Solved Problems in Classical Electrodynamics and Theory of Relativity

This book is intended for undergraduate and graduate students in physics, engineering, astronomy, applied mathematics and for researchers working in related subjects. It is an excellent study tool for those students who would like to work independently on more electrodynamics problems in order to deepen their understanding and problem solving skills.

The book discusses main concepts and techniques related to Maxwell's equations, potentials and fields (including Liénard-Wiechert potentials), electromagnetic waves, and the interaction and dynamics of charged point particles. It also includes content on magnetohydrodynamics and plasma, radiation and antennas, special relativity, relativistic kinematics, relativistic dynamics and relativistic-covariant dynamics and general theory of relativity. It contains a wide range of problems, ranging from electrostatics and magnetostatics to the study of the stability of dynamical systems, field theories and black hole orbiting. The book even contains interdisciplinary problems from the fields of electronics, elementary particle theory, antenna design.

Detailed, step-by step calculations are presented, meeting the need for a thorough understanding of the reasoning and steps of the calculations by all students, regardless of their level of training. Additionally, numerical solutions are also proposed and accompanied by adjacent graphical representations and even multiple methods of solving the same problem.

It is structured in a coherent and unified way, having a deep didactic character, being thus oriented towards a university environment, where the transmission of knowledge in a logical, unified and coherent way is essential. It teaches students how to think about and how to approach solving electrodynamics problems.

- Contains a wide range of problems and applications from the fields of electrodynamics and the theory of special relativity.
- Presents numerical solutions to problems involving nonlinearities.
- Details command lines specific to Mathematica software dedicated to both analytical and numerical calculations, which allows readers to obtain the numerical solutions as well as the related graphical representations.

Daniel Radu is a Lecturer in the Faculty of Physics, Al.I.Cuza University, Romania.
Ioan Merches is Professor Emeritus in the Faculty of Physics, Al.I.Cuza University, Romania.

Solved Problems in Classical Electrodynamics and Theory of Relativity

Daniel Radu and Ioan Merches

 CRC Press
Taylor & Francis Group
Boca Raton London New York

CRC Press is an imprint of the
Taylor & Francis Group, an **informa** business

Designed cover image: Shutterstock_1393409009

First edition published 2023
by CRC Press
2385 NW Executive Center Drive, Suite 320, Boca Raton FL 33431

and by CRC Press
4 Park Square, Milton Park, Abingdon, Oxon, OX14 4RN

CRC Press is an imprint of Taylor & Francis Group, LLC

© 2023 Daniel Radu and Ioan Merches

ISBN: 978-1-0325-1495-6 (hbk)
ISBN: 978-1-0325-1509-0 (pbk)
ISBN: 978-1-0034-0260-2 (ebk)

DOI: 10.1201/9781003402602

Typeset in CMR10
by KnowledgeWorks Global Ltd.

Contents

Preface

According to Albert Einstein, "The formulation of the problem is often more essential than its solution, which may be merely a matter of mathematical or experimental skill". We do not want to comment on this quote but just urge the reader to reflect upon this judgment while solving our proposed problems.

This collection contains a number of eighty-eight solved problems, distributed on twelve chapters, consistent with the system of organization of the material taught to our students within the course of *Electrodynamics and Theory of Relativity*. Besides, to help the students endeavour, we added six appendices concerning mathematical formalism used in problem-solving.

Over time, Electrodynamics has been approached in various ways, corresponding to the level of the research instrument. For example, at the end of nineteenth century Maxwell's theory regarding production and propagation of the electromagnetic field has been studied mostly phenomenologically; this is quite natural if we consider that in Maxwell's time, the electron had not yet been discovered, so that at that time the structure of matter was not known as we know it today (as a matter of fact, the electron was discovered by J.J. Thomson, in 1897, just eighteen years after the Maxwell's death). At the beginning of the twentieth century, the emergence of the Special − and then General − Theory of Relativity, together with Quantum Mechanics and many other discoveries, arrises the necessity of a new procedure of investigation, based on tensor calculus and Riemannian geometry. Such a mathematical formalism allows the relativist-covariant treatment of Electrodynamics in flat spaces (such as the Minkowski Universe), or curved spaces (*e.g.*, the Riemann space). An exhaustive approach of the Electromagnetic Field Theory requires the use of Quantum Field Theory, but this would go beyond the limits of this book.

The problems are inserted − as far as possible − according to their difficulty and dedication. In our opinion, the collection could be useful to both the students who attend the physical faculty courses and to those studying theoretical physics as chemists, or preparing as future engineers, etc.

At the end of most of the problems, the result is physically interpreted. In same cases, for the sake of clarity, the problems contain at the beginning a preparatory part, helping the reader to easily understand the solving of the problem. We have to mention that obtaining and interpreting the solutions, in some cases, lead to non-linear and/or transcendent equations, in which cases, the numerical solutions have been discussed by means of software specialized in analytical and numerical calculation, that is *Mathematica*.

A few words about how the problems, formulas and graphical representations in this book are numbered.

Problems are numbered like this (an example): 8.3 Problem No. 61. The number 8 represents the number of the chapter to which the problem belongs, three is the sequence number of the problem within the chapter, and sixty-one is actually the real number of the problem, as all the problems in the book are numbered continuously, in order, starting with the number 1 and ending with the number 88, which is the number of the last problem in the collection. So the number of the problems does not take into account the number of the chapter to which a problem belongs. In brief, in the case of the given example, the third problem in the chapter 8 is actually the sixty-first problem of the collection.

Numbered formulas carry three numbers separated by two dots − for example, Eq. (4.7.141) − as follows: the first figure is given by the number of the chapter in which the formula is presented (4, in our example), the second figure (7, in the chosen example) is given by the number of order of the problem − to which that formula belongs − from the current chapter (for instance, in the seventh problem from chapter 4, all the formulas that carry a number will be numbered so that the first two figures − from the series of three − will be 4 and 7, that is, all these formulas will have numbers starting with 4.7: (4.7.139), (4.7.140), (4.7.141), ..., up to the last numbered formula from the seventh problem of chapter 4, which is formula (4.7.169)), and the third figure is given by the order number of the formula within the respective chapter (in other words, all the numbered formulas from a chapter are numbered from the first − which has 1 as the third digit, whatever the chapter − to the last, in ascending order, regardless of the problem to which the respective formula belongs). For an easier understanding, let's give a concrete example: The fifth problem from chapter 8 is Problem No. 63 (the sixty-third problem in the collection). It contains seven numbered formulas, from (8.5.26) to (8.5.32). This means that in chapter 8, up to the fifth problem in this chapter, there are four problems, from Problem No. 59 to Problem No. 62, containing twenty-five numbered formulas, from formula (8.1.1) up to formula (8.4.25).

Many of the problems contain one or more figures (graphical representations), which are intended to help the understanding of the reasoning or to facilitate the interpretation of the results, but the problem collection also contains "purely theoretical" problems, which do not require any helpful or explanatory figures. The graphical representations have been numbered with two figures separated by a dot (*e.g.*, Fig. 1.3). The first number in the figure's "code" is given by the number of the chapter to which the figure belongs (in our example, 1), and the second number indicates the order number of the figure in that chapter (3, in the chosen example), so the figures are numbered in order (starting with number 1 within each chapter − this being the second number of the figure "code") only within the same chapter, even if not all problems are accompanied by supporting/explanatory figures.

As we have mentioned above, at the end of this problem collection with solutions, we added six appendices containing mathematical complements absolutely necessary in the study of Electrodynamics by means of applications: tensor calculus, vector analysis, Dirac's delta distribution, Green's function, etc.

Although to some readers the organization of the book's content may not seem quite appropriate (e.g., some readers might comment on the fact that chapters where the treatment is Lorentz invariant appear interspersed with chapters that are treated non-Lorentz invariant, etc.), nevertheless the order of presentation of the notions respects a certain logic, on which we do not want to insist here. Indeed, some chapters contain problems treated in a relativistic-covariant way, but others do not. For example, chapter 5 dealing with electromagnetic radiation theory is fully Lorentz invariant, but the next chapter is not, and Relativity is not explicitly introduced until chapter 8. Such observations might seem pertinent, but we point out that the material has been organized according to the analytical syllabus of the course we teach to our students, and the polemic on this issue is purely subjective.

In compiling this problem collection, we have tried to ensure an equilibrium between the "standard" problems which are popular in many physics faculties/departments around the world, and lesser known problems, which occur quite rarely or not at all in the collections of problems dedicated to the same subject.

Many of the standard problems which are often addressed to students in exams have been previously published in other collections of problems, in most cases being difficult to determine when and where a certain problem was first presented. Because of this, it is very difficult, if not impossible, to know who is the real author of such a problem. Being, however, "famous" problems, we decided to include some of them in this book.

Even if sometimes it may seem boring to solve many of these problems (so-called "standard"), most often, this effort is worth making, since more than half of such problems are usually given in the electrodynamics examinations. From this point of view, we think that this way we come to the aid of students, our endeavour to include such problems being welcome.

We recognize with reluctance but are aware of the fact that all mistakes produced in formulating the problems and solutions is the sole responsibility of the authors. We tried to offer detailed solutions and to closely follow the thread of the calculation, even if sometimes the way to the solution was not at all difficult. Obviously, we cannot claim that our solutions are the best or that there are no errors in the book; we hope that possible errors be only editorial ones.

So, we will be grateful to the readers for any comments, pertinent criticism or alternative solutions offered to improve an eventual new edition of the book. In this respect, the authors want to propose here the most difficult exercise of the present collection of problems, that is, the eighty-ninth problem of the book, whose statement is as follows: "Find all the errors in this book and report them to the authors".

To the end of this forward, here is another quote belonging to Albert Einstein: "*There comes a point in your life when you need to stop reading other people's books and write your own*". We followed this precious advice, and we wrote not only this book but also another one on the same subject (a monograph that appeared in 2016 at Springer), the present book being an "applicative companion" for the first.

The authors

1

Electrostatics

1.1 Problem No. 1

Determine the potential $V(r)$ and the electric field intensity $\vec{E}(r)$ in an arbitrary point situated inside and outside of a sphere of radius R, uniformly charged with the electric charge Q.

Solution

Method 1: The fundamental problem of Electrostatics

Following the general method of solving a problem of Electrostatics, the electrostatic field potential $V(\vec{r})$ at an arbitrary point $P(\vec{r})$ in the three-dimensional Euclidean space results as the solution of the Poisson equation

$$\Delta V = \begin{cases} -\rho/\varepsilon_0, & \text{if } P(\vec{r}) \text{ is inside the sphere,} \\ 0, & \text{if } P(\vec{r}) \text{ is outside the sphere,} \end{cases} \qquad (1.1.1)$$

with corresponding boundary conditions. Here $\rho = \dfrac{3Q}{4\pi R^3}$ is the spatial charge density inside the sphere.

Due to the spherical symmetry of the problem, we will use the Laplacian in spherical coordinates

$$\Delta V = \frac{1}{r^2}\left\{ \frac{\partial}{\partial r}\left(r^2\frac{\partial V}{\partial r}\right) + \frac{1}{\sin\theta}\left[\frac{\partial}{\partial\theta}\left(\sin\theta\frac{\partial V}{\partial\theta}\right) + \frac{1}{\sin\theta}\frac{\partial^2 V}{\partial\varphi^2}\right]\right\}. \qquad (1.1.2)$$

Because of the same reason, the potential V does not depend on angular variables, so $V(r,\theta,\varphi) \to V(r) \equiv V$, and, as a consequence, in Eq. (1.1.2) remains only the radial part

$$\Delta V = \frac{1}{r^2}\left[\frac{\partial}{\partial r}\left(r^2\frac{\partial V}{\partial r}\right)\right] \equiv \frac{1}{r^2}\left[\frac{d}{dr}\left(r^2\frac{dV}{dr}\right)\right]. \qquad (1.1.3)$$

Case 1. The potential V and the electric field intensity $\vec{E}$ inside the sphere.

Denoting by $V_{in}(r)$ and $E_{in}(r)$, the potential and the modulus of the electric field intensity inside the sphere, one follows from Eqs. (1.1.1) and (1.1.3) that

$$\frac{1}{r^2}\left[\frac{d}{dr}\left(r^2\frac{dV_{in}}{dr}\right)\right] = -\frac{\rho}{\varepsilon_0}, \tag{1.1.4}$$

and, by integration,

$$r^2\frac{dV_{in}}{dr} = -\frac{\rho}{\varepsilon_0}\frac{r^3}{3} + C_1, \tag{1.1.5}$$

where C_1 is an arbitrary integration constant (to be determined at the right time). In view of Eq. (1.1.5), we still have

$$\frac{dV_{in}}{dr} = -\frac{\rho r}{3\varepsilon_0} + \frac{C_1}{r^2}, \tag{1.1.5$'$}$$

and, by a new integration,

$$V_{in}(r) = -\frac{\rho r^2}{6\varepsilon_0} - \frac{C_1}{r} + C_2, \tag{1.1.6}$$

where C_2 is a new arbitrary integration constant.

The electric field intensity $\vec{E} = -\nabla V$ follows directly from Eq. (1.1.5$'$):

$$E_{in} = -\frac{dV_{in}}{dr} = \frac{\rho r}{3\varepsilon_0} - \frac{C_1}{r^2}. \tag{1.1.7}$$

Case 2. The potential V and the electric field intensity $\vec{E}$ at an arbitrary point outside the sphere.

Denote the potential and modulus of the electric field intensity outside the sphere by $V_e(r)$ and $E_e(r)$, respectively. The Poisson equation (1.1.1) corresponding to this choice, then writes

$$\Delta V_e(r) = 0, \tag{1.1.8}$$

or, in view of Eq. (1.1.3),

$$\frac{1}{r^2}\left[\frac{d}{dr}\left(r^2\frac{dV_e}{dr}\right)\right] = 0, \tag{1.1.9}$$

leading to

$$r^2\frac{dV_e}{dr} = C_3,$$

and so

$$\frac{dV_e}{dr} = \frac{C_3}{r^2}.$$

By integration, one follows that the potential of the electrostatic field outside the sphere is

$$V_e = -\frac{C_3}{r} + C_4, \tag{1.1.10}$$

while the electric field intensity writes

$$E_e(r) = -\frac{dV_e}{dr} = -\frac{C_3}{r^2}. \tag{1.1.11}$$

The arbitrary integration constants C_1, C_2, C_3 and C_4 can be determined by imposing the boundary conditions

$$V_e(r \to \infty) = 0, \tag{1.1.12}$$

$$V_{in}(r \to 0) = \text{finite}, \tag{1.1.13}$$

and the continuity conditions as well,

$$V_{in}(R) = V_e(R), \tag{1.1.14}$$

$$E_{in}(R) = E_e(R). \tag{1.1.15}$$

According to Eqs. (1.1.10) and (1.1.12), $C_4 = 0$, while following Eqs. (1.1.6) and (1.1.13), $C_1 = 0$. The remaining conditions expressed by Eqs. (1.1.14) and (1.1.15) give C_2 and C_3 as being

$$C_2 = \frac{\rho R^2}{2\varepsilon_0}, \tag{1.1.16}$$

$$C_3 = -\frac{\rho R^3}{3\varepsilon_0}. \tag{1.1.17}$$

Introducing all the constants of integration in Eqs. (1.1.6), (1.1.7), (1.1.10) and (1.1.11), one finds the expressions for the potential and modulus of the electric field intensity, inside and, respectively, outside the sphere (which is uniformly charged with the electric charge Q):

$$\begin{cases} V_{in}(r) \equiv V(r)\Big|_{r \leq R} = \dfrac{\rho R^2}{2\varepsilon_0}\left(1 - \dfrac{r^2}{3R^2}\right) = \dfrac{3Q}{8\pi\varepsilon_0 R}\left(1 - \dfrac{r^2}{3R^2}\right), \\[4mm] E_{in}(r) \equiv E(r)\Big|_{r \leq R} = \dfrac{\rho r}{3\varepsilon_0} = \dfrac{Q}{4\pi\varepsilon_0}\dfrac{r}{R^3}, \end{cases} \tag{1.1.18}$$

respectively,

$$\begin{cases} V_e \equiv V(r)\Big|_{r > R} = \dfrac{\rho R^3}{3\varepsilon_0 r} = \dfrac{Q}{4\pi\varepsilon_0 r}, \\[4mm] E_e \equiv E(r)\Big|_{r > R} = \dfrac{\rho R^3}{3\varepsilon_0 r^2} = \dfrac{Q}{4\pi\varepsilon_0 r^2}. \end{cases} \tag{1.1.19}$$

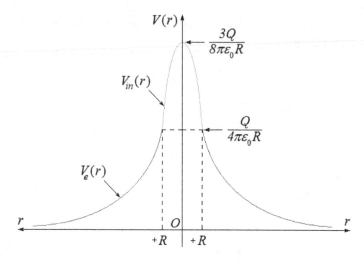

FIGURE 1.1
Graphical representation of the electrostatic potential V as a function of the radial distance r.

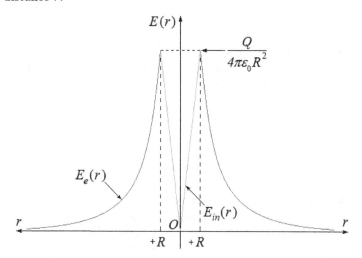

FIGURE 1.2
Graphical representation of the electrostatic field E as a function of the radial distance r.

The results given by Eqs. (1.1.18) and (1.1.19) are graphically represented in Figs. 1.1 and 1.2.

Method 2: The Gauss's Flux Theorem

We will use the same notations as above, *i.e.*, $E(r \leq R) \equiv E_{in}(r)$, $E(r > R) \equiv E_e(r)$, $V(r \leq R) \equiv V_{in}(r)$ and $V(r > R) \equiv V_e(r)$. The same notation will be used for all the other quantities that appear in solving the problem.

If the first method determines the potential first and then the electric field intensity, this second method determines the electric field intensity first and then the potential. Both methods appeal to the same relationship between the field and potential, which is always valid in Electrostatics, namely $\vec{E} = -\nabla V$.

Case 1. The potential V and the electric field intensity $\vec{E}$ inside the sphere.

In this case, the Gauss's flux theorem (or the Gauss's integral law) writes

$$\oint_{(S_{in})} \vec{E}_{in}(r) \cdot d\vec{S}_{in} = \frac{q_{in}}{\varepsilon_0}, \qquad (1.1.20)$$

or, having in view the spherical symmetry of the problem (E_{in} is constant on the whole surface of any sphere of a given radius r),

$$E_{in}(r)\left(4\pi r^2\right) = \frac{\rho 4\pi r^3}{3\varepsilon_0}, \qquad (1.1.21)$$

where $\rho = \frac{3Q}{4\pi R^3}$. From Eq. (1.1.21) we get

$$E_{in}(r) = \frac{\rho r}{3\varepsilon_0}, \qquad (1.1.22)$$

and then, from $\vec{E} = -\nabla V$, which, for spherical symmetry simply writes $E(r) = -\frac{dV}{dr}$, we have in this case

$$V_{in}(r) = -\int E_{in}(r)dr = -\int \frac{\rho r}{3\varepsilon_0}dr = -\frac{\rho r^2}{6\varepsilon_0} + K_1, \qquad (1.1.23)$$

where K_1 is an arbitrary integration constant, determinable from the continuity condition expressed by Eq. (1.1.14).

The result expressed by Eq. (1.1.22) – which is identical to that expressed by Eq. $(1.1.18)_2$, as is normal – was obtained very easily and in a straightforward manner, without the need to determine any arbitrary integration constant.

Case 2. The potential V and the electric field intensity $\vec{E}$ at an arbitrary point $P(r > R)$ outside the sphere.

In this case, according to Gauss's flux theorem for electrostatic field, the potential and the field intensity generated by the uniformly distributed electrical charge inside the sphere of radius R, at an arbitrary point $P(r > R)$ outside the sphere are identical to those generated by a point charge situated in the center of the sphere, and whose charge is equal to the total charge contained inside the sphere, that is Q. Indeed, in this case, we can write

$$\oint_{(S_e)} \vec{E}_e(r) \cdot d\vec{S}_e = \frac{Q}{\varepsilon_0}, \qquad (1.1.24)$$

or,

$$E_e(r)\left(4\pi r^2\right) = \frac{Q}{\varepsilon_0},\qquad(1.1.25)$$

which gives

$$E_e(r) = \frac{Q}{4\pi\varepsilon_0 r^2},\qquad(1.1.26)$$

i.e., the result expressed by Eq. $(1.1.19)_2$ was re-obtained directly (without the need of finding of any arbitrary integration constant), as is natural. Obviously,

$$V_e(r) = -\int E_e(r)dr = -\int \frac{Q}{4\pi\varepsilon_0 r^2}dr = \frac{Q}{4\pi\varepsilon_0 r} + K_2,\qquad(1.1.27)$$

where K_2 is a new arbitrary integration constant, determinable from the boundary condition expressed by Eq. (1.1.12). It results immediately that $K_2 = 0$, so that the relation $(1.1.19)_1$ is very easily re-obtained, as we expected.

Now, using the condition expressed by Eq. (1.1.14), the constant K_1 can very quickly be found; in a jiffy, it results $K_1 = \frac{3Q}{8\pi\varepsilon_0 R}$, so that the relation $(1.1.18)_1$ is re-obtained, as is normal.

Note the simplicity (in the case of this second method only two arbitrary integration constants were needed, and not four, as in the case of the first method) and also the elegance of this method, which thus proves to be much more appropriate than the first. In fact, this is the case with all problems with "high" symmetry: the method of Gauss's flux theorem is preferable to the "classical" method, which involves solving the Poisson's or Laplace's equations, as the case may be.

1.2   Problem No. 2

A disk of radius R is uniformly charged with electricity of superficial density σ = const. Determine the electrostatic field produced by the disk at an arbitrary point situated on its axis.

Solution

The field potential at an arbitrary point of z-axis (see Fig. 1.3) is given by

$$V(0,0,z>0) = \frac{1}{4\pi\varepsilon_0}\int\limits_{(S)} \frac{dq}{r} = \frac{1}{4\pi\varepsilon_0}\int\limits_{(S)} \frac{\sigma dS}{r}$$

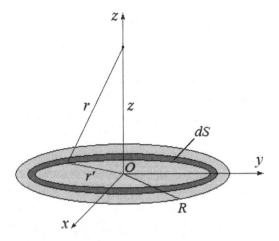

FIGURE 1.3
A uniformly charged disk.

$$= \frac{1}{4\pi\varepsilon_0} \int_0^R \frac{\sigma 2\pi r' dr'}{r} = \frac{\sigma}{2\varepsilon_0} \int_0^R \frac{r' dr'}{\sqrt{r'^2 + z^2}}$$

$$= \frac{\sigma}{2\varepsilon_0} \sqrt{r'^2 + z^2} \Big|_0^R = \frac{\sigma}{2\varepsilon_0} \left(\sqrt{R^2 + z^2} - z \right). \quad (1.2.28)$$

For symmetry reasons, the potential has to be the same for $z < 0$, that is

$$V(0, 0, z < 0) = \frac{\sigma}{2\varepsilon_0} \left(\sqrt{R^2 + z^2} + z \right). \quad (1.2.29)$$

According to Eqs. (1.2.28) and (1.2.29), the potential V has a singular behaviour at the point $z = 0$. This fact is also put into evidence by the graphic representation in Fig. 1.4.

For $z < 0$, the graph has a positive slope since the function $V(r)$ is increasing, while for $z > 0$ the slope is negative, because $V(r)$ is decreasing. Consequently, the potential $V(z)$ exhibits a sudden change of the slope sign at $z = 0$. According to the well-known relation between potential and the field intensity $\vec{E} = -\nabla V$, we conclude that at the point $z = 0$ the field intensity suffers a jump (has a discontinuity). For $z = 0$,

$$V(z)\Big|_{z=0} = \frac{\sigma R}{2\varepsilon_0}, \quad (1.2.30)$$

while for $z \gg R$, the series expansion gives

$$\sqrt{R^2 + z^2} - z = z \left[\sqrt{1 + \left(\frac{R}{z} \right)^2} - 1 \right] \cong z \left(1 + \frac{1}{2} \frac{R^2}{z^2} - 1 \right) = \frac{R^2}{2z}. \quad (1.2.31)$$

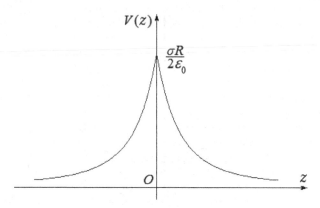

FIGURE 1.4
Graphical representation of the electrostatic potential V as a function of the distance z.

According to Eqs. (1.2.28) and (1.2.31), in the approximation $z \gg R$ the potential writes

$$V(z \gg R) = \frac{\sigma}{2\varepsilon_0} \frac{R^2}{2z} = \frac{q}{4\pi\varepsilon_0 z}, \qquad (1.2.32)$$

a result that was to be expected, since for $z \gg R$ the disk appears as being punctiform.

Since $V = V(z)$ (*i.e.*, V does not depend on x and y), and $\vec{E} = -\nabla V$, one follows that $\vec{E} = (0, 0, E_z)$, with $E_z = -\frac{dV}{dz}$, that is

$$E_{z+} = E_z(z > 0) = -\frac{dV(z > 0)}{dz} = \frac{\sigma}{2\varepsilon_0}\left[1 - \frac{z}{\sqrt{z^2 + R^2}}\right], \qquad (1.2.33)$$

and

$$E_{z-} = E_z(z < 0) = -\frac{dV(z < 0)}{dz} = -\frac{\sigma}{2\varepsilon_0}\left[1 + \frac{z}{\sqrt{z^2 + R^2}}\right]. \qquad (1.2.34)$$

In the limit $z = 0$, the previous two relations become

$$\lim_{\substack{z \to 0 \\ z > 0}} E_z = \lim_{z \to 0} E_{z+} = \frac{\sigma}{2\varepsilon_0}, \qquad (1.2.35)$$

and

$$\lim_{\substack{z \to 0 \\ z < 0}} E_z = \lim_{z \to 0} E_{z-} = -\frac{\sigma}{2\varepsilon_0}. \qquad (1.2.36)$$

According to Eqs. (1.2.35) and (1.2.36), the values of a field intensity produced by an electrostatic charged disk, at the point $z = 0$, are identical to the field intensity values of an infinite plane uniformly charged with the electric charge of superficial density σ, situated in vacuum.

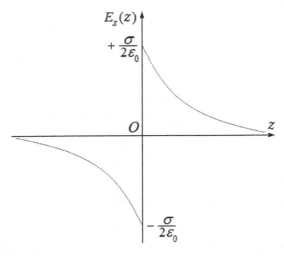

FIGURE 1.5
Jump of the electrostatic field while passing from one side of the disk to the other.

This result is correct, since for $z \to 0$ the disk behaves like an infinite plane. In addition,

$$(E_{z+} - E_{z-})\Big|_{z=0} = \frac{\sigma}{2\varepsilon_0} - \left(-\frac{\sigma}{2\varepsilon_0}\right) = \frac{\sigma}{\varepsilon_0}, \qquad (1.2.37)$$

which means that at the point $z = 0$, when passing from one part of the disk to the other, the electric field suffers a jump of value $\frac{\sigma}{\varepsilon_0}$ (see Fig. 1.5).

1.3   Problem No. 3

Determine the shape of the equipotential surfaces of an electrostatic field produced by a charge uniformly distributed ($\lambda = $ const.) along a rectilinear wire of length $2c$.

Solution

By a convenient choice of the coordinate axes, the potential V_P of the field created by the electrized wire, at an arbitrary point P, is

$$V_P = k_e \int\limits_{-c}^{+c} \frac{dq}{r} = k_e \lambda \int\limits_{-c}^{+c} \frac{d\xi}{r} \;; \quad k_e = \frac{1}{4\pi\varepsilon_0}, \qquad (1.3.38)$$

where $d\xi$ is a length element of the wire (see Fig. 1.6).

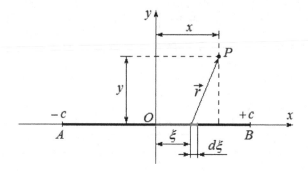

FIGURE 1.6
A uniformly distributed charge q along a straight wire of length $2c$.

Using the obvious formula $r = \left[(x - \xi)^2 + y^2\right]^{1/2}$ and making the substitution $u = x - \xi + r$, one can write

$$\frac{dr}{d\xi} = -\frac{x - \xi}{r} = 1 - \frac{u}{r}. \tag{1.3.39}$$

On the other hand, since $dr = du + d\xi$, we still have

$$\frac{d\xi}{r} = -\frac{du}{u}, \tag{1.3.40}$$

so that

$$V_P = -k_e \lambda \int_{u_1}^{u_2} \frac{du}{u} = k_e \ln \frac{x + c + r_1}{x - c + r_2}. \tag{1.3.41}$$

The shape of the equipotential surfaces is found by equaling V_P to a constant K. Denoting $x + c = x_1$, $x - c = x_2$, we then have

$$x_1 + r_1 = K(x_2 + r_2). \tag{1.3.42}$$

"Processing" conveniently the relations

$$r_1^2 = x_1^2 + y^2; \quad r_2^2 = x_2^2 + y^2,$$

and using Eq. (1.3.42), we still have

$$r_2 - x_2 = K(r_1 - x_1). \tag{1.3.43}$$

By virtue of Eqs. (1.3.42) and (1.3.43), we finally obtain

$$r_1 + r_2 = 2c\frac{K + 1}{K - 1} = 2a \, (= const.). \tag{1.3.44}$$

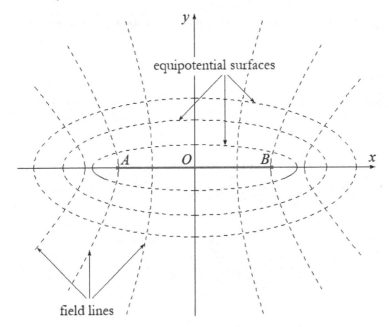

FIGURE 1.7
Plane projection of the field lines and equipotential surfaces for a uniformly distributed charge q along a straight wire of length $2c$.

Consequently, the locus of the points that satisfy the condition $V_P = const.$ (k fixed) is an ellipse with foci at the wire's ends. Rotating the figure around the x-axis and giving to K different values, we obtain *homophocal revolution ellipsoids, with foci at the points A and B*. The field lines are orthogonal to the equipotential surfaces at each point of the field (see Fig. 1.7).

One can easily show that $K = \frac{a+c}{a-c}$, so that, finally we arrive at

$$V_P = \frac{\lambda}{4\pi\varepsilon_0} \ln\frac{a+c}{a-c}. \tag{1.3.45}$$

Particular cases. If the points A and B approach until they overlap, the ellipsoids degenerate into concentric spheres, while if they are moved away to infinity, one obtains coaxial cylinders with the wire as axis.

1.4   Problem No. 4

The potential of the electrostatic field produced in vacuum by a dipole of moment $\vec{p} = const.$ is

$$V = \frac{1}{4\pi\varepsilon_0}\frac{\vec{p}\cdot\vec{r}}{r^3}. \tag{1.4.46}$$

Determine the electric field intensity $\vec{E}$ and the electric field lines of the dipole electrostatic field.

Solution

The electric field intensity of the dipole electrostatic field is given by

$$\vec{E} = -\operatorname{grad} V = -\frac{1}{4\pi\varepsilon_0}\nabla\left(\frac{\vec{p}\cdot\vec{r}}{r^3}\right).$$ (1.4.47)

Using the relation (B.3.55), we can write

$$\nabla\left(\frac{\vec{p}\cdot\vec{r}}{r^3}\right) = \frac{1}{r^3}\nabla(\vec{p}\cdot\vec{r}) + (\vec{p}\cdot\vec{r})\,\nabla\left(\frac{1}{r^3}\right)$$

$$= \frac{1}{r^3}\Big[\underbrace{\vec{p}\times(\nabla\times\vec{r})}_{=0} + \underbrace{\vec{r}\times(\nabla\times\vec{p})}_{=0} + (\vec{p}\cdot\nabla)\vec{r}$$

$$+ \underbrace{(\vec{r}\cdot\nabla)\vec{p}}_{=0}\Big] + (\vec{p}\cdot\vec{r})\left(-\frac{3\vec{r}}{r^5}\right)$$

$$= \frac{1}{r^3}(\vec{p}\cdot\nabla)\vec{r} - \frac{3(\vec{p}\cdot\vec{r})\vec{r}}{r^5} = \frac{\vec{p}}{r^3} - \frac{3(\vec{p}\cdot\vec{r})\vec{r}}{r^5},$$

where we have considered the fact that $\vec{p} = const.$ and, also,

$$(\vec{p}\cdot\nabla)\vec{r} = \left(p_i\frac{\partial}{\partial x_i}\right)(x_k\vec{u}_k) = p_i\delta_{ik}\vec{u}_k = p_i\vec{u}_i = \vec{p}.$$

Thus, according to Eq. (1.4.47), the electric field intensity $\vec{E}$ of the dipole electrostatic field is given by

$$\vec{E} = -\operatorname{grad} V = \frac{1}{4\pi\varepsilon_0}\left[-\frac{\vec{p}}{r^3} + \frac{3(\vec{p}\cdot\vec{r})\vec{r}}{r^5}\right].$$ (1.4.48)

Keeping in mind the symmetry of the problem, we will use a reference frame with spherical coordinates, having the polar axis along $\vec{p}$. Projecting the vector $\vec{E}$ on the axes of the local trihedron, one obtains

$$E_r = \frac{1}{4\pi\varepsilon_0}\frac{2p\cos\theta}{r^3}, \quad E_\theta = \frac{1}{4\pi\varepsilon_0}\frac{p\sin\theta}{r^3}, \quad E_\varphi = 0.$$ (1.4.49)

By integrating the differential equations of the field lines,

$$\frac{dr}{E_r} = \frac{r\,d\theta}{E_\theta} = \frac{r\sin\theta\,d\varphi}{E_\varphi},$$

one then obtains

$$\frac{1}{r}dr = \frac{E_r}{E_\theta}d\theta = 2\cot\theta\,d\theta \;\Rightarrow\; \ln r = 2\ln|\sin\theta| + \ln C_1 \;\Rightarrow\; r = C_1\sin^2\theta,$$

as well as

$$d\varphi = 0 \;\Rightarrow\; \varphi = C_2,$$

where C_1 and C_2 are arbitrary integration constants.

1.5 Problem No. 5

A family of nonintersecting surfaces is given by the equation $u(x, y, z) = const$. Find the condition which has to be satisfied by the function u, so that the family surfaces could be the equipotential surfaces of the field created by a system of conductors.

Solution

To satisfy the required condition, the potential $V(x, y, z)$ must depend on coordinates by means of function $u(x, y, z)$. Consequently, we must have, on one side

$$V(x, y, z) = f(u), \qquad (1.5.50)$$

and, on the other, $V(x, y, z)$ must be a solution of the Laplace equation

$$\Delta V(x, y, z) = 0. \qquad (1.5.51)$$

In view of Eq. (1.5.50), we then have

$$\nabla V = f'(u)\nabla u, \qquad (1.5.52)$$

as well as

$$\Delta V = (\nabla \cdot \nabla)V = \nabla \cdot (\nabla V) = \nabla \cdot \left(f'(u)\nabla u\right) = f'(u)\nabla \cdot (\nabla u)$$
$$+ (\nabla u) \cdot \nabla f'(u) = f'(u)\left(\nabla \cdot \nabla\right)u + f''(u)\left(\nabla u\right) \cdot \left(\nabla u\right)$$
$$= f'(u)\Delta u + f''(u)\left(\nabla u\right)^2. \qquad (1.5.53)$$

According to Eqs. (1.5.51) and (1.5.53), we can write

$$\frac{\Delta u}{(\nabla u)^2} = -\frac{f''(u)}{f'(u)} = -\frac{d}{du}\left[\ln f'(u)\right] \equiv F(u). \qquad (1.5.54)$$

So, $F(u) = \dfrac{\Delta u}{(\nabla u)^2}$, and the searched potential is then found by means of integration. Indeed, we have

$$-d\left[\ln f'(u)\right] = F(u)\,du \;\Rightarrow\; \ln f'(u) = -\int F(u)du + \ln A$$
$$\Rightarrow f'(u) = A\,e^{-\int F(u)du}.$$

Thus, a new integration gives the searched solution as being

$$f(u)\left[\,= V(x, y, z)\right] = \int f'(u)du = A\int e^{-\int F(u)du}du + B,$$

where A and B are two arbitrary integration constants.

1.6 Problem No. 6

Show that the electric quadrupole momentum tensor of a homogeneous charge distribution with axial symmetry has a single essential distinct component, and calculate the quadrupole component of the potential of the given charge distribution.

Solution

The symmetry of the problem clearly suggests the usage of the cylindrical coordinates ρ, φ, z. Taking Oz as the symmetry axis and observing that – due to the symmetry – the electric charge spatial density does not depend on angle φ [$\rho_e = \rho_e(\rho, z)$], we have by definition

$$p_{ik} = \int\limits_{(\mathcal{D})} \rho_e(\rho, z)\big(3x_i x_k - r^2 \delta_{ik}\big) d\tau, \qquad (1.6.55)$$

where $\mathcal{D}$ is the 3D-domain that contains the homogeneous charge distribution.

Calculating the six components of the symmetric tensor p_{ik}, we obtain for the non-diagonal components,

$$p_{12} \equiv p_{xy} = 3 \int \rho_e(\rho, z)\rho^3 d\rho dz \int_0^{2\pi} \sin\varphi \cos\varphi \, d\varphi = 0,$$

$$p_{23} \equiv p_{yz} = 3 \int \rho_e(\rho, z)\rho^2 z d\rho dz \int_0^{2\pi} \sin\varphi d\varphi = 0,$$

$$p_{31} \equiv p_{zx} = 3 \int \rho_e(\rho, z)\rho^2 z d\rho dz \int_0^{2\pi} \cos\varphi d\varphi = 0.$$

Since $x = \rho\cos\varphi$, $y = \rho\sin\varphi$, $r^2 = \rho^2 + z^2$, the diagonal components are given by

$$p_{33} \equiv p_{zz} = 2\pi \int \rho_e(\rho, z)\left(2z^2 - \rho^2\right) \rho d\rho dz = p,$$

$$p_{11} \equiv p_{xx} = \int \rho_e(\rho, z)\left[3x^2 - \left(\rho^2 + z^2\right)\right]\rho d\rho d\varphi dz$$

$$= -\pi \int \rho_e(\rho, z)\left(2z^2 - \rho^2\right)\rho d\rho dz = -\frac{p}{2},$$

$$p_{22} \equiv p_{yy} = \int \rho_e(\rho, z)\left[3y^2 - \left(\rho^2 + z^2\right)\right]\rho d\rho d\varphi dz = p_{xx} = -\frac{p}{2},$$

which finalise the answer to the first part of the problem. As can easily be seen,

$$\text{Tr}(p_{ik}) = \sum_{i=1}^{3} p_{ii} = 0,$$

which is quite normal, because the tensor p_{ik} was built from the beginning to have this property. The quadrupole component of the potential is determined by means of the formula (see Ref. [15]):

$$V^{(2)} = \frac{k_e}{6} p_{ik} \left(\frac{3x_i x_k}{r^5} - \frac{\delta_{ik}}{r^3} \right); \quad k_e = \frac{1}{4\pi\varepsilon_0}. \tag{1.6.56}$$

Using the above determined components of the quadrupole electric moment, we then have

$$\begin{aligned} V^{(2)} &= \frac{k_e}{6} \left[p_{11} \left(\frac{3x^2}{r^5} - \frac{1}{r^3} \right) \right. \\ &\quad + p_{22} \left(\frac{3y^2}{r^5} - \frac{1}{r^3} \right) + p_{33} \left(\frac{3z^2}{r^5} - \frac{1}{r^3} \right) \Bigg] \\ &= k_e \frac{1}{6} \frac{p}{r^3} \left[\frac{3z^2}{r^2} - \frac{3}{2} \left(\frac{x^2 + y^2}{r^2} \right) \right]. \end{aligned}$$

Since $x^2 + y^2 = r^2 - z^2$, $z = r\cos\theta$, we finally obtain

$$V^{(2)} = \frac{1}{4\pi\varepsilon_0} \frac{1}{2} \frac{p}{r^3} P_2(\cos\theta), \tag{1.6.57}$$

where

$$P_2(\cos\theta) = \frac{3\cos^2\theta - 1}{2}$$

is the Legendre polynomial of degree two.

1.7 Problem No. 7

Consider a parallelepipedic box of dimensions a, b, c, corresponding to the three axes x, y, and, respectively, z. All faces of the box are maintained at zero potential, except for the face $z = c$, on which the potential is $V = V(x,y)$. Determine the potential $V(x,y,z)$ inside the box.

Solution

The problem symmetry requires a Cartesian coordinate system. Without restricting the problem generality, let us take the origin of the frame in one of the box corners, with coordinate axes "leaving" that point (see Fig. 1.8). The searched potential is then the solution of the Laplace equation

$$\frac{\partial^2 V}{\partial x^2} + \frac{\partial^2 V}{\partial y^2} + \frac{\partial^2 V}{\partial z^2} = 0. \tag{1.7.58}$$

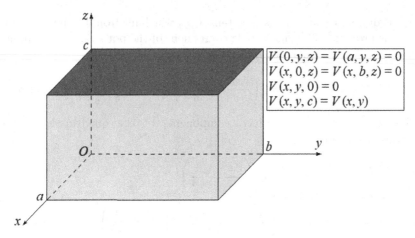

FIGURE 1.8
Illustration of boundary conditions for the electrostatic potential on the walls of the parallelepipedic box of sides a, b and c.

To solve this equation we will use the Fourier method (variable separation method) and look for the solution $V(x, y, z)$ of the form

$$V(x, y, z) = X(x)Y(y)Z(z). \tag{1.7.59}$$

According to the "standard procedure", one introduces Eq. (1.7.59) in Eq. (1.7.58) and the result is divided by $X(x)Y(y)Z(z) \equiv V(x, y, z)$. The result is

$$\frac{1}{X(x)} \frac{\partial^2 X(x)}{\partial x^2} + \frac{1}{Y(y)} \frac{\partial^2 Y(y)}{\partial y^2} + \frac{1}{Z(z)} \frac{\partial^2 Z(z)}{\partial z^2} = 0. \tag{1.7.60}$$

This relation is satisfied only if each term on the l.h.s. equals a constant. These (real) constants can be conveniently chosen, so that the algebraic sum of the three constants equals zero; for instance,

$$\frac{1}{X(x)} \frac{\partial^2 X(x)}{\partial x^2} = -\lambda^2, \tag{1.7.61}$$

$$\frac{1}{Y(y)} \frac{\partial^2 Y(y)}{\partial y^2} = -\mu^2, \tag{1.7.62}$$

$$\frac{1}{Z(z)} \frac{\partial^2 z(z)}{\partial z^2} = \chi^2, \tag{1.7.63}$$

with

$$\chi^2 = \lambda^2 + \mu^2. \tag{1.7.64}$$

The general form of the solutions of Eqs. (1.7.61)−(1.7.63) is

$$X(x) = A_1 e^{i\lambda x} + B_1 e^{-i\lambda x}, \tag{1.7.65}$$

$$Y(y) = A_2 e^{i\mu y} + B_2 e^{-i\mu y}, \tag{1.7.66}$$

$$Z(z) = A_3 e^{\chi z} + B_3 e^{-\chi z}, \tag{1.7.67}$$

or, equivalently,

$$X(x) = M_1 \cos(\lambda x) + N_1 \sin(\lambda x), \tag{1.7.68}$$

$$Y(y) = M_2 \cos(\mu y) + N_2 \sin(\mu y), \tag{1.7.69}$$

$$Z(z) = M_3 \cosh(\chi z) + N_3 \sinh(\chi z), \tag{1.7.70}$$

where the arbitrary integration constants M_i and N_i, $i = \overline{1,3}$, are determined by means of the boundary conditions. First, the boundary conditions

$$V(0, y, z)\big[= X(0)Y(y)Z(z)\big] = 0, \quad \forall y \in \big[0, b\big], \ z \in \big[0, c\big],$$

$$V(x, 0, z)\big[= X(x)Y(0)Z(z)\big] = 0, \quad \forall x \in \big[0, a\big], \ z \in \big[0, c\big],$$

$$V(x, y, 0)\big[= X(x)Y(y)Z(0)\big] = 0, \quad \forall x \in \big[0, a\big], \ y \in \big[0, b\big],$$

demand $M_i = 0$, $i = \overline{1,3}$, so that $X(x)$, $Y(y)$ and $Z(z)$ receive a simpler form, namely

$$X(x) = N_1 \sin(\lambda x), \tag{1.7.71}$$

$$Y(y) = N_2 \sin(\mu y), \tag{1.7.72}$$

$$Z(z) = N_3 \sinh(\chi z). \tag{1.7.73}$$

Secondly, from the boundary conditions

$$V(a, y, z) = \big[X(a)Y(y)Z(z)\big] = 0, \quad \forall y \in \big[0, b\big], \ z \in \big[0, c\big],$$

and

$$V(x, b, z) = \big[X(x)Y(b)Z(z)\big] = 0, \quad \forall x \in \big[0, a\big], \ z \in \big[0, c\big],$$

one follows that λ and μ must be of the form

$$\lambda \equiv \lambda_n = \frac{n\pi}{a}, \quad n \in \mathbb{Z}, \tag{1.7.74}$$

$$\mu \equiv \mu_m = \frac{m\pi}{b}, \quad m \in \mathbb{Z}. \tag{1.7.75}$$

According to Eq. (1.7.64), we then have

$$\chi_{n,m} = \pi \sqrt{\frac{n^2}{a^2} + \frac{m^2}{b^2}}, \quad n, m \in \mathbb{Z}. \tag{1.7.76}$$

The above considerations show that the potential $V(x, y, z)$ has to be written as a double series of the form

$$V(x, y, z) = \sum_{n=1}^{\infty} \sum_{m=1}^{\infty} A_{nm} \sin(\lambda_n x) \sin(\mu_m y) \sinh(\chi_{nm} z), \tag{1.7.77}$$

where the set of quantities $A_{nm} = (N_1 N_2 N_3)_{nm}$ is going to be determined from the condition $V(x, y, c) = V(x, y)$, where $V(x, y)$ is known:

$$V(x, y) = \sum_{n=1}^{\infty} \sum_{m=1}^{\infty} A_{nm} \sin(\lambda_n x) \sin(\mu_m y) \sinh(\chi_{nm} c). \tag{1.7.78}$$

This is a double Fourier series for $V(x, y)$. Multiplying Eq. (1.7.78) by $\sin(\lambda_p x) \sin(\mu_q y)$ and integrating over x and y, one obtains

$$\int_0^a dx \int_0^b dy \, V(x, y) \sin(\lambda_p x) \sin(\mu_q y)$$

$$= \int_0^a dx \int_0^b dy \sum_{n=1}^{\infty} \sum_{m=1}^{\infty} A_{nm} \sin(\lambda_n x) \sin(\mu_m y)$$

$$\times \sin(\lambda_p x) \sin(\mu_q y) \sinh(\chi_{nm} c) = \sum_{n=1}^{\infty} \sum_{m=1}^{\infty} A_{nm} \sinh(\chi_{nm} c)$$

$$\times \int_0^a dx \int_0^b dy \sin(\lambda_n x) \sin(\lambda_p x) \sin(\mu_m y) \sin(\mu_q y)$$

$$= \frac{ab}{4} \sum_{n=1}^{\infty} \sum_{m=1}^{\infty} A_{nm} \sinh(\chi_{nm} c) \delta_{np} \delta_{mq} = \frac{ab}{4} \sinh(\chi_{pq} c) A_{pq}, \tag{1.7.79}$$

where the following orthogonality relations have been used:

$$\int_0^a \sin(\lambda_m x) \sin(\lambda_n x) dx = \frac{a}{2} \delta_{mn},$$

$$\int_0^b \sin(\mu_m y) \sin(\mu_n y) dy = \frac{b}{2} \delta_{mn}.$$

In view of Eq. (1.7.79), one can write

$$A_{pq} = \frac{4}{ab \sinh(\chi_{pq} c)} \int_0^a dx \int_0^b dy \, V(x, y) \sin(\lambda_p x) \sin(\mu_q y). \tag{1.7.80}$$

Introducing Eq. (1.7.80) into Eq. (1.7.77), we can write the final solution of the problem as

$$V(x, y, z) = \frac{4}{ab} \sum_{n=1}^{\infty} \sum_{m=1}^{\infty} \left[\frac{\sin(\lambda_n x) \sin(\mu_m y) \sinh(\chi_{nm} z)}{\sinh(\chi_{nm} c)} \right.$$

$$\left. \times \int_0^a dx \int_0^b dy \, V(x, y) \sin(\lambda_n x) \sin(\mu_m y) \right]. \tag{1.7.81}$$

In particular, if $V(x, y) = V_0 = const.$, according to Eq. (1.7.80) one follows that

$$
\begin{aligned}
A_{pq} &= \frac{4}{ab \sinh(\chi_{pq}c)} \int_0^a dx \int_0^b dy\, V(x,y) \sin(\lambda_p x) \sin(\mu_q y) \\
&= \frac{4V_0}{ab \sinh(\chi_{pq}c)} \int_0^a dx \int_0^b dy\, \sin(\lambda_p x) \sin(\mu_q y) \\
&= \frac{4V_0}{ab \sinh(\chi_{pq}c)} \frac{1}{\lambda_p \mu_q} \left[1 - \cos(p\pi)\right]\left[1 - \cos(q\pi)\right] \\
&= \frac{4V_0}{pq\,\pi^2 \sinh(\chi_{pq}c)} \times \begin{cases} 4, & \text{if } p,q = 2k+1, \quad k \in \mathbb{Z}, \\ 0, & \text{if } p,q = 2k, \quad k \in \mathbb{Z}, \end{cases}
\end{aligned}
$$

which finally gives

$$
\begin{aligned}
V(x,y,z) &= \frac{16V_0}{\pi^2} \sum_{n=1}^{\infty} \sum_{m=1}^{\infty} \frac{(2m-1)^{-1}(2n-1)^{-1}}{\sinh(\chi_{n'm'}c)} \\
&\quad \times \sin(\lambda_{n'}x) \sin(\mu_{m'}y) \sinh(\chi_{n'm'}z),
\end{aligned} \tag{1.7.82}
$$

where $n' = 2n - 1$ and $m' = 2m - 1$, $m, n \in \mathbb{Z}$.

1.8   Problem No. 8

A rectilinear, infinitely long wire of negligible section, uniformly charged with the electric charge Λ per unit length, is inserted in an infinitely metal tube with rectangular section ($0 \le x \le a$, $0 \le y \le b$), so that the wire is parallel to the edges of the tube. Let x_0 and y_0 be the wire coordinates in a cross section of the tube. Determine the electric potential in the tube cross section, if its walls are grounded.

Solution

Consider, for the beginning, that the electric charge of the wire is uniformly distributed, in a very thin prismatic column, the rectangular section having its faces parallel to the tube walls (see Fig. 1.9).

The spatial electric charge density in the tube cross section can be written as

$$
\rho(x,y) = \begin{cases} \frac{\Lambda}{4hk}, & \text{for } x_0 - h < x < x_0 + h, \ y_0 - k < y < y_0 + k, \\ 0, & \text{for any other values of } x \text{ and } y. \end{cases} \tag{1.8.83}
$$

Obviously, we will proceed to the limit at the right time, by directing h and k to zero. Since the charge density distributed in the cross section of the

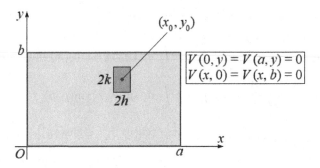

FIGURE 1.9
Cross-section through the grounded rectangular metal tube, showing the boundary conditions for the electrostatic potential on the walls of the tube, and the uniform distribution of electric charge in the prismatic column section having sides $2h$ and $2k$.

tube is known, the problem requires integration of the Poisson equation in the cross section of the tube:

$$\frac{\partial^2 V}{\partial x^2} + \frac{\partial^2 V}{\partial y^2} = -\frac{1}{\varepsilon_0}\rho(x,y), \tag{1.8.84}$$

with boundary conditions

$$\begin{cases} V(0,y) = V(a,y) = 0, \\ V(x,0) = V(x,b) = 0. \end{cases} \tag{1.8.85}$$

The boundary conditions expressed by Eq. (1.8.85) suggests a solution for Eq. (1.8.84) of the form

$$V(x,y) = \sum_{n=1}^{\infty} u_n(x) \sin\left(\frac{n\pi}{b}y\right), \tag{1.8.86}$$

since, this way, the condition expressed by Eq. $(1.8.85)_2$ for $y = 0$ and $y = b$ (regardless of the values of variable x) is automatically satisfied.

Demanding for Eq. (1.8.84) to be satisfied by solution given by Eq. (1.8.86), one obtains

$$\sum_{n=1}^{\infty}\left[u_n''(x) - \frac{n^2\pi^2}{b^2}u_n(x)\right]\sin\left(\frac{n\pi}{b}y\right) = -\frac{1}{\varepsilon_0}\rho(x,y). \tag{1.8.87}$$

According to the orthogonality relation

$$\int_0^b \sin\left(\frac{n\pi}{b}y\right)\sin\left(\frac{m\pi}{b}y\right)dy = \frac{b}{2}\delta_{nm},$$

and using Eq. (1.8.87), one obtains the following ordinary differential equation of order two for $u_n(x)$:

$$u_n''(x) - \frac{n^2\pi^2}{b^2}u_n(x) = -\frac{2}{b\varepsilon_0}\int_0^b \rho(x,\eta)\sin\left(\frac{n\pi}{b}\eta\right)d\eta\,. \qquad (1.8.88)$$

The general solution of this equation can be obtained by means of the Lagrange's method of variation of parameters. The homogeneous equation attached to the non-homogeneous Eq. (1.8.88) is

$$\frac{d^2u_n(x)}{dx^2} - \frac{n^2\pi^2}{b^2}u_n(x) = 0\,,$$

and, since the solutions of the attached characteristic equation are real, it has the following fundamental system of solutions:

$$\begin{cases} u_n^{(1)}(x) = \sinh\left(\frac{n\pi}{b}x\right)\,, \\ u_n^{(2)}(x) = \cosh\left(\frac{n\pi}{b}x\right)\,. \end{cases}$$

According to the Lagrange's method of variation of parameters, the general solution of the non-homogeneous equation then is

$$u_n(x) = u_n^{(1)}(x)\int A_n'(x)\,dx + u_n^{(2)}(x)\int B_n'(x)\,dx\,, \qquad (1.8.89)$$

where $A_n'(x)$ and $B_n'(x)$ are the solutions of the algebraic system

$$\begin{cases} u_n^{(1)}(x)A_n'(x) + u_n^{(2)}(x)B_n'(x) = 0\,, \\ \dfrac{du_n^{(1)}}{dx}A_n'(x) + \dfrac{du_n^{(2)}}{dx}B_n'(x) \\ \qquad = -\dfrac{2}{b\varepsilon_0}\displaystyle\int_0^b \rho(x,\eta)\sin\left(\frac{n\pi}{b}\eta\right)d\eta\,, \end{cases}$$

that is

$$\begin{cases} \sinh\left(\dfrac{n\pi}{b}x\right)A_n'(x) + \cosh\left(\dfrac{n\pi}{b}x\right)B_n'(x) = 0\,, \\ \dfrac{n\pi}{b}\cosh\left(\dfrac{n\pi}{b}x\right)A_n'(x) + \dfrac{n\pi}{b}\sinh\left(\dfrac{n\pi}{b}x\right)B_n'(x) \\ \qquad = -\dfrac{2}{b\varepsilon_0}\displaystyle\int_0^b \rho(x,\eta)\sin\left(\frac{n\pi}{b}\eta\right)d\eta\,. \end{cases}$$

The first equation of the above system yields

$$B_n'(x) = -A_n'(x)\tanh\left(\frac{n\pi}{b}x\right)\,,$$

in which case the second equation of the same system becomes

$$\frac{n\pi}{b} A'_n(x) = -\frac{2}{b\varepsilon_0} \cosh\left(\frac{n\pi}{b}x\right) \int_0^b \rho(x,\eta) \sin\left(\frac{n\pi}{b}\eta\right) d\eta, \qquad (1.8.90)$$

where the identity of the hyperbolic functions trigonometry $\cosh^2 \psi - \sinh^2 \psi = 1$, $\forall \psi \in \mathbb{R}$ has been used. Equation (1.8.90) then yields

$$A'_n(x) = -\frac{2}{n\pi\varepsilon_0} \cosh\left(\frac{n\pi}{b}x\right) \int_0^b \rho(x,\eta) \sin\left(\frac{n\pi}{b}\eta\right) d\eta,$$

so that

$$A_n(x) = \int A'_n(x)dx = -\frac{2}{n\pi\varepsilon_0} \int \left[\int_0^b \rho(x,\eta) \sin\left(\frac{n\pi}{b}\eta\right) d\eta\right] \cosh\left(\frac{n\pi}{b}x\right) dx$$

$$= -\frac{2}{n\pi\varepsilon_0} \int_0^x \left[\int_0^b \rho(\xi,\eta) \sin\left(\frac{n\pi}{b}\eta\right) d\eta\right] \cosh\left(\frac{n\pi}{b}\xi\right) d\xi + \mathcal{A}_n,$$

where $\mathcal{A}_n$ are pure arbitrary integration constants. In this case

$$B'_n(x) = -A'_n(x) \tanh\left(\frac{n\pi}{b}x\right)$$

$$= \frac{2}{n\pi\varepsilon_0} \sinh\left(\frac{n\pi}{b}x\right) \int_0^b \rho(x,\eta) \sin\left(\frac{n\pi}{b}\eta\right) d\eta, \qquad (1.8.91)$$

which leads to

$$B_n(x) = \int B'_n(x)\, dx = \frac{2}{n\pi\varepsilon_0} \int \left[\int_0^b \rho(x,\eta) \sin\left(\frac{n\pi}{b}\eta\right) d\eta\right] \sinh\left(\frac{n\pi}{b}x\right) dx$$

$$= \frac{2}{n\pi\varepsilon_0} \int_0^x \left[\int_0^b \rho(\xi,\eta) \sin\left(\frac{n\pi}{b}\eta\right) d\eta\right] \sinh\left(\frac{n\pi}{b}\xi\right) d\xi + \mathcal{B}_n,$$

where $\mathcal{B}_n$ are also true integration constants. Both $\mathcal{A}_n$ and $\mathcal{B}_n$ are determined by means of the boundary conditions for $u_n(x)$, namely $u_n(0) = 0$ and $u_n(a) = 0$.

If $\int A'_n(x)dx$ and $\int B'_n(x)dx$ are now introduced into Eq. (1.8.89), one obtains the following expression for the general solution of Eq. (1.8.88):

$$u_n(x) = u_n^{(1)}(x) \int A'_n(x)\, dx + u_n^{(2)}(x) \int B'_n(x)\, dx$$

$$= \sinh\left(\frac{n\pi}{b}x\right) \left\{ -\frac{2}{n\pi\varepsilon_0} \int \left[\int_0^b \rho(x,\eta) \sin\left(\frac{n\pi}{b}\eta\right) d\eta\right] \right.$$

$$\left. \times \cosh\left(\frac{n\pi}{b}x\right) dx \right\} + \cosh\left(\frac{n\pi}{b}x\right) \left\{ \frac{2}{n\pi\varepsilon_0} \int \left[\int_0^b \rho(x,\eta)\right.\right.$$

(a)   (b)

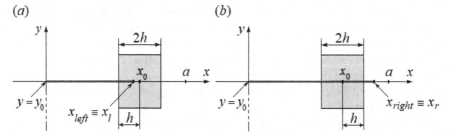

FIGURE 1.10
An auxiliary construction used to explain the solution to problem, concerning
the process of passing to the limit $h \to 0$.

$$\left. \times \sin\left(\frac{n\pi}{b}\eta\right) d\eta \right] \sinh\left(\frac{n\pi}{b}x\right) dx \right\}$$

$$= \mathcal{A}_n \sinh\left(\frac{n\pi}{b}x\right) + \mathcal{B}_n \cosh\left(\frac{n\pi}{b}x\right) - \frac{2}{n\pi\varepsilon_0}\sinh\left(\frac{n\pi}{b}x\right)$$

$$\times \int_0^x \int_0^b \rho(\xi,\eta) \sin\left(\frac{n\pi}{b}\eta\right) \cosh\left(\frac{n\pi}{b}\xi\right) d\xi d\eta + \frac{2}{n\pi\varepsilon_0}$$

$$\times \cosh\left(\frac{n\pi}{b}x\right) \int_0^x \int_0^b \rho(\xi,\eta) \sin\left(\frac{n\pi}{b}\eta\right) \sinh\left(\frac{n\pi}{b}\xi\right) d\xi d\eta$$

$$= \mathcal{A}_n \sinh\left(\frac{n\pi}{b}x\right) + \mathcal{B}_n \cosh\left(\frac{n\pi}{b}x\right) - \frac{2}{n\pi\varepsilon_0} \int_0^x \int_0^b \rho(\xi,\eta)$$

$$\times \sin\left(\frac{n\pi}{b}\eta\right) \left[\sinh\left(\frac{n\pi}{b}x\right) \cosh\left(\frac{n\pi}{b}\xi\right) \right.$$

$$\left. - \sinh\left(\frac{n\pi}{b}\xi\right) \cosh\left(\frac{n\pi}{b}x\right) \right] d\xi d\eta$$

$$= \mathcal{A}_n \sinh\left(\frac{n\pi}{b}x\right) + \mathcal{B}_n \cosh\left(\frac{n\pi}{b}x\right)$$

$$- \frac{2}{n\pi\varepsilon_0} \int_0^x \int_0^b \rho(\xi,\eta) \sinh\left[\frac{n\pi}{b}(x-\xi)\right] \sin\left(\frac{n\pi}{b}\eta\right) d\xi d\eta. \quad (1.8.92)$$

For $x < x_0$, the last term vanishes by going to the limit (h can be supposed
small enough so that $x < x_0 - h$ and then, $\rho = 0$ according to Eq. (1.8.83)).
Indeed, for $x < x_0$ (*i.e.*, for x situated to the left of x_0 – see the Fig. 1.10.a),
for any value of x, with $x < x_0$, that is, for any length x_s which cannot be
greater than x_0, when going to the limit $h \to 0$, x_s will always remain outside
of the rectangle of width $2h$ (remember that the integration is performed from
0 to $x < x_0$ (or, more precisely, from 0 to x_s) and, when the rectangle shrinks,
as a result of going to the limit, x_s will always remain outside the rectangle,
no matter how big is $x_s < x_0$).

On the contrary, for $x > x_0$, no matter how long the "integration segment" would be (*i.e.*, no matter how big is $x_d < a$ − see the Fig. 1.10.*b*), when going to the limit $h \to 0$ (when the rectangle shrinks more and more until − at the limit − it becomes the straight line segment of equation $x = x_0$, of length $2k$ and oriented perpendicular to x-axis) there will always be a small rectangle of width $2h$ and height $2k$ for which $\rho(\xi, \eta) = \frac{\Lambda}{4hk} \neq 0$, situated inside the integration interval. In other words, $\forall x_d$, with $x_0 < x_d < a$, $\exists h < x_d - x_0 \neq 0$. Then, for $x > x_0$, supposing that h is small enough to have $x_0 + h < x$, and taking into account Eq. (1.8.83), we get

$$I(h,k) \equiv \int_0^x \int_0^b \rho(\xi,\eta) \sinh\left[\frac{n\pi}{b}(x-\xi)\right] \sin\left(\frac{n\pi}{b}\eta\right) d\xi d\eta$$

$$= \frac{\Lambda}{4hk} \int_{x_0-h}^{x_0+h} \int_{y_0-k}^{y_0+k} \sinh\left[\frac{n\pi}{b}(x-\xi)\right] \sin\left(\frac{n\pi}{b}\eta\right) d\xi d\eta$$

$$= \frac{\Lambda}{4hk} \int_{x_0-h}^{x_0+h} \sinh\left[\frac{n\pi}{b}(x-\xi)\right] d\xi \int_{y_0-k}^{y_0+k} \sin\left(\frac{n\pi}{b}\eta\right) d\eta$$

$$= \frac{\Lambda}{4hk} \frac{b^2}{(n\pi)^2} \left\{-\cosh\left[\frac{n\pi}{b}(x-\xi)\right]\right\}_{x_0-h}^{x_0+h} \left[-\cos\left(\frac{n\pi}{b}\eta\right)\right]_{y_0-k}^{y_0+k}$$

$$= \frac{\Lambda b^2}{4hk(n\pi)^2} \left\{\cosh\left[\frac{n\pi}{b}(x-x_0+h)\right] - \cosh\left[\frac{n\pi}{b}(x-x_0-h)\right]\right\}$$

$$\times \left\{\cos\left[\frac{n\pi}{b}(y_0-k)\right] - \cos\left[\frac{n\pi}{b}(y_0+k)\right]\right\}. \tag{1.8.93}$$

Going to the limit $(h,k) \to 0$ in Eq. (1.8.93), one obtains

$$\lim_{\substack{h\to 0 \\ k\to 0}} I(h,k) = \lim_{\substack{h\to 0 \\ k\to 0}} \int_0^x \int_0^b \rho(\xi,\eta) \sinh\left[\frac{n\pi}{b}(x-\xi)\right] \sin\left(\frac{n\pi}{b}\eta\right) d\xi d\eta$$

$$= \lim_{\substack{h\to 0 \\ k\to 0}} \left\{\frac{\Lambda b^2}{4hk(n\pi)^2} \left\{\cosh\left[\frac{n\pi}{b}(x-x_0+h)\right] - \cosh\left[\frac{n\pi}{b}\right.\right.\right.$$

$$\left.\left.\times(x-x_0-h)\right]\right\} \left\{\cos\left[\frac{n\pi}{b}(y_0-k)\right] - \cos\left[\frac{n\pi}{b}(y_0+k)\right]\right\}\right\}$$

$$= \frac{\Lambda b^2}{4(n\pi)^2} \lim_{h\to 0} \frac{\cosh\left[\frac{n\pi}{b}(x-x_0+h)\right] - \cosh\left[\frac{n\pi}{b}(x-x_0-h)\right]}{h}$$

$$\times \lim_{k\to 0} \frac{\cos\left[\frac{n\pi}{b}(y_0-k)\right] - \cos\left[\frac{n\pi}{b}(y_0+k)\right]}{k}$$

$$= \frac{\Lambda b^2}{4(n\pi)^2} \left\{\lim_{h\to 0} \frac{2\sinh\left[\frac{n\pi}{b}(x-x_0)\right] \sinh\left(\frac{n\pi}{b}h\right)}{h}\right\}$$

$$\times \left[\lim_{k\to 0} \frac{2\sin\left(\frac{n\pi}{b}y_0\right) \sin\left(\frac{n\pi}{b}k\right)}{k}\right] = \frac{\Lambda b^2}{(n\pi)^2} \sinh\left[\frac{n\pi}{b}(x-x_0)\right]$$

$$\times \sin\left(\frac{n\pi}{b}y_0\right)\left[\lim_{h\to 0}\frac{\sinh\left(\frac{n\pi}{b}h\right)}{h}\right]\left[\lim_{k\to 0}\frac{\sin\left(\frac{n\pi}{b}k\right)}{k}\right]$$

$$= \frac{\Lambda b^2}{(n\pi)^2}\sinh\left[\frac{n\pi}{b}(x-x_0)\right]\sin\left(\frac{n\pi}{b}y_0\right)$$

$$\times \left[\frac{n\pi}{b}\underbrace{\lim_{h\to 0}\frac{\sinh\left(\frac{n\pi}{b}h\right)}{\frac{n\pi}{b}h}}_{=1}\right]\left[\frac{n\pi}{b}\underbrace{\lim_{k\to 0}\frac{\sinh\left(\frac{n\pi}{b}k\right)}{\frac{n\pi}{b}k}}_{=1}\right]$$

$$= \Lambda\sinh\left[\frac{n\pi}{b}(x-x_0)\right]\sin\left(\frac{n\pi}{b}y_0\right).$$

The quantities $u_n(x)$ are then given by the formula

$$u_n(x) = \mathcal{A}_n\sinh\left(\frac{n\pi}{b}x\right) + \mathcal{B}_n\cosh\left(\frac{n\pi}{b}x\right)$$

$$+ \begin{cases} 0, & \text{for } x < x_0, \\ -\frac{2\Lambda}{n\pi\varepsilon_0}\sinh\left[\frac{n\pi}{b}(x-x_0)\right]\sin\left(\frac{n\pi}{b}y_0\right), & \text{for } x > x_0. \end{cases} \quad (1.8.94)$$

The condition $u_n(0) = 0$ yields $\mathcal{B}_n = 0$, while the condition $u_n(a) = 0$ leads to

$$\mathcal{A}_n = \frac{2\Lambda}{n\pi\varepsilon_0}\left[\sinh\left(\frac{n\pi}{b}a\right)\right]^{-1}\sinh\left[\frac{n\pi}{b}(a-x_0)\right]\sin\left(\frac{n\pi}{b}y_0\right). \quad (1.8.95)$$

Introducing now $\mathcal{A}_n$ and $\mathcal{B}_n$ into Eq. (1.8.94), one obtains the final form of $u_n(x)$ and, by means of Eq. (1.8.86), the answer to the problem (*i.e.*, the solution of Eq. (1.8.84)):

$$V(x,y) = \begin{cases} \dfrac{2\Lambda}{\pi\varepsilon_0}\displaystyle\sum_{n=1}^{\infty}\left[n\sinh\left(\frac{n\pi}{b}a\right)\right]^{-1}\sinh\left[\frac{n\pi}{b}(a-x_0)\right]\sin\left(\frac{n\pi}{b}y_0\right) \\ \times\sinh\left(\frac{n\pi}{b}x\right)\sin\left(\frac{n\pi}{b}y\right), \quad \text{for } x < x_0, \\[2mm] \dfrac{2\Lambda}{\pi\varepsilon_0}\displaystyle\sum_{n=1}^{\infty}\left[n\sinh\left(\frac{n\pi}{b}a\right)\right]^{-1}\sin\left[\frac{n\pi}{b}(a-x_0)\right]\sin\left(\frac{n\pi}{b}y_0\right) \\ \times\sinh\left(\frac{n\pi}{b}x\right)\sin\left(\frac{n\pi}{b}y\right) - \dfrac{2\Lambda}{\pi\varepsilon_0}\displaystyle\sum_{n=1}^{\infty}\frac{1}{n} \\ \times\sinh\left[\frac{n\pi}{b}(x-x_0)\right]\sin\left(\frac{n\pi}{b}y_0\right)\sin\left(\frac{n\pi}{b}y\right), \quad \text{for } x > x_0. \end{cases}$$

If in $V(x,y)$ for $x > x_0$ one considers the identity

$$\sinh(\alpha-\beta)\sinh(\gamma) + \sinh(\beta-\gamma)\sinh(\alpha) + \sinh(\gamma-\alpha)\sinh(\beta) = 0,$$

$$\forall\,\alpha,\beta,\gamma\in\mathbb{R}, \quad (1.8.96)$$

where $\alpha = \dfrac{n\pi}{b}a$, $\beta = \dfrac{n\pi}{b}x_0$, $\gamma = \dfrac{n\pi}{b}x$, then the result can be written in the following simpler form:

$$V(x,y) = \begin{cases} \dfrac{2\Lambda}{\pi\varepsilon_0}\displaystyle\sum_{n=1}^{\infty}\left[n\sinh\left(\dfrac{n\pi}{b}y_0\right)\right]^{-1}\sinh\left[\dfrac{n\pi}{b}(a-x_0)\right]\sin\left(\dfrac{n\pi}{b}a\right) \\ \quad\times\sinh\left(\dfrac{n\pi}{b}x\right)\sin\left(\dfrac{n\pi}{b}y\right), \quad \text{for } x < x_0, \\[4mm] \dfrac{2\Lambda}{\pi\varepsilon_0}\displaystyle\sum_{n=1}^{\infty}\left[n\sinh\left(\dfrac{n\pi}{b}a\right)\right]^{-1}\sinh\left(\dfrac{n\pi}{b}x_0\right)\sin\left(\dfrac{n\pi}{b}y_0\right) \\ \quad\times\sinh\left[\dfrac{n\pi}{b}(a-x)\right]\sin\left(\dfrac{n\pi}{b}y\right), \quad \text{for } x > x_0. \end{cases}$$

1.9 Problem No. 9

An electric point charge q is placed at the point $P_0(x_0, y_0, z_0)$ situated inside a parallelepipedic box $(0 \le x \le a, \ 0 \le y \le b, \ 0 \le z \le c)$. The walls of the box are conductive and grounded. Determine the electrostatic potential inside the box.

Solution

To avoid the difficult work with exact mathematical expression of the volume charge density of a single point charge distribution, we will first consider that the electric charge q is uniformly distributed in a small parallelepiped of dimensions $2h \times 2k \times 2l$, whose sides are parallel to the box walls, and then go to the limit when simultaneously $h \to 0$, $k \to 0$, $l \to 0$. In this situation, the electric charge density $\rho(x, y, z)$ writes (see Fig. 1.11) as follows:

$$\rho(x,y,z) = \begin{cases} \dfrac{q}{8hkl}, & \text{for } x_0 - h < x < x_0 + h, \\ & y_0 - k < y < y_0 + k, \ z_0 - l < z < z_0 + l, \\ 0, & \text{for the other values of } x, y, \text{ and } z, \end{cases} \quad (1.9.97)$$

while the electrostatic field potential is obtained as the solution of the Poisson's equation

$$\frac{\partial^2 V}{\partial x^2} + \frac{\partial^2 V}{\partial y^2} + \frac{\partial^2 V}{\partial z^2} = -\frac{1}{\varepsilon_0}\rho(x,y,z), \qquad (1.9.98)$$

with the boundary conditions

$$\begin{cases} V(0,y,z) = V(a,y,z) = 0, & \forall \ y, z \text{ inside the box,} \\ V(x,0,y) = V(x,b,y) = 0, & \forall \ x, z \text{ inside the box,} \\ V(x,y,0) = V(x,y,c) = 0, & \forall \ x, y \text{ inside the box,} \end{cases}$$

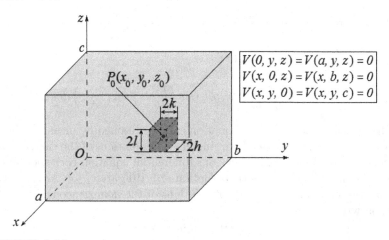

FIGURE 1.11
Schematic representation of the conductive and grounded parallelepipedic box, showing the boundary conditions for the electrostatic potential on the walls of the box.

where, due to the rectangular symmetry of the problem, the Laplacian was expressed in Cartesian coordinates.

In view of the above given boundary conditions, the solution of Eq. (1.9.98) is searched in the form

$$V(x, y, z) = \sum_{m=1}^{\infty} \sum_{n=1}^{\infty} u_{mn}(x) \sin\left(\frac{m\pi}{b}y\right) \sin\left(\frac{n\pi}{c}z\right), \qquad (1.9.99)$$

which automatically ensures the cancellation of the potential on the walls $y = 0$, $y = b$, $z = 0$ and $z = c$. By imposing that the potential given by Eq. (1.9.99) verifies Eq. (1.9.98), one obtains

$$\sum_{m=1}^{\infty} \sum_{n=1}^{\infty} \left[\frac{d^2 u_{mn}(x)}{dx^2} - \alpha_{mn}^2 u_{mn}(x)\right]$$
$$\times \sin\left(\frac{m\pi}{b}y\right) \sin\left(\frac{n\pi}{c}z\right) = -\frac{1}{\varepsilon_0}\rho(x, y, z), \qquad (1.9.100)$$

where $\alpha_{mn} = \pi\sqrt{\dfrac{m^2}{b^2} + \dfrac{n^2}{c^2}}$. In view of the orthogonality relations

$$\int_0^b \sin\left(\frac{n\pi}{b}y\right) \sin\left(\frac{m\pi}{b}y\right) dy = \frac{b}{2}\delta_{mn}$$

and

$$\int_0^c \sin\left(\frac{n\pi}{c}z\right) \sin\left(\frac{m\pi}{c}z\right) dz = \frac{c}{2}\delta_{mn},$$

one follows from Eq. (1.9.100) that

$$\frac{d^2 u_{mn}(x)}{dx^2} - \alpha_{mn}^2 u_{mn}(x) = -\frac{4}{bc\varepsilon_0}$$

$$\times \int_0^b \int_0^c \rho(x, \eta, \zeta) \sin\left(\frac{m\pi}{b}\eta\right) \sin\left(\frac{n\pi}{c}\zeta\right) d\eta d\zeta. \quad (1.9.101)$$

The general solution of this equation can be obtained by means of the Lagrange's method of variation of parameters. The solutions of the characteristic equation attached to the homogeneous differential equation corresponding to the non-homogeneous differential equation (1.9.101) are $r_{1,2} = \pm\alpha_{mn}$, meaning that the fundamental system of solutions of the homogeneous differential equation writes

$$\begin{cases} u_{mn}^{(1)}(x) = \sinh(\alpha_{mn}x), \\ u_{mn}^{(2)}(x) = \cosh(\alpha_{mn}x). \end{cases}$$

According to the Lagrange's method of variation of parameters, the general solution of the non-homogeneous differential equation then is

$$u_{mn}(x) = u_{mn}^{(1)}(x) \int A_{mn}'(x)dx + u_{mn}^{(2)}(x) \int B_{mn}'(x)dx, \quad (1.9.102)$$

where $A_{mn}'(x)$ and $B_{mn}'(x)$ are the solutions of the algebraic system

$$\begin{cases} u_{mn}^{(1)}(x)A_{mn}'(x) + u_{mn}^{(2)}(x)B_{mn}'(x) = 0, \\ \dfrac{du_{mn}^{(1)}}{dx}A_{mn}'(x) + \dfrac{du_{mn}^{(2)}}{dx}B_{mn}'(x) \\ = -\dfrac{4}{bc\varepsilon_0} \int_0^b \int_0^c \rho(x, \eta, \zeta) \sin\left(\dfrac{m\pi}{b}\eta\right) \sin\left(\dfrac{n\pi}{c}\zeta\right) d\eta d\zeta, \end{cases}$$

that is

$$\begin{cases} \sinh(\alpha_{mn}x)A_{mn}'(x) + \cosh(\alpha_{mn}x)B_{mn}'(x) = 0, \\ \alpha_{mn}\cosh(\alpha_{mn}x)A_{mn}'(x) + \alpha_{mn}\sinh(\alpha_{mn}x)B_{mn}'(x) \\ = -\dfrac{4}{bc\varepsilon_0} \int_0^b \int_0^c \rho(x, \eta, \zeta) \sin\left(\dfrac{m\pi}{b}\eta\right) \sin\left(\dfrac{n\pi}{c}\zeta\right) d\eta d\zeta. \end{cases}$$

The first equation of the above system gives

$$B_{mn}'(x) = -A_{mn}'(x)\tanh(\alpha_{mn}x),$$

in which case, the second equation of the same system leads to

$$\alpha_{mn}A_{mn}'(x) = -\frac{4\cosh(\alpha_{mn}x)}{bc\varepsilon_0} \int_0^b \int_0^c \rho(x, \eta, \zeta)$$

$$\times \sin\left(\frac{m\pi}{b}\eta\right) \sin\left(\frac{n\pi}{c}\zeta\right) d\eta d\zeta, \quad (1.9.103)$$

where the identity $\cosh^2 \psi - \sinh^2 \psi = 1$, $\forall\, \psi \in \mathbb{R}$ has been considered. Given that $A'_{mn}(x) = dA_{mn}(x)/dx$, Eq. (1.9.103) yields

$$A_{mn}(x) = -\int_0^x \left[\int_0^b \int_0^c \rho(\xi,\eta,\zeta) \sin\left(\frac{m\pi}{b}\eta\right) \sin\left(\frac{n\pi}{c}\zeta\right) d\eta d\zeta \right]$$
$$\times \frac{4}{bc\varepsilon_0\alpha_{mn}} \cosh(\alpha_{mn}\xi)d\xi + \mathcal{A}_{mn}, \qquad (1.9.104)$$

where $\mathcal{A}_{mn}$ are true, arbitrary, integration constants. Then

$$B'_{mn}(x) = -A'_{mn}(x)\tanh(\alpha_{mn}x) = \frac{4\sinh(\alpha_{mn}x)}{bc\varepsilon_0\alpha_{mn}}$$
$$\times \int_0^b \int_0^c \rho(x,\eta,\zeta) \sin\left(\frac{m\pi}{b}\eta\right) \sin\left(\frac{n\pi}{c}\zeta\right) d\eta d\zeta, \quad (1.9.105)$$

from which we obtain

$$B_{mn}(x) = \int_0^x \left[\int_0^b \int_0^c \rho(\xi,\eta,\zeta) \sin\left(\frac{m\pi}{b}\eta\right) \sin\left(\frac{n\pi}{c}\zeta\right) d\eta d\zeta \right]$$
$$\times 4(bc\varepsilon_0\alpha_{mn})^{-1}\sinh(\alpha_{mn}\xi)d\xi + \mathcal{B}_{mn}, \qquad (1.9.106)$$

where $\mathcal{B}_{mn}$ are, also, true arbitrary integration constants. Both $\mathcal{A}_{mn}$ and $\mathcal{B}_{mn}$ are determined by means of the boundary conditions for $u_{mn}(x)$: $u_{mn}(0) = 0$ and $u_{mn}(a) = 0$.

If $\int A'_{mn}(x)dx$ and $\int B'_{mn}(x)dx$ are introduced into Eq. (1.9.102), one finds the following expression for the general solution of Eq. (1.9.100):

$$u_{mn}(x) = u_{mn}^{(1)}(x)\int A'_{mn}(x)dx + u_{mn}^{(2)}(x)\int B'_{mn}(x)dx$$
$$= \sinh(\alpha_{mn}x)\Bigg\{ -\frac{4}{bc\varepsilon_0\alpha_{mn}}\int_0^x \left[\int_0^b \int_0^c \rho(\xi,\eta,\zeta) \right.$$
$$\times \sin\left(\frac{m\pi}{b}\eta\right) \sin\left(\frac{n\pi}{c}\zeta\right) d\eta d\zeta \Bigg] \cosh(\alpha_{mn}\xi)d\xi$$
$$+\mathcal{A}_{mn}\Bigg\} + \cosh(\alpha_{mn}x)\Bigg\{ \frac{4}{bc\varepsilon_0\alpha_{mn}}\int_0^x \left[\int_0^b \int_0^c \rho(\xi,\eta,\zeta) \right.$$
$$\times \sin\left(\frac{m\pi}{b}\eta\right) \sin\left(\frac{n\pi}{c}\zeta\right) d\eta d\zeta \Bigg] \sinh(\alpha_{mn}\xi)d\xi + \mathcal{B}_{mn}\Bigg\}$$
$$= \mathcal{A}_{mn}\sinh(\alpha_{mn}x) + \mathcal{B}_{mn}\cosh(\alpha_{mn}x) - \frac{4\sinh(\alpha_{mn}x)}{bc\varepsilon_0\alpha_{mn}}$$
$$\times \int_0^x \int_0^b \int_0^c \rho(\xi,\eta,\zeta) \sin\left(\frac{m\pi}{b}\eta\right) \sin\left(\frac{n\pi}{c}\zeta\right) \cosh(\alpha_{mn}\xi)d\xi d\eta d\zeta$$
$$+ \frac{4\cosh(\alpha_{mn}x)}{bc\varepsilon_0\alpha_{mn}}\int_0^x \int_0^b \int_0^c \rho(\xi,\eta,\zeta) \sin\left(\frac{m\pi}{b}\eta\right) \sin\left(\frac{n\pi}{c}\zeta\right)$$
$$\times \sinh(\alpha_{mn}\xi)d\xi d\eta d\zeta = \mathcal{A}_{mn}\sinh(\alpha_{mn}x) + \mathcal{B}_{mn}\cosh(\alpha_{mn}x)$$

$$- \frac{4}{bc\varepsilon_0 \alpha_{mn}} \int_0^x \int_0^b \int_0^c \rho(\xi, \eta, \zeta) \sin\left(\frac{m\pi}{b}\eta\right)$$

$$\times \sin\left(\frac{n\pi}{c}\zeta\right) \Big[\sinh(\alpha_{mn}x) \cosh(\alpha_{mn}\xi)$$

$$- \cosh(\alpha_{mn}x) \sinh(\alpha_{mn}\xi) \Big] d\xi \, d\eta \, d\zeta = \mathcal{A}_{mn} \sinh(\alpha_{mn}x)$$

$$+ \mathcal{B}_{mn} \cosh(\alpha_{mn}x) - \frac{4}{bc\varepsilon_0 \alpha_{mn}} \int_0^x \int_0^b \int_0^c \rho(\xi, \eta, \zeta)$$

$$\times \sinh\left[\alpha_{mn}(x - \xi)\right] \sin\left(\frac{m\pi}{b}\eta\right) \sin\left(\frac{n\pi}{c}\zeta\right) d\xi \, d\eta \, d\zeta. \tag{1.9.107}$$

For $x < x_0$, when going to the limit, the last term vanishes (we can suppose h small enough to have $x < x_0 - h$, and then $\rho = 0$, according to Eq. (1.9.97); see the previous problem for a more detailed justification). For $x > x_0$, in view of Eq. (1.9.97), we can write

$$I(h, k, l) \equiv \int_0^x \int_0^b \int_0^c \rho(\xi, \eta, \zeta) \sinh\left[\alpha_{mn}(x - \xi)\right] \sin\left(\frac{m\pi}{b}\eta\right)$$

$$\times \sin\left(\frac{n\pi}{c}\zeta\right) d\xi \, d\eta \, d\zeta = \frac{q}{8hkl} \int_{x_0-h}^{x_0+h} \int_{y_0-k}^{y_0+k} \int_{z_0-l}^{z_0+l}$$

$$\times \sinh\left[\alpha_{mn}(x - \xi)\right] \sin\left(\frac{m\pi}{b}\eta\right) \sin\left(\frac{n\pi}{c}\zeta\right) d\xi \, d\eta \, d\zeta = \frac{q}{8hkl}$$

$$\times \int_{x_0-h}^{x_0+h} \sinh\left[\alpha_{mn}(x - \xi)\right] d\xi \int_{y_0-k}^{y_0+k} \sin\left(\frac{m\pi}{b}\eta\right) d\eta \int_{z_0-l}^{z_0+l}$$

$$\times \sin\left(\frac{n\pi}{c}\zeta\right) d\zeta = -\frac{q}{8hkl} \frac{bc}{\pi^2 mn\alpha_{mn}} \Big\{ \cosh\left[\alpha_{mn}(x - \xi)\right] \Big\}_{x_0-h}^{x_0+h}$$

$$\times \left[\cos\left(\frac{m\pi}{b}\eta\right)\right]_{y_0-k}^{y_0+k} \left[\cos\left(\frac{n\pi}{c}\zeta\right)\right]_{z_0-l}^{z_0+l}$$

$$= \frac{qbc}{8hklmn\pi^2\alpha_{mn}} \Big\{ \cosh\left[\alpha_{mn}(x - x_0 + h)\right]$$

$$- \cosh\left[\alpha_{mn}(x - x_0 - h)\right] \Big\} \left\{ \cos\left[\frac{m\pi}{b}(y_0 + k)\right] - \cos\left[\frac{m\pi}{b}(y_0 - k)\right] \right\}$$

$$\times \left\{ \cos\left[\frac{n\pi}{c}(z_0 + l)\right] - \cos\left[\frac{n\pi}{c}(z_0 - l)\right] \right\},$$

so that

$$\lim_{\substack{h \to 0 \\ k \to 0 \\ l \to 0}} I(h, k, l) = \frac{qbc}{\pi^2 mn\alpha_{mn}} \sinh\left[\alpha_{mn}(x - x_0)\right]$$

$$\times \sin\left(\frac{m\pi}{b}y_0\right) \sin\left(\frac{n\pi}{c}z_0\right) \lim_{h \to 0} \frac{\sinh(\alpha_{mn}h)}{h}$$

$$\times \lim_{k \to 0} \frac{\sin\left(\frac{m\pi}{b}k\right)}{k} \lim_{l \to 0} \frac{\sin\left(\frac{n\pi}{c}l\right)}{l} = q \sinh\left[\alpha_{mn}(x - x_0)\right]$$

$$\times \sin\left(\frac{m\pi}{b}y_0\right)\sin\left(\frac{n\pi}{c}z_0\right)\lim_{h\to 0}\frac{\sinh(\alpha_{mn}h)}{\alpha_{mn}h}$$

$$\times \lim_{k\to 0}\frac{\sin\left(\frac{m\pi}{b}k\right)}{\frac{m\pi}{b}k}\lim_{l\to 0}\frac{\sin\left(\frac{n\pi}{c}l\right)}{\frac{n\pi}{c}l}$$

$$= q\sinh\left[\alpha_{mn}(x-x_0)\right]\sin\left(\frac{m\pi}{b}y_0\right)\sin\left(\frac{n\pi}{c}z_0\right).$$

Therefore, as a result of integration and transition to the limit, one obtains the functions $u_{mn}(x)$ as

$$u_{mn}(x) = \mathcal{A}_{mn}\sinh(\alpha_{mn}x) + \mathcal{B}_{mn}\cosh(\alpha_{mn}x)$$

$$+ \begin{cases} 0, & \text{for } x < x_0, \\[2mm] -\dfrac{4q}{bc\varepsilon_0}\dfrac{\sinh\left[\alpha_{mn}(x-x_0)\right]\sin\left(\frac{m\pi}{b}y_0\right)}{\alpha_{mn}} \\[4mm] \times\sin\left(\dfrac{n\pi}{c}z_0\right), & \text{for } x > x_0. \end{cases} \qquad (1.9.108)$$

The condition $u_{mn}(0) = 0$ gives $\mathcal{B}_{mn} = 0$, while $u_{mn}(a) = 0$ leads to

$$\mathcal{A}_{mn} = \frac{4q}{\alpha_{mn}bc\varepsilon_0}\frac{\sinh\left[\alpha_{mn}(x-x_0)\right]\sin\left(\frac{m\pi}{b}y_0\right)\sin\left(\frac{n\pi}{c}z_0\right)}{\sinh(\alpha_{mn}a)}. \qquad (1.9.109)$$

If $\mathcal{A}_{mn}$ and $\mathcal{B}_{mn}$ are now introduced into Eq. (1.9.108), one obtains the final form of the functions $u_{mn}(x)$. If these functions are inserted into Eq. (1.9.99), we are left with the problem result (*i.e.*, the solution of Eq. (1.9.98)):

$$V(x,y,z) = \begin{cases} \dfrac{4q}{bc\varepsilon_0}\displaystyle\sum_{m=1}^{\infty}\sum_{n=1}^{\infty}\dfrac{\sinh\left[\alpha_{mn}(a-x_0)\right]\sin\left(\frac{m\pi}{b}y_0\right)\sin\left(\frac{n\pi}{c}z_0\right)}{\alpha_{mn}\sinh(\alpha_{mn}a)} \\[4mm] \times\sinh(\alpha_{mn}x)\sin\left(\dfrac{m\pi}{b}y\right)\sin\left(\dfrac{n\pi}{c}z\right), & \text{for } x < x_0, \\[6mm] \dfrac{4q}{bc\varepsilon_0}\displaystyle\sum_{m=1}^{\infty}\sum_{n=1}^{\infty}\dfrac{\sinh\left[\alpha_{mn}(a-x_0)\right]\sin\left(\frac{m\pi}{b}y_0\right)\sin\left(\frac{n\pi}{c}z_0\right)}{\alpha_{mn}\sinh(\alpha_{mn}a)} \\[4mm] \times\sinh(\alpha_{mn}x)\sin\left(\dfrac{m\pi}{b}y\right)\sin\left(\dfrac{n\pi}{c}z\right) \\[4mm] -\dfrac{4q}{bc\varepsilon_0}\displaystyle\sum_{m=1}^{\infty}\sum_{n=1}^{\infty}\dfrac{\sinh\left[\alpha_{mn}(x-x_0)\right]\sin\left(\frac{m\pi}{b}y_0\right)}{\alpha_{mn}} \\[4mm] \times\sin\left(\frac{n\pi}{c}z_0\right)\sin\left(\dfrac{m\pi}{b}y\right)\sin\left(\dfrac{n\pi}{c}z\right), & \text{for } x > x_0, \end{cases}$$

or, in a simpler form,

$$V(x, y, z) =$$

$$\begin{cases} \dfrac{4q}{bc\varepsilon_0} \displaystyle\sum_{m=1}^{\infty} \sum_{n=1}^{\infty} \dfrac{\sinh\left[\alpha_{mn}(a - x_0)\right] \sin\left(\dfrac{m\pi}{b} y_0\right) \sin\left(\dfrac{n\pi}{c} z_0\right)}{\alpha_{mn} \sinh(\alpha_{mn} a)} \\ \times \sinh(\alpha_{mn} x) \sin\left(\dfrac{m\pi}{b} y\right) \sin\left(\dfrac{n\pi}{c} z\right), \quad \text{for } x < x_0, \\[2em] \dfrac{4q}{bc\varepsilon_0} \displaystyle\sum_{m=1}^{\infty} \sum_{n=1}^{\infty} \dfrac{\sinh(\alpha_{mn} x_0) \sin\left(\dfrac{m\pi}{b} y_0\right) \sin\left(\dfrac{n\pi}{c} z_0\right)}{\alpha_{mn} \sinh(\alpha_{mn} a)} \\ \times \sinh\left[\alpha_{mn}(a - x)\right] \sin\left(\dfrac{m\pi}{b} y\right) \sin\left(\dfrac{n\pi}{c} z\right), \quad \text{for } x > x_0, \end{cases} \qquad (1.9.110)$$

where in Eq. (1.9.110), for $x > x_0$, the following identity has been considered:

$$\sinh(\alpha - \beta) \sinh \gamma + \sinh(\beta - \gamma) \sinh \alpha + \sinh(\gamma - \alpha) \sinh \beta = 0,$$
$$\forall\, \alpha, \beta, \gamma\, \in \mathbb{R},$$

written for $\alpha = \alpha_{mn} a$, $\beta = \alpha_{mn} x_0$, $\gamma = \alpha_{mn} x$.

2

Magnetostatics

2.1 Problem No. 10

Show that a uniform magnetostatic field, $\vec{B}_0$, admits the vector potential $\vec{A} = \frac{1}{2}\vec{B}_0 \times \vec{r}$.

2.1.1 Solution

First method. By means of a straight calculation, we have

$$\nabla \times \vec{A} = \frac{1}{2}\nabla \times (\vec{B}_0 \times \vec{r}) = \frac{1}{2}\varepsilon_{ijk}\partial_j(\vec{B}_0 \times \vec{r})_k \vec{u}_i$$

$$= \frac{1}{2}\varepsilon_{ijk}\partial_j(\varepsilon_{klm}B_{0l}x_m)\vec{u}_i = \frac{1}{2}\varepsilon_{ijk}\varepsilon_{klm}B_{0l}(\partial_j x_m)\vec{u}_i$$

$$= \frac{1}{2}\varepsilon_{ijk}\varepsilon_{klm}B_{0l}\delta_{jm}\vec{u}_i = \frac{1}{2}\varepsilon_{imk}\varepsilon_{klm}B_{0l}\vec{u}_i = \frac{1}{2}\varepsilon_{imk}\varepsilon_{lmk}B_{0l}\vec{u}_i$$

$$= \frac{1}{2}2\delta_{il}B_{0l}\vec{u}_i = B_{0i}\vec{u}_i = \vec{B}_0,$$

where the fact that $\vec{B}_0$ is a magnetostatic field ($B_{0i} = const.$, $\forall\, i = 1, 2, 3$) has been used.

Second method. Making allowance for the identity expressed by Eq. (B.3.56) in **Appendix B**, we can write

$$\nabla \times \vec{A} = \nabla \times \left(\frac{1}{2}\vec{B}_0 \times \vec{r}\right) = \frac{1}{2}\nabla \times (\vec{B}_0 \times \vec{r})$$

$$= \frac{1}{2}\left[\vec{B}_0\, \nabla \cdot \vec{r} - \vec{r}\, \nabla \cdot \vec{B}_0 + (\vec{r} \cdot \nabla)\vec{B}_0\right.$$

$$\left. -(\vec{B}_0 \cdot \nabla)\vec{r}\right] = \frac{1}{2}(3\vec{B}_0 - \vec{B}_0) = \vec{B}_0,$$

where the following results have been taken into account:

$$\nabla \cdot \vec{r} \equiv \mathrm{div}\, \vec{r} = 3;$$

$$\nabla \cdot \vec{B}_0 \equiv \mathrm{div}\, \vec{B}_0 = 0 \text{ (the local form of Gauss's flux theorem)};$$

$$(\vec{r} \cdot \nabla)\vec{B}_0 = \left(x_i \frac{\partial}{\partial x_i} \right)(B_{0j}\vec{u}_j) = \vec{u}_j x_i \frac{\partial B_{0j}}{\partial x_i} = 0;$$

$$(\vec{B}_0 \cdot \nabla)\vec{r} = \left(B_{0i} \frac{\partial}{\partial x_i} \right)(x_j\vec{u}_j) = \vec{u}_j B_{0i} \frac{\partial x_j}{\partial x_i} = \vec{u}_j B_{0i}\delta_{ij} = \vec{B}_0.$$

The result $\left(x_i \dfrac{\partial}{\partial x_i} \right)(B_{0j}\vec{u}_j) = \vec{u}_j \left(x_i \dfrac{\partial}{\partial x_i} \right) B_{0j} = \vec{u}_j x_i \dfrac{\partial B_{0j}}{\partial x_i}$ written above is based on the fact that in a Cartesian frame, $\vec{u}_j = \mathrm{const.}$, $\forall\, j = \overline{1,3}$, while the fact that $B_{0j} = \mathrm{const.}$, $\forall\, j = 1, 2, 3$, ensures the validity of $\dfrac{\partial B_{0j}}{\partial x_i} = 0$.

2.2   Problem No. 11

Find the magnetostatic scalar potential for the field produced by an infinite straight current of intensity I.

Solution

Since

$$\nabla \times \vec{H} = \nabla \times (\mathrm{grad}\, V_m) = 0, \tag{2.2.1}$$

everywhere around the current, one follows that a scalar potential V_m can be defined, so that

$$\vec{H} = -\nabla V_m. \tag{2.2.2}$$

In this case, the field lines of the magnetic field are circles situated in planes orthogonal to the wire, with their centers on the wire, and the magnitude of the field is constant along a field line (see Fig. 2.1). Applying the Ampère's law

$$\oint_{(C)} \vec{H} \cdot \vec{dl} = I_{int},$$

where I_{int} is the intensity of the current surrounded by the integration contour C, one obtains

$$H_r = H_z = 0, \tag{2.2.3}$$

$$H_\varphi = \frac{I}{2\pi r}, \tag{2.2.4}$$

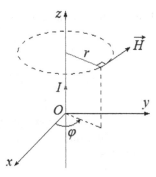

FIGURE 2.1
Schematic representation of the infinite, rectilinear current I along the z-axis.

where r, φ, z are the cylindrical coordinates, with the z-axis oriented along the current. On the other hand, if relation (2.2.2) is written in cylindrical coordinates, one results

$$H_r = -\frac{\partial V_m}{\partial r}, \qquad (2.2.5)$$

$$H_\varphi = -\frac{1}{r}\frac{\partial V_m}{\partial \varphi}, \qquad (2.2.6)$$

$$H_z = -\frac{\partial V_m}{\partial z}. \qquad (2.2.7)$$

According to Eqs. (2.2.4) and (2.2.6), the magnetostatic scalar potential V_m is given by

$$V_m = -\frac{I}{2\pi}\varphi. \qquad (2.2.8)$$

For univocity, we can consider $0 \le \varphi \le 2\pi$. Therefore, the equipotential surfaces are meridian planes, $\varphi = K$, where K is an arbitrary constant.

2.3 Problem No. 12

a) Find the expression of the vector potential for a magnetic field created by an infinite straight current of intensity I.

b) Find the induction $\vec{B}$ of a magnetostatic field produced in vacuum by a straight steady current of intensity I and length $2L$.

Solution

a) The problem symmetry requires the use of cylindrical coordinates ρ, φ, z. Without restricting the generality of the problem, we shall orient the z-axis along the current. In this case, the magnetic induction vector $\vec{B}$ has the components

$$B_\rho = B_z = 0, \qquad (2.3.9)$$

and

$$B_\varphi = \frac{\mu_0 I}{2\pi\rho}. \qquad (2.3.10)$$

Since the only non-zero component of the current is oriented along the z-axis, one follows that $A_\rho = A_\varphi = 0$, in which case the relation between $\vec{B}$ and $\vec{A}$, written in cylindrical coordinates,

$$\vec{B} = \nabla \times \vec{A} = \vec{e}_\rho \left(\frac{1}{\rho}\frac{\partial A_z}{\partial \varphi} - \frac{\partial A_\varphi}{\partial z} \right) + \vec{e}_\varphi \left(\frac{\partial A_\rho}{\partial z} - \frac{\partial A_z}{\partial \rho} \right)$$
$$+ \vec{e}_z \left[\frac{1}{\rho}\frac{\partial(\rho A_\varphi)}{\partial \rho} - \frac{\partial A_\rho}{\partial \varphi} \right]$$

becomes

$$\vec{B} = \frac{1}{\rho}\frac{\partial A_z}{\partial \varphi}\,\vec{e}_\rho - \frac{\partial A_z}{\partial \rho}\,\vec{e}_\varphi,$$

which leads to

$$B_\rho = \frac{1}{\rho}\frac{\partial A_z}{\partial \varphi}, \qquad (2.3.11)$$

$$B_\varphi = -\frac{\partial A_z}{\partial \rho}, \qquad (2.3.12)$$

$$B_z = 0. \qquad (2.3.13)$$

According to Eqs. (2.3.10) and (2.3.12), the only non-zero component of the vector potential is

$$A_z = -\frac{\mu_0 I}{2\pi}\ln\rho. \qquad (2.3.14)$$

b) This time we shall choose, for convenience, a Cartesian system of coordinates $Oxyz$, with z-axis oriented along the current and the coordinate origin at the center of the wire (see Fig. 2.2). The vector potential of a continuous distribution of steady currents at the point defined by the position vector $\vec{r}$, is given by

$$\vec{A}(\vec{r}) = \frac{\mu_0}{4\pi} \int\limits_{(D')} \frac{\vec{j}(\vec{r}')\,d\tau'}{|\vec{r} - \vec{r}'|}, \qquad (2.3.15)$$

where D' is the three-dimensional domain occupied by currents. Since

$$\vec{j}\,d\tau' = \vec{j}S'\,dl' = j\,S'\,\vec{dl}' = I\,\vec{dl}',$$

in our case, we have

$$\vec{A}(\vec{r}) = \frac{\mu_0 I}{4\pi} \int_{-L}^{+L} \frac{\vec{dl}'}{R}, \qquad (2.3.16)$$

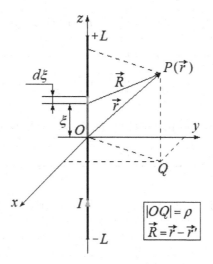

FIGURE 2.2
The geometry of the finite current I along the symmetric interval $[-L, L]$ of the z-axis.

where $I\vec{dl}'$ is an infinitesimal current element with

$$\vec{dl}' = (dx', dy', dz') = (0, 0, d\xi),$$

while $R = \sqrt{(z - \xi)^2 + \rho^2}$. If Eq. (2.3.16) is projected on the coordinate axes, one obtains

$$\begin{cases} A_x = 0, \\ A_y = 0, \\ A_z = \dfrac{\mu_0 I}{4\pi} \displaystyle\int_{-L}^{+L} \dfrac{d\xi}{R}. \end{cases} \tag{2.3.17}$$

To perform integration in Eq. (2.3.17) one introduces a new integration variable u by means of substitution $z - \xi = u$. Since $R = \sqrt{\rho^2 + u^2}$ and $d\xi = -du$ (the coordinates x, y and z of the "observation" point $P(\vec{r})$ remain unchanged when ξ runs from $-L$ to $+L$, to cover the entire length of the finite current), we have

$$\begin{aligned} A_z &= \frac{\mu_0 I}{4\pi} \int_{z-L}^{z+L} \frac{du}{\sqrt{\rho^2 + u^2}} \\ &= \frac{\mu_0 I}{4\pi} \left(\operatorname{arsinh}\frac{z + L}{\rho} - \operatorname{arsinh}\frac{z - L}{\rho} \right). \end{aligned} \tag{2.3.18}$$

Due to the cylindrical symmetry of the problem (here, the cylindrical coordinates are denoted by ρ, φ and z – see Fig. 2.2), we shall use Eq. (2.3.18) in order to determine the induction $\vec{B}$ of the magnetostatic field. Because $A_\rho = 0$, $A_\varphi = 0$ and A_z do not depend on φ we have $B_\rho = B_z = 0$, the only non vanishing component of $\vec{B}$ being

$$B_\varphi = \frac{\partial A_\rho}{\partial z} - \frac{\partial A_z}{\partial \rho} = -\frac{\partial A_z}{\partial \rho}.$$

Using Eq. (2.3.18) and performing the required derivative, we finally obtain

$$B_\varphi = \frac{\mu_0 I}{4\pi \rho} \left[\frac{z + L}{\sqrt{\rho^2 + (z + L)^2}} - \frac{z - L}{\sqrt{\rho^2 + (z - L)^2}} \right]. \qquad (2.3.19)$$

As it can be observed, for $L \to \infty$, $(z \neq 0)$, one re-find the well-known relation (2.3.10),

$$B_\varphi = \frac{\mu_0 I}{2\pi \rho},$$

which is valid for an infinite, straight and steady current of intensity I.

2.4   Problem No. 13

A thin conducting wire in the shape of an ellipse, with semi-major axis a and eccentricity e, carries a steady current I. Determine the magnetic field intensity at the center of the ellipse.

Solution

The ellipse of equation $\dfrac{x^2}{a^2} + \dfrac{y^2}{b^2} = 1$ has the following parametric equations:

$$\begin{cases} x = a \cos\varphi, \\ y = b \sin\varphi, \quad \varphi \in [-\pi, \pi], \end{cases} \qquad (2.4.20)$$

where a and $b < a$ are the ellipse semi-axes, while the eccentricity is given by $e = \sqrt{1 - b^2/a^2}$.

The magnetic field intensity $\vec{H}$ at the center of the ellipse is

$$\vec{H} = \frac{I}{4\pi} \oint_{(ellipse)} \frac{d\vec{l} \times \vec{r}}{r^3}, \qquad (2.4.21)$$

where $d\vec{l} = (dx, dy, 0)$ is an arc element of the ellipse, while $\vec{r}$ is the position vector of the ellipse center with respect to the origin of $d\vec{l}$, *i.e.*, $\vec{r} = (-x, -y, 0)$ (see Fig. 2.3).

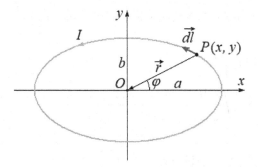

FIGURE 2.3
A stationary electric current I passing through a thin wire of an elliptic form.

Then, for the vector product $\vec{dl} \times \vec{r}$ one obtains

$$\vec{dl} \times \vec{r} = \begin{vmatrix} \vec{i} & \vec{j} & \vec{k} \\ dx & dy & 0 \\ -x & -y & 0 \end{vmatrix} = \vec{k}\,(-y\,dx + x\,dy), \tag{2.4.22}$$

showing that the only non-zero component of the magnetic field at the center of the ellipse is normal to the ellipse surface, and oriented in the direction of the z-axis, that is

$$H_z = \vec{H} \cdot \vec{k} = \frac{I}{4\pi} \oint_{(ellipse)} \frac{(\vec{dl} \times \vec{r}) \cdot \vec{k}}{r^3} = \frac{I}{4\pi} \oint_{(ellipse)} \frac{-y\,dx + x\,dy}{r^3}. \tag{2.4.23}$$

According to parameterization given by Eq. (2.4.20), we can write

$$\begin{aligned} r^2 &= x^2 + y^2 \\ &= a^2 \cos^2\varphi + b^2 \sin^2\varphi = a^2(1 - \sin^2\varphi) + b^2 \sin^2\varphi \\ &= a^2 \left[1 - \left(\frac{a^2 - b^2}{a^2} \right) \sin^2\varphi \right] = a^2 \left[1 - \left(1 - \frac{b^2}{a^2} \right) \sin^2\varphi \right] \\ &= a^2(1 - e^2 \sin^2\varphi), \end{aligned} \tag{2.4.24}$$

so that

$$r = a\sqrt{1 - e^2 \sin^2\varphi}. \tag{2.4.25}$$

Next, by differentiation of Eq. (2.4.20), one obtains

$$\begin{cases} dx = -a\,\sin\varphi\,d\varphi, \\ dy = b\,\cos\varphi\,d\varphi, \end{cases}$$

which, together with Eqs. (2.4.20) and (2.4.23) leads to

$$
\begin{aligned}
H_z &= \frac{I}{4\pi} \oint_{(ellipse)} \frac{-y\,dx + x\,dy}{r^3} \\
&= \frac{I}{4\pi} \int_{-\pi}^{\pi} \frac{\left[(-b\sin\varphi)(-a\sin\varphi) + a\cos\varphi\, b\cos\varphi\right]}{a^3\left(1 - e^2\sin^2\varphi\right)^{3/2}} \, d\varphi \\
&= \frac{bI}{4\pi a^2} \int_{-\pi}^{\pi} \frac{d\varphi}{\left(1 - e^2\sin^2\varphi\right)^{3/2}} \\
&= \frac{I\sqrt{1 - e^2}}{4\pi a} \int_{-\pi}^{\pi} \frac{d\varphi}{\left(1 - e^2\sin^2\varphi\right)^{3/2}} \\
&= \frac{I\sqrt{1 - e^2}}{2\pi a} \int_{0}^{\pi} \frac{d\varphi}{\left(1 - e^2\sin^2\varphi\right)^{3/2}} \\
&= \frac{I\sqrt{1 - e^2}}{\pi a} \int_{0}^{\pi/2} \frac{d\varphi}{\left(1 - e^2\sin^2\varphi\right)^{3/2}}.
\end{aligned} \tag{2.4.26}
$$

To write the penultimate equality in Eq. (2.4.26), we took into account that the integrand is an even function and the integration interval is symmetric with respect to the origin, while in the last equality we appealed to the periodicity of the function $\sin^2\varphi$. To solve the last integral in Eq. (2.4.26), we make use of the change of variable

$$
1 - e^2\sin^2\varphi = \frac{1 - e^2}{1 - e^2\sin^2\alpha}. \tag{2.4.27}
$$

According to Eq. (2.4.27), we have

$$
\sin\varphi = \frac{\cos\alpha}{\sqrt{1 - e^2\sin^2\alpha}} \tag{2.4.28}
$$

and

$$
\cos\varphi = \sqrt{1 - e^2} \frac{\sin\alpha}{\sqrt{1 - e^2\sin^2\alpha}}. \tag{2.4.29}
$$

By differentiating Eq. (2.4.28), one finds

$$
\begin{aligned}
\cos\varphi\, d\varphi &= \frac{-\sin\alpha\sqrt{1 - e^2\sin^2\alpha}\, d\alpha + \dfrac{e^2\sin\alpha\cos^2\alpha\, d\alpha}{\sqrt{1 - e^2\sin^2\alpha}}}{1 - e^2\sin^2\alpha} \\
&= \frac{-\sin\alpha(1 - e^2)}{(1 - e^2\sin^2\alpha)^{3/2}} \, d\alpha,
\end{aligned}
$$

and, by means of Eq. (2.4.29),

$$d\varphi = \frac{-\sin\alpha(1 - e^2)}{\cos\varphi(1 - e^2\sin^2\alpha)^{3/2}}\,d\alpha$$

$$= \frac{-\sin\alpha(1 - e^2)\sqrt{1 - e^2\sin^2\alpha}}{\sqrt{1 - e^2}\sin\alpha(1 - e^2\sin^2\alpha)^{3/2}}\,d\alpha$$

$$= -\frac{\sqrt{1 - e^2}}{1 - e^2\sin^2\alpha}\,d\alpha. \tag{2.4.30}$$

By using of Eqs. (2.4.27) and (2.4.30), Eq. (2.4.26) leads to

$$H_z = \frac{I\sqrt{1 - e^2}}{\pi a}\int_0^{\pi/2}\frac{d\varphi}{(1 - e^2\sin^2\varphi)^{3/2}}$$

$$= -\frac{I\sqrt{1 - e^2}}{\pi a}\int_{\pi/2}^0\frac{(1 - e^2\sin^2\alpha)^{3/2}\sqrt{1 - e^2}\,d\alpha}{(1 - e^2)^{3/2}(1 - e^2\sin^2\alpha)}$$

$$= \frac{I}{\pi a\sqrt{1 - e^2}}\int_0^{\pi/2}\sqrt{1 - e^2\sin^2\alpha}\,d\alpha = \frac{I\,E(e)}{\pi a\sqrt{1 - e^2}},$$

where

$$E(e) = \int_0^{\pi/2}\sqrt{1 - e^2\sin^2\alpha}\,d\alpha$$

is the complete elliptic integral of the second kind, written in Legendre's trigonometric form.

To conclude, the magnetic field intensity at the center of the elliptic wire, of semi-major axis a and eccentricity e, is oriented perpendicular to the plan of the wire, and its value is

$$H_z = \frac{I\,E(e)}{\pi a\sqrt{1 - e^2}} = \frac{I\,E(e)}{\pi b}. \tag{2.4.31}$$

2.5 Problem No. 14

Determine the vector potential $\vec{A}$ of the magnetic field created by a circular steady current of radius R and intensity I.

Solution

Consider a coordinate system with its origin at the wire center and the z-axis orthogonal to current plane (see Fig. 2.4). Due to the symmetry of the problem, it is convenient to work in cylindrical coordinates. Let $P(\rho, \varphi, z)$ be the observation point and P' its projection on the wire plane. Consider an

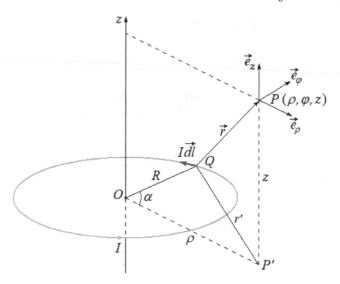

FIGURE 2.4
A stationary electric current I passing through the plane circular loop of radius R.

arbitrary point Q on the wire and $I\vec{dl}$ an infinitesimal element of current with its origin at Q. The $\Delta PP'Q$ triangle is right-angled, with its right-angle at P'. The local trihedron situated at the observation point, formed by the unit vectors $\vec{e}_\rho, \vec{e}_\varphi, \vec{e}_z$ is shown in Fig. 2.4.

The vector potential is given by

$$\vec{A} = \frac{\mu_0 I}{4\pi} \oint_{(wire)} \frac{\vec{dl}}{r}, \qquad (2.5.32)$$

where r is the distance between points P and Q. From Fig. 2.4 it follows that

$$r^2 = r'^2 + z^2 = R^2 + \rho^2 - 2\rho R \cos\alpha + z^2. \qquad (2.5.33)$$

The components of vector $\vec{A}$ on the axes of local trihedron are found by a scalar multiplication of Eq. (2.5.32) with the corresponding unit vectors. Since $\vec{dl} \perp \vec{e}_z$, one follows that

$$A_z = \vec{A} \cdot \vec{e}_z = 0. \qquad (2.5.34)$$

Observing Fig. 2.5, we also have

$$\vec{dl} \cdot \vec{e}_\rho = dl \cos\left(\alpha + \frac{\pi}{2}\right) = dl \sin\alpha, \qquad (2.5.35)$$

so that

$$A_\rho = \vec{A} \cdot \vec{e}_\rho = \frac{\mu_0 I}{4\pi} \oint_{(wire)} \frac{\vec{dl} \cdot \vec{e}_\rho}{r}. \qquad (2.5.36)$$

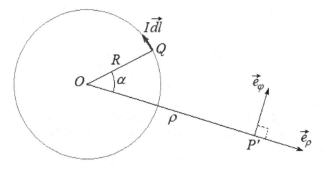

FIGURE 2.5
An auxiliary geometrical construction showing the relative positioning of elementary stationary electric current $I\vec{dl}$ and the unit vectors $\vec{e}_\rho$ and $\vec{e}_\varphi$.

Since $dl = R\,d\alpha$, one obtains

$$A_\rho = \frac{\mu_0 IR}{4\pi} \int_{-\pi}^{+\pi} \frac{\sin\alpha\,d\alpha}{r} = 0, \qquad (2.5.37)$$

because the integrand is an odd function of α, and the integration interval is symmetrical to the origin.

Finally, taking into account that $\vec{dl} \cdot \vec{e}_\varphi = dl\cos\alpha = R\cos\alpha\,d\alpha$, the azimuthal component of the vector potential $\vec{A}$ is

$$A_\varphi = \vec{A} \cdot \vec{e}_\varphi = \frac{\mu_0 I}{4\pi} \oint_{(wire)} \frac{\vec{dl} \cdot \vec{e}_\varphi}{r}$$

$$= \frac{\mu_0 IR}{4\pi} \int_{-\pi}^{+\pi} \frac{\cos\alpha\,d\alpha}{r} = \frac{\mu_0 IR}{2\pi} \int_0^\pi \frac{\cos\alpha\,d\alpha}{r}, \qquad (2.5.38)$$

where we took into account that the integrand is an even function of α, and the integration interval is symmetrical to the origin.
The change of variable $\alpha = 2\psi$ leads to the following expression for A_φ:

$$A_\varphi = A_\varphi(\rho, z)$$

$$= \frac{\mu_0 IR}{\pi} \int_0^{\pi/2} \frac{(2\cos^2\psi - 1)\,d\psi}{(\rho^2 + R^2 + z^2 - 2\rho R\cos 2\psi)^{1/2}}$$

$$= \frac{\mu_0 IR}{\pi} \int_0^{\pi/2} \frac{(2\cos^2\psi - 1)\,d\psi}{\left[(\rho + R)^2 + z^2 - 4\rho R\cos^2\psi\right]^{1/2}}$$

$$= \frac{\mu_0 I\xi}{2\pi} \sqrt{\frac{R}{\rho}} \int_0^{\pi/2} \frac{(2\cos^2\psi - 1)\,d\psi}{(1 - \xi^2\cos^2\psi)^{1/2}}, \qquad (2.5.39)$$

where $\xi^2 = \dfrac{4\rho R}{(\rho + R)^2 + z^2}$. As can be seen,

$$\xi^2 \le \frac{4\rho R}{(\rho + R)^2} = 1 - \left(\frac{\rho - R}{\rho + R}\right)^2 \le 1, \qquad (2.5.40)$$

so that $0 \le \xi \le 1$.

The last integral in Eq. (2.5.39) can be expressed by means of the complete elliptic integrals of the first and second kind, written in Legendre's trigonometric form,

$$K(\xi) = \int_0^{\pi/2} \frac{d\psi}{\sqrt{1 - \xi^2 \sin^2 \psi}}, \qquad (2.5.41)$$

and, respectively,

$$E(\xi) = \int_0^{\pi/2} \sqrt{1 - \xi^2 \sin^2 \psi}\, d\psi, \quad (0 \le \xi \le 1). \qquad (2.5.42)$$

Using the elliptic integrals given by Eqs. (2.5.41) and (2.5.42), and integrating, one obtains

$$A_\varphi(\rho, z) = \frac{\mu_0 I}{\pi \xi} \sqrt{\frac{R}{\rho}} \left[\left(1 - \frac{\xi^2}{2}\right) K(\xi) - E(\xi) \right]. \qquad (2.5.43)$$

To prove this result, we first perform in Eq. (2.5.39) the change of variable $\cos \psi = \sin \theta$, which gives

$$\int_0^{\pi/2} \frac{(2\cos^2 \psi - 1)}{(1 - \xi^2 \cos^2 \psi)^{1/2}}\, d\psi = \int_0^{\pi/2} \frac{(2\sin^2 \theta - 1)}{\sqrt{1 - \xi^2 \sin^2 \theta}}\, d\theta,$$

leading to

$$\int_0^{\pi/2} \frac{(2\cos^2 \psi - 1)}{(1 - \xi^2 \cos^2 \psi)^{1/2}}\, d\psi = \int_0^{\pi/2} \frac{(2\sin^2 \theta - 1)\, d\theta}{\sqrt{1 - \xi^2 \sin^2 \theta}}$$

$$= -\frac{2}{\xi^2} \int_0^{\pi/2} \frac{-\xi^2 \sin^2 \theta}{\sqrt{1 - \xi^2 \sin^2 \theta}}\, d\theta - \int_0^{\pi/2} \frac{d\theta}{\sqrt{1 - \xi^2 \sin^2 \theta}}$$

$$= -\frac{2}{\xi^2} \int_0^{\pi/2} \frac{(1 - \xi^2 \sin^2 \theta - 1)}{\sqrt{1 - \xi^2 \sin^2 \theta}}\, d\theta - \int_0^{\pi/2} \frac{d\theta}{\sqrt{1 - \xi^2 \sin^2 \theta}}$$

$$= -\frac{2}{\xi^2} \int_0^{\pi/2} \sqrt{1 - \xi^2 \sin^2 \theta}\, d\theta + \left(\frac{2}{\xi^2} - 1\right) \int_0^{\pi/2} \frac{d\theta}{\sqrt{1 - \xi^2 \sin^2 \theta}}$$

$$= -\frac{2}{\xi^2} E(\xi) + \left(\frac{2}{\xi^2} - 1\right) K(\xi).$$

Consequently,

$$A_\varphi(\rho, z) = \frac{\mu_0 I \xi}{2\pi} \sqrt{\frac{R}{\rho}} \int_0^{\pi/2} \frac{(2\cos^2\psi - 1)}{(1 - \xi^2 \cos^2\psi)^{1/2}} \, d\psi$$

$$= \frac{\mu_0 I \xi}{2\pi} \sqrt{\frac{R}{\rho}} \left[-\frac{2}{\xi^2} E(\xi) + \left(\frac{2}{\xi^2} - 1 \right) K(\xi) \right]$$

$$= \frac{\mu_0 I}{\pi \xi} \sqrt{\frac{R}{\rho}} \left[\left(1 - \frac{\xi^2}{2} \right) K(\xi) - E(\xi) \right],$$

which finally verifies Eq. (2.5.43).

2.6 Problem No. 15

By means of relation $\vec{B} = \nabla \times \vec{A}$, calculate the magnetic field induction created by the circular wire met in the previous problem.

Solution

Using the curl "operator" in cylindrical coordinates ρ, φ, z, we have (see the relation (D.3.34)):

$$\vec{B} = \nabla \times \vec{A} = \vec{e}_\rho \left(\frac{1}{\rho} \frac{\partial A_z}{\partial \varphi} - \frac{\partial A_\varphi}{\partial z} \right) + \vec{e}_\varphi \left(\frac{\partial A_\rho}{\partial z} - \frac{\partial A_z}{\partial \rho} \right)$$

$$+ \vec{e}_z \left[\frac{1}{\rho} \frac{\partial(\rho A_\varphi)}{\partial \rho} - \frac{\partial A_\rho}{\partial \varphi} \right]. \quad (2.6.44)$$

Since $A_\rho = A_z = 0$, one follows that

$$B_\rho = -\frac{\partial A_\varphi}{\partial z}, \quad (2.6.45)$$

$$B_\varphi = 0, \quad (2.6.46)$$

$$B_z = \frac{1}{\rho} \frac{\partial}{\partial \rho}(\rho A_\varphi), \quad (2.6.47)$$

where, as it has been proved in the previous problem,

$$A_\varphi(\rho, z) = \frac{\mu_0 I}{\pi \xi} \sqrt{\frac{R}{\rho}} \left[\left(1 - \frac{\xi^2}{2} \right) K(\xi) - E(\xi) \right], \quad (2.6.48)$$

with

$$\xi^2 = \frac{4\rho R}{(\rho + R)^2 + z^2}. \tag{2.6.49}$$

Since A_φ depends implicitly on variable z by means of ξ, the relation (2.6.45) can also be written as

$$B_\rho = -\frac{\partial A_\varphi}{\partial \xi} \frac{\partial \xi}{\partial z}, \tag{2.6.50}$$

where the derivative $\dfrac{\partial \xi}{\partial z}$ can be easily calculated by using Eq. (2.6.49); we have

$$\frac{\partial \xi}{\partial z} = -\frac{2z\sqrt{\rho R}}{\left[(\rho + R)^2 + z^2\right]^{3/2}} = -\frac{z\xi^2}{4\rho R}. \tag{2.6.51}$$

In addition,

$$\frac{\partial \xi}{\partial \rho} = \sqrt{\frac{R}{\rho}} \, \frac{R^2 + z^2 - \rho^2}{\left[(\rho + R)^2 + z^2\right]^{3/2}} = \frac{\xi}{2\rho} - \frac{(\rho + R)\xi^3}{4R\rho}. \tag{2.6.52}$$

We are left now with calculation of the derivatives

$$\frac{\partial E(\xi)}{\partial \xi} \equiv \frac{dE(\xi)}{d\xi}$$

and

$$\frac{\partial K(\xi)}{\partial \xi} \equiv \frac{dK(\xi)}{d\xi},$$

where $E(\xi)$ and $K(\xi)$ are given by

$$E(\xi) = \int_0^{\pi/2} \sqrt{1 - \xi^2 \sin^2 \psi} \, d\psi$$

and

$$K(\xi) = \int_0^{\pi/2} \frac{d\psi}{\sqrt{1 - \xi^2 \sin^2 \psi}}, \qquad 0 \le \xi \le 1.$$

By using the formula that gives the derivative of a parameter-dependent integral of the form

$$F(y) = \int_{a(y)}^{b(y)} f(x, y) dx,$$

that is

$$\frac{dF(y)}{dy} = \int_{a(y)}^{b(y)} \frac{\partial f(x, y)}{\partial y} dx + b'(y) f\big(b(y), y\big) - a'(y) f\big(a(y), y\big), \tag{2.6.53}$$

where $f(x, y)$ must have continuous partial derivatives with respect to y, one gets

$$\frac{dK(\xi)}{d\xi} = \int_0^{\pi/2} \frac{\partial}{\partial \xi} \left(\frac{1}{\sqrt{1 - \xi^2 \sin^2 \psi}} \right) d\psi$$

$$= \int_0^{\pi/2} \frac{\xi \sin^2 \psi \, d\psi}{\left(1 - \xi^2 \sin^2 \psi \right)^{3/2}}$$

$$= \int_0^{\pi/2} \left(-\frac{1}{\xi} \right) \frac{-\xi^2 \sin \psi}{\left(1 - \xi^2 \sin^2 \psi \right) \sqrt{1 - \xi^2 \sin^2 \psi}} d\psi$$

$$= \int_0^{\pi/2} \left(-\frac{1}{\xi} \right) \left[\frac{1 - \xi^2 \sin^2 \psi}{\left(1 - \xi^2 \sin^2 \psi \right) \sqrt{1 - \xi^2 \sin^2 \psi}} \right.$$

$$\left. - \frac{1}{\left(1 - \xi^2 \sin^2 \psi \right)^{3/2}} \right] d\psi$$

$$= -\frac{1}{\xi} K(\xi) + \frac{1}{\xi} \int_0^{\pi/2} \frac{d\psi}{\left(1 - \xi^2 \sin^2 \psi \right)^{3/2}} \tag{2.6.54}$$

and

$$\frac{dE(\xi)}{d\xi} = \int_0^{\pi/2} \frac{\partial}{\partial \xi} \left(\sqrt{1 - \xi^2 \sin^2 \psi} \right) d\psi$$

$$= \int_0^{\pi/2} \frac{-\xi \sin^2 \psi}{\sqrt{1 - \xi^2 \sin^2 \psi}} d\psi$$

$$= \int_0^{\pi/2} \frac{1}{\xi} \left(\sqrt{1 - \xi^2 \sin^2 \psi} - \frac{1}{\sqrt{1 - \xi^2 \sin^2 \psi}} \right) d\psi$$

$$= \frac{1}{\xi} \left[E(\xi) - K(\xi) \right]. \tag{2.6.55}$$

In order to express the last integral in Eq. (2.6.54) in terms of the elliptic integrals $K(\xi)$ and $E(\xi)$, we shall use the following change of variables

$$1 - \xi^2 \sin^2 \psi = \frac{1 - \xi^2}{1 - \xi^2 \sin^2 \theta}. \tag{2.6.56}$$

The relation (2.6.56) then yields

$$\sin^2 \psi = \frac{1}{\xi^2} \left(1 - \frac{1 - \xi^2}{1 - \xi^2 \sin^2 \theta} \right) = \frac{\cos^2 \theta}{1 - \xi^2 \sin^2 \theta},$$

that is

$$\sin \psi = \frac{\cos \theta}{\sqrt{1 - \xi^2 \sin^2 \theta}}, \tag{2.6.57}$$

and

$$\cos \psi = \frac{\sqrt{1 - \xi^2} \sin \theta}{\sqrt{1 - \xi^2 \sin^2 \theta}}. \tag{2.6.58}$$

Next, from Eq. (2.6.57) one obtains

$$\cos \psi \, d\psi = \frac{-\sin \theta \left(1 - \xi^2 \sin^2 \theta\right) + \xi^2 \cos^2 \theta \sin \theta}{\left(1 - \xi^2 \sin^2 \theta\right) \sqrt{1 - \xi^2 \sin^2 \theta}} \, d\theta$$

$$= -\frac{\left(1 - \xi^2\right) \sin \theta}{\left(1 - \xi^2 \sin^2 \theta\right)^{3/2}} \, d\theta,$$

and, by means of Eq. (2.6.58),

$$\frac{\sqrt{1 - \xi^2} \sin \theta}{\sqrt{1 - \xi^2 \sin^2 \theta}} \, d\psi = -\frac{\left(1 - \xi^2\right) \sin \theta}{\left(1 - \xi^2 \sin^2 \theta\right)^{3/2}} \, d\theta,$$

leading to

$$d\psi = -\frac{\sqrt{1 - \xi^2}}{1 - \xi^2 \sin^2 \theta} \, d\theta. \tag{2.6.59}$$

The last integral in Eq. (2.6.54) then becomes

$$\int_0^{\pi/2} \frac{d\psi}{\left(1 - \xi^2 \sin^2 \psi\right)^{3/2}}$$

$$= -\int_{\pi/2}^0 \left(\frac{1 - \xi^2 \sin^2 \theta}{1 - \xi^2}\right)^{3/2} \frac{\sqrt{1 - \xi^2}}{1 - \xi^2 \sin^2 \theta} \, d\theta$$

$$= \frac{1}{1 - \xi^2} \int_0^{\pi/2} \sqrt{1 - \xi^2 \sin^2 \theta} \, d\theta = \frac{1}{1 - \xi^2} E(\xi),$$

so that

$$\frac{dK(\xi)}{d\xi} = \frac{1}{\xi} \left[\frac{E(\xi)}{1 - \xi^2} - K(\xi)\right]. \tag{2.6.60}$$

Introducing these results into Eq. (2.6.50), the radial component of the magnetic induction writes

$$B_\rho = -\frac{\partial A_\varphi}{\partial \xi} \frac{\partial \xi}{\partial z} = -\frac{\partial}{\partial \xi} \left\{\frac{\mu_0 I}{\pi \xi} \sqrt{\frac{R}{\rho}} \left[\left(1 - \frac{\xi^2}{2}\right) K(\xi) - E(\xi)\right]\right\} \frac{\partial \xi}{\partial z}$$

$$= -\frac{\partial}{\partial \xi} \left\{\frac{\mu_0 I}{\pi \xi} \sqrt{\frac{R}{\rho}} \left[\left(1 - \frac{\xi^2}{2}\right) K(\xi) - E(\xi)\right]\right\} \left(-\frac{z \xi^2}{4 \rho R}\right)$$

$$= \frac{\mu_0 I}{4\pi} \sqrt{\frac{R}{\rho}} \frac{z \xi^3}{\rho R} \frac{\partial}{\partial \xi} \left\{\frac{1}{\xi} \left[\left(1 - \frac{\xi^2}{2}\right) K(\xi) - E(\xi)\right]\right\}$$

$$= \frac{\mu_0 I}{4\pi} \sqrt{\frac{R}{\rho}} \frac{z\xi^3}{\rho R} \frac{d}{d\xi} \left[\left(\frac{1}{\xi} - \frac{\xi}{2} \right) K(\xi) - E(\xi) \right]$$

$$= \frac{\mu_0 I}{4\pi} \sqrt{\frac{R}{\rho}} \frac{z\xi^3}{\rho R} \left[\left(-\frac{1}{\xi^2} - \frac{1}{2} \right) K(\xi) + \left(\frac{1}{\xi} - \frac{\xi}{2} \right) \frac{dK(\xi)}{d\xi} \right.$$

$$\left. + \frac{1}{\xi^2} E(\xi) - \frac{1}{\xi} \frac{dE(\xi)}{d\xi} \right] = \frac{\mu_0 I}{4\pi} \sqrt{\frac{R}{\rho}} \frac{z\xi^3}{\rho R} \left\{ \left(-\frac{1}{\xi^2} - \frac{1}{2} \right) K(\xi) \right.$$

$$\left. + \frac{1}{\xi} \left(\frac{1}{\xi} - \frac{\xi}{2} \right) \left[\frac{E(\xi)}{1 - \xi^2} - K(\xi) \right] + \frac{1}{\xi^2} E(\xi) - \frac{1}{\xi^2} \left[E(\xi) - K(\xi) \right] \right\}$$

$$= \frac{\mu_0 I}{4\pi} \sqrt{\frac{R}{\rho}} \frac{z\xi^3}{\rho R} \left[-\frac{1}{\xi^2} K(\xi) + E(\xi) \frac{1}{1 - \xi^2} \left(\frac{1}{\xi^2} - \frac{1}{2} \right) \right]$$

$$= \frac{\mu_0 I}{4\pi} \sqrt{\frac{R}{\rho}} \frac{z\xi}{\rho R} \left[-K(\xi) + \frac{1}{2} E(\xi) \frac{2 - \xi^2}{1 - \xi^2} \right],$$

or, in view of Eq. (2.6.49),

$$B_\rho = \frac{\mu_0 I}{2\pi} \frac{z}{\rho \sqrt{(\rho + R)^2 + z^2}} \left[-K(\xi) + \frac{\rho^2 + R^2 + z^2}{(\rho - R)^2 + z^2} E(\xi) \right].$$

Finally, by using Eqs. (2.6.48), (2.6.49), (2.6.52), (2.6.55) and (2.6.60), the axial component B_z of the magnetic induction, given by Eq. (2.6.47), is determined as follows:

$$B_z = \frac{1}{\rho} \frac{\partial}{\partial \rho} (\rho A_\varphi) = \frac{1}{\rho} A_\varphi + \frac{\partial A_\varphi}{\partial \rho}$$

$$= \frac{1}{\rho} \frac{\mu_0 I}{\pi \xi} \sqrt{\frac{R}{\rho}} \left[\left(1 - \frac{\xi^2}{2} \right) K(\xi) - E(\xi) \right]$$

$$+ \frac{\partial}{\partial \rho} \left\{ \frac{\mu_0 I}{\pi \xi} \sqrt{\frac{R}{\rho}} \left[\left(1 - \frac{\xi^2}{2} \right) K(\xi) - E(\xi) \right] \right\}$$

$$= \frac{\mu_0 I}{\pi \xi} \sqrt{\frac{R}{\rho^3}} \left[\left(1 - \frac{\xi^2}{2} \right) K(\xi) - E(\xi) \right]$$

$$- \frac{1}{2} \frac{\mu_0 I}{\pi \xi} \sqrt{\frac{R}{\rho^3}} \left[\left(1 - \frac{\xi^2}{2} \right) K(\xi) - E(\xi) \right]$$

$$+ \frac{\mu_0 I}{\pi} \sqrt{\frac{R}{\rho}} \frac{\partial \xi}{\partial \rho} \frac{\partial}{\partial \xi} \left\{ \left[\left(1 - \frac{\xi^2}{2} \right) K(\xi) - E(\xi) \right] \frac{1}{\xi} \right\}$$

$$= \frac{1}{2} \frac{\mu_0 I}{\pi \xi} \sqrt{\frac{R}{\rho}} \left[\left(1 - \frac{\xi^2}{2} \right) K(\xi) - E(\xi) \right] + \frac{\mu_0 I}{\pi} \sqrt{\frac{R}{\rho}} \left[\frac{\xi}{2\rho} \right.$$

$$-\frac{(\rho+R)}{4\rho R}\xi^3\right] \frac{\partial}{\partial\xi}\left[\left(\frac{1}{\xi}-\frac{\xi}{2}\right)K(\xi)-\frac{1}{\xi}E(\xi)\right]$$

$$=\frac{\mu_0 I}{2\pi\xi}\sqrt{\frac{R}{\rho^3}}\left[\left(1-\frac{\xi^2}{2}\right)K(\xi)-E(\xi)\right]+\frac{\mu_0 I}{\pi}\sqrt{\frac{R}{\rho}}$$

$$\times\left[\frac{\xi}{2\rho}-\frac{(\rho+R)\xi^3}{4R\rho}\right]\left[\left(\frac{1}{\xi}-\frac{\xi}{2}\right)\frac{\partial K(\xi)}{\partial\xi}\right.$$

$$\left.-\left(\frac{1}{\xi^2}+\frac{1}{2}\right)K(\xi)-\frac{1}{\xi}\frac{\partial E(\xi)}{\partial\xi}+E(\xi)\frac{1}{\xi^2}\right]$$

$$=\frac{\mu_0 I}{2\pi\xi}\sqrt{\frac{R}{\rho^3}}\left[\left(1-\frac{\xi^2}{2}\right)K(\xi)-E(\xi)\right]+\frac{\mu_0 I}{2\pi\xi}\sqrt{\frac{R}{\rho^3}}$$

$$\times\left[\xi^2-\frac{(\rho+R)\xi^4}{2R}\right]\left\{\left(\frac{1}{\xi^2}-\frac{1}{2}\right)\left[\frac{E(\xi)}{1-\xi^2}-K(\xi)\right]\right.$$

$$\left.-\left(\frac{1}{\xi^2}+\frac{1}{2}\right)K(\xi)-\frac{1}{\xi^2}\left[E(\xi)-K(\xi)\right]+\frac{1}{\xi^2}E(\xi)\right\}$$

$$=\frac{\mu_0 I}{2\pi\xi}\sqrt{\frac{R}{\rho^3}}\left[\frac{\rho\xi^2}{2R}K(\xi)-\frac{2\rho\xi^2-(\rho+R)\xi^4}{4R(1-\xi^2)}E(\xi)\right],$$

or, by taking into account Eq. (2.6.49),

$$B_z=\frac{\mu_0 I}{2\pi}\frac{1}{\sqrt{(\rho+R)^2+z^2}}\left[K(\xi)-\frac{\rho^2-R^2+z^2}{(\rho-R)^2+z^2}E(\xi)\right].$$

2.7   Problem No. 16

Determine the field lines of the magnetic field created by a circular loop of radius R, through which flows a steady current I.

Solution

Since in cylindrical coordinates the line element $\vec{dl}$ has the components $d\rho$, $\rho d\varphi$, dz, the differential equations of the field lines are written as

$$\frac{d\rho}{B_\rho}=\frac{\rho d\varphi}{B_\varphi}=\frac{dz}{B_z}. \tag{2.7.61}$$

According to the previous problem, the components of the magnetic induction generated by a circular current of radius R are

$$B_\rho=-\frac{\partial A_\varphi}{\partial z},\quad B_\varphi=0,\quad B_z=\frac{1}{\rho}\frac{\partial}{\partial\rho}(\rho A_\varphi), \tag{2.7.62}$$

where A_φ is given by Eq. (2.6.48). In view of Eqs. (2.7.61) and (2.7.62), one follows that

$$\frac{d\rho}{-\dfrac{\partial A_\varphi}{\partial z}} = \frac{\rho\,d\varphi}{0} = \frac{dz}{\dfrac{1}{\rho}\dfrac{\partial}{\partial \rho}\left(\rho A_\varphi\right)}. \tag{2.7.63}$$

As can be easily seen, a first integral is given by

$$\varphi = const. = C_1, \tag{2.7.64}$$

which shows that the field lines are plane curves, contained in the meridian planes. Another first integral is obtained from the remaining equation, which can be written as

$$\frac{\partial}{\partial \rho}\left(\rho A_\varphi\right)d\rho + \frac{\partial\left(\rho A_\varphi\right)}{\partial z}dz = 0. \tag{2.7.65}$$

Since $A_\varphi = A_\varphi(\rho, z)$, it follows from Eq. (2.7.65) that $d\left(\rho A_\varphi\right) = 0$, that is

$$\rho A_\varphi = const. = C_2. \tag{2.7.66}$$

The curves expressed by this equation are some ovals, surrounding the wire through which flows the current, being symmetric with respect to the wire plane. The field lines are completely determined by the relations (2.7.64) and (2.7.66).

2.8   Problem No. 17

Show that far enough from the center of a circular (let R be the radius of the circle) conducting wire, through which flows a steady current of intensity I, the vector potential $\vec{A}$ of the generated magnetostatic field can be written in the form

$$\vec{A}_\varphi = \frac{\mu_0}{4\pi}\frac{\vec{m} \times \vec{r}_0}{r_0^3},$$

where $\vec{r}_0$ is the radius vector of the observation point (with respect to the center of the wire), and $\vec{m} = IS\vec{n} = m\vec{n}$ is the magnetic moment "of the wire", $S = \pi R^2$ being the surface of the "equivalent" magnetic sheet, while $\vec{n}$ is the unit vector orthogonal to this surface.

Solution

Let θ be the angle between the vector $\vec{r}_0$ and z-axis (see Fig. 2.6). Analysing the figure, one can easily observe that

$$\begin{cases} \rho = r_0 \sin\theta, \\ z = r_0 \cos\theta, \\ r_0^2 = \rho^2 + z^2. \end{cases} \tag{2.8.67}$$

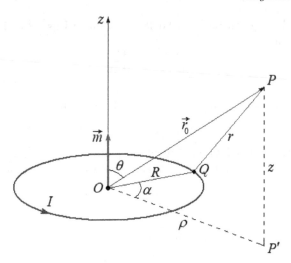

FIGURE 2.6
Graphical representation of the point $P(\vec{r}_0)$ where the vector potential $\vec{A}$ of
the magnetostatic field created by a circular steady current is determined.

Since $\Delta PQP'$ is a right triangle, with the right angle at P', we can write

$$r^2 = z^2 + |P'Q|^2 = r_0^2 + R^2 - 2Rr_0 \sin\theta\cos\alpha. \qquad (2.8.68)$$

The only non-zero component of the vector potential is the azimuthal component A_φ which, according to relation (2.6.50) it is given by

$$A_\varphi = \vec{A} \cdot \vec{e}_\varphi = \frac{\mu_0 I}{4\pi} \oint_{wire} \frac{\vec{dl} \cdot \vec{e}_\varphi}{r}$$

$$= \frac{\mu_0 IR}{4\pi} \int_{-\pi}^{+\pi} \frac{\cos\alpha \, d\alpha}{r} = \frac{\mu_0 IR}{2\pi} \int_0^\pi \frac{\cos\alpha \, d\alpha}{r},$$

or, in view of Eq. (2.8.68),

$$A_\varphi = \frac{\mu_0 IR}{2\pi r_0} \int_0^\pi \frac{\cos\alpha \, d\alpha}{\sqrt{1 - \dfrac{2R}{r_0}\sin\theta\cos\alpha + \dfrac{R^2}{r_0^2}}}. \qquad (2.8.69)$$

At great distances from the wire, that is for $R/r_0 \ll 1$, one can use the
approximation

$$\frac{1}{\sqrt{1 - \dfrac{2R}{r_0}\sin\theta\cos\alpha + \dfrac{R^2}{r_0^2}}} \simeq \frac{1}{\sqrt{1 - \dfrac{2R}{r_0}\sin\theta\cos\alpha}}$$

$$\simeq 1 + \frac{R}{r_0}\sin\theta\cos\alpha,$$

in which case Eq. (2.8.69) leads to

$$A_\varphi = \frac{\mu_0 I R}{2\pi r_0} \int_0^\pi \left(1 + \frac{R}{r_0}\sin\theta\cos\alpha\right)\cos\alpha\, d\alpha$$

$$= \frac{\mu_0 I R^2 \sin\theta}{2\pi r_0^2}\int_0^\pi \cos^2\alpha\, d\alpha = \frac{\mu_0 I R^2}{4r_0^2}\sin\theta = \frac{\mu_0}{4\pi}\frac{m\,\sin\theta}{r_0^2}, \quad (2.8.70)$$

where $m = IS = I\pi R^2$ is the modulus of the magnetic moment of the wire, $\vec{m}$ being oriented along the normal to the circuit surface, that is along z-axis. Since the angle between the vectors $\vec{r}_0$ and $\vec{m}$ is θ, we can write

$$|\vec{A}_\varphi| = \frac{\mu_0}{4\pi}\frac{|\vec{m}\times\vec{r}_0|}{r_0^3},$$

or, in vector form,

$$\vec{A}_\varphi = \frac{\mu_0}{4\pi}\frac{\vec{m}\times\vec{r}_0}{r_0^3}. \quad (2.8.71)$$

2.9   Problem No. 18

Determine the magnetic moment of a homogeneous, non-magnetic, filled in sphere of radius R, uniformly charged with total electric charge Q. The sphere rotates uniformly ($\omega = const.$) about an axis passing through its center.

Solution

By rotation about the z-axis, the infinitesimal quantity of charge

$$dQ = \rho\, dV = \rho r^2 \sin\theta dr d\theta d\varphi$$

generates an "elementary" steady current of intensity

$$dI = \frac{dQ}{T} = \frac{\omega}{2\pi}dQ,$$

equivalent to that of an "elementary" conducting circular wire (an elementary circular loop) of radius $R_{ec} = r\sin\theta$ (see Fig. 2.7, where the volume element dV, corresponding to the infinitesimal charge dQ, was represented exceedingly large for clarity reasons), whose infinitesimal magnetic moment is

$$dm_z = S_z dI = \pi R_{ec}^2 dI = \pi r^2 \sin^2\theta dI = \pi r^2 \sin^2\theta\frac{dQ}{T}$$

$$= \frac{1}{2}\omega\rho r^4 \sin^3\theta\, dr\, d\theta\, d\varphi. \quad (2.9.72)$$

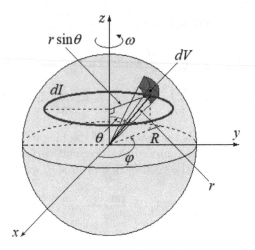

FIGURE 2.7
An infinitesimal volume element dV of the non-magnetic, uniformly charged sphere of radius R which rotates uniformly about an axis passing through its center. The figure also shows the "elementary" circular current dI created by the infinitesimal charge dQ (situated inside dV) that moves around the z-axis, on the circle of radius $r \sin \theta$.

By integration, one obtains

$$m_z = \frac{\rho \omega}{2} \int_0^R r^4 dr \int_0^\pi \sin^3 \theta \, d\theta \int_0^{2\pi} d\varphi = \frac{\rho \omega}{2} \frac{R^5}{5} \left(\int_0^\pi \sin^3 \theta \, d\theta \right) 2\pi$$

$$= \frac{\pi R^5 \rho \omega}{5} \int_0^\pi \sin^3 \theta \, d\theta. \qquad (2.9.73)$$

Since

$$\int_0^\pi \sin^3 \theta \, d\theta = \int_0^\pi \sin^2 \theta \sin \theta \, d\theta = \int_0^\pi \sin^2 \theta \, d(-\cos \theta)$$

$$= -(\sin^2 \theta \cos \theta) \Big|_0^\pi + 2 \int_0^\pi \sin \theta \cos^2 \theta d\theta = 2 \int_0^\pi \sin \theta \cos^2 \theta d\theta$$

$$= -2 \int_0^\pi \cos^2 \theta \, d(\cos \theta) = \left(-\frac{2}{3} \cos^3 \theta \right) \Big|_0^\pi = \frac{4}{3}, \qquad (2.9.74)$$

one then follows that the magnetic moment of the homogeneous, rotating, charged sphere, is

$$m_z = \frac{1}{5} \frac{4\pi R^3}{3} \rho \omega R^2 = \frac{1}{5} Q \omega R^2. \qquad (2.9.75)$$

2.10 Problem No. 19

Show that the field lines of the magnetic field created in vacuum by a plane filiform current, of an arbitrary shape, are symmetrical curves with respect to the plane of the circuit.

Solution

Without restricting the generality of the problem, let us consider the circuit as being situated in the xy-plane. Let $A(\xi, \eta, 0)$ be an arbitrary point of the circuit, and $P(x, y, z)$ a point of observation (see Fig. 2.8). According to the Biot-Savart law, the magnetic field intensity at the point P is

$$\vec{H} = \frac{I}{4\pi} \oint \frac{\vec{dl} \times \vec{r}}{r^3}, \tag{2.10.76}$$

where $\vec{r}$ is the radius vector of the point P with respect to A: $\vec{r} = \vec{r}(x - \xi, y - \eta, z)$, $r = \sqrt{(x - \xi)^2 + (y - \eta)^2 + z^2}$. If $\vec{dl} = (d\xi, d\eta, 0)$ is an infinitesimal element of length along the circuit, then the vector product appearing in formula (2.10.76) writes

$$\vec{dl} \times \vec{r} = \begin{vmatrix} \vec{i} & \vec{j} & \vec{k} \\ d\xi & d\eta & 0 \\ x - \xi & y - \eta & z \end{vmatrix} = \vec{i}z\,d\eta - \vec{j}z\,d\xi + \vec{k}[(y - \eta)d\xi - (x - \xi)d\eta],$$

and formula (2.10.76) gives the following components for $\vec{H}$:

$$H_x(x, y, z) = \frac{I}{4\pi} \oint \frac{z\,d\eta}{\left[(x - \xi)^2 + (y - \eta)^2 + z^2\right]^{3/2}}, \tag{2.10.77}$$

$$H_y(x, y, z) = -\frac{I}{4\pi} \oint \frac{z\,d\xi}{\left[(x - \xi)^2 + (y - \eta)^2 + z^2\right]^{3/2}}, \tag{2.10.78}$$

$$H_z(x, y, z) = \frac{I}{4\pi} \oint \frac{(y - \eta)d\xi - (x - \xi)d\eta}{\left[(x - \xi)^2 + (y - \eta)^2 + z^2\right]^{3/2}}. \tag{2.10.79}$$

Let us now consider the point $P'(x, y, -z)$, which is symmetric to the point $P(x, y, z)$ with respect to the circuit plane. Then, according to Eqs. (2.10.77), (2.10.78) and (2.10.79), we have

$$H_x(x, y, -z) = -H_x(x, y, z), \tag{2.10.80}$$

$$H_y(x, y, -z) = -H_y(x, y, z), \tag{2.10.81}$$

$$H_z(x, y, -z) = H_z(x, y, z). \tag{2.10.82}$$

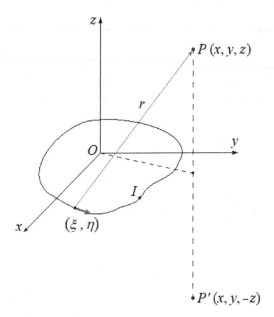

FIGURE 2.8
A stationary electric current I passing through a plane filiform wire of an arbitrary shape.

The field lines of the magnetic field are the integral curves of the following system of differential equations:

$$\frac{dx}{H_x(x,y,z)} = \frac{dy}{H_y(x,y,z)} = \frac{dz}{H_z(x,y,z)} \equiv d\zeta. \qquad (2.10.83)$$

Let

$$x = f(\zeta), \ y = g(\zeta), \ z = h(\zeta), \quad (\zeta_1 < \zeta < \zeta_2), \qquad (2.10.84)$$

be a solution of the system (2.10.83), given in a parametric form (*i.e.*, the parametric equations of a portion Γ of a field line, the parameter being denoted here by ζ).

In view of the symmetry relations (2.10.80), (2.10.81) and (2.10.82), one can easily observe that

$$x = f(\zeta) = f(-\zeta),$$
$$y = g(\zeta) = g(-\zeta),$$
$$z = -h(\zeta) = -h(-\zeta), \quad (-\zeta_2 < \zeta < -\zeta_1),$$

are well-defined curves according to solution given by Eq. (2.10.84) and satisfy the system (2.10.83). They represent the portion of the field line Γ', which is symmetric to Γ with respect to the circuit plane.

2.11   Problem No. 20

Show that the interaction force between two circuits through which flow steady currents satisfies the action and reaction principle, while the interaction force between two elements of circuit does not satisfy this principle.

Solution

From the very beginning we specify that the two circuits C_1 and C_2 (see Fig. 2.9), through which flow the steady (uniform and constant) currents I_1 and I_2, are supposed to be fixed.

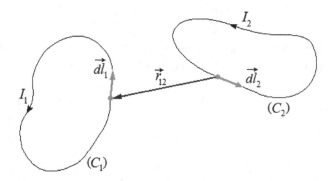

FIGURE 2.9
Two interacting current-carrying loops.

The elementary/infinitesimal force with which the field produced by the elementary/infinitesimal current $I_2\vec{dl}_2$, belonging to the circuit C_2, acts upon the elementary current $I_1\vec{dl}_1$ of the circuit C_1 is given by

$$d\vec{F}_{12} = I_1\vec{dl}_1 \times d\vec{B}_2$$

$$= \frac{\mu_0}{4\pi} I_1\vec{dl}_1 \times \frac{I_2\vec{dl}_2 \times \vec{r}_{12}}{r_{12}^3} \equiv k_m I_1\vec{dl}_1 \times \frac{I_2\vec{dl}_2 \times \vec{r}_{12}}{r_{12}^3}, \quad (2.11.85)$$

where the formula of the Laplace force and Biot-Savart law were used (in this order). To be very rigorous, mathematically speaking, the infinitesimal force $d\vec{F}_{12}$ in the above formula is in fact a squared infinitesimal force, because it writes as a product of two infinitesimal quantities (it is an infinitesimal quantity "of second order"); so, more correctly we should write $d^2\vec{F}_{12}$ instead of $d\vec{F}_{12}$.

If we develop the double vector product in Eq. (2.11.85) then we can write

$$d^2\vec{F}_{12} = k_m I_1 I_2 \left[\frac{\vec{dl}_2 \left(\vec{r}_{12} \cdot \vec{dl}_1 \right) - \vec{r}_{12} \left(\vec{dl}_1 \cdot \vec{dl}_2 \right)}{r_{12}^3} \right]. \qquad (2.11.86)$$

In the same way one calculates the elementary force with which the current element $I_1 \vec{dl}_1$ acts upon $I_2 \vec{dl}_2$, namely

$$d^2\vec{F}_{21} = k_m I_1 I_2 \left[\frac{\vec{r}_{12} \left(\vec{dl}_1 \cdot \vec{dl}_2 \right) - \vec{dl}_1 \left(\vec{r}_{12} \cdot \vec{dl}_2 \right)}{r_{12}^3} \right]. \qquad (2.11.87)$$

As can be easily seen,

$$d^2\vec{F}_{12} \neq -d^2\vec{F}_{21}, \qquad (2.11.88)$$

which demonstrates the second requirement of the problem.

Let us first calculate the force acting by the current element $I_2 \vec{dl}_2$ upon the whole circuit C_1; denote this force by $d\vec{F}_{12}^{(C_1)}$. We have

$$d\vec{F}_{12}^{(C_1)} = k_m I_1 I_2 \left[\oint\limits_{(C_1)} \frac{\vec{dl}_2 \left(\vec{r}_{12} \cdot \vec{dl}_1 \right)}{r_{12}^3} - \oint\limits_{(C_1)} \frac{\vec{r}_{12} \left(\vec{dl}_1 \cdot \vec{dl}_2 \right)}{r_{12}^3} \right]$$

$$= k_m I_1 I_2 \left[\vec{dl}_2 \oint\limits_{(C_1)} \frac{\vec{r}_{12} \cdot \vec{dl}_1}{r_{12}^3} - \oint\limits_{(C_1)} \frac{\vec{r}_{12} \left(\vec{dl}_1 \cdot \vec{dl}_2 \right)}{r_{12}^3} \right].$$

(Note that this force is not the same with that acting by the whole circuit C_2 upon the current element $I_1 \vec{dl}_1$, which is given by

$$d\vec{F}_{12}^{(C_2)} = k_m I_1 I_2 \left[\oint\limits_{(C_2)} \frac{\vec{dl}_2 \left(\vec{r}_{12} \cdot \vec{dl}_1 \right)}{r_{12}^3} - \oint\limits_{(C_2)} \frac{\vec{r}_{12}(\vec{dl}_1 \cdot \vec{dl}_2)}{r_{12}^3} \right].) \qquad (2.11.89)$$

But

$$\oint\limits_{(C_1)} \frac{\vec{r}_{12} \cdot \vec{dl}_1}{r_{12}^3} = - \oint\limits_{(C_1)} \nabla_1 \left(\frac{1}{r_{12}} \right) \cdot \vec{dl}_1$$

$$= - \oint\limits_{(C_1)} d \left(\frac{1}{r_{12}} \right) = 0; \quad \left(\nabla_1 \equiv \frac{\partial}{\partial \vec{l}_1} \right),$$

so that

$$d\vec{F}_{12}^{(C_1)} = - k_m I_1 I_2 \oint\limits_{(C_1)} \frac{\vec{r}_{12} \left(\vec{dl}_1 \cdot \vec{dl}_2 \right)}{r_{12}^3}, \qquad (2.11.90)$$

and then, the (total) force with which the whole circuit C_2 acts upon the whole circuit C_1 is

$$\vec{F}_{12} = \oint_{(C_2)} d\vec{F}_{12}^{(C_1)} = -k_m I_1 I_2 \oint_{(C_2)} \oint_{(C_1)} \frac{\vec{r}_{12}\left(\vec{dl}_1 \cdot \vec{dl}_2\right)}{r_{12}^3}. \qquad (2.11.91)$$

Let us now calculate the force acting by the current element $I_1\vec{dl}_1$ upon the whole circuit C_2; denote this force by $d\vec{F}_{21}^{(C_2)}$. This time we have

$$d\vec{F}_{21}^{(C_2)} = k_m I_1 I_2 \left[-\oint_{(C_2)} \frac{\vec{dl}_1\left(\vec{r}_{12} \cdot \vec{dl}_2\right)}{r_{12}^3} + \oint_{(C_2)} \frac{\vec{r}_{12}\left(\vec{dl}_1 \cdot \vec{dl}_2\right)}{r_{12}^3} \right]$$

$$= k_m I_1 I_2 \left[-\vec{dl}_1 \oint_{(C_2)} \frac{\vec{r}_{12} \cdot \vec{dl}_2}{r_{12}^3} + \oint_{(C_2)} \frac{\vec{r}_{12}\left(\vec{dl}_1 \cdot \vec{dl}_2\right)}{r_{12}^3} \right].$$

(Note again that this force is not the same with that acting by the whole circuit C_1 upon the current element $I_2\vec{dl}_2$, which is given by

$$d\vec{F}_{21}^{(C_1)} = k_m I_1 I_2 \left[-\oint_{(C_1)} \frac{\vec{dl}_1\left(\vec{r}_{12} \cdot \vec{dl}_2\right)}{r_{12}^3} + \oint_{(C_1)} \frac{\vec{r}_{12}(\vec{dl}_1 \cdot \vec{dl}_2)}{r_{12}^3} \right].) \qquad (2.11.92)$$

Because

$$\oint_{(C_2)} \frac{\vec{r}_{12} \cdot \vec{dl}_2}{r_{12}^3} = -\oint_{(C_2)} \nabla_1\left(\frac{1}{r_{12}}\right) \cdot \vec{dl}_2 = \oint_{(C_2)} \nabla_2\left(\frac{1}{r_{12}}\right) \cdot \vec{dl}_2$$

$$= \oint_{(C_2)} d\left(\frac{1}{r_{12}}\right) = 0; \quad \left(\nabla_2 \equiv \frac{\partial}{\partial \vec{l}_2}\right),$$

the force $d\vec{F}_{21}^{(C_2)}$ becomes

$$d\vec{F}_{21}^{(C_2)} = k_m I_1 I_2 \oint_{(C_2)} \frac{\vec{r}_{12}\left(\vec{dl}_1 \cdot \vec{dl}_2\right)}{r_{12}^3}, \qquad (2.11.93)$$

and then, the (total) force with which the whole circuit C_1 acts upon the whole circuit C_2 is

$$\vec{F}_{21} = \oint_{(C_1)} d\vec{F}_{21}^{(C_2)} = k_m I_1 I_2 \oint_{(C_1)} \oint_{(C_2)} \frac{\vec{r}_{12}\left(\vec{dl}_1 \cdot \vec{dl}_2\right)}{r_{12}^3}, \qquad (2.11.94)$$

which means that

$$\vec{F}_{21} = -\vec{F}_{12}, \tag{2.11.95}$$

as it has been expected. With that, the first requirement of the problem has also been demonstrated.

In order to draw a general conclusion, let us first systematise the results obtained so far. Therefore, in view of the above, the following eight forces can be identified:

1) the ("double/squared") infinitesimal force acting by the elementary current $I_1\vec{dl}_1$ upon the elementary current $I_2\vec{dl}_2$: $d^2\vec{F}_{21}$ (formula (2.11.87));

2) the ("double/squared") infinitesimal force acting by the elementary current $I_2\vec{dl}_2$ upon the elementary current $I_1\vec{dl}_1$: $d^2\vec{F}_{12}$ (formula (2.11.86));

3) the ("single/simple") infinitesimal force acting by the elementary current $I_1\vec{dl}_1$ upon the whole circuit C_2: $d\vec{F}_{21}^{(C_2)}$ (formula (2.11.93));

4) the ("single/simple") infinitesimal force acting by the whole circuit C_2 upon the elementary current $I_1\vec{dl}_1$: $d\vec{F}_{12}^{(C_2)}$ (formula (2.11.89));

5) the ("single/simple") infinitesimal force acting by the elementary current $I_2\vec{dl}_2$ upon the whole circuit C_1: $d\vec{F}_{12}^{(C_1)}$ (formula (2.11.90));

6) the ("single/simple") infinitesimal force acting by the whole circuit C_1 upon the elementary current $I_2\vec{dl}_2$: $d\vec{F}_{21}^{(C_1)}$ (formula (2.11.92));

7) the force acting by the whole circuit C_1 upon the whole circuit C_2: $\vec{F}_{21}$ (formula (2.11.94));

8) the force acting by the whole circuit C_2 upon the whole circuit C_1: $\vec{F}_{12}$ (formula (2.11.91));

Obviously, from the point of view of the validity of the action and reaction principle, we can only study the relationships that exist between forces that have the same order of "infinitesimality"/differentiability, as follows (cases which have no relevance in terms of the principle of action and reaction have been excluded):

a) 1) with 2);

b) 3) with 4);

c) 5) with 6);

d) 7) with 8);

The comparisons corresponding to the cases a) and d) have already been analyzed. A comparative analysis between the formulas corresponding to cases b) and c), which can be done simply and quickly, shows that in none of these cases is the principle of action and reaction satisfied.

3

Energy of Electrostatic and Magnetostatic Fields: Electromagnetic Induction

3.1 Problem No. 21

An electric charge Q is uniformly distributed inside a cone whose hight h equals the base radius. The cone rotates around its symmetry axis with the constant angular velocity ω. At the top of the cone is placed a particle with the internal magnetic moment $\vec{\mu}$. Calculate the magnetic energy of interaction between the particle and the cone.

Solution

The magnetic energy of interaction between the particle and the cone is the same − as an expression − with that corresponding to a magnetic dipole or a permanent magnetic moment $\vec{\mu}$, situated in an external magnetic field,

$$W = -\vec{\mu} \cdot \vec{B}, \tag{3.1.1}$$

where the induction $\vec{B}$ of the external magnetic field can be variable over time.

This well-known result for the potential energy of a dipole with magnetic moment $\vec{\mu}$ shows that the dipole tends by itself to orient parallel to the field, in order to reach a position of minimal potential energy.

In reality, the energy W given by Eq. (3.1.1) is not the total energy of the dipole of magnetic moment $\vec{\mu}$ placed in the external field $\vec{B}$, because in order to bring the dipole in the field $\vec{B}$ it must be spent a mechanical work to keep constant the current $\vec{j}$ which produces $\vec{\mu}$. Even if finally the state is stationary, initially there is a transient state, when the field is time variable. In fact, when an object is placed in a magnetic field with fixed sources, if initially the fields are $\vec{H}_0$ and $\vec{B}_0$, while after inserting the object they become $\vec{H}$ and $\vec{B}$, the variation of the magnetic energy is given by

$$W = \frac{1}{2} \int\limits_{(V)} \left(\vec{B} \cdot \vec{H}_0 - \vec{H} \cdot \vec{B}_0 \right) dV. \tag{3.1.2}$$

DOI: 10.1201/9781003402602-3

If the object is introduced in an empty space, the energy given by Eq. (3.1.2) can be expressed in terms of magnetization (the magnetic moment per unit volume) as

$$W = \frac{1}{2} \int\limits_{(V)} \vec{M} \cdot \vec{B} \, dV, \qquad (3.1.3)$$

where the integration is extended over the object volume.

The difference between Eq. (3.1.3) (or Eqs. (3.1.2)) and (3.1.1) for a permanent magnetic moment in an external magnetic field (apart of factor $1/2$ which appears due to the supposed linear relationship between $\vec{M}$ and $\vec{B}$) comes from the fact that the quantity in Eq. (3.1.3) is the total energy necessary to "produce" the given configuration, while Eq. (3.1.1) includes only the mechanical work performed to bring the permanent magnetic moment in the field, but not the energy needed to create and preserve it permanently. For this reason, the energy given by Eq. (3.1.1) is also called *potential energy* and, obviously, does not coincide with the total energy. The difference was explained above.

Therefore, in order to find the energy W given by Eq. (3.1.1), the induction of the external magnetic field must be determined at the point where the particle is, *i.e.*, at the top of the cone. This magnetic field is produced by the currents generated by the uniform circular motion of the electric charges which fill uniformly the cone.

An infinitesimal volume element of the cone, $dV = rdrdzd\varphi$, contains the infinitesimal quantity of electric charge (see Fig. 3.1):

$$dq\left(= d^3q\right) = \rho_e dV = \frac{Q}{V_{cone}} rdrdzd\varphi = \frac{3Q}{\pi h^3} rdrdzd\varphi = \frac{3Q\omega}{\pi h^3} rdrdzdt, \ (3.1.4)$$

where ρ_e is the electric charge spatial density inside the cone. It was also taken into account that the angular velocity of rotation of the cone around its axis is $\omega = d\varphi/dt$. (The notation in parentheses d^3q is rigorously mathematically correct and highlights the fact that it is a third order differential quantity, being the product of three total differentials.)

The uniform motion of rotation of this "elementary" electric charge generates an (infinitesimal) electric current of intensity

$$I\left(= d^2I\right) = \frac{dq}{dt}\left(= \frac{d^3q}{dt}\right) = \frac{3Q\omega}{\pi h^3} rdrdz, \qquad (3.1.5)$$

which produces at the top of the cone an infinitesimal magnetic field whose induction is given by the Biot-Savart law,

$$d\vec{B}\left(= d^3\vec{B}\right) = \frac{\mu_0}{4\pi} \frac{I\vec{dl} \times \vec{R}}{R^3}\left(= \frac{\mu_0}{4\pi} \frac{d^2I \, \vec{dl} \times \vec{R}}{R^3}\right), \qquad (3.1.6)$$

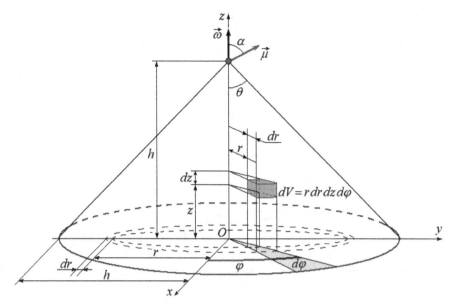

FIGURE 3.1
A uniformly distributed charge Q inside a uniformly rotating cone that has at the top a particle with the internal magnetic moment $\vec{\mu}$.

where $\vec{R}$ is the position vector of the top of the cone (*i.e.*, the point in which the field is going to be calculated) with respect to the "origin" of the infinitesimal current element $I d\vec{l} \left(= d^2 I d\vec{l} \right)$ (see Fig. 3.2).

As it can be observed in Fig. 3.2, the vector $d\vec{B} \left(= d^3 \vec{B} \right)$ has only two non-zero components in a frame with cylindrical coordinates (obviously, the problem symmetry is cylindrical), because the vector $d^2 I d\vec{l} \times \vec{R}$ has no component in the azimuthal direction (it is perpendicular to the arc element $r d\varphi$). In addition, due to the symmetry of the problem, after integration over the angular coordinate φ, the resulting component of $d\vec{B} \left(= \int_{(\varphi)} d^3 \vec{B} = d^2 \vec{B} \right)$ on the radial direction cancels. Indeed, for any value of the angular variable φ, the radial component of $d\vec{B} (= d^3 \vec{B})$ will be opposed to a quantity equal in modulus but of opposite sense, namely that corresponding to the value $\varphi + \pi$ of the integration variable (remember that integration is, at the limit, a summation which, here, is a vector summation). Therefore, as a result of integration over $I d\vec{l} \left(= d^2 I d\vec{l} \right)$ (in fact, after the angular variable φ) remains only the axial component of the magnetic induction

$$B_z \left(= d^2 B_z \right) = |\vec{B}| \cos\beta = B\frac{r}{R}. \tag{3.1.7}$$

For now, the problem lies in calculation of the magnetic induction produced by a circular current (a circular loop or ring) of radius r, at a point situated

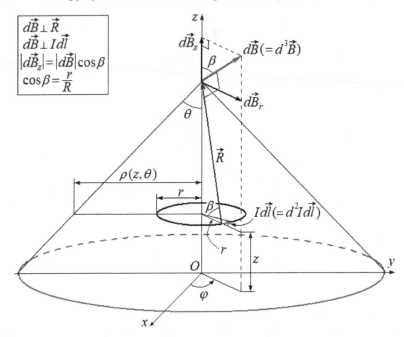

FIGURE 3.2
The infinitesimal magnetic field $d\vec{B}$ created at the top of the cone by the infinitesimal circular current element $I\vec{dl}$ of radius r.

on the positive semi-axis of the ring, at the height $\sqrt{R^2 - r^2}$ with respect to the horizontal plane of the ring. According to Biot-Savart law, the current element $I\vec{dl}$ of the ring of radius r, through which flows a steady current of intensity I (see Fig. 3.3) produces at the point P the elementary magnetic field of induction

$$d\vec{B} = \frac{\mu_0}{4\pi} \frac{I\vec{dl} \times \vec{R}}{R^3},$$

while the total magnetic field produced by the whole circular loop at the point P is

$$\vec{B} = \frac{\mu_0}{4\pi} \int_{(C)} \frac{I\vec{dl} \times \vec{R}}{R^3} = \frac{\mu_0}{4\pi} \int_0^{2\pi} \frac{Ir d\varphi (\hat{u}_\varphi \times \vec{R})}{R^3} = \frac{\mu_0}{4\pi} \frac{Ir\hat{u}_B}{R^2} \int_0^{2\pi} d\varphi$$

$$= \frac{\mu_0}{2} \frac{Ir}{R^2} \hat{u}_B,$$

where $\hat{u}_\varphi$ is the unit vector of the azimuthal direction, and $\hat{u}_B = \hat{u}_\varphi \times \hat{u}_R$ is the unit vector of the vector $I\vec{dl} \times \vec{R}$ (see Fig. 3.3).

Taking into account the above observation regarding the cancellation of the resulting radial component of $\vec{B}$, it follows that the induction of the resultant

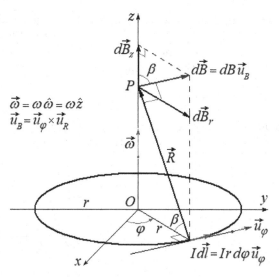

FIGURE 3.3
Detailed image allowing easy observation of $d\vec{B}$ vector spatial orientation.

magnetic field, created by the circular loop at the point P is

$$\left|\vec{B}_{res}\right| = \vec{B} \cdot \hat{z} = \frac{\mu_0}{2}\frac{Ir}{R^2}(\hat{u}_B \cdot \hat{z}) = \frac{\mu_0}{2}\frac{Ir}{R^2}\cos\beta,$$

having the same direction as z-axis,

$$\vec{B}_{res} = \frac{\mu_0}{2}\frac{Ir\cos\beta}{R^2}\hat{z} = \frac{\mu_0}{2}\frac{Ir\cos\beta}{R^2}\hat{\omega}, \qquad (3.1.8)$$

where $\hat{\omega} = \vec{\omega}/\omega\,(=\hat{z})$ is the angular velocity unit vector, which here coincides with the unit vector of the z-axis.

Using this result, one can write the induction of the "resultant" magnetic field, produced at the top of the cone by the current (in fact, the current element) $I\left(=d^2I\right) = \left(3Q\omega/\pi h^3\right)r dr dz$, which is an elementary circular loop crossed by the steady current d^2I:

$$d^2\vec{B}_{res} = \frac{\mu_0}{2}\frac{d^2Ir\cos\beta}{R^2}\hat{\omega} = \hat{\omega}\frac{\mu_0}{2}\frac{3Q\omega r^3}{\pi h^3 R^3}dr dz. \qquad (3.1.9)$$

The resulting total magnetic field produced by the whole cone (by all current elements $I d\vec{l}\left(=d^2I d\vec{l}\right)$, which are *sui-generis* elementary circular loops), is then obtained by summing up all the contributions brought by all these infinitesimal current elements (which, for each elementary individual ring they represent self-standing steady currents). Mathematically, the resulting magnetic field is obtained by integration over all possible values of the variables intervening in $d^2\vec{B}_{res}$, namely r (from 0 to ρ, where ρ is the radius of the rings tangent to the side wall of the cone), and z (from 0 to h), this way being considered all current elements contained inside the cone. When integrating, must also be considered the dependence of R on the two variables, which are not independent, but satisfy the cone equation,

$$x^2 + y^2 = \rho^2 = (h - z)^2 \tan^2 \theta, \tag{3.1.10}$$

where 2θ is the angular opening of the cone (see Fig. 3.1).

Since $R = \sqrt{r^2 + (h - z)^2}$, the integration over $r = r(\rho)$ gives

$$d\vec{B}_{res} = \int_{(r)} d^2\vec{B}_{res} = \frac{\mu_0}{2}\frac{3Q\omega}{\pi h^3}dz\hat{\omega}\int_0^\rho \frac{r^3}{\left[r^2 + (h-z)^2\right]^{3/2}}dr$$

$$= \frac{\mu_0}{2}\frac{3Q\omega}{\pi h^3}dz\hat{\omega}\left[\sqrt{r^2 + (h-z)^2} + \frac{(h-z)^2}{\sqrt{r^2 + (h-z)^2}}\right]_{r=0}^{r=\rho}$$

$$= \frac{\mu_0}{2}\frac{3Q\omega}{\pi h^3}dz\hat{\omega}\frac{\left[\sqrt{\rho^2 + (h-z)^2} - (h-z)\right]^2}{\sqrt{\rho^2 + (h-z)^2}}$$

$$= \frac{\mu_0}{2}\frac{3Q\omega}{\pi h^3}dz\hat{\omega}\frac{\left[\sqrt{(h-z)^2\tan^2\theta + (h-z)^2} - (h-z)\right]^2}{\sqrt{(h-z)^2\tan^2\theta + (h-z)^2}}$$

$$= \frac{\mu_0}{2}\frac{3Q\omega}{\pi h^3}dz\hat{\omega}\frac{(1-\cos\theta)^2}{\cos\theta}(h-z).$$

Finally, integration over z of the last relation gives the induction of the resultant magnetic field, generated by all the elementary currents, at the top of the cone,

$$\vec{B}_{res} = \int_{(z)} d\vec{B}_{res}(z) = \frac{\mu_0}{2}\frac{3Q\omega}{\pi h^3}\hat{\omega}\frac{(1-\cos\theta)^2}{\cos\theta}\int_0^h (h-z)dz$$

$$= \frac{\mu_0}{2}\frac{3Q\omega}{\pi h^3}\hat{\omega}\frac{(1-\cos\theta)^2}{\cos\theta}\left(hz - \frac{z^2}{2}\right)_{z=0}^{z=h}$$

$$= \frac{\mu_0}{2}\frac{3Q\omega}{\pi h^3}\hat{\omega}\frac{(1-\cos\theta)^2}{\cos\theta}\frac{h^2}{2} = \frac{\mu_0}{4\pi}\frac{3Q\vec{\omega}}{h}\frac{(1-\cos\theta)^2}{\cos\theta}$$

$$= \frac{\mu_0}{4\pi}\frac{3Q\vec{\omega}}{h}\frac{\left(1 - \dfrac{1}{\sqrt{2}}\right)^2}{\dfrac{1}{\sqrt{2}}} = \frac{\mu_0}{4\pi}\frac{3\sqrt{2} - 4}{2}\frac{3Q}{h}\vec{\omega}.$$

To conclude, the magnetic energy of interaction between the particle with internal magnetic moment $\vec{\mu}$ and the magnetic field generated by the charged, moving cone is

$$W = -\vec{\mu} \cdot \vec{B} \left(= -\vec{\mu} \cdot \vec{B}_{res} \right) = (-\vec{\mu}) \cdot \left(\frac{\mu_0}{4\pi} \frac{3\sqrt{2} - 4}{2} \frac{3Q\vec{\omega}}{h} \right)$$

$$= \frac{\mu_0}{4\pi} \frac{3Q(4 - 3\sqrt{2})}{2h} (\vec{\mu} \cdot \vec{\omega}) = \frac{3(4 - 3\sqrt{2})\mu\mu_0 Q \cos\alpha}{8\pi h}, \qquad (3.1.11)$$

where α is the angle between vectors $\vec{\omega}$ (the angular velocity of uniform rotation of the cone around its axis), and $\vec{\mu}$ (the internal magnetic moment of the particle at the top of the cone – see Fig. 3.1).

3.2 Problem No. 22

Consider an arbitrary (but bounded) distribution of electric charges in vacuum and let W_0 be the energy of its electrostatic field. If the space is occupied by a dielectric (which can be non-homogeneous, and even anisotropic), the energy of the electrostatic field of this distribution becomes W. Show that

$$\delta W = W - W_0 = -\frac{1}{2} \int \vec{P} \cdot \vec{E}_0 \, d\tau, \qquad (3.2.12)$$

where $\vec{P}$ is the polarization vector, and $\vec{E}_0$ the intensity of the initial electrostatic field.

Solution

As it is known, the energy of the electrostatic field in vacuum is given by

$$W_0 = \frac{1}{2} \int \vec{E}_0 \cdot \vec{D}_0 \, d\tau, \qquad (3.2.13)$$

while in the presence of a dielectric it is

$$W = \frac{1}{2} \int \vec{E} \cdot \vec{D} \, d\tau. \qquad (3.2.14)$$

Since the charge distributions in vacuum and dielectric are the same, the fields $\vec{E}_0$ and $\vec{D}_0$ satisfy the well-known relations

$$\nabla \times \vec{E}_0 = 0, \quad \nabla \cdot \vec{D}_0 = \rho, \qquad (3.2.15)$$

and, similarly, for the fields $\vec{E}$ and $\vec{D}$:

$$\nabla \times \vec{E} = 0, \quad \nabla \cdot \vec{D} = \rho. \qquad (3.2.16)$$

We then have

$$\delta W = W - W_0 = \frac{1}{2} \int \left(\vec{E} \cdot \vec{D} - \vec{E}_0 \cdot \vec{D}_0 \right) d\tau$$

$$= \frac{1}{2} \int \left[\vec{E} \cdot \left(\vec{D} - \vec{D}_0 \right) + \vec{E}_0 \cdot \left(\vec{D} - \vec{D}_0 \right) - \vec{P} \cdot \vec{E}_0 \right] d\tau, \quad (3.2.17)$$

where, between the square brackets, we added and subtracted the expression $\vec{P} \cdot \vec{E}_0$. Indeed, since $\vec{D}_0 = \varepsilon_0 \vec{E}_0$ and $\vec{D} = \varepsilon_0 \vec{E} + \vec{P}$, we have

$$-\vec{E} \cdot \vec{D}_0 + \vec{E}_0 \cdot \vec{D} = \vec{P} \cdot \vec{E}_0,$$

and so

$$\vec{E} \cdot \left(\vec{D} - \vec{D}_0 \right) + \vec{E}_0 \cdot \left(\vec{D} - \vec{D}_0 \right) - \vec{P} \cdot \vec{E}_0 = \vec{E} \cdot \vec{D} - \vec{E}_0 \cdot \vec{D}_0.$$

It can be shown that

$$\int \vec{E} \cdot \left(\vec{D} - \vec{D}_0 \right) d\tau = 0.$$

Indeed, since $\vec{E} = -\text{grad}\, V$, we can write

$$\vec{E} \cdot \left(\vec{D} - \vec{D}_0 \right) = -\text{grad}\, V \cdot \left(\vec{D} - \vec{D}_0 \right) = -\text{div} \left[V \left(\vec{D} - \vec{D}_0 \right) \right]$$

$$+ V \underbrace{\text{div} \left(\vec{D} - \vec{D}_0 \right)}_{= \rho - \rho_0 = 0} = -\text{div} \left[V \left(\vec{D} - \vec{D}_0 \right) \right].$$

Therefore,

$$\int\limits_{(D)} \vec{E} \cdot \left(\vec{D} - \vec{D}_0 \right) d\tau = - \int\limits_{(D)} \text{div} \left[V \left(\vec{D} - \vec{D}_0 \right) \right] d\tau$$

$$= - \oint\limits_{(S_D)} V \left(\vec{D} - \vec{D}_0 \right) \cdot d\vec{S} \to 0,$$

because – recalling that the charge distribution is bounded – V decreases as $\frac{1}{r}$ when $r \to \infty$, while the field vectors $\vec{D}$ and $\vec{D}_0$ decrease as $\frac{1}{r^2}$, so that the product $\left(\vec{D} - \vec{D}_0 \right) \cdot d\vec{S}$ vary as $r^2 \times \frac{1}{r} \times \frac{1}{r^2} = \frac{1}{r}$ and tends to zero when r tends to ∞.

Similarly, one can also show that

$$\int \vec{E}_0 \cdot \left(\vec{D} - \vec{D}_0 \right) d\tau \to 0.$$

Under these circumstances, relation (3.2.17) takes the form

$$\delta W = W - W_0 = -\frac{1}{2} \int \vec{P} \cdot \vec{E}_0 \, d\tau,$$

which gives the answer to the problem.

3.3 Problem No. 23

A bounded distribution of electric currents of density[1] $\vec{j}(x, y, z)$ produces in vacuum a magnetic field whose energy is W_0. In a magnetizable medium (non-ferromagnetic) the same current distribution produces a field whose total energy is W. Show that the contribution of the medium to the value of energy is

$$\delta W = W - W_0 = \frac{1}{2} \int \vec{M} \cdot \vec{B}_0 \, d\tau, \tag{3.3.18}$$

where $\vec{M}$ is the magnetization vector of the medium, and $\vec{B}_0$ is the magnetic induction in vacuum.

Solution

For the beginning, let us consider the magnetic field produced by the distribution in vacuum. In this case, we have

$$\nabla \times \vec{H} = \vec{j}, \quad \nabla \cdot \vec{B}_0 = 0, \quad \vec{B}_0 = \nabla \times \vec{A}_0, \quad \vec{B}_0 = \mu_0 \vec{H}_0, \tag{3.3.19}$$

and the field energy is

$$W_0 = \frac{1}{2} \int \vec{H}_0 \cdot \vec{B}_0 \, d\tau. \tag{3.3.20}$$

The corresponding set of equations that can be written for the field produced by the same distribution of currents in the considered magnetizable medium (isotropic or anisotropic, but non-ferromagnetic) is

$$\nabla \times \vec{H} = \vec{i}, \quad \nabla \cdot \vec{B} = 0, \quad \vec{B} = \nabla \times \vec{A}, \quad \vec{M} = \frac{1}{\mu_0}\vec{B} - \vec{H}, \tag{3.3.21}$$

as well as

$$W = \frac{1}{2} \int \vec{H} \cdot \vec{B} \, d\tau. \tag{3.3.22}$$

The contribution of the medium to the energy is then given by

$$\delta W = W - W_0 = \frac{1}{2} \int \left(\vec{H} \cdot \vec{B} - \vec{H}_0 \cdot \vec{B}_0 \right) d\tau. \tag{3.3.23}$$

Taking into account the expression of magnetization (see Eq. $(3.3.21)_4$) and

---

[1]The *current density* $\vec{j}$ is defined through the relation $I = \int_{(S)} \vec{j} \cdot d\vec{S}$. Thus, the modulus of the current density is numerically equal to the intensity of the current passing a surface of unit area, orthogonal to the direction of the displacement of charges. By definition, $\vec{j}$ is oriented along the direction of displacement of positive charges.

the relation between $\vec{H}_0$ and $\vec{B}_0$ (formula $(3.3.19)_4$), the integrand of Eq. (3.3.23) can be transformed as follows:

$$
\begin{aligned}
\vec{H} \cdot \vec{B} &- \vec{H}_0 \cdot \vec{B}_0 \\
&= \left(\vec{H} - \vec{H}_0 \right) \cdot \vec{B} + \left(\vec{H} - \vec{H}_0 \right) \cdot \vec{B}_0 + \vec{H}_0 \cdot \vec{B} - \vec{H} \cdot \vec{B}_0 \\
&= \left(\vec{H} - \vec{H}_0 \right) \cdot \vec{B} + \left(\vec{H} - \vec{H}_0 \right) \cdot \vec{B}_0 + \frac{1}{\mu_0} \vec{B}_0 \cdot \vec{B} - \vec{H} \cdot \vec{B}_0 \\
&= \left(\vec{H} - \vec{H}_0 \right) \cdot \vec{B} + \left(\vec{H} - \vec{H}_0 \right) \cdot \vec{B}_0 + \vec{M} \cdot \vec{B}_0.
\end{aligned}
\tag{3.3.24}
$$

In this case, relation (3.3.23) becomes

$$
\begin{aligned}
\delta W = W - W_0 &= \frac{1}{2} \int (\vec{H} \cdot \vec{B} - \vec{H}_0 \cdot \vec{B}_0)\, d\tau \\
&= \frac{1}{2} \int (\vec{H} - \vec{H}_0) \cdot \vec{B}\, d\tau + \frac{1}{2} \int (\vec{H} - \vec{H}_0) \cdot \vec{B}_0\, d\tau \\
&\quad + \frac{1}{2} \int \vec{M} \cdot \vec{B}_0\, d\tau.
\end{aligned}
\tag{3.3.25}
$$

Therefore, in order to solve the problem, it remains to be shown that the first two integrals of the r.h.s. of the last equality in Eq. (3.3.25) are equal to zero. Indeed, in view of Eqs. (3.3.19), (3.3.21) and the vector identity

$$
\nabla \cdot (\vec{u} \times \vec{v}) = \vec{v} \cdot \nabla \times \vec{u} - \vec{u} \cdot \nabla \times \vec{v},
$$

we can write

$$
\begin{aligned}
\int \left(\vec{H} - \vec{H}_0 \right) \cdot \vec{B}\, d\tau &= \int \left(\vec{H} - \vec{H}_0 \right) \cdot \nabla \times \vec{A}\, d\tau \\
&= \int_{(D)} \nabla \cdot \left[\vec{A} \times \left(\vec{H} - \vec{H}_0 \right) \right] d\tau + \int_{(D)} \vec{A} \cdot \underbrace{\nabla \times \left(\vec{H} - \vec{H}_0 \right)}_{=\vec{\jmath}-\vec{\jmath}=0} d\tau \\
&= \int_{(D)} \nabla \cdot \left[\vec{A} \times \left(\vec{H} - \vec{H}_0 \right) \right] d\tau = \oint_{(S_D)} \left[\vec{A} \times \left(\vec{H} - \vec{H}_0 \right) \right] \cdot d\vec{S} \to 0,
\end{aligned}
$$

where we took into account that the distribution of the currents is the same, both in vacuum and magnetizable non-ferromagnetic medium $\left(\nabla \times \vec{H}_0 = \nabla \times \vec{H} = \vec{\jmath} \right)$ and, in addition (the distribution of the electric currents being bounded), for $r \to \infty$, $|\vec{A}| \sim \frac{1}{r}$ and $|\vec{H} - \vec{H}_0| \sim \frac{1}{r^2}$. Since $dS \sim r^2$ when $r \to \infty$, the product $\left[\vec{A} \times \left(\vec{H} - \vec{H}_0 \right) \right] \cdot d\vec{S}$ varies as $\frac{1}{r} \cdot \frac{1}{r^2} \cdot r^2 = \frac{1}{r} \to 0$.

In a very similar way it is demonstrated the fact that the integral

$$
\int \left(\vec{H} - \vec{H}_0 \right) \cdot \vec{B}_0\, d\tau
$$

in Eq. (3.3.25) vanishes. This way, the problem is fully solved.

3.4 Problem No. 24

Show that the presence of material bodies in an electromagnetic field can be fully described by introducing in the field equations of an additional charge density, $-\nabla \cdot \vec{P}$, and an additional current density, $\dfrac{\partial \vec{P}}{\partial t} + \nabla \times \vec{M}$, where $\vec{P}$ is the electric polarization of the medium, and $\vec{M}$ is the magnetization.

Solution

As well-known, the electric field intensity $\vec{E}$, the electric induction (electric displacement field) $\vec{D}$ and the polarization (or polarization density) $\vec{P}$ are connected by the relation

$$\vec{D} = \varepsilon_0 \vec{E} + \vec{P}, \tag{3.4.26}$$

while the magnetic field intensity $\vec{H}$, the magnetic induction $\vec{B}$ and the magnetization field $\vec{M}$ are related by

$$\vec{B} = \mu_0 \vec{H} + \mu_0 \vec{M}. \tag{3.4.27}$$

The equations describing the field are Maxwell's four fundamental equations:

– the electro-magnetic induction law: $\nabla \times \vec{E} = -\dfrac{\partial \vec{B}}{\partial t}$;

– the magneto-electric induction law: $\nabla \times \vec{H} = \vec{j} + \dfrac{\partial \vec{D}}{\partial t}$;

– Gauss's theorem for the electric field: $\nabla \cdot \vec{D} = \rho$;

– Gauss's theorem for the magnetic field: $\nabla \cdot \vec{B} = 0$.

The first and the last equation are also known as *Maxwell's source-free equations*, while the second and the third equations are called *Maxwell's source equations*. Here $\rho(\vec{r}, t)$ is the spatial density of the electric charges, and $\vec{j}(\vec{r}, t)$ is the current density.

Obviously, Maxwell's equations without sources remain unchanged if some bodies are introduced in the electromagnetic field, but the other two equations undergo appropriate modifications.

Let us begin with Gauss's flux theorem, written in its local (differential) form,

$$\nabla \cdot \vec{D} = \rho.$$

So, in view of Eq. (3.4.26), we have

$$\nabla \cdot \left(\varepsilon_0 \vec{E} + \vec{P} \right) = \rho,$$

or

$$\nabla \cdot \vec{E} = \frac{\rho}{\varepsilon_0} - \frac{\nabla \cdot \vec{P}}{\varepsilon_0} \equiv \frac{\rho}{\varepsilon_0} + \frac{\rho_{add}}{\varepsilon_0} = \frac{\rho + \rho_{add}}{\varepsilon_0}, \tag{3.4.28}$$

where we have introduced the obvious notation $\rho_{add} \equiv -\nabla \cdot \vec{P}$. Comparing Eq. (3.4.28) with Gauss's flux theorem written for vacuum,

$$\nabla \cdot \vec{E} = \frac{\rho}{\varepsilon_0},$$

it can be easily seen that the presence of the bodies in the electromagnetic field lead to "appearance" of an additional electric charge density $\rho_{add} = -\nabla \cdot \vec{P}$.

Proceeding in a similar way with the other Maxwell's source equation

$$\nabla \times \vec{H} = \vec{j} + \frac{\partial \vec{D}}{\partial t},$$

we obtain

$$\nabla \times \left(\frac{\vec{B}}{\mu_0} - \vec{M} \right) = \vec{j} + \frac{\partial \left(\varepsilon_0 \vec{E} + \vec{P} \right)}{\partial t}, \tag{3.4.29}$$

where we used the relation (3.4.27) to express $\vec{H}$, and $\vec{D}$ "has been taken" from Eq. (3.4.26). Equation (3.4.29) can also be written as

$$\nabla \times \vec{B} - \varepsilon_0 \mu_0 \frac{\partial \vec{E}}{\partial t} = \mu_0 \vec{j} + \mu_0 \left(\frac{\partial \vec{P}}{\partial t} + \nabla \times \vec{M} \right) \equiv \mu_0 \vec{j} + \mu_0 \vec{j}_{add} = \mu_0 \left(\vec{j} + \vec{j}_{add} \right).$$

If this equation is compared with the magneto-electric induction law for vacuum,

$$\nabla \times \vec{B} - \varepsilon_0 \mu_0 \frac{\partial \vec{E}}{\partial t} = \mu_0 \vec{j},$$

one observes the "appearance" of an additional electric current,

$$\vec{j}_{add} = \frac{\partial \vec{P}}{\partial t} + \nabla \times \vec{M},$$

when a body is introduced in the electromagnetic field.

3.5   Problem No. 25

A mathematical pendulum of length l, made of a conductive material, moves so that its lower end slips without friction on a support in the form of a circular arc, whose contact surface is also made of a conductive material. The pendulum arm of mass m is rigid and moves in a static magnetic field of

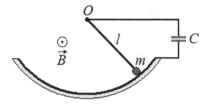

FIGURE 3.4
The electric circuit closing through the pendulum of length l.

induction $\vec{B}$, oriented perpendicular on any point of the pendulum arm. An end of the conductive material and the support point of the pendulum arm are connected by means of an ideal capacitor of capacitance C. The electric circuit (of variable length, due to the pendulum motion) closes through the arm of the pendulum, as in Fig. 3.4. Determine the pendulum period, by neglecting the electrical resistance and the circuit inductance.

Solution

During the infinitesimal time interval dt the pendulum arm of length l covers (on the circle circumference described by the pendulum lower end) the distance $ds = l\,d\alpha$ and "sweeps" the surface of area $dS = l\,ds/2 = l^2 d\alpha/2$ (see Fig. 3.5).

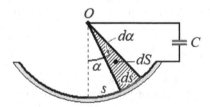

FIGURE 3.5
The elementary area swept by the pendulum rod (of length l) in the infinitesimal time interval dt, in which the body of mass m covers the elementary arc ds.

Since the magnetic field is orthogonal at any point to this surface, the elementary magnetic flux through dS is $d\Phi = \vec{B} \cdot d\vec{S} = B\,dS = Bl^2 d\alpha/2$. Since this flux is time variable, it induces in the pendulum arm the electromotive force (the minus sign is due to the Lenz's law):

$$e = -\frac{d\Phi}{dt} = -\frac{Bl^2}{2}\frac{d\alpha}{dt} = -\frac{\omega Bl^2}{2}, \qquad (3.5.30)$$

whose modulus is $E = |e| = Bl^2\omega/2$.

This voltage generates in the circuit shown in Fig. 3.4 an electric current of intensity

$$I = \frac{dq}{dt} = \frac{d}{dt}(CE) = C\frac{dE}{dt} = \frac{CBl^2}{2}\frac{d\omega}{dt} = \frac{CBl^2}{2}\frac{d^2\alpha}{dt^2}. \tag{3.5.31}$$

According to Laplace's law, on the pendulum arm (which is moving in the magnetic field of induction $\vec{B}$ and through which flows the current I) is exercised the force

$$\left|\vec{F}_{em}\right| = \left|I\,\vec{l}\times\vec{B}\right| = IBl = \frac{CBl^3}{2}\frac{d^2\alpha}{dt^2}, \tag{3.5.32}$$

having direction and sense shown in Fig. 3.6, and the application point at the middle of the segment $|OA|$.

The pendulum motion is produced by both the electromagnetic force, of momentum

$$\left|\vec{M}_{em}\right| \equiv M_{em} = \left|\vec{r}_B \times \vec{F}_{em}\right| = \frac{l}{2}F_{em} = \frac{CB^2l^4}{4}\frac{d^2\alpha}{dt^2}, \tag{3.5.33}$$

where $\vec{r}_B$ is the position vector of the application point of the force $\vec{F}_{em}$ with respect to the center of rotation (the point O), and the gravitational force, whose moment has the modulus

$$\left|\vec{M}_g\right| \equiv M_g = \left|\vec{r}_A \times \vec{F}_g\right| = \left|\vec{r}_A \times m\vec{g}\right| = lmg\sin\alpha, \tag{3.5.34}$$

where $\vec{r}_A$ is the position vector of the point of application of the gravitational force with respect to the same point. The differential equation of motion then is

$$J\frac{d^2\alpha}{dt^2} = -M_g - M_{em} = -mgl\sin\alpha - \frac{CB^2l^4}{4}\frac{d^2\alpha}{dt^2}, \tag{3.5.35}$$

where $J = ml^2$ is the moment of inertia of the pendulum arm with respect to the point O. The moments of both forces have been taken with minus sign, since both are "return"/"coming back" forces. Equation (3.5.35) can also be written in the form

$$\frac{d^2\alpha}{dt^2}\left(ml^2 + \frac{CB^2l^4}{4}\right) + mgl\sin\alpha = 0. \tag{3.5.36}$$

In the limit of small angles, $\sin\alpha \simeq \alpha$, and Eq. (3.5.36) becomes

$$\frac{d^2\alpha}{dt^2}\left(ml^2 + \frac{CB^2l^4}{4}\right) + mgl\,\alpha = 0. \tag{3.5.37}$$

By introducing the notation

$$\omega^2 \equiv \frac{mgl}{ml^2 + \dfrac{CB^2l^4}{4}} = \frac{1}{\dfrac{l}{g} + \dfrac{CB^2l^3}{4mg}} = \frac{1}{\dfrac{l}{g}\left(1 + \dfrac{CB^2l^2}{4m}\right)},$$

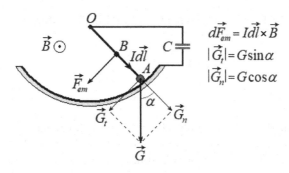

FIGURE 3.6
Graphical representation of the forces acting on the pendulum in Fig. 3.4.

Eq. (3.5.38) rewrites in the form

$$\frac{d^2\alpha}{dt^2} + \omega^2\alpha = 0,$$ (3.5.38)

which is the well-known differential equation of the linear harmonic oscillator, the oscillation period being

$$T = \frac{2\pi}{\omega} = 2\pi\sqrt{\frac{l}{g}\left(1 + \frac{CB^2l^2}{4m}\right)}.$$ (3.5.39)

The term $CB^2l^2/4m$ can be considered as a "contribution" brought by the electromagnetic effects to those of the gravitational field, since, as well-known, the period of motion of the mathematical pendulum in gravitational field, in the case of oscillations of small amplitudes (isochronous oscillations), is given by

$$T = 2\pi\sqrt{\frac{l}{g}}.$$ (3.5.40)

3.6 Problem No. 26

The Lagrangian of a charged particle of mass m and electric charge q that moves with the velocity $\vec{v}$ in the electromagnetic field $(\vec{E}, \vec{B})$ is

$$L = \frac{1}{2}m\vec{v}^2 - qV + q\vec{v}\cdot\vec{A},$$

where V is the scalar potential, and $\vec{A}$ is the vector potential of the field. Show that, if the magnetic field is constant and homogeneous (i.e., $\vec{A} = (\vec{B}\times\vec{r})/2$), and weak (*i.e.*, $\mathcal{O}(\vec{B}^2) = 0$), then, in the first approximation, the energy of a magnetic dipole of magnetic moment $\vec{\mu}$ placed in the exterior field $(\vec{E}, \vec{B})$ is given by the following formula:

$$W_m = -\vec{\mu} \cdot \vec{B}.$$

Solution

Because the particle moves freely (*i.e.*, there are no constraints to limit its motion) we can choose as generalized coordinates, just the Cartesian coordinates x_i, $i = \overline{1,3}$. Thus, the generalized momentum p_i canonically conjugated to the generalized coordinate x_i is

$$p_i = \frac{\partial L}{\partial \dot{x}_i} = \frac{\partial L}{\partial v_i} = mv_i + qA_i, \qquad (3.6.41)$$

so, the Hamiltonian of the particle is

$$H = p_i \dot{x}_i - L = p_i v_i - L = m\vec{v}^2 + q\,\vec{v} \cdot \vec{A} - \frac{1}{2}m\vec{v}^2 + qV$$

$$- q\,\vec{v} \cdot \vec{A} = \frac{1}{2}m\vec{v}^2 + qV = \frac{1}{2m}m^2\vec{v}^2 + qV = \frac{1}{2m}(p_i - qA_i)$$

$$\times (p_i - qA_i) + qV = \frac{1}{2m}\vec{p}^2 + \frac{1}{2m}q^2\vec{A}^2 - \frac{q}{m}\vec{p} \cdot \vec{A} + qV$$

$$= \frac{1}{2m}\vec{p}^2 + qV - \frac{q}{2}\vec{v} \cdot (\vec{B} \times \vec{r}) + \frac{q^2}{8m}|\vec{B} \times \vec{r}|^2$$

$$\simeq \frac{1}{2m}\vec{p}^2 + qV - \frac{q}{2}\vec{v} \cdot (\vec{B} \times \vec{r}) \equiv H_0 + H_{\text{int}},$$

where

$$H_0 = \frac{1}{2m}\vec{p}^2 + qV \qquad (3.6.42)$$

is the energy of the magnetic dipole in the absence of the magnetic field, while

$$H_{\text{int}} = -\frac{q}{2}\vec{v} \cdot (\vec{B} \times \vec{r}) \qquad (3.6.43)$$

is the "supplementary" energy of the magnetic dipole, which is due to the presence of the magnetic field. In order to show that

$$H_{\text{int}} = -\vec{\mu} \cdot \vec{B}, \qquad (3.6.44)$$

we will use the expression of the areolar velocity of the moving particle. During the finite time interval Δt, the position vector $\vec{r}$ of the particle sweeps the aria

$$\overrightarrow{\Delta S} = \frac{1}{2}\vec{r} \times \overrightarrow{\Delta r},$$

so the areolar velocity is

$$\frac{\overrightarrow{\Delta S}}{\Delta t} = \frac{1}{2}\vec{r} \times \vec{v}, \qquad (3.6.45)$$

and thus

$$H_{\text{int}} \equiv W_m = -\frac{q}{2}\,\vec{v}\cdot\left(\vec{B}\times\vec{r}\right) = -q\vec{B}\cdot\left(\frac{1}{2}\,\vec{r}\times\vec{v}\right) = -q\vec{B}\cdot\frac{\overrightarrow{\Delta S}}{\Delta t}$$

$$= -\vec{B}\cdot\left(\frac{q}{\Delta t}\,\overrightarrow{\Delta S}\right) = -\vec{B}\cdot\left(I\,\overrightarrow{\Delta S}\right) = -\vec{B}\cdot\vec{\mu} = -\vec{\mu}\cdot\vec{B},$$

which concludes the proof.

4

Stationary and Quasi-stationary Currents

4.1   Problem No. 27

Calculate the current distribution (in a stationary regime) in a rectangular metal plate, of thickness h and electric conductivity λ. The wires through which enters and exits the current of intensity I are applied at the centers of the two opposite sides of the plate.

Solution

In stationary regime, the equation of continuity leads to

$$\operatorname{div} \vec{j} = -\frac{\partial \rho}{\partial t} = 0. \tag{4.1.1}$$

Integrating over the volume V of the plate, bordered by surface S, we have

$$\int_V \operatorname{div} \vec{j} \, d\tau = \oint_S \vec{j} \cdot \vec{n} \, dS = 0,$$

or

$$\left(\vec{j}_2 - \vec{j}_1 \right) \cdot \vec{n} = j_{2n} - j_{1n} = 0. \tag{4.1.2}$$

According to Ohm's law,

$$\vec{j} = \lambda \vec{E} = -\lambda \nabla V, \tag{4.1.3}$$

and thus, Eq. (4.1.1) leads to

$$\Delta V = \frac{\partial^2 V}{\partial x^2} + \frac{\partial^2 V}{\partial y^2} = 0. \tag{4.1.4}$$

For the beginning, we suppose that the input and output electrodes are disposed on small surfaces of dimension $2k$, characterized by $k \ll b$ (see Fig. 4.1). Here k is a constant which we will make tend to zero at the right time.

In order to write the boundary conditions, we observe that $j_{2y} = j_{1y} = 0$, therefore

$$j_y \big|_{y=-b} = j_y \big|_{y=+b} = -\lambda \frac{\partial V}{\partial y} \bigg|_{y=-b} = -\lambda \frac{\partial V}{\partial y} \bigg|_{y=+b} = 0.$$

DOI: 10.1201/9781003402602-4

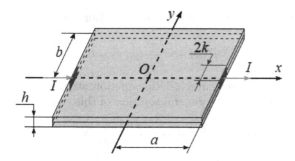

FIGURE 4.1

A rectangular metallic plate of thickness h and surface $2a \times 2b$, which is traversed along the side of length $2a$ by an electric current of intensity I. The figure shows the input and output electrodes of dimensions $h \times 2k$ which have been considered as having a finite surface to avoid working with δ-Dirac distribution.

Since the current flows along the x-axis, we also have

$$
j_x\big|_{x=-a} = j_x\big|_{x=+a} = -\lambda\frac{\partial V}{\partial x}\bigg|_{x=-a} = -\lambda\frac{\partial V}{\partial x}\bigg|_{x=+a} \equiv g(y) = \begin{cases} \frac{I}{2hk}, & |y| < k, \\ 0, & |y| > k. \end{cases}
$$

Therefore, the boundary conditions for Eq. (4.1.4) are written as follows:

$$
\begin{cases} \left(\dfrac{\partial V}{\partial y}\right)_{y=-b} = \left(\dfrac{\partial V}{\partial y}\right)_{y=+b} = 0, \\[2mm] \left(\dfrac{\partial V}{\partial x}\right)_{x=-a} = \left(\dfrac{\partial V}{\partial x}\right)_{x=+a} = f(y) = \begin{cases} -\dfrac{I}{2hk\lambda}, & |y| \le k, \\ 0, & |y| > k. \end{cases} \end{cases} \tag{4.1.5}
$$

Given the problem symmetry with respect to x-axis, the potential $V(x, y)$ must be an even function of y, that is a series of the form

$$
V(x, y) = \sum_{n=0}^{\infty} u_n(x) \cos\left(\frac{n\pi}{b}y\right). \tag{4.1.6}
$$

As can be seen, this choice satisfies the boundary conditions at the points $y = \pm b$. Also, since

$$
\frac{\partial^2 V}{\partial x^2} = \sum_{n=0}^{\infty} u_n''(x) \cos\left(\frac{n\pi}{b}y\right),
$$

and

$$
\frac{\partial^2 V}{\partial y^2} = -\sum_{n=0}^{\infty} \frac{n^2\pi^2}{b^2} u_n(x) \cos\left(\frac{n\pi}{b}y\right),
$$

one obtains from Eq. (4.1.4) the following equation for $u_n(x)$:

$$u_n''(x) - \frac{n^2\pi^2}{b^2}u_n(x) = 0. \tag{4.1.7}$$

Since the solutions of the characteristic equation attached to the differential equation (4.1.7) are real, the general solution of this equation writes

$$u_n(x) = a_n \sinh\left(\frac{n\pi}{b}x\right) + b_n \cosh\left(\frac{n\pi}{b}x\right), \quad n \in \mathbb{Z}, \ n > 0. \tag{4.1.8}$$

For $n = 0$, Eq. (4.1.7) becomes $u_0''(x) = 0$, with the solution

$$u_0 = a_0 x + b_0. \tag{4.1.9}$$

All constants b_n must be zero, since, for symmetry reasons, the line $x = 0$ is − obviously − equipotential and, without restricting the generality of the problem, we can suppose that the potential equals zero on it. Therefore, for the "singular" solution $u_0(x)$, we will ask that

$$u_0(x = 0) = 0 \ \Rightarrow \ u_0(x) = a_0 x,$$

while for the general solution $u_n(x)\big|_{n>0}$ we have

$$u_n(x = 0) = 0 \Rightarrow b_n = 0 \Rightarrow u_n(x) = a_n \sinh\left(\frac{n\pi}{b}x\right), \quad n > 0.$$

The final solution of Eq. (4.1.4) therefore is

$$V = V(x,y) = a_0 x + \sum_{n=1}^{\infty} a_n \sinh\left(\frac{n\pi}{b}x\right)\cos\left(\frac{n\pi}{b}y\right). \tag{4.1.10}$$

The boundary conditions for $x = \pm a$ lead to

$$\frac{\partial V}{\partial x}\bigg|_{x=-a} = \frac{\partial V}{\partial x}\bigg|_{x=+a}$$

$$= a_0 + \frac{\pi}{b}\sum_{n=1}^{\infty} n a_n \cosh\left(\frac{n\pi}{b}a\right)\cos\left(\frac{n\pi}{b}y\right) = f(y).$$

If this equation is integrated over y from 0 to b, one obtains

$$\int_0^b f(y)dy = \int_0^b \left[a_0 + \frac{\pi}{b}\sum_{n=1}^{\infty} n a_n \cosh\left(\frac{n\pi}{b}a\right)\cos\left(\frac{n\pi}{b}y\right)\right] dy$$

$$= a_0 b + \frac{\pi}{b}\sum_{n=1}^{\infty} n a_n \cosh\left(\frac{n\pi}{b}a\right)\underbrace{\int_0^b \cos\left(\frac{n\pi}{b}y\right)dy}_{=0} = a_0 b,$$

which leads to

$$a_0 = \frac{1}{b} \int_0^b f(y)dy = \frac{1}{b}\left[\int_0^k f(y)dy + \int_k^b f(y)dy\right]$$

$$= \frac{1}{b}\left[-\frac{I}{2hk\lambda}\int_0^k dy + \int_k^b 0\,dy\right] = -\frac{I}{2hb\lambda}. \qquad (4.1.11)$$

To determine the rest of the constants a_n, $n = 1, 2, 3, ...$, we will multiply the relation

$$a_0 + \frac{\pi}{b}\sum_{n=1}^{\infty} na_n \cosh\left(\frac{n\pi}{b}a\right)\cos\left(\frac{n\pi}{b}y\right) = f(y),$$

(that was found above) by $\cos\left(\frac{m\pi}{b}y\right)$ and then we will integrate over y from 0 to b, taking into account the corresponding orthogonality relation. So, we have

$$\int_0^b \left[a_0 + \frac{\pi}{b}\sum_{n=1}^{\infty} na_n \cosh\left(\frac{n\pi}{b}a\right)\cos\left(\frac{n\pi}{b}y\right)\right]\cos\left(\frac{m\pi}{b}y\right)dy$$

$$= \int_0^b f(y)\cos\left(\frac{m\pi}{b}y\right)dy,$$

or

$$a_0 \underbrace{\int_0^b \cos\left(\frac{m\pi}{b}y\right)dy}_{=0} + \frac{\pi}{b}\sum_{n=1}^{\infty} na_n \cosh\left(\frac{n\pi}{b}a\right)$$

$$\times \underbrace{\int_0^b \cos\left(\frac{n\pi}{b}y\right)\cos\left(\frac{m\pi}{b}y\right)dy}_{=\frac{b}{2}\delta_{mn}}$$

$$= \int_0^b f(y)\cos\left(\frac{m\pi}{b}y\right)dy = \int_0^k f(y)\cos\left(\frac{m\pi}{b}y\right)dy$$

$$+ \int_k^b f(y)\cos\left(\frac{m\pi}{b}y\right)dy,$$

or, still

$$\frac{\pi}{2}\sum_{n=1}^{\infty}\left[na_n \cosh\left(\frac{n\pi}{b}a\right)\delta_{mn}\right] = \frac{\pi}{2}ma_m \cosh\left(\frac{m\pi}{b}a\right)$$

$$= -\frac{I}{2hk\lambda}\int_0^k \cos\left(\frac{m\pi}{b}y\right)dy + \underbrace{\int_k^b 0\cos\left(\frac{m\pi}{b}y\right)dy}_{=0}$$

$$= -\frac{I}{2hk\lambda}\frac{b}{m\pi}\left[\sin\left(\frac{m\pi}{b}y\right)\right]_{y=0}^{y=k} = -\frac{I}{2hk\lambda}\frac{b}{m\pi}\sin\left(\frac{m\pi}{b}k\right),$$

leading to

$$a_m = -\frac{I}{\pi m h \lambda} \frac{1}{\cosh\left(\dfrac{m\pi}{b}a\right)} \frac{\sin\left(\dfrac{m\pi}{b}k\right)}{\dfrac{m\pi}{b}k}.$$

At the limit $k \to 0$, this becomes

$$a_n = -\frac{I}{\pi n h \lambda} \frac{1}{\cosh\left(\dfrac{n\pi}{b}a\right)} \lim_{k\to 0} \frac{\sin\left(\dfrac{n\pi}{b}k\right)}{\dfrac{n\pi}{b}k}$$

$$= -\frac{I}{\pi n h \lambda} \frac{1}{\cosh\left(\dfrac{n\pi}{b}a\right)}, \quad n = 1, 2, 3, \dots. \qquad (4.1.12)$$

This way, all the constants a_n, $n = 0, 1, 2, 3, \dots$, have been determined, and the final expression for the potential $V(x, y)$ follows by introducing a_0 and a_n given by Eqs. (4.1.11) and (4.1.12), respectively, into Eq. (4.1.10). One obtains

$$V(x, y) = -\frac{I}{2hb\lambda}\left[x + \frac{2b}{\pi}\sum_{n=1}^{\infty} \frac{\sinh\left(\dfrac{n\pi}{b}x\right)\cos\left(\dfrac{n\pi}{b}y\right)}{n\cosh\left(\dfrac{n\pi}{b}a\right)} \right]. \qquad (4.1.13)$$

According to the local form of Ohm's law, the current distribution in the plate is given by

$$\vec{j} = -\lambda\,\mathrm{grad}\,V,$$

or, by components

$$\begin{cases} j_x(x, y) = -\lambda\dfrac{\partial V(x, y)}{\partial x}, \\[2mm] j_y(x, y) = -\lambda\dfrac{\partial V(x, y)}{\partial y}, \\[2mm] j_z(x, y) = -\lambda\dfrac{\partial V(x, y)}{\partial z} = 0, \end{cases}$$

where $V = V(x, y)$ is given by Eq. (4.1.13). Some simple calculations finally lead to

$$\begin{cases} j_x(x, y) = \dfrac{I}{2hb} + \dfrac{I}{hb}\sum_{n=1}^{\infty} \dfrac{\cosh\left(\dfrac{n\pi}{b}x\right)\cos\left(\dfrac{n\pi}{b}y\right)}{\cosh\left(\dfrac{n\pi}{b}a\right)}, \\[4mm] j_y(x, y) = -\dfrac{I}{hb}\sum_{n=1}^{\infty} \dfrac{\sinh\left(\dfrac{n\pi}{b}x\right)\sin\left(\dfrac{n\pi}{b}y\right)}{\cosh\left(\dfrac{n\pi}{b}a\right)}, \\[4mm] j_z(x, y) = -\lambda\dfrac{\partial V(x, y)}{\partial z} = 0. \end{cases}$$

4.2   Problem No. 28

Calculate the current distribution (in stationary regime) for a metallic circular plate of radius a and conductivity λ. The electrodes through which the current (of intensity I) enters and leaves the plate are applied in two diametrically opposed points (see Fig. 4.2).

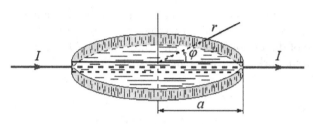

FIGURE 4.2

A circular metallic plate of radius a and thickness h. The electric current of intensity I enters and leaves the plate in two diametrically opposed points.

Solution

Since the symmetry of the problem requires usage of the plane polar coordinates r and φ, inside the disk, the potential $V(r,\varphi)$ satisfies Laplace's equation

$$r\frac{\partial}{\partial r}\left(r\frac{\partial V}{\partial r}\right)+\frac{\partial^2 V}{\partial \varphi^2}=0, \quad 0\le \varphi < 2\pi, \tag{4.2.14}$$

where Laplace's operator in plane polar coordinates has been considered as emerging from its expression in cylindrical coordinates, in which the contribution of z-coordinate has been formally removed (as well known, the plane polar coordinates can be obtained from cylindrical coordinates setting formally $z = 0$):

$$(\Delta *)_{cyl}=\frac{1}{\rho}\left[\frac{\partial}{\partial \rho}\left(\rho\frac{\partial *}{\partial \rho}\right)+\frac{\partial}{\partial \varphi}\left(\frac{1}{\rho}\frac{\partial *}{\partial \varphi}\right)+\frac{\partial}{\partial z}\left(\rho\frac{\partial *}{\partial z}\right)\right]$$

$$\xrightarrow{\rho\to r\ +\ \text{elimination of the term in } z}$$

$$(\Delta *)_{pl.pol.}=\frac{1}{r}\left[\frac{\partial}{\partial r}\left(r\frac{\partial *}{\partial r}\right)+\frac{\partial}{\partial \varphi}\left(\frac{1}{r}\frac{\partial *}{\partial \varphi}\right)\right]=\frac{1}{r}\frac{\partial}{\partial r}\left(r\frac{\partial *}{\partial r}\right)+\frac{1}{r^2}\frac{\partial^2 *}{\partial \varphi^2}.$$

The Laplace's equation in plane polar coordinates is then written as follows:

$$\Delta V=\frac{1}{r}\frac{\partial}{\partial r}\left(r\frac{\partial V}{\partial r}\right)+\frac{1}{r^2}\frac{\partial^2 V}{\partial \varphi^2}=0,$$

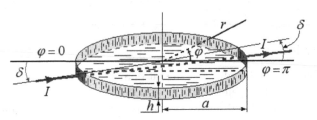

FIGURE 4.3
The input and output electrodes through which the electric current of intensity
I enters and leaves the plate. The two electrodes are extended on a finite very
small surface of dimensions $h \times a\delta$, and they are considered as having a finite
surface to avoid working with δ−Dirac distribution.

or, equivalently,

$$r\frac{\partial}{\partial r}\left(r\frac{\partial V}{\partial r}\right) + \frac{\partial^2 V}{\partial \varphi^2} = 0.$$

In order to avoid the use of Dirac's delta distribution, we will consider for
the beginning that the electrodes are not pointy applied, but are "extended"
on a finite (very small) surface: from 0 to δ (at the entrance of the current into
the plate) and from π to $\pi + \delta$ (at the exit) − regarding the angular coordinate
− and on the whole thickness of the plate, h, concerning the second essential
dimension of the plate (see Fig. 4.3). At the end, we will calculate the limit
for $\delta \to 0$. The boundary conditions therefore are

$$\left.\frac{\partial V}{\partial r}\right|_{r=a} = f(\varphi) = \begin{cases} -\dfrac{I}{\lambda h a \delta}, & 0 < \varphi < \delta, \\[2mm] +\dfrac{I}{\lambda h a \delta}, & \pi < \varphi < \pi + \delta, \\[2mm] 0, & \text{for the other angles.} \end{cases} \tag{4.2.15}$$

The symmetry of the problem requires the potential to be an even function
of φ; let this function be of the form

$$V(r, \varphi) = \sum_{n=0}^{\infty} u_n(r) \cos n\varphi. \tag{4.2.16}$$

By asking function $V(r, \varphi)$ given by Eq. (4.2.16) to verify Laplace's equation (4.2.14), an Euler-type differential equation will result for the unknown
$u_n(r)$, namely

$$r^2 u_n''(r) + r u_n'(r) - n^2 u_n(r) = 0. \tag{4.2.17}$$

The solution of this ordinary differential equation is of the form

$$u_n(r) = a_n r^n + b_n r^{-n}, \quad n > 0. \tag{4.2.18}$$

For $n = 0$, equation (4.2.17) becomes

$$ru_n''(r) + u_n'(r) = 0,$$

and has the solution

$$u_0(r) = a_0 + b_0 \ln r. \tag{4.2.19}$$

As one can be easily observed, the solutions (4.2.18) and (4.2.19) diverge for $r = 0$. To eliminate this divergence, we must consider $b_n = 0$, $n = 0, 1, 2, \dots$. Consequently, the potential $V(r, \varphi)$ (given by Eq. (4.2.16)) receives the simpler form

$$V(r, \varphi) = \sum_{n=0}^{\infty} a_n r^n \cos n\varphi. \tag{4.2.20}$$

Since $V(r, \varphi) = a_0 = \text{const.}$, for $n = 0$ the solution of the problem can be written as

$$V(r, \varphi) = \sum_{n=1}^{\infty} a_n r^n \cos n\varphi + \text{const.}, \tag{4.2.21}$$

where the constant "const." can be determined by a convenient choice of the reference point (which, by convention, is associated with the value zero of the potential).

The coefficients a_n, $n = 1, 2, 3, \dots$, are determined from the boundary condition

$$\frac{\partial V}{\partial r}\bigg|_{r=a} = f(\varphi) = \sum_{n=1}^{\infty} n\, a_n a^{n-1} \cos n\varphi. \tag{4.2.22}$$

Multiplying Eq. (4.2.22) by $\cos m\varphi$ and integrating over φ from 0 to 2π, one obtains

$$\int_0^{2\pi} f(\varphi) \cos m\varphi \, d\varphi = \sum_{n=1}^{\infty} n\, a_n a^{n-1} \int_0^{2\pi} \cos n\varphi \cos m\varphi \, d\varphi$$

$$= \pi \sum_{n=1}^{\infty} n\, a_n a^{n-1} \delta_{nm} = \pi m a_m a^{m-1},$$

where the orthogonality relation

$$\int_0^{2\pi} \cos n\varphi \cos m\varphi \, d\varphi = \pi \delta_{nm}$$

has been used. Therefore,

$$a_n = \frac{1}{n\pi a^{n-1}}$$

$$\times \int_0^{2\pi} f(\varphi) \cos n\varphi d\varphi = \frac{1}{n\pi a^{n-1}} \left[\int_0^{\delta} \left(-\frac{I}{\lambda h a \delta} \right) \cos n\varphi d\varphi \right.$$

$$\left. + \underbrace{\int_{\delta}^{\pi} 0 \cdot \cos n\varphi d\varphi}_{=0} + \int_{\pi}^{\pi+\delta} \frac{I}{\lambda h a \delta} \cos n\varphi d\varphi + \underbrace{\int_{\pi+\delta}^{2\pi} 0 \cdot \cos n\varphi d\varphi}_{=0} \right]$$

$$= \frac{1}{n\pi a^{n-1}} \left[\left(-\frac{I}{\lambda h a \delta} \right) \frac{1}{n} (\sin n\varphi) \Big|_{\varphi=0}^{\varphi=\delta} + \frac{I}{\lambda h a \delta} \frac{1}{n} (\sin n\varphi) \Big|_{\varphi=\pi}^{\varphi=\pi+\delta} \right]$$

$$= \frac{1}{n\pi a^{n-1}} \left[\frac{-I}{\lambda h a} \frac{\sin n\delta}{n\delta} + \frac{I}{\lambda h a} \frac{\cos n\pi \sin n\delta}{n\delta} \right]$$

$$= \frac{1}{n\pi a^{n}} \frac{I}{h\lambda} \left[-1 + (-1)^{n} \right] \frac{\sin n\delta}{n\delta}. \tag{4.2.23}$$

Going now to the limit $\delta \to 0$, one obtains

$$a_n = \frac{1}{n\pi a^n} \frac{I}{\lambda h} \left[(-1)^n - 1 \right], \quad n = 1, 2, 3, \ldots . \tag{4.2.24}$$

If the constants from Eq. (4.2.24) are introduced into Eq. (4.2.21), we find

$$V(r, \varphi) = \frac{I}{\pi \lambda h} \sum_{n=1}^{\infty} \left[(-1)^n - 1 \right] \frac{1}{n} \left(\frac{r}{a} \right)^n \cos n\varphi + \text{const.}. \tag{4.2.25}$$

But, for $r < a$, we have

$$r^2 - 2ar \cos \varphi + a^2 = a^2 \left(\frac{r^2}{a^2} - 2\frac{r}{a} \cos \varphi + 1 \right)$$

$$= a^2 \left(1 - \frac{r}{a} e^{i\varphi} \right) \left(1 - \frac{r}{a} e^{-i\varphi} \right),$$

where we have used the relation $e^{i\varphi} + e^{-i\varphi} = 2\cos \varphi$. Taking the logarithm of the square root of the last formula, one finds

$$\ln \left(r^2 - 2ar \cos \varphi + a^2 \right)^{1/2}$$

$$= \ln a + \frac{1}{2} \ln \left(1 - \frac{r}{a} e^{i\varphi} \right) + \frac{1}{2} \ln \left(1 - \frac{r}{a} e^{-i\varphi} \right)$$

$$= \ln a - \frac{1}{2} \sum_{n=1}^{\infty} \frac{1}{n} \left(\frac{r}{a} \right)^n e^{in\varphi} - \frac{1}{2} \sum_{n=1}^{\infty} \frac{1}{n} \left(\frac{r}{a} \right)^n e^{-in\varphi}$$

$$= \ln a - \sum_{n=1}^{\infty} \frac{1}{n} \left(\frac{r}{a} \right)^n \cos n\varphi, \tag{4.2.26}$$

where we have used the following Mac-Laurin series expansion:

$$\ln(1 - x) = - \left(\frac{x}{1} + \frac{x^2}{2} + \frac{x^3}{3} + \ldots \right) = - \sum_{n=1}^{\infty} \frac{x^n}{n}.$$

This power series is convergent only if $|x| < 1$. Obviously, in order to obtain relation (4.2.26), x has to be taken as $x = \frac{r}{a} e^{i\varphi}$ and, respectively, $x = \frac{r}{a} e^{-i\varphi}$; these expressions satisfy the convergence condition only if $r < a$ (since $\left| e^{\pm i\varphi} \right| = 1$, $\forall \varphi \in \mathbb{R}$), and we have considered this requirement from the very beginning.

The Mac-Laurin series expansion of

$$\ln(1 + x) = \frac{x}{1} - \frac{x^2}{2} + \frac{x^3}{3} - \frac{x^4}{4} + \dots = -\sum_{n=1}^{\infty} (-1)^n \frac{x^n}{n}$$

leads – in a perfectly analogous way – to

$$\ln(r^2 + 2ar\cos\varphi + a^2)^{1/2} = \ln a - \sum_{n=1}^{\infty} (-1)^n \frac{1}{n} \left(\frac{r}{a}\right)^n \cos n\varphi. \qquad (4.2.27)$$

Therefore we can write

$$\sum_{n=1}^{\infty} [(-1)^n - 1] \frac{1}{n} \left(\frac{r}{a}\right)^n \cos n\varphi = \ln a - \ln(r^2 + 2ar\cos\varphi + a^2)^{1/2}$$

$$+ \ln(r^2 - 2ar\cos\varphi + a^2)^{1/2} - \ln a = \ln \frac{(r^2 - 2ar\cos\varphi + a^2)^{1/2}}{(r^2 + 2ar\cos\varphi + a^2)^{1/2}}$$

$$= \frac{1}{2}\ln \frac{(r^2 - 2ar\cos\varphi + a^2)}{(r^2 + 2ar\cos\varphi + a^2)},$$

and then

$$V(r,\varphi) = \frac{I}{2\pi\lambda h} \ln\left(\frac{r^2 - 2ar\cos\varphi + a^2}{r^2 + 2ar\cos\varphi + a^2}\right) + \text{const.}, \qquad (4.2.28)$$

or

$$V(r,\varphi) = \frac{I}{\pi\lambda h} \ln\frac{r_1}{r_2} + \text{const.}, \qquad (4.2.29)$$

where r_1 and r_2 are the distances to the two electrodes (see Fig. 4.4):

$$\begin{cases} r_1^2 = r^2 + a^2 - 2ar\cos\varphi, \\ r_2^2 = r^2 + a^2 - 2ar\cos(\pi - \varphi) = r^2 + a^2 + 2ar\cos\varphi. \end{cases}$$

According to the local form of Ohm's law, the current distribution in the plate is then given by

$$\vec{j} = -\lambda \operatorname{grad} V,$$

or, by components

$$\begin{cases} j_r(r,\varphi) = -\lambda \dfrac{\partial V(r,\varphi)}{\partial r}, \\ j_\varphi(r,\varphi) = -\dfrac{\lambda}{r} \dfrac{\partial V(r,\varphi)}{\partial \varphi}, \end{cases}$$

where $V(r,\varphi)$ is given by Eq. (4.2.28). Several simple calculations lead, finally, to

$$\begin{cases} j_r(r,\varphi) = \dfrac{2I}{\pi a h} \dfrac{(r^2 - a^2)\cos\varphi}{4r^2\cos^2\varphi - \left(\frac{r^2}{a} + a\right)^2}, \\ j_\varphi(r,\varphi) = \dfrac{2I}{\pi a h} \dfrac{(r^2 + a^2)\sin\varphi}{4r^2\cos^2\varphi - \left(\frac{r^2}{a} + a\right)^2}. \end{cases}$$

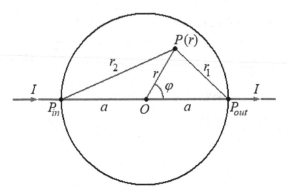

FIGURE 4.4
Some point $P(r, \varphi)$ in the plane of the circular metallic plate, at which the potential $V(r, \varphi)$ is determined. r_1 and r_2 are the distances (in the plane of the plate) between the point P and the two electrodes.

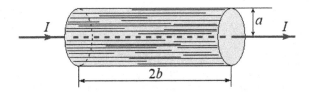

FIGURE 4.5
A cylindrical conductor of radius a and length $2b$. The two electrodes through which the current of intensity I enters and leaves the conductor are applied at the centres of the two bases.

4.3   Problem No. 29

A stationary current of intensity I enters in a cylindrical conductor of radius a and length $2b$ through one of the bases and comes out from the other. Determine the current distribution in the cylinder, supposing that the electrodes are applied at the centers of the bases.

Solution

The problem symmetry requires, obviously, the use of cylindrical coordinates. Let us consider the z-axis along the axis of the cylindrical conductor (see Fig. 4.5).

The Laplace's equation in cylindrical coordinates r, φ and z is written as (see **Appendix D**):

$$\frac{1}{r}\frac{\partial}{\partial r}\left(r\frac{\partial V}{\partial r}\right) + \frac{1}{r^2}\frac{\partial^2 V}{\partial \varphi^2} + \frac{\partial^2 V}{\partial z^2} = 0. \qquad (4.3.30)$$

Since, for symmetry reasons, the potential V does not depend on φ, the solution of the Laplace's equation will be of the form $V = V(r, z)$, and will satisfy the "simplified" Laplace's equation

$$\frac{1}{r}\frac{\partial}{\partial r}\left(r\frac{\partial V}{\partial r}\right) + \frac{\partial^2 V}{\partial z^2} = 0. \tag{4.3.30'}$$

We suppose that the electrodes are small disks of radius h (see Fig. 4.6) and, at the right time, we will consider the limit $h \to 0$.

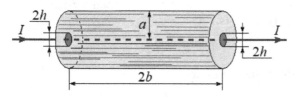

FIGURE 4.6
The input and output electrodes through which the electric current of intensity I enters and leaves the cylindrical conductor. The two electrodes are very small disks of radius h, and they are considered as having a finite surface to avoid working with δ−Dirac distribution.

Taking into account the local form of Ohm's law $\left(\vec{j} = \lambda\vec{E} = -\lambda\operatorname{grad} V\right)$, as well as the interface condition for the current density $(j_{2n} - j_{1n} = 0)$, where $\vec{n}$ is the unit vector of the normal to the separation surface, oriented from medium 1 towards medium 2, the boundary conditions are written as follows:

$$\left.\frac{\partial V}{\partial z}\right|_{z=-b} = \left.\frac{\partial V}{\partial z}\right|_{z=+b} = f(r) = \begin{cases} -\dfrac{I}{\lambda\pi h^2}, & r < h, \\ 0, & r > h, \end{cases} \tag{4.3.31}$$

and

$$\left.\frac{\partial V}{\partial r}\right|_{r=a} = 0. \tag{4.3.32}$$

In order to integrate Eq. (4.3.30') we will use the Fourier method (the variable separation method), *i.e.*, we look for a solution of the form

$$V(r, z) = R(r)Z(z), \tag{4.3.33}$$

and we require that solution given by Eq. (4.3.33) verifies Eq. (4.3.30'). Therefore, one obtains

$$R'' + \frac{1}{r}R' + m^2 R = 0 \tag{4.3.34}$$

and

$$Z'' - m^2 Z = 0, \tag{4.3.35}$$

where m^2 is the constant we have introduced when separating the variables.

Equation (4.3.34) is just the equation of Bessel functions, of order zero and argument mr. Its non-singular solution is $R(r) = J_0(mr)$, while the boundary condition for $r = a$ gives the possible values of m. Indeed, Eqs. (4.3.32) and (4.3.33) show that we must have

$$m J_0'(ma) = 0, \qquad (4.3.36)$$

or

$$m J_1(ma) = 0.$$

Therefore, ma has to be a root of the equation $x J_1(x) = 0$. This equation admits an infinity of roots,

$$\mu_0(=0) < \mu_1 < \mu_2 < \cdots < \mu_i < \cdots$$

and the possible values of m are

$$m \equiv m_i = \frac{\mu_i}{a}, \ i = 0, 1, 2, \dots .$$

Equation (4.3.35) then becomes

$$Z'' - \left(\frac{\mu_i}{a}\right)^2 Z = 0, \qquad (4.3.37)$$

and has the solution

$$Z(z) = c_i \sinh\left(\frac{\mu_i z}{a}\right) + d_i \cosh\left(\frac{\mu_i z}{a}\right), \quad i > 0. \qquad (4.3.38)$$

Considering the fact that $\mu_0 = 0$, for $i = 0$ from Eq. (4.3.37) one simply obtains $Z(z) = c_0 z + d_0$. The general solution of Eq. (4.3.30') can be then represented by the series

$$V(r, z) = c_0 z + \sum_{i=1}^{\infty} \left[c_i \sinh\left(\frac{\mu_i z}{a}\right) + d_i \cosh\left(\frac{\mu_i z}{a}\right)\right] J_0\left(\frac{\mu_i r}{a}\right) + \text{const.},$$

$$(4.3.39)$$

with $J_1(\mu_i) = 0$, $i = 1, 2, 3, \dots .$

For symmetry reasons, the section $z = 0$ represents an equipotential surface. If this "level" is chosen as the reference level for the potential, all the coefficients d_i, $i = 1, 2, 3, \dots$, cancel and $V(r, z)$ receives the simpler expression

$$V(r, z) = c_0 z + \sum_{i=1}^{\infty} c_i \sinh\left(\frac{\mu_i z}{a}\right) J_0\left(\frac{\mu_i r}{a}\right). \qquad (4.3.39')$$

This way, only the constants c_i, $i = 0, 1, 2, \dots$, remained undetermined. In order to find them, we will use the boundary conditions expressed by Eq. (4.3.31) which are written

$$\left.\frac{\partial V}{\partial z}\right|_{z=\pm b} = f(r) = c_0 + \sum_{i=1}^{\infty} c_i \frac{\mu_i}{a} \cosh\left(\frac{\mu_i b}{a}\right) J_0\left(\frac{\mu_i r}{a}\right). \qquad (4.3.40)$$

Multiplying both members of this equation by r and integrating from 0 to a, we obtain

$$\int_0^a rf(r)dr = c_0 \int_0^a rdr + \sum_{i=1}^{\infty} c_i \frac{\mu_i}{a} \cosh\left(\frac{\mu_i b}{a}\right) \int_0^a r J_0\left(\frac{\mu_i r}{a}\right) dr,$$

or

$$\int_0^h rf(r)dr + \int_h^a rf(r)dr = c_0\frac{a^2}{2} + \sum_{i=1}^{\infty} c_i \frac{\mu_i}{a} \cosh\left(\frac{\mu_i b}{a}\right)\left[\frac{a^2}{\mu_i}J_1(\mu_i)\right],$$

which gives

$$-\frac{I}{\pi h^2 \lambda}\frac{h^2}{2} = c_0\frac{a^2}{2} \quad \Rightarrow \quad c_0 = -\frac{I}{\pi a^2 \lambda}, \qquad (4.3.41)$$

where we took into consideration the relation (4.3.31) and the fact that $J_1(\mu_i) = 0$.

To find the rest of the constants c_i, $i = 1, 2, ...$, we multiply Eq. (4.3.40) by $r J_0\left(\frac{\mu_j r}{a}\right)$ and integrate again from 0 to a. It follows that

$$\int_0^a rf(r) J_0\left(\frac{\mu_j r}{a}\right) dr = c_0 \int_0^a r J_0\left(\frac{\mu_j r}{a}\right) dr$$

$$+ \sum_{i=1}^{\infty} c_i \frac{\mu_i}{a} \cosh\left(\frac{\mu_i b}{a}\right) \int_0^a r J_0\left(\frac{\mu_i r}{a}\right) J_0\left(\frac{\mu_j r}{a}\right) dr. \qquad (4.3.42)$$

But

$$\int_0^a r J_0\left(\frac{\mu_j r}{a}\right) dr = \frac{a^2}{\mu_j}J_1(\mu_j) = 0,$$

(because $J_1(\mu_j) = 0$) and, in addition, we have the orthogonality condition of the Bessel functions,

$$\int_0^a r J_\nu\left(\frac{\mu_i r}{a}\right) J_\nu\left(\frac{\mu_j r}{a}\right) dr = \frac{a^2}{2}\left[J_\nu^2(\mu_i) + J_{\nu-1}(\mu_i)J_{\nu+1}(\mu_i)\right]\delta_{ij}.$$

Customising the above relation for the zero-order Bessel functions ($\nu = 0$) and considering that in our case μ_j are solutions of the equation $J_1(\mu_j) = 0$, we have

$$\int_0^a r J_0\left(\frac{\mu_i r}{a}\right) J_0\left(\frac{\mu_j r}{a}\right) dr = \frac{a^2}{2}J_0^2(\mu_i)\delta_{ij},$$

so that Eq. (4.3.42) becomes

$$\int_0^a rf(r) J_0\left(\frac{\mu_j r}{a}\right) dr = \sum_{i=1}^{\infty} c_i \frac{\mu_i}{a} \cosh\left(\frac{\mu_i b}{a}\right)\frac{a^2}{2}J_0^2(\mu_i)\delta_{ij}$$

$$= c_j \frac{\mu_j a}{2} \cosh\left(\frac{\mu_j b}{a}\right) J_0^2(\mu_j),$$

or, if we take into account that

$$\int_0^a rf(r)J_0\left(\frac{\mu_j r}{a}\right) dr = \int_0^h rf(r)J_0\left(\frac{\mu_j r}{a}\right) dr$$

$$+ \int_h^a rf(r)J_0\left(\frac{\mu_j r}{a}\right) dr = \int_0^h rf(r)J_0\left(\frac{\mu_j r}{a}\right) dr$$

$$= -\frac{I}{\pi h^2 \lambda}\int_0^h rJ_0\left(\frac{\mu_j r}{a}\right) dr = -\frac{I}{\pi h^2 \lambda}\frac{ah}{\mu_j}J_1\left(\frac{\mu_j h}{a}\right),$$

one obtains

$$-\frac{I}{\pi h^2 \lambda}\frac{ah}{\mu_j}J_1\left(\frac{\mu_j h}{a}\right) = c_j\frac{\mu_j a}{2}\cosh\left(\frac{\mu_j b}{a}\right)J_0^2(\mu_j) \Rightarrow$$

$$c_j = -\frac{2IJ_1\left(\dfrac{\mu_j h}{a}\right)}{\pi h \lambda \mu_j^2 \cosh\left(\dfrac{\mu_j b}{a}\right)J_0^2(\mu_j)}, \quad j = 1,2,3,\dots. \tag{4.3.43}$$

Taking now $h \to 0$ in Eq. (4.3.43), it follows that for $i = 1,2,3,\dots$, we have

$$c_i = -\frac{2I}{\pi a \lambda \cosh\left(\dfrac{\mu_i b}{a}\right)\mu_i J_0^2(\mu_i)}\underbrace{\lim_{h \to 0}\frac{J_1\left(\dfrac{\mu_i h}{a}\right)}{\dfrac{\mu_i h}{a}}}_{=1/2}$$

$$= -\frac{I}{\pi a \lambda \cosh\left(\dfrac{\mu_i b}{a}\right)\mu_i J_0^2(\mu_i)}. \tag{4.3.44}$$

Substituting the constant c_0 from Eq. (4.3.41), and c_i, $i = 1,2,3,\dots$, from Eq. (4.3.44) into Eq. (4.3.39′), we get

$$V(r,z) = -\frac{I}{\pi a \lambda}\left[\frac{z}{a} + \sum_{i=1}^{\infty}\frac{J_0\left(\dfrac{\mu_i r}{a}\right)\sinh\left(\dfrac{\mu_i z}{a}\right)}{\mu_i J_0^2(\mu_i)\cosh\left(\dfrac{\mu_i b}{a}\right)}\right]. \tag{4.3.45}$$

Finally, the three components of the current density $\vec{j}$ are

$$\begin{cases} j_r = -\lambda\dfrac{\partial V(r,z)}{\partial r}, \\[2mm] j_\varphi = -\dfrac{\lambda}{r}\dfrac{\partial V(r,z)}{\partial \varphi} = 0, \\[2mm] j_z = -\lambda\dfrac{\partial V(r,z)}{\partial z}, \end{cases}$$

with $V = V(r, z)$ given by Eq. (4.3.45). It is now easy to obtain the answer to the problem; it is given by the following relations:

$$
\begin{cases}
j_r(r, z) = -\dfrac{I}{\pi a^2} \displaystyle\sum_{i=1}^{\infty} \dfrac{J_1\left(\frac{\mu_i r}{a}\right) \sinh\left(\frac{\mu_i z}{a}\right)}{J_0^2(\mu_i) \cosh\left(\frac{\mu_i b}{a}\right)}, \\[4mm]
j_\varphi(r, z) = 0, \\[2mm]
j_z(r, z) = \dfrac{I}{\pi a^2}\left[1 + \displaystyle\sum_{i=1}^{\infty} \dfrac{J_0\left(\frac{\mu_i r}{a}\right) \cosh\left(\frac{\mu_i z}{a}\right)}{J_0^2(\mu_i) \cosh\left(\frac{\mu_i b}{a}\right)} \right].
\end{cases}
$$

4.4   Problem No. 30

A capacitor with capacitance C_1 is connected in parallel with a coil of inductance L_1 and the whole system is connected in series with another coil of inductance L_2 and a second capacitor of capacitance C_2, as shown in Fig. 4.7. Using Kirchhoff's laws deduced within the Lagrangian formalism based on the analogy between mechanical and electrical systems, find the differential equation satisfied by the electric charge $q(t)$ existent on the armatures/plates of the capacitor C_2, knowing that at the initial moment $q(0) = q_0$, while the capacitor of capacitance C_1 was uncharged ($q_1(0) = 0$).

Solution

The analogy between the electric circuits containing resistors, coils, capacitors and electromotive forces (voltage sources), on the one side, and the mechanical systems of particles, on the other, was first identified by James Clerk Maxwell. This analogy allows application of the Lagrangian formalism in the study of electrical systems. Since resistors are circuit dissipative elements, and the electric resistance together with conductance (the inverse of electrical resistance) have as correspondent − within the above described analogy − the coefficient of the friction force (which is a dissipative force), in order to settle the basic relations of the analogy in question we have to appeal to the Lagrangian formalism for mechanical systems, which includes non-potential forces of negative power (dissipative forces).

As known from Analytical Mechanics, the Lagrangian formalism for systems in which act non-potential forces is based on Lagrange's equations

$$
\frac{d}{dt}\left(\frac{\partial L}{\partial \dot{q}_k} \right) - \frac{\partial L}{\partial q_k} = \tilde{\Phi}_k, \quad k = \overline{1, n}, \tag{4.4.46}
$$

where q_k are the generalized coordinates, $L = T - V$ is the Lagrangian of the

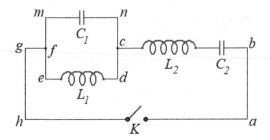

FIGURE 4.7
A DC circuit containing a capacitor C_1 connected in parallel with a coil L_1, the whole system being connected in series with another coil L_2 and a second capacitor C_2.

system defined only for potential forces, $\vec{F}_i = -\text{grad}_i V$, $i = \overline{1,N}$, T is the kinetic energy of the system, and

$$\tilde{\Phi}_k = \sum_{i=1}^{N} \vec{\tilde{F}}_i \cdot \frac{\partial \vec{r}_i}{\partial q_k}, \ k = \overline{1,n},$$

are the non-potential generalized forces. We also have to mention that N represents the number of particles in the system, n means the number of effective degrees of freedom of the system, $\vec{F}_i$, $i = \overline{1,N}$, are the potential forces (which "derive" from the mechanical potential V) and $\vec{\tilde{F}}_i$ are the non-potential forces.

In the space of real coordinates, the infinitesimal mechanical work of non-potential forces writes

$$\delta \tilde{W} = \sum_{i=1}^{N} \vec{\tilde{F}}_i \cdot \delta \vec{r}_i,$$

while in the configuration space (the space of generalized coordinates q_k, $k = \overline{1,n}$) it is given by

$$\delta \tilde{W} = \sum_{k=1}^{n} \tilde{\Phi}_k \delta q_k = \sum_{k=1}^{n} \sum_{i=1}^{N} \vec{\tilde{F}}_i \cdot \frac{\partial \vec{r}_i}{\partial q_k} \delta q_k.$$

By means of this mechanical work, one can write the power $\tilde{P}$ of the non-potential forces as

$$\tilde{P} \equiv \frac{\delta \tilde{W}}{\delta t} = \sum_{i=1}^{N} \vec{\tilde{F}}_i \cdot \vec{v}_i = \sum_{k=1}^{n} \tilde{\Phi}_k \dot{q}_k. \quad (4.4.47)$$

The non-potential forces whose power is negative, $\tilde{P} < 0$, are called *dissipative forces*; a well-known example in this respect is the *friction force*. A special exemple of non-potential forces is given by the *gyroscopic forces* (those for which the power $\tilde{P}$ is zero).

If the velocities of the particles are not very large, then the friction forces are proportional to the velocities,

$$\vec{\tilde{F}}_i \equiv \vec{F}_i^f = -k\vec{v}_i, \quad i = \overline{1,N}, \ k > 0, \tag{4.4.48}$$

in which case there exists a scalar function $\mathcal{T}$ whose expression is

$$\mathcal{T} = \frac{1}{2}k\sum_{i=1}^{N}|\vec{v}_i|^2, \tag{4.4.49}$$

so that formally one can write

$$\vec{F}_i^f = -\frac{\partial \mathcal{T}}{\partial \vec{v}_i} = -\nabla_{\vec{v}_i}\mathcal{T}, \quad i = \overline{1,N}, \tag{4.4.50}$$

where $\nabla_{\vec{v}_i}$ formally designates the partial derivative with respect to velocity $\vec{v}_i$. The scalar function $\mathcal{T}$ defined this way is called *Rayleigh's dissipation function.* For scleronomic systems, it is a quadratic and homogeneous function of the generalized velocities $\dot{q}_k$,

$$\mathcal{T} = \frac{1}{2}\sum_{j=1}^{n}\sum_{k=1}^{n}C_{jk}\dot{q}_j\dot{q}_k, \tag{4.4.51}$$

which is a form similar to the kinetic energy for scleronomic systems,

$$T = \frac{1}{2}\sum_{j=1}^{n}\sum_{k=1}^{n}a_{jk}\dot{q}_j\dot{q}_k.$$

The physical significance of $\mathcal{T}$ can be found by writing the power of the friction forces

$$\tilde{P}^f = \sum_{i=1}^{N}\vec{F}_i^f \cdot \vec{v}_i = -k\sum_{i=1}^{N}|\vec{v}_i|^2 = -2\mathcal{T},$$

which means that $\mathcal{T}$ represents one half of the power developed by the friction forces.

The generalized forces associated to the friction forces are

$$\tilde{\Phi}_k^f = \sum_{i=1}^{N}\vec{F}_i^f \cdot \frac{\partial \vec{r}_i}{\partial q_k} = -\frac{\partial \mathcal{T}}{\partial \dot{q}_k}, \quad k = \overline{1,n}, \tag{4.4.52}$$

in which case the Lagrange equations of the second kind for the systems in which are also present non-potential forces given by Eq. (4.4.46), write as follows:

$$\frac{d}{dt}\left(\frac{\partial \mathrm{L}}{\partial \dot{q}_k}\right) - \frac{\partial \mathrm{L}}{\partial q_k} + \frac{\partial \mathcal{T}}{\partial \dot{q}_k} = 0, \quad k = \overline{1,n}. \tag{4.4.53}$$

Let us now go back to the problem of analogy between a mechanical system of particles and the electrical circuits, and establish the two fundamental Kirchhoff's laws within the Lagrangian formalism.

1°. **Kirchhoff's loop** (or **mesh**) **rule** (or **Kirchhoff's second rule,** or **Kirchhoff's second law**)

The analogy between mechanical systems of particles and electrical circuits leads to the following correspondence between mechanical and electrical systems:

Generalized coordinate $q_k(t)$	$\longleftrightarrow$	Electric charge $q(t)$
Generalized velocity $\dot{q}_k(t)$	$\longleftrightarrow$	Electric current $I(t)$
External periodical force F	$\longleftrightarrow$	Electromotive force $\mathcal{E}(t)$
Mass m	$\longleftrightarrow$	Inductance L
Elastic constant k	$\longleftrightarrow$	Elastance S
Damping force constant r	$\longleftrightarrow$	Electric resistance R
Kinetic energy T	$\longleftrightarrow$	Magnetic energy W_{mag}
Potential energy V	$\longleftrightarrow$	Electric energy W_{el}

According to the previously mentioned analogy, an electric circuit can be watched as an oscillating mechanical system with n degrees of freedom, subject to two potential and one non-potential (dissipative) forces. The potential energy of the system (circuit) is

$$V = \frac{1}{2} \sum_{i=1}^{n} \sum_{j=1}^{n} S_{ij} q_i q_j - \sum_{i=1}^{n} q_i \mathcal{E}_i(t), \qquad (4.4.54)$$

where

$$\frac{1}{2} \sum_{i=1}^{n} \sum_{j=1}^{n} S_{ij} q_i q_j = W_{el}$$

is the electric energy of the circuit (without taking into account its generating sources), the coefficients S_{ij} – called *elastances* – represent the reverse of capacitances: $[S_{ij}] = [C_{ij}^{-1}]$, while $\sum_{i=1}^{n} q_i \mathcal{E}_i(t)$ represents the total electric energy supplied by the electromotive forces (voltage sources) of the circuit. The quantities C_{ij} are the *influence coefficients*, and for $i = j$, $C_{ii} \equiv C_i$ are called *capacitance coefficients* or *capacitances*. The upper index n of the summation symbol appearing in Eq. (4.4.54) represents the number of conductors of the circuit. The magnetic energy of the circuit is

$$W_{mag} = \frac{1}{2} \sum_{i=1}^{n} \sum_{j=1}^{n} M_{ij} \dot{q}_i \dot{q}_j, \qquad (4.4.55)$$

where M_{ij} is the *mutual inductance* of the electrical net's loops i and j, $i, j = \overline{1, n}$. For $i = j$, $M_{ii} \equiv L_i$ is called *self-inductance* of the net's loop i. We draw attention on the fact that this time n represents the number of loops of the circuit. This is important because, even if in the final expression of the net's loops it has been used the same n for all terms, when this law is written for a loop which is part of a more complicated net, n takes different values for each term of the sum since, in general, the current $I_k = \dot{q}_k$ is not the same for all branches of the considered loop.

Since resistors are dissipative elements of the circuit, the Rayleigh's dissipative function is written in this case as

$$\mathcal{T} = \frac{1}{2} \sum_{i=1}^{n} \sum_{j=1}^{n} R_{ij} \dot{q}_i \dot{q}_j, \qquad (4.4.56)$$

where the coefficients R_{ij} are constant and closely related to the circuit resistances. For $i = j$, $R_{ii} \equiv R_i$ is the *electrical resistance* of the loop i of the circuit.

By means of these elements, in view of the previously presented correspondences, one can build up the Lagrangian of the electric circuit, namely

$$\mathrm{L}(= T - \mathrm{V}) = W_{mag} - W_{el}$$

$$= \frac{1}{2} \sum_{i=1}^{n} \sum_{j=1}^{n} (M_{ij} \dot{q}_i \dot{q}_j - S_{ij} q_i q_j) + \sum_{i=1}^{n} q_i \mathcal{E}_i(t). \quad (4.4.57)$$

In agreement with the same correspondence, using Eqs. (4.4.56) and (4.4.57), the Lagrange's equations of the second kind (4.4.53) are

$$\sum_{k=1}^{n} (M_{ik} \ddot{q}_k + R_{ik} \dot{q}_k + S_{ik} q_k) = \mathcal{E}_i(t), \quad i = \overline{1, n}, \qquad (4.4.58)$$

because

$$\frac{\partial \mathrm{L}}{\partial \dot{q}_k} = \frac{\partial}{\partial \dot{q}_k} \left[\frac{1}{2} \sum_{i=1}^{n} \sum_{j=1}^{n} (M_{ij} \dot{q}_i \dot{q}_j - S_{ij} q_i q_j) + \sum_{i=1}^{n} q_i \mathcal{E}_i(t) \right]$$

$$= \frac{\partial}{\partial \dot{q}_k} \left(\frac{1}{2} \sum_{i=1}^{n} \sum_{j=1}^{n} M_{ij} \dot{q}_i \dot{q}_j \right) = \frac{1}{2} \left[\sum_{i=1}^{n} \sum_{j=1}^{n} \left(M_{ij} \frac{\partial \dot{q}_i}{\partial \dot{q}_k} \dot{q}_j + M_{ij} \dot{q}_i \frac{\partial \dot{q}_j}{\partial \dot{q}_k} \right) \right]$$

$$= \frac{1}{2} \left[\sum_{i=1}^{n} \sum_{j=1}^{n} (M_{ij} \delta_{ik} \dot{q}_j + M_{ij} \dot{q}_i \delta_{jk}) \right] = \frac{1}{2} \left[\sum_{j=1}^{n} M_{kj} \dot{q}_j + \sum_{i=1}^{n} M_{ik} \dot{q}_i \right]$$

$$= \sum_{j=1}^{n} M_{kj} \dot{q}_j = \sum_{i=1}^{n} M_{ki} \dot{q}_i,$$

where, in the second term of the sum in the last parenthesis, the summation index has been conveniently changed and the symmetry property of mutual inductances, $M_{jk} = M_{kj}$, has been used. Then,

$$\frac{d}{dt}\left(\frac{\partial \mathrm{L}}{\partial \dot{q}_k}\right) = \frac{d}{dt}\left(\sum_{i=1}^{n} M_{ki}\dot{q}_i\right) = \sum_{i=1}^{n} M_{ki}\ddot{q}_i.$$

Also,

$$\frac{\partial \mathrm{L}}{\partial q_k} = \frac{\partial}{\partial q_k}\left[\frac{1}{2}\sum_{i=1}^{n}\sum_{j=1}^{n}\left(M_{ij}\dot{q}_i\dot{q}_j - S_{ij}q_iq_j\right) + \sum_{i=1}^{n}q_i\mathcal{E}_i(t)\right]$$

$$= \frac{\partial}{\partial q_k}\left(-\frac{1}{2}\sum_{i=1}^{n}\sum_{j=1}^{n}S_{ij}q_iq_j + \sum_{i=1}^{n}q_i\mathcal{E}_i(t)\right)$$

$$= -\frac{1}{2}\sum_{i=1}^{n}\sum_{j=1}^{n}\left(S_{ij}\frac{\partial q_i}{\partial q_k}q_j + S_{ij}q_i\frac{\partial q_j}{\partial q_k}\right) + \sum_{i=1}^{n}\frac{\partial q_i}{\partial q_k}\mathcal{E}_i(t)$$

$$-\frac{1}{2}\sum_{i=1}^{n}\sum_{j=1}^{n}\left(S_{ij}\delta_{ik}q_j + S_{ij}q_i\delta_{jk}\right) + \sum_{i=1}^{n}\delta_{ik}\mathcal{E}_i(t)$$

$$= -\frac{1}{2}\sum_{j=1}^{n}S_{kj}q_j - \frac{1}{2}\sum_{i=1}^{n}S_{ik}q_i + \mathcal{E}_k(t)$$

$$= -\sum_{i=1}^{n}S_{ik}q_i + \mathcal{E}_k(t) = -\sum_{i=1}^{n}S_{ki}q_i + \mathcal{E}_k(t),$$

where the same simple mathematical artifices as for calculation of derivative $\frac{\partial \mathrm{L}}{\partial \dot{q}_k}$ have been used. We also have

$$\frac{\partial \mathcal{T}}{\partial \dot{q}_k} = \frac{\partial}{\partial \dot{q}_k}\left(\frac{1}{2}\sum_{i=1}^{n}\sum_{j=1}^{n}R_{ij}\dot{q}_i\dot{q}_j\right)$$

$$= \frac{1}{2}\sum_{i=1}^{n}\sum_{j=1}^{n}\left(R_{ij}\frac{\partial \dot{q}_i}{\partial \dot{q}_k}\dot{q}_j + R_{ij}\dot{q}_i\frac{\partial \dot{q}_j}{\partial \dot{q}_k}\right)$$

$$= \frac{1}{2}\sum_{i=1}^{n}\sum_{j=1}^{n}(R_{ij}\delta_{ik}\dot{q}_j + R_{ij}\dot{q}_i\delta_{jk})$$

$$= \frac{1}{2}\sum_{j=1}^{n}R_{kj}\dot{q}_j + \frac{1}{2}R_{ik}\dot{q}_i = \sum_{i=1}^{n}R_{ik}\dot{q}_i = \sum_{i=1}^{n}R_{ki}\dot{q}_i,$$

where, again, the useful mentioned above mathematical artifices have been applied. Replacing these results into Eq. (4.4.53), one obtains

$$\frac{d}{dt}\left(\frac{\partial L}{\partial \dot{q}_k}\right) - \frac{\partial L}{\partial q_k} + \frac{\partial \mathcal{T}}{\partial \dot{q}_k} = 0 = \sum_{i=1}^{n} M_{ki}\ddot{q}_i + \sum_{i=1}^{n} S_{ki}q_i - \mathcal{E}_k(t)$$

$$+ \sum_{i=1}^{n} R_{ki}\dot{q}_i = \sum_{i=1}^{n} (M_{ki}\ddot{q}_i + R_{ki}\dot{q}_i + S_{ki}q_i) - \mathcal{E}_k(t), \quad k = \overline{1,n},$$

or, equivalently,

$$\sum_{k=1}^{n} (M_{ik}\ddot{q}_k + R_{ik}\dot{q}_k + S_{ik}q_k) = \mathcal{E}_i(t), \quad i = \overline{1,n},$$

i.e., nothing else but relation (4.4.58), which is the *Kirchhoff's mesh rule* (or *Kirchhoff's second law*) for the loop i. Since $\dot{q}_k = I_k$, (where I_k is the current through loop k, being considered the same in any point of this net's loop), an equivalent form of this law is

$$\sum_{k=1}^{n}\left(M_{ik}\frac{dI_k}{dt} + R_{ik}I_k + S_{ik}\int I_k dt\right) = \mathcal{E}_i(t), \quad i = \overline{1,n}. \qquad (4.4.59)$$

Obviously, when this law is effectively applied, one must take care of the fact that the currents are different from one branch of the net to the other, if this is part of a more complex circuit.

2°. Kirchhoff's point rule (or **Kirchhoff's junction** (or **nodal**) **rule**, or **Kirchhoff's first law**)

Consider a node (junction) of an electric circuit, where n net's branches are meeting. If the voltage drop on all branches is the same, namely $U(t)$, then the correlation between the quantities associated to mechanical systems and electric circuits – in this case – are:

Generalized coordinate q_k $\longleftrightarrow$ Electric voltage $U(r)$

Generalized velocity $\dot{q}(t)$ $\longleftrightarrow$ Time derivative of the voltage $\frac{dU(t)}{dt}$

External force F $\longleftrightarrow$ Time derivative of the current $\frac{dI(t)}{dt}$

Mass m $\longleftrightarrow$ Capacitance C

Elastic constant k $\longleftrightarrow$ Inverse of the inductance $\mathcal{L} = 1/L$

Constant of the damping force r $\longleftrightarrow$ Conductance $G = 1/R$

Kinetic energy T $\longleftrightarrow$ $\frac{1}{2}\sum_{i=1}^{n}\sum_{k=1}^{n} C_{ik}\dot{U}_i\dot{U}_k$

Potential energy V $\longleftrightarrow$ $\frac{1}{2}\sum_{i=1}^{n}\sum_{k=1}^{n} \mathcal{L}_{ik}U_iU_k - \sum_{i=1}^{n} U_i\frac{dI_i}{dt}$

Dissipative Rayleigh function $\mathcal{T}$ $\longleftrightarrow$ $\frac{1}{2}\sum_{i=1}^{n}\sum_{k=1}^{n} G_{ik}\dot{U}_i\dot{U}_k$

Taking into account these connections, the Lagrangian of the system writes

$$L(= T - V) = \frac{1}{2}\sum_{i=1}^{n}\sum_{k=1}^{n}\left(C_{ik}\dot{U}_i\dot{U}_k - \mathcal{L}_{ik}U_iU_k\right) + \sum_{i=1}^{n} U_i\frac{dI_i}{dt}. \qquad (4.4.60)$$

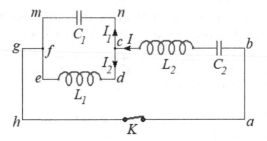

FIGURE 4.8
Same circuit as in Fig. 4.7, this time highlighting the loops and currents flowing
through the circuit.

Using the dissipative Rayleigh's function and the Lagrange's equations of the
second kind (4.4.53), one obtains

$$\sum_{k=1}^{n}\left(C_{ik}\ddot{U}_k + G_{ik}\dot{U}_k + \mathcal{L}_{ik}U_k\right) = \frac{dI_i}{dt}, \quad i = \overline{1,n}, \tag{4.4.61}$$

and integrating with respect to time,

$$\sum_{k=1}^{n}\left(C_{ik}\dot{U}_k + G_{ik}U_k + \mathcal{L}_{ik}\int U_k dt\right) = I_i(t), \quad i = \overline{1,n}. \tag{4.4.62}$$

This relation expresses the *Kirchhoff's first law* for the net's nodes (more
precisely, for the node i of the net).

For the net's loops *abcdefgha* and *abcnmfgha* of the circuit shown in Fig.
4.8, the law expressed by Eq. (4.4.59) writes

$$L_2\frac{dI}{dt} + \frac{1}{C_2}\int I\,dt + L_1\frac{dI_2}{dt} = 0 \tag{4.4.63}$$

and

$$L_2\frac{dI}{dt} + \frac{1}{C_2}\int I\,dt + \frac{1}{C_1}\int I_1\,dt = 0, \tag{4.4.64}$$

respectively.

The Kirchhoff's first law for one of the two net's nodes of the circuit can
be easier written if the circuit is drawn as in Fig. 4.9. As easily seen, on the
branches fc and ed of the net, the voltage drop $U(t)$ is the same, while in the
node c the current $I(t)$ is entering. Under these conditions can be applied the
law expressed by Eq. (4.4.62), written as

$$C_1\frac{dU}{dt} + \frac{1}{L_1}\int U\,dt = I(t). \tag{4.4.65}$$

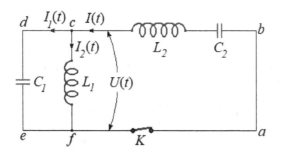

FIGURE 4.9
Same circuit as in Fig. 4.7, redesigned for easier writing of Kirchhoff's laws.

But, as well known, the voltage drop on a capacitor is $U = q/C$, where q is the electric charge on the plates of the capacitor, and C is its capacitance. Writing this relation for the capacitor C_1 from Fig. 4.9 and taking then its time derivative, one obtains

$$U = \frac{q_1}{C_1} \Rightarrow \frac{dU}{dt} = \frac{1}{C_1}\frac{dq_1}{dt} = \frac{1}{C_1}I_1(t)$$

$$\Rightarrow I_1(t) = C_1\frac{dU}{dt}. \tag{4.4.66}$$

Similarly, the voltage drop U on a coil with inductance L is $U = L\,dI/dt$. Therefore, for the coil with inductance L_1 of the circuit shown in Fig. 4.9 an analogous discussion is true, so that

$$U(t) = L_1\frac{dI_2}{dt} \Rightarrow \frac{dI_2}{dt} = \frac{U(t)}{L_1}$$

$$\Rightarrow I_2(t) = \frac{1}{L_1}\int U\,dt. \tag{4.4.67}$$

Introducing Eqs. (4.4.66) and (4.4.67) into Eq. (4.4.65), one follows that

$$I_1(t) + I_2(t) = I(t), \tag{4.4.68}$$

which is exactly the Kirchhoff's first law (applied for the node c of the circuit drawn in Fig. 4.9), written in the "classical" form. We must not forget that this relation has been consequently deduced by means of the Lagrangian formalism, as required by the problem statement.

Relations (4.4.63), (4.4.64) and (4.4.68) lead in a simple and rapid way to the solution of the problem, if we take into account that $I = dq/dt$, $I_1 = dq_1/dt$ and $I_2 = dq_2/dt$, where q is the electric charge on the plates of the capacitor of capacitance C_2, and q_1 is the charge on the plates of capacitor with capacitance C_1. Indeed, relations (4.4.63) and (4.4.64) rewrite as follows:

$$L_2\frac{d^2q}{dt^2} + \frac{q}{C_2} + L_1\frac{d^2q_2}{dt^2} = 0 \tag{4.4.69}$$

and

$$L_2 \frac{d^2q}{dt^2} + \frac{q}{C_2} + \frac{q_1}{C_1} = 0, \qquad (4.4.70)$$

respectively. It we take the second derivative with respect to time of the last relation, one obtains

$$L_2 \frac{d^4q}{dt^4} + \frac{1}{C_2}\frac{d^2q}{dt^2} + \frac{1}{C_1}\frac{d^2q_1}{dt^2} = 0 \Rightarrow$$

$$\frac{d^2q_1}{dt^2} = -C_1\left(L_2\frac{d^4q}{dt^4} + \frac{1}{C_2}\frac{d^2q}{dt^2}\right),$$

which, added to d^2q_2/dt^2 from Eq. (4.4.69), leads to

$$\frac{d^2q_1}{dt^2} + \frac{d^2q_2}{dt^2} = -C_1\left(L_2\frac{d^4q}{dt^4} + \frac{1}{C_2}\frac{d^2q}{dt^2}\right) - \frac{1}{L_1}\left(L_2\frac{d^2q}{dt^2} + \frac{q}{C_2}\right).$$

Taking now into account Eq. (4.4.68), we finally obtain

$$\frac{d^2q}{dt^2} = -C_1\left(L_2\frac{d^4q}{dt^4} + \frac{1}{C_2}\frac{d^2q}{dt^2}\right) - \frac{1}{L_1}\left(L_2\frac{d^2q}{dt^2} + \frac{q}{C_2}\right), \qquad (4.4.71)$$

or

$$L_1L_2C_1C_2\frac{d^4q}{dt^4} + (L_1C_1 + L_2C_2 + L_1C_2)\frac{d^2q}{dt^2} + q = 0, \qquad (4.4.72)$$

which can be written as

$$\frac{d^4q}{dt^4} + 2A\frac{d^2q}{dt^2} + Bq = 0, \qquad (4.4.73)$$

where

$$A \equiv \frac{L_1C_1 + L_2C_2 + L_1C_2}{2L_1L_2C_1C_2} > 0, \qquad (4.4.74)$$

and

$$B \equiv \frac{1}{L_1L_2C_1C_2} > 0. \qquad (4.4.75)$$

Equation (4.4.73) is an ordinary, homogeneous, differential equation with constant coefficients, for the unknown $q = q(t)$. The attached characteristic equation is

$$r^4 + 2Ar^2 + B = 0 \qquad (4.4.76)$$

and has the solutions

$$(r)_{1,2,3,4} = \pm\sqrt{-A \pm \sqrt{A^2 - B}}. \qquad (4.4.77)$$

Then the time variation of the electric charge $q(t)$ on the plates of the capacitor having the capacitance C_2 is

$$q(t) = K_1 e^{t\sqrt{-A+\sqrt{A^2-B}}} + K_2 e^{t\sqrt{-A-\sqrt{A^2-B}}}$$

$$+ K_3 e^{-t\sqrt{-A+\sqrt{A^2-B}}} + K_4 e^{-t\sqrt{-A-\sqrt{A^2-B}}}, \qquad (4.4.78)$$

where K_i, $i = \overline{1,4}$, are four arbitrary integration constants, which can be determined by means of the initial conditions.

If we want to find the final solution of Eq. (4.4.73), first of all we have to analyze the nature of the solutions of Eq. (4.4.76). Introducing the notations

$$a \equiv L_1 C_1, \quad b \equiv L_2 C_2, \quad c \equiv L_1 C_2,$$

then, since $a > 0$, $b > 0$ and $c > 0$, it immediately follows that

$$A^2 - B = \frac{1}{4a^2 b^2} \left[(a - b)^2 + 2c(a + b) + c^2 \right] > 0,$$

so that $\sqrt{A^2 - B} \in \mathbb{R}$, but, given relations (4.4.74) and (4.4.75), it follows that $\sqrt{A^2 - B} < A$, which implies that $-A \pm \sqrt{A^2 - B} < 0$, so $\sqrt{-A \pm \sqrt{A^2 - B}} \in \mathbb{C}$, and so, all exponents in Eq. (4.4.78) are purely imaginary, so the relation (4.4.78) can also be written in the following (more suggestive) form:

$$q(t) = K_1 e^{i\omega_1 t} + K_2 e^{i\omega_2 t} + K_3 e^{-i\omega_1 t} + K_4 e^{-i\omega_2 t}, \tag{4.4.79}$$

where

$$\omega_1 \equiv \sqrt{A - \sqrt{A^2 - B}}, \quad \omega_2 \equiv \sqrt{A + \sqrt{A^2 - B}}. \tag{4.4.80}$$

The initial conditions

$$q(0) = q_0, \quad I(0) = \frac{dq}{dt}(0) = 0, \quad \frac{d^2 q}{dt^2}(0) = 0, \quad \frac{d^3 q}{dt^3}(0) = 0,$$

lead to the following system of equations for the integration constants K_i, $i = \overline{1,4}$:

$$\begin{cases} q_0 = K_1 + K_2 + K_3 + K_4 \\ 0 = \omega_1(K_1 - K_3) - \omega_2(K_4 - K_2) \\ 0 = \omega_1^2(K_1 + K_3) + \omega_2^2(K_2 + K_4) \\ 0 = \omega_1^3(K_3 - K_1) - \omega_2^3(K_2 - K_4) \end{cases} \tag{4.4.81}$$

with the solutions

$$K_1 = K_3 = -\frac{q_0}{2} \frac{\omega_2^2}{\omega_1^2 - \omega_2^2}, \quad K_2 = K_4 = \frac{q_0}{2} \frac{\omega_1^2}{\omega_1^2 - \omega_2^2}.$$

With these constants into Eq. (4.4.79) we get

$$q(t) = \frac{q_0}{2} \frac{1}{\omega_1^2 - \omega_2^2} \left[\omega_1^2 \left(e^{-i\omega_2 t} + e^{i\omega_2 t} \right) - \omega_2^2 \left(e^{-i\omega_1 t} + e^{i\omega_1 t} \right) \right]$$

$$= q_0 \frac{\omega_1^2}{\omega_1^2 - \omega_2^2} \left[\cos \omega_2 t - \left(\frac{\omega_2}{\omega_1} \right)^2 \cos \omega_1 t \right]. \tag{4.4.82}$$

The dependence $q = q(t)$ given by the relation (4.4.82) is dictated by the ratio in which the two pulsations, ω_1 and ω_2, lie. For relatively close values of the two pulsations, the dependence $q = q(t)$ shows a behaviour similar to the beat phenomenon — see Fig. 4.10, which was drawn using the following command lines, written using the Mathematica 5.0 software:

Needs["Graphics'Colors'"]

$q_0 = 1 * 10^{-5};$

$L_1 = 4 * \pi * 10^{-4};$

$L_2 = 6 * \pi * 10^{-4};$

$C_1 = 2 * 10^{-6};$

$C_2 = 4 * 10^{-6};$

$A = (L_1 * C_1 + L_2 * C_2 + L_1 * C_2)/(2 * L_1 * L_2 * C_1 * C_2);$

$B = 1/(L_1 * L_2 * C_1 * C_2);$

$\omega_1 = \sqrt{A - \sqrt{A^2 - B}};$

$\omega_2 = \sqrt{A + \sqrt{A^2 - B}};$

$q[t\_] := q_0 * \frac{\omega_1^2}{\omega_1^2 - \omega_2^2} * (\text{Cos}[\omega_2 * t] - (\omega_2/\omega_1)^2 * \text{Cos}[\omega_1 * t]);$

$t_i = 0.0;$

$t_f = 0.45 * 10^{-1};$

gr = Plot[$q[t]$, {t, t_i, t_f}, PlotStyle → Blue,

AxesLabel → "$t(s)$", "$q(C)$"}]

Print["$q = q(t)$ for", t_i, "s ≤ t ≤ ", t_f "s"];

$q = q(t)$ for $0\,s \le t \le 0.045\,s$

For significantly different values of the two pulsations the same dependence has predominantly a simple periodic behaviour — see Fig. 4.11, which was drawn using the following command lines, written using the same Mathematica 5.0 software:

Needs["Graphics'Colors'"]

$q_0 = 1 * 10^{-5};$

$L_1 = 4 * \pi * 10^{-4};$

$L_2 = 6 * \pi * 10^{-2};$

$C_1 = 2 * 10^{-6};$

$C_2 = 4 * 10^{-2};$

$A = (L_1 * C_1 + L_2 * C_2 + L_1 * C_2)/(2 * L_1 * L_2 * C_1 * C_2);$

$B = 1/(L_1 * L_2 * C_1 * C_2);$

$\omega_1 = \sqrt{A - \sqrt{A^2 - B}};$

$\omega_2 = \sqrt{A + \sqrt{A^2 - B}};$

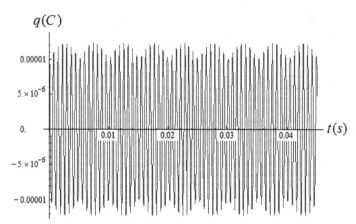

FIGURE 4.10
Graphical representation of the electric charge on the plates of the capacitor C_2 as a function of time, $q = q(t)$, for relatively close values of the two pulsations ω_1 and ω_2 in formula (4.4.82). The figure highlights a behaviour similar to the phenomenon of beats/beating.

$q_p[t_-] := q_0 * \frac{\omega_1^2}{\omega_1^2 - \omega_2^2} * (\text{Cos}[\omega_2 * t] - (\omega_2/\omega_1)^2 * \text{Cos}[\omega_1 * t]);$

$t_i = 0.0;$

$t_f = 0.45 * 25;$

$\text{grp} = \text{Plot}[q_p[t], \{t, t_i, t_f\}, \text{PlotStyle} \rightarrow \text{Green},$

$\text{AxesLabel} \rightarrow \text{``}t(s)\text{''}, \text{``}q_p(C)\text{''}\}]$

$\text{Print}[\text{``}q_p = q_p(t) \text{ for''}, t_i, \text{``}s \leq t \leq \text{''}, t_f \text{ ``s''}];$

$q_p = q_p(t) \text{ for } 0 s \leq t \leq 11.25 s$

and in the limiting case, when the two pulsations tend to be equal, because

$$\lim_{\substack{\omega_2 \to \omega_1 \\ (\omega_1 \equiv \omega)}} \frac{\omega_1^2}{\omega_1^2 - \omega_2^2} \left[\cos \omega_2 t - \left(\frac{\omega_2}{\omega_1} \right)^2 \cos \omega_1 t \right] = \cos \omega t + \frac{1}{2} \omega t \sin \omega t,$$

the graph of dependency in question shows like that in Fig. 4.12, being obtained by using the following command lines, written using Mathematica 5.0 software:

$\text{Needs}[\text{``Graphics'Colors'''}]$

$q_0 = 1 * 10^{-5};$

$L_1 = 4 * \pi * 10^{-4};$

$L_2 = 6 * \pi * 10^{-4};$

$C_1 = 2 * 10^{-6};$

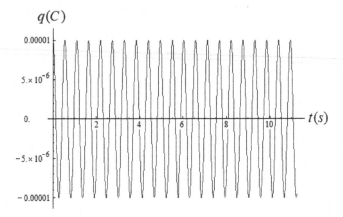

FIGURE 4.11
Graphical representation of the electric charge on the plates of the capacitor C_2 as a function of time, $q = q(t)$, for significantly different values of the two pulsations ω_1 and ω_2 in formula (4.4.82).

$C_2 = 4 * 10^{-6}$;

$A = (L_1 * C_1 + L_2 * C_2 + L_1 * C_2)/(2 * L_1 * L_2 * C_1 * C_2)$;

$B = 1/(L_1 * L_2 * C_1 * C_2)$;

$\omega = \sqrt{A - \sqrt{A^2 - B}}$;

$q_l[t_-] := q_0 * (\text{Cos}[\omega * t] + 1/2 * \omega * t*\text{Sin}[\omega * t])$;

$t_i = 0.0$;

$t_f = 0.45 * 10^{-1}$;

$\text{grl} = \text{Plot}[q_l[t], \{t, t_i, t_f\}, \text{PlotStyle} \rightarrow \text{Red},$

$\text{AxesLabel} \rightarrow \text{"}t(s)\text{"}, \text{"}q_l(C)\text{"}\}]$

$\text{Print}[\text{"}q_l = q_l(t) \text{ for"}, t_i, \text{"s} \leq t \leq \text{"}, t_f \text{ "s"}]$;

$q_l = q_l(t)$ for $0\,s \leq t \leq 0.045\,s$

We conclude this discussion here with the remark that in the general case of an arbitrary ratio between the two pulsations, the dependence $q = q(t)$ shows a behaviour intermediate to the significant cases that have been briefly studied above.

4.5   Problem No. 31

A power supply with electromotive force (voltage) $\mathcal{E}$ and internal resistance R is connected in parallel with a coil of inductance L and a capacitor of

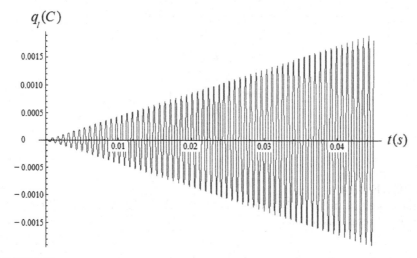

FIGURE 4.12
Graphical representation of the electric charge on the plates of the capacitor C_2 as a function of time, $q = q(t)$, when the two pulsations ω_1 and ω_2 in formula (4.4.82) tend to be equal.

capacitance C, as shown in Fig. 4.13. Determine the time dependence of the electric current through the (electrical) source, $I = I(t)$, which appears in the circuit after closing the switch K, using the Lagrangian formalism based on the analogy between mechanical systems of material points and the electric circuits. The coil and the capacitor are supposed to be ideal (without internal resistance).

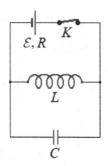

FIGURE 4.13
A parallel circuit containing a coil L, a capacitor C, and a DC power supply with electromotive force $\mathcal{E}$ and internal resistance R.

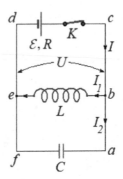

FIGURE 4.14
Same circuit as in Fig. 4.13 redesigned for highlighting the loops and currents
flowing through the circuit.

Solution

Applying Kirchhoff's second law (for the loops of a net) deduced in the previous problem,

$$\sum_{k=1}^{n} \left(M_{ik}\frac{dI_k}{dt} + R_{ik}I_k + S_{ik}\int I_k dt \right) = \mathcal{E}_i(t), \quad i = \overline{1, n},$$

to the net's loop $bcdeb$ (see Fig. 4.14), one obtains

$$L\frac{dI_1}{dt} + RI = \mathcal{E}, \tag{4.5.83}$$

and if the same law is applied to the loop $acdfa$, the result is

$$RI + \frac{1}{C}\int I_2 dt = \mathcal{E}. \tag{4.5.84}$$

The first Kirchhoff's law (concerning the net nodes/junctions),

$$\sum_{k=1}^{n} \left(C_{ik}\dot{U}_k + G_{ik}U_k + \mathcal{L}_{ik}\int U_k dt \right) = I_i(t), \quad i = \overline{1, n},$$

provides the supplementary relation

$$C\frac{dU}{dt} + \frac{1}{L}\int U dt = I(t), \tag{4.5.85}$$

where U is the voltage on the coil (the same as that on the capacitor – see
Fig. 4.14). Since the voltage on any capacitor is $U = \frac{q}{C}$, where q is the electric
charge on the plates of the capacitor and C its capacitance, one can write for
the capacitor in Fig. 4.14 the following relation:

$$U = \frac{q_2}{C} \Rightarrow \frac{dU}{dt} = \frac{1}{C}\frac{dq_2}{dt} = \frac{1}{C}I_2(t) \Rightarrow I_2(t) = C\frac{dU}{dt}. \tag{4.5.86}$$

Next, the (same) voltage U on the coil of inductance L is $U = L dI_1/dt$, where, as seen in Fig. 4.14, I_1 is the current flowing through the coil. It then follows that

$$\frac{dI_1}{dt} = \frac{U}{L} \Rightarrow I_1(t) = \frac{1}{L} \int U \, dt. \tag{4.5.87}$$

Introducing Eqs. (4.5.86) and (4.5.87) into Eq. (4.5.85), it follows that

$$I_2(t) + I_1(t) = I(t), \tag{4.5.88}$$

which is the net's nodes law (the first Kirchhoff's law) applied to the node b of the circuit given in Fig. 4.14, in a well-known (classical) form. Relations (4.5.83), (4.5.84) and (4.5.88) lead to the solution of the problem. Thus, taking the second derivative of Eq. (4.5.84) with respect to time, we have

$$R \frac{d^2 I}{dt^2} + \frac{1}{C} \frac{dI_2}{dt} = 0,$$

leading to

$$\frac{dI_2}{dt} = -RC \frac{d^2 I}{dt^2}.$$

Next, using Eq. (4.5.83), it follows that

$$\frac{dI_1}{dt} = -\frac{R}{L} I + \frac{1}{L} \mathcal{E}.$$

In this case, the problem requirement, which is the dependence $I = I(t)$, follows as a solution of the differential equation obtained by adding the above obtained results $\left(\frac{dI_2}{dt} \text{ and } \frac{dI_1}{dt} \right)$ and using the relation (4.5.88), derived once with respect to time; the result is

$$\frac{dI_2}{dt} + \frac{dI_1}{dt} \left(= \frac{dI}{dt} \right) = -RC \frac{d^2 I}{dt^2} - \frac{R}{L} I + \frac{1}{L} \mathcal{E},$$

or

$$\frac{d^2 I}{dt^2} + \frac{1}{RC} \frac{dI}{dt} + \frac{1}{LC} I = \frac{1}{RLC} \mathcal{E}, \tag{4.5.89}$$

which is a non-homogeneous ordinary differential equation, with constant coefficients for the unknown $I = I(t)$. The general solution of such an equation is obtained as a sum of general solution of the homogeneous equation, and a particular solution of the non-homogeneous equation. A particular solution of the non-homogeneous equation can be obtained by the general method (the Lagrange's method of variation of parameters) or, more simply, but which may not always apply, looking for a solution to the form of the term that gives the non-homogeneity (in our case, a constant).

The homogeneous equation attached to the non-homogeneous equation (4.5.89) is

$$\frac{d^2 I}{dt^2} + \frac{1}{RC} \frac{dI}{dt} + \frac{1}{LC} I = 0.$$

The characteristic equation attached to this equation is

$$r^2 + \frac{1}{RC}r + \frac{1}{LC} = 0,$$

and has the roots

$$(r)_{1,2} = \frac{-\frac{1}{RC} \pm \sqrt{\frac{1}{R^2C^2} - \frac{4}{LC}}}{2} = -\frac{1}{2RC}\left(1 \mp \sqrt{1 - \frac{4R^2C}{L}}\right)$$

$$= -\frac{1}{2RC} \pm \sqrt{\frac{1}{4R^2C^2} - \frac{1}{LC}}. \tag{4.5.90}$$

The general solution of the homogeneous equation then is

$$I_0(t) = Ae^{r_1 t} + Be^{r_2 t}$$

$$= Ae^{-\frac{t}{2RC}}e^{+t\sqrt{\frac{1}{4R^2C^2} - \frac{1}{LC}}} + Be^{-\frac{t}{2RC}}e^{-t\sqrt{\frac{1}{4R^2C^2} - \frac{1}{LC}}}$$

$$= Ae^{-\frac{t}{2RC}}e^{i\omega t} + Be^{-\frac{t}{2RC}}e^{-i\omega t}, \tag{4.5.91}$$

where $\omega^2 \equiv \frac{1}{LC} - \frac{1}{4R^2C^2}$, while A and B are two arbitrary constants of integration.

Looking now for a particular solution of the non-homogeneous equation (4.5.89) of the form $I_p(t) = K_0 = const.$, and requiring that this solution verifies the non-homogeneous equation, one results

$$I_p(t) = K_0 = \frac{\mathcal{E}}{R}.$$

Therefore, the general solution of the non-homogeneous equation writes as

$$I(t) = I_0(t) + I_p(t) = e^{-\frac{t}{2RC}}\left(Ae^{i\omega t} + Be^{-i\omega t}\right) + \frac{\mathcal{E}}{R}, \tag{4.5.92}$$

where the constants A and B are determined by the initial conditions: $I(0) = 0$ and $\frac{dI}{dt}(0) = 0$. Thus, the following algebraic system results for the unknowns A and B:

$$\begin{cases} A + B + \dfrac{\mathcal{E}}{R} = 0, \\ (A - B)\left(I\omega - \dfrac{1}{2RC}\right) = 0. \end{cases}$$

On the assumption that $\left(I\omega - \dfrac{1}{2RC}\right) \neq 0$, the above system has the solutions $A = B = -\frac{\mathcal{E}}{2R}$. By introducing this last result in Eq. (4.5.92), one obtains the final form of the problem solution,

$$I(t) = \frac{\mathcal{E}}{R} - \frac{\mathcal{E}}{2R}e^{-\frac{t}{2RC}}\left(e^{i\omega t} + e^{-i\omega t}\right)$$

$$= \frac{\mathcal{E}}{R} - \frac{\mathcal{E}}{2R}e^{-\frac{t}{2RC}}2\cos\omega t = \frac{\mathcal{E}}{R}\left(1 - e^{-\frac{t}{2RC}}\cos\omega t\right), \tag{4.5.93}$$

which is graphically depicted in Figs. 4.15–4.19, by means of the following command lines, written using Mathematica 5.0 software:

Needs["Graphics'Colors'"]
$R = 200;$
$ES = 10;$
$L_1 = 4 * \pi * 10^{-4};$
$CA = 10^{-6};$
$$\omega_1 = \sqrt{\frac{1}{L * CA} - \frac{1}{4 * R^2 * CA^2}};$$
$CE[t_-] := \dfrac{ES}{R} * (1 - e^{\frac{-t}{2 * R * CA}} * \text{Cos}[\omega * t]);$
$t_i = 0.0;$
$t_f = 1.75 * 10^{-3};$
gr1 = Plot[$CE[t]$, $\{t, t_i, t_f\}$, PlotStyle → Red,
AxesLabel → {"$t(s)$", "$I(A)$"}]
Print["$I = I(t)$ for", t_i, "s ≤ t ≤ ", t_f "s"];
gr2 = Plot[$\dfrac{ES}{R} * \left(1 - e^{\frac{-t}{2 * R * CA}}\right)$, $\{t, t_i, t_f\}$, PlotStyle → Black,
AxesLabel → {"$t(s)$", "$I_{cr+}(A)$"}]
Print["$I_{cr+} = I_{cr+}(t)$ for", t_i, "s ≤ t ≤ ", t_f "s"];
gr3 = Plot[$\dfrac{ES}{R} * \left(1 + e^{\frac{-t}{2 * R * CA}}\right)$, $\{t, t_i, t_f\}$, PlotStyle → Black,
AxesLabel → {"$t(s)$", "$I_{cr-}(A)$"}]
Print["$I_{cr-} = I_{cr-}(t)$ for", t_i, "s ≤ t ≤ ", t_f "s"];
Show[$gr1, gr2, gr3$];

On a single graph, for $0\,s \leq t \leq 0.00175\,s$, the three curves shown in Figs. 4.15–4.17, look as in Fig. 4.18, while for $0\,s \leq t \leq 0.01\,s$, the same graph looks like in Fig. 4.19.

Here are the notations used above:

$$I_{cr+} = \frac{\mathcal{E}}{R}\left(1 - e^{-\frac{t}{2RC}}\right), \qquad (4.5.94)$$

that is $I_{cr+} = I\big|_{\cos \omega t = +1}$, $\forall\, t > 0$, which implies

$$\omega^2 = \frac{1}{LC} - \frac{1}{4R^2C^2} = 0 \;\Leftrightarrow\; R = R_{cr} \equiv \frac{1}{2}\sqrt{\frac{L}{C}}$$

(the critical value of the electric resistance) and

$$I_{cr-} = \frac{\mathcal{E}}{R}\left(1 + e^{-\frac{t}{2RC}}\right), \qquad (4.5.95)$$

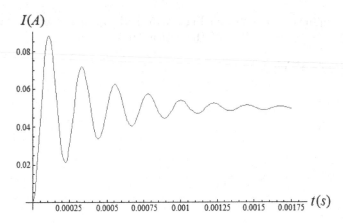

FIGURE 4.15
Graphical representation of the "total" electric current given by Eq. (4.5.93)
as a function of time, $I = I(t)$.

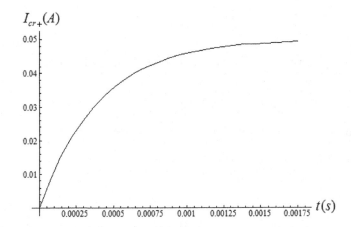

FIGURE 4.16
Graphical representation of the electric current given by Eq. (4.5.94) as a
function of time, $I_{cr+} = I_{cr+}(t)$.

that is $I_{cr-} = I\big|_{\cos\omega t=-1}$, $\forall t > 0$, which graphically represents the symmet-
rical of I_{cr+} with respect to the straight line

$$I_\infty = \lim_{t \to \infty} I(t) = \frac{\mathcal{E}}{R}.$$

For the considered values of the electromotive force (voltage) $\mathcal{E} = 10\,\mathrm{V}$, its
internal resistance, $R = 200\,\Omega$, the coil inductance, $L = 4\pi \times 10^{-7}\mathrm{H}$, and the
capacitance of the capacitor, $C = 10^{-6}\mathrm{F} = 1\,\mu\mathrm{F}$, the last figure (represented
for $t_f = 10^{-2}\mathrm{s} = 10\,\mathrm{ms}$) shows that the transient regime disappears almost
completely after a time shorter than $0.006\,\mathrm{s} = 6\,\mathrm{ms}$.

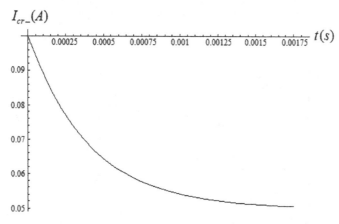

FIGURE 4.17
Graphical representation of the electric current given by Eq. (4.5.95) as a function of time, $I_{cr-} = I_{cr-}(t)$.

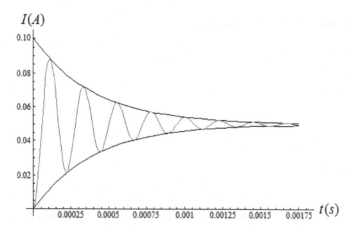

FIGURE 4.18
Graphical representation on the same figure of the three currents, $I = I(t)$, $I_{cr+} = I_{cr+}(t)$ and $I_{cr-} = I_{cr-}(t)$ for $0\,s \le t \le 0.00175\,s$.

4.6 Problem No. 32

Determine the time variation of the electric charge $q = q(t)$ on the plates of a capacitor in a series RLC circuit, powered by a source with constant electromotive force ($E = $ const.). The circuit elements are considered ideal and $R = $ const., $L = $ const., $C = $ const.

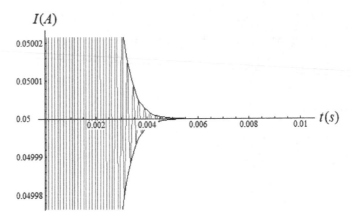

FIGURE 4.19
Graphical representation on the same figure of the three currents, $I = I(t)$, $I_{cr+} = I_{cr+}(t)$ and $I_{cr-} = I_{cr-}(t)$ for $0\,s \le t \le 0.01\,s$.

Complements: useful theorems

Theorem 1. Consider the nth-order homogeneous differential equation with real constant coefficients $a_k \in \mathbb{R}$, $k = 0, 1, 2, ..., n$:

$$a_0 y^{(n)} + a_1 y^{(n-1)} + a_2 y^{(n-2)} + ... + a_{n-3} y^{(3)} + a_{n-2} y'' + a_{n-1} y' + a_n y = 0. \tag{4.6.96}$$

If the characteristic equation attached to Eq. (4.6.96), *i.e.*,

$$a_0 r^n + a_1 r^{n-1} + a_2 r^{n-2} + ... + a_{n-2} r^2 + a_{n-1} r + a_n = 0,$$

has complex roots

$$\alpha_1 + i\beta_1, \ \alpha_2 + i\beta_2, \ ..., \ \alpha_p + i\beta_p,$$

(and, because the coefficients a_k, $k = 0, 1, 2, ..., n$, are real numbers, the characteristic equation also admits the complex-conjugate roots

$$\alpha_1 - i\beta_1, \ \alpha_2 - i\beta_2, \ ..., \ \alpha_p - i\beta_p)$$

of multiplicity orders m_1, m_2, ..., m_p, and real roots

$$\gamma_1, \ \gamma_2, \ ..., \ \gamma_q,$$

of multiplicity orders s_1, s_2, ..., s_q, then the general solution of the homogeneous differential equation (4.6.96) is

$$y(x) = \sum_{k=1}^{p} e^{\alpha_k x} \left[P_{m_k-1}(x) \cos \beta_k x + Q_{m_k-1}(x) \sin \beta_k x \right]$$

$$+ \sum_{h=1}^{q} e^{\gamma_h x} R_{s_h-1}(x), \tag{4.6.97}$$

where $P_{m_k-1}(x)$, $Q_{m_k-1}(x)$ and $R_{s_h-1}(x)$ are arbitrary polynomials in x of degrees $m_k - 1$, $m_k - 1$ and $s_h - 1$, respectively. Besides, the following relationship holds:

$$2(m_1 + m_2 + ... + m_p) + s_1 + s_2 + ... + s_q = n,$$

or, in a more compact writing,

$$2 \sum_{m_k=1}^{p} m_k + \sum_{s_h=1}^{q} s_h = n.$$

Theorem 2 (Lagrange's method of variation of parameters). Consider the nth-order non-homogeneous differential equation with real constant coefficients

$$a_0(x)y^{(n)} + a_1(x)y^{(n-1)} + a_2(x)y^{(n-2)} + ... + a_{n-1}(x)y' + a_n(x)y = f(x), \tag{4.6.98}$$

with $a_0(x)$, $a_1(x)$, $a_2(x)$, ..., $a_{n-1}(x)$, $a_n(x)$, $f(x)$ − continuous functions and $a_0(x) \neq 0$ on the $\mathbb{R}$−interval $[a, b]$ and let $\{y_1, y_2, ..., y_n\}$ be a fundamental system of solutions on $[a, b]$ of the homogeneous differential equation attached to Eq. (4.6.98):

$$a_0(x)y^{(n)} + a_1(x)y^{(n-1)} + a_2(x)y^{(n-2)} + ... + a_{n-1}(x)y' + a_n(x)y = 0.$$

Then, a particular solution of the non-homogeneous equation (4.6.98) on $[a, b]$ is given by

$$y_p(x) = y_1 \int C_1'(x)dx + y_2 \int C_2'(x)dx + ... + y_n \int C_n'(x)dx, \tag{4.6.99}$$

where $\{C_1'(x), C_2'(x), ..., C_n'(x)\}$ is the solution of the algebraic system

$$\begin{cases} y_1 C_1'(x) + y_2 C_2'(x) + ... + y_n C_n'(x) = 0, \\ y_1' C_1'(x) + y_2' C_2'(x) + ... + y_n' C_n'(x) = 0, \\ y_1'' C_1'(x) + y_2'' C_2'(x) + ... + y_n'' C_n'(x) = 0, \\ \vdots \\ y_1^{(n-2)} C_1'(x) + y_2^{(n-2)} C_2'(x) + ... + y_n^{(n-2)} C_n'(x) = 0, \\ y_1^{(n-1)} C_1'(x) + y_2^{(n-1)} C_2'(x) + ... + y_n^{(n-1)} C_n'(x) = \dfrac{f(x)}{a_0(x)}. \end{cases} \tag{4.6.100}$$

If all the integrals in Eq. (4.6.99) are performed, each integral introducing an arbitrary integration constant (let A_1, A_2, ..., A_n, be these n constants), *i.e.*,

$$\begin{cases} \int C_1'(x)dx = \varphi_1(x) + A_1, \\ \int C_2'(x)dx = \varphi_2(x) + A_2, \\ \vdots \\ \int C_{n-1}'(x)dx = \varphi_{n-1}(x) + A_{n-1}, \\ \int C_n'(x)dx = \varphi_n(x) + A_n, \end{cases}$$

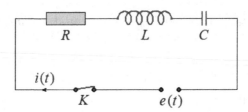

FIGURE 4.20
A series RLC circuit powered by a generic source $e = e(t)$.

then the *general solution of the non-homogeneous differential equation* (4.6.98) has the following (final) form:

$$y(x) = y_1\varphi_1 + y_2\varphi_2 + \ldots + y_n\varphi_n + A_1 y_1 + A_2 y_2 + \ldots + A_n y_n. \quad (4.6.101)$$

Solution

Let us first consider the general case when the voltage supply is variable. Then, the differential equation which governs the series RLC circuit shown in Fig. 4.20 can be found judging in two different ways:
1) The sum of the voltage drops on the three circuit elements must be equal to the electromotive force of the ideal source:

$$u_R + u_L + u_C = e,$$

or,

$$Ri + L\frac{di}{dt} + \frac{q}{C} = e, \quad (4.6.102)$$

where

$$q(t) = \int i(t)dt.$$

2) Considering that actually the coil is an inductance through which a variable electric current flows, according to the law of electromagnetic induction, it will be "equivalent" to a voltage source with (self-induced) electromotive force

$$e_L = -L\frac{di}{dt},$$

where the Lenz's law has been taken into account. Thus the sum of the voltage drops on the "consumer" elements will be equal to the sum of the electromotive forces of the circuit (which "feeds" the circuit):

$$u_R + u_C = e + e_L,$$

or,

$$Ri + \frac{q}{C} = e - L\frac{di}{dt} \quad \Leftrightarrow \quad Ri + \frac{q}{C} + L\frac{di}{dt} = e,$$

that is the same equation obtained by the first method, as natural.
Taking into account that $i = dq/dt$, Eq. (4.6.102) becomes

$$R\frac{dq}{dt} + L\frac{d^2q}{dt^2} + \frac{1}{C}q = e,$$

or

$$\frac{d^2q}{dt^2} + \frac{R}{L}\frac{dq}{dt} + \frac{1}{LC}q = \frac{e}{L}. \qquad (4.6.103)$$

Note that by taking the total derivative with respect to time of Eq.
(4.6.102), it is obtained an equation similar to Eq. (4.6.103) but for the intensity of the electric current in the circuit,

$$\frac{d^2i}{dt^2} + \frac{R}{L}\frac{di}{dt} + \frac{1}{LC}i = \frac{1}{L}\frac{de}{dt}. \qquad (4.3.103')$$

Equations (4.6.103) and (4.3.103') are non-homogeneous second-order differential equations with real constant coefficients. If the electromotive force e
is constant (DC power supply), then Eq. (4.3.103') is a homogeneous second-
order differential equation with real constant coefficients.

As stated in the problem statement, the series RLC circuit is powered by
a DC voltage source, $e \equiv E = \text{const.}$ (see Fig. 4.21).

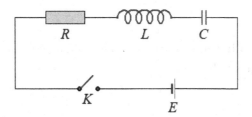

FIGURE 4.21
A series RLC circuit powered by a DC source with constant electromotive
force E.

Thus, the problem turns to solving the non-homogeneous second-order
differential equation (4.6.103), for which purpose we will use the Lagrange's
method of variation of parameters. The characteristic equation attached to
the corresponding homogeneous differential equation is

$$r^2 + \frac{R}{L}r + \frac{1}{LC} = 0, \qquad (4.6.104)$$

with the solutions

$$r_{1,2} = \frac{-\frac{R}{L} \pm \sqrt{\frac{R^2}{L^2} - \frac{4}{LC}}}{2} = -\frac{R}{2L} \pm \sqrt{\frac{R^2}{4L^2} - \frac{1}{LC}}.$$

Introducing the suggestive notations

$$\tau \equiv \frac{2L}{R}$$

(since the ratio $\frac{L}{R}$ has dimensions of time) and

$$\omega^2 \equiv \frac{R^2}{4L^2} - \frac{1}{LC},$$

(since the ratios $\frac{R}{L}$ and $\frac{1}{\sqrt{LC}}$ are measured in s^{-1}), we have the following three possibilities:

a) $\dfrac{R^2}{4L^2} - \dfrac{1}{LC} > 0$. In this case, the solutions of quadratic equation (4.6.104) are real and distinct:

$$r_1 = -\frac{1}{\tau} + \omega$$

and

$$r_2 = -\frac{1}{\tau} - \omega,$$

b) $\dfrac{R^2}{4L^2} - \dfrac{1}{LC} = 0$. In this case, the solutions of the characteristic equation are real and equal:

$$r_1 = r_2 \left(= -\frac{R}{2L} \right) = -\frac{1}{\tau}.$$

c) $\dfrac{R^2}{4L^2} - \dfrac{1}{LC} < 0$. In this case, the solutions of quadratic equation (4.6.104) are complex-conjugated:

$$r_1 = -\frac{1}{\tau} + i\omega$$

and

$$r_2 = -\frac{1}{\tau} - i\omega.$$

Therefore, the free operating mode of a series RLC circuit (corresponding to the situation where the circuit is not supplied by any electromotive force, i.e., $E = 0$), and which can occur only when initially (before closing the switch K) the capacitor is charged with the "initial" electric charge $q_0 = q(t_0 = 0)$, can be of three kinds:

$\mathbf{a_1}$) *overdamped response* (overdamped regime of operation or *aperiodic operating regime*), corresponding to the situation $\frac{R^2}{4L^2} - \frac{1}{LC} > 0$, or $\frac{R^2}{4L^2} > \frac{1}{LC}$, or, still, $R > R_{cr}$ where

$$R_{cr} \equiv 2\sqrt{\frac{L}{C}}$$

is called the *critical resistance* of the circuit;

$\mathbf{b_1}$) *critically damped response* (critically damped regime of operation or *critically aperiodic operating regime*), corresponding to the case $\frac{R^2}{4L^2} - \frac{1}{LC} = 0$, that is $R = R_{cr}$, and

c_1) *underdamped response* (underdamped regime of operation or *periodic operating regime*), corresponding to the case $\frac{R^2}{4L^2} - \frac{1}{LC} < 0$, or $R < R_{cr}$.

According to relation (4.6.97), the fundamental system of solutions of the homogeneous differential equation attached to the non-homogeneous differential equation (4.6.103) therefore will be written as:

a_2) in the case of overdamped response:

$$\begin{cases} q_1(t) = e^{\left(-\frac{1}{\tau} + \omega\right)t}, \\ q_2(t) = e^{\left(-\frac{1}{\tau} - \omega\right)t}; \end{cases} \tag{4.6.105}$$

b_2) in the case of critically damped response:

$$\begin{cases} q_1(t) = t\,e^{-\frac{t}{\tau}}, \\ q_2(t) = e^{-\frac{t}{\tau}}; \end{cases} \tag{4.6.106}$$

c_2) in the case of underdamped response:

$$\begin{cases} q_1(t) = e^{-\frac{t}{\tau}} \cos \omega t, \\ q_2(t) = e^{-\frac{t}{\tau}} \sin \omega t. \end{cases} \tag{4.6.107}$$

Case a)

In this case, for the concrete situation of our problem, the system (4.6.100) writes as follows:

$$\begin{cases} C_1'(t)e^{\left(-\frac{1}{\tau}+\omega\right)t} + C_2'(t)e^{-\left(\frac{1}{\tau}+\omega\right)t} = 0, \\ \left(-\frac{1}{\tau}+\omega\right)C_1'(t)e^{\left(-\frac{1}{\tau}+\omega\right)t} - \left(-\frac{1}{\tau}+\omega\right)C_2'(t)e^{-\left(-\frac{1}{\tau}+\omega\right)t} = \frac{E}{L}. \end{cases} \tag{4.6.108}$$

Multiplying the first equation of the above system by $\left(\frac{1}{\tau}+\omega\right)$ and adding the resulting equation member by member to the second equation of the system, we have

$$2\omega C_1'(t)e^{\left(-\frac{1}{\tau}+\omega\right)t} = \frac{E}{L},$$

which yields

$$C_1'(t) = \frac{E}{2\omega L}e^{\left(\frac{1}{\tau}-\omega\right)t}, \tag{4.6.109}$$

so that

$$C_1(t) = \int C_1'(t)\,dt = \frac{E}{2\omega L}\int e^{\left(-\frac{1}{\tau}+\omega\right)t}dt$$

$$= \frac{E}{2\omega L}\frac{\tau}{1-\omega\tau}e^{\left(\frac{1}{\tau}-\omega\right)t} + \mathcal{C}_1, \tag{4.6.110}$$

where $\mathcal{C}_1$ is a true arbitrary constant of integration. Then, the first equation of the system (4.6.108) gives

$$C_2'(t) = -C_1'(t)e^{2\omega t},$$

or, using Eq. (4.6.109),

$$C_2'(t) = -\frac{E}{2\omega L}e^{\left(\frac{1}{\tau}-\omega\right)t}e^{2\omega t} = -\frac{E}{2\omega L}e^{\left(\frac{1}{\tau}+\omega\right)t},$$

which gives

$$C_2(t) = \int C_2'(t)\, dt = -\frac{E}{2\omega L}\int e^{\left(\frac{1}{\tau}+\omega\right)t}$$

$$= -\frac{E}{2\omega L}\frac{\tau}{1+\omega\tau}e^{\left(\frac{1}{\tau}+\omega\right)t} + C_2, \qquad (4.6.111)$$

where C_2 is also a true, arbitrary, integration constant. Both constants C_1 and C_2 can be determined by means of the initial conditions:

$$\begin{cases} q(t_0 = 0) = q_0, \\ \dot{q}(t_0 = 0)[\equiv i(t_0 = 0)] = 0, \end{cases} \qquad (4.6.112)$$

if at the initial moment $t_0 = 0$ the capacitor was charged with the electric charge q_0, or,

$$\begin{cases} q(t_0 = 0) = 0. \\ \dot{q}(t_0 = 0)\left(\equiv i(t_0 = 0)\right) = 0, \end{cases} \qquad (4.6.113)$$

if before closing the switch K the capacitor was uncharged.

Therefore, in this first case, in agreement with relation (4.6.101) met in the Lagrange's method of variation of parameters, the general solution of the non-homogeneous differential equation (4.6.103) is

$$q(t) = C_1\, e^{-\left(\frac{1}{\tau}-\omega\right)t} + C_2\, e^{-\left(\frac{1}{\tau}+\omega\right)t} + \frac{E}{2\omega L}\left(\frac{\tau}{1-\omega\tau} - \frac{\tau}{1+\omega\tau}\right)$$

$$= C_1\, e^{-\left(\frac{1}{\tau}-\omega\right)t} + C_2\, e^{-\left(\frac{1}{\tau}+\omega\right)t} + \frac{\tau^2}{1-\omega^2\tau^2}\frac{E}{L}. \qquad (4.6.114)$$

Using the initial conditions expressed by Eq. (4.6.112), we find

$$\begin{cases} C_1 + C_2 + \dfrac{\tau^2}{1-\omega^2\tau^2}\dfrac{E}{L} = q_0, \\ C_1 + C_2 + \omega\tau(C_1 - C_2) = 0, \end{cases}$$

or, in a more convenient form

$$\begin{cases} C_1 + C_2 = q_0 - \dfrac{\tau^2}{1-\omega^2\tau^2}\dfrac{E}{L}, \\ C_1 - C_2 = \dfrac{q_0}{\omega\tau} - \dfrac{\tau}{\omega(1-\omega^2\tau^2)}\dfrac{E}{L}. \end{cases} \qquad (4.6.115)$$

Adding member by member the two equations of this system, we obtain

$$C_1 = \frac{1}{2}\left(q_0 - \frac{\tau^2}{1-\omega^2\tau^2}\frac{E}{L}\right)\left(1 + \frac{1}{\omega\tau}\right)$$

$$= \frac{1+\omega\tau}{2\omega\tau}q_0 - \frac{1}{2\omega}\frac{\tau}{1-\omega\tau}\frac{E}{L},$$

and, by subtracting the second equation from the first in Eq. (4.6.115), we still have

$$C_2 = \frac{1}{2}\left(q_0 - \frac{\tau^2}{1-\omega^2\tau^2}\frac{E}{L}\right)\left(1 - \frac{1}{\omega\tau}\right)$$

$$= -\frac{1-\omega\tau}{2\omega\tau}q_0 + \frac{1}{2\omega}\frac{\tau}{1+\omega\tau}\frac{E}{L}.$$

Introducing these results into Eq. (4.6.114), one obtains the final solution to the problem in this case (the overdamped response or the overdamped regime of operation of the circuit):

$$q(t) = \left(\frac{1+\omega\tau}{2\omega\tau}q_0 - \frac{1}{2\omega}\frac{\tau}{1-\omega\tau}\frac{E}{L}\right)e^{-\left(\frac{1}{\tau}-\omega\right)t}$$

$$- \left(\frac{1-\omega\tau}{2\omega\tau}q_0 - \frac{1}{2\omega}\frac{\tau}{1+\omega\tau}\frac{E}{L}\right)e^{-\left(\frac{1}{\tau}+\omega\right)t}$$

$$+ \frac{\tau^2}{1-\omega^2\tau^2}\frac{E}{L} = \frac{q_0 e^{-\frac{t}{\tau}}}{2\omega\tau}\left[(1+\omega\tau)e^{\omega t} - (1-\omega\tau)e^{-\omega t}\right]$$

$$- \frac{\tau}{2\omega}\frac{E}{L}e^{-\frac{t}{\tau}}\left(\frac{e^{\omega t}}{1-\omega\tau} - \frac{e^{-\omega t}}{1+\omega\tau}\right) + \frac{\tau^2}{1-\omega^2\tau^2}\frac{E}{L}$$

$$= \frac{q_0 e^{-\frac{t}{\tau}}}{2\omega\tau}\left[e^{\omega t} - e^{-\omega t} + \omega\tau\left(e^{\omega t} + e^{-\omega t}\right)\right]$$

$$- \frac{1}{2\omega\tau}\frac{\tau^2}{1-\omega^2\tau^2}\frac{E}{L}e^{-\frac{t}{\tau}}\left[e^{\omega t} - e^{-\omega t} + \omega\tau\left(e^{\omega t} + e^{-\omega t}\right)\right]$$

$$+ \frac{\tau^2}{1-\omega^2\tau^2}\frac{E}{L} = \frac{q_0 e^{-\frac{t}{\tau}}}{\omega\tau}\left(\sinh\omega t + \omega\tau\cosh\omega t\right)$$

$$- \frac{1}{\omega\tau}\frac{\tau^2}{1-\omega^2\tau^2}\frac{E}{L}e^{-\frac{t}{\tau}}\left(\sinh\omega t + \omega\tau\cosh\omega t\right) + \frac{\tau^2}{1-\omega^2\tau^2}\frac{E}{L}$$

$$= q_0\frac{1}{\omega\tau}e^{-\frac{t}{\tau}}\left(\sinh\omega t + \omega\tau\cosh\omega t\right)$$

$$+ \frac{\tau^2}{1-\omega^2\tau^2}\frac{E}{L}\left[1 - \frac{e^{-\frac{t}{\tau}}}{\omega\tau}\left(\sinh\omega t + \omega\tau\cosh\omega t\right)\right],$$

or, observing that according to Eq. (4.6.115) the quantity

$$\frac{\tau^2}{1-\omega^2\tau^2}\frac{E}{L}$$

has the dimension of an electric charge (indeed,

$$\frac{\tau^2}{1-\omega^2\tau^2}\frac{E}{L} = \frac{4L^2}{R^2}\frac{1}{1 - \left(\frac{R^2}{4L^2} - \frac{1}{LC}\right)\frac{4L^2}{R^2}}\frac{E}{L}$$

$$= \frac{4L^2}{R^2}\frac{R^2LC}{4L^2}\frac{E}{L} = EC \equiv Q_0),$$

therefore it can be suggestively denoted by

$$\frac{\tau^2}{1 - \omega^2\tau^2}\frac{E}{L} \equiv Q_0,$$

and we finally have

$$q(t) = q_0 e^{-\frac{t}{\tau}}\left(\cosh\omega t + \frac{1}{\omega\tau}\sinh\omega t\right)$$

$$+ Q_0\left[1 - e^{-\frac{t}{\tau}}\left(\cosh\omega t + \frac{1}{\omega\tau}\sinh\omega t\right)\right]$$

$$= Q_0 + (q_0 - Q_0)\,e^{-\frac{t}{\tau}}\left(\cosh\omega t + \frac{1}{\omega\tau}\sinh\omega t\right). \quad (4.6.116)$$

In the particular case $q_0 = 0$ (meaning that the capacitor is not initially charged), the solution to the problem for the overdamped regime of operation is

$$q(t) = Q_0\left[1 - e^{-\frac{t}{\tau}}\left(\cosh\omega t + \frac{1}{\omega\tau}\sinh\omega t\right)\right]. \quad (4.6.117)$$

This last case is graphically represented in Fig. 4.22. The shape of the characteristic curve justifies the name of overdamped (or "aperiodic") given to this regime.

Below there are the command lines written using Mathematica 5.0 software to plot the time variation $q = q(t)$ in the particular case $q_0 = 0$ for the overdamped regime.

$R = 10^4$;

$L = 100$;

$CC = 10 * 10^{-6}$;

$EE = 200$;

$Q_0 = CC * EE$;

$\tau = \dfrac{2 * L}{R}$;

$\omega = \sqrt{\dfrac{R^2}{4 * L^2} - \dfrac{1}{L * CC}}$;

$q[t_-] := Q_0 - Q_0 * e^{-\frac{t}{\tau}}(\text{Cosh}[\omega * t] + \dfrac{1}{\omega * \tau}\text{Sinh}[\omega * t])$;

$\text{Plot}[q[t], (t, 0, 0.57)]$

For $q_0 \neq 0$, the graphic representation of the time dependence $q = q(t)$ is given in Fig. 4.23.

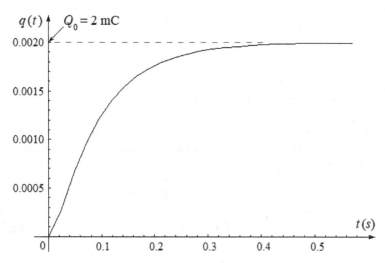

FIGURE 4.22
Graphical representation of time variation of the electric charge $q = q(t)$ on the plates of a capacitor in a DC powered series RLC circuit, for the overdamped regime of operation, in the particular case when the capacitor is not initially charged.

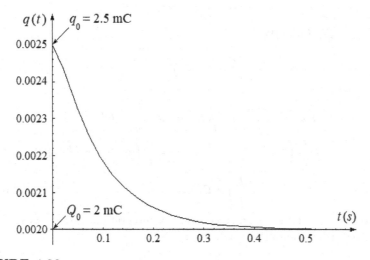

FIGURE 4.23
Graphical representation of time variation of the electric charge $q = q(t)$ on the plates of a capacitor in a DC powered series RLC circuit, for the overdamped regime of operation, in the particular case when the capacitor is initially charged with electric charge q_0.

Case b)

In this case, the system (4.6.100) writes as follows:

$$\begin{cases} C_1'(t)te^{-\frac{t}{\tau}} + C_2'(t)e^{-\frac{t}{\tau}} = 0, \\ \left(1 - \dfrac{t}{\tau}\right) C_1'(t)e^{-\frac{t}{\tau}} - \dfrac{1}{\tau}C_2'(t)e^{-\frac{t}{\tau}} = \dfrac{E}{L}. \end{cases} \tag{4.6.118}$$

Let us now multiply the first equation of the system (4.6.118) by $\frac{1}{\tau}$ and add the resulting equation, member by member, to the second equation of the system. The result is

$$C_1'(t)e^{-\frac{t}{\tau}} = \frac{E}{L},$$

or

$$C_1' = \frac{E}{L}e^{\frac{t}{\tau}}, \tag{4.6.119}$$

so that

$$C_1(t) = \int C_1'(t)dt = \frac{E}{L}\int e^{\frac{t}{\tau}}dt = \frac{\tau E}{L}e^{\frac{t}{\tau}} + \mathcal{C}_1, \tag{4.6.120}$$

where $\mathcal{C}_1$ is a true arbitrary constant of integration, which can be determined by means of the initial conditions.

The first equation of the system (4.6.118) leads to

$$C_2'(t) = -tC_1'(t),$$

or, by using Eq. (4.6.119),

$$C_2'(t) = -\frac{E}{L}t\,e^{\frac{t}{\tau}},$$

so that we have

$$\begin{aligned} C_2(t) &= \int C_2'(t)dt = -\frac{E}{L}\int t\,e^{\frac{t}{\tau}}dt = -\frac{E}{L}\int t\,d\left(\tau e^{\frac{t}{\tau}}\right) \\ &= -\frac{E}{L}\left(t\tau e^{\frac{t}{\tau}} - \tau\int e^{\frac{t}{\tau}}dt\right) = -\frac{E}{L}\left(t\tau e^{\frac{t}{\tau}} - \tau^2 e^{\frac{t}{\tau}}\right) + \mathcal{C}_2 \\ &= \tau^2 e^{\frac{t}{\tau}}\frac{E}{L}\left(1 - \frac{t}{\tau}\right) + \mathcal{C}_2, \end{aligned} \tag{4.6.121}$$

where $\mathcal{C}_2$ is a new true arbitrary constant of integration which, as $\mathcal{C}_1$, can be determined using the initial conditions:

$$\begin{cases} q(t_0 = 0) = q_0, \\ \dot{q}(t_0 = 0)\big(\equiv i(t_0 = 0)\big) = 0, \end{cases} \tag{4.6.122}$$

if at the initial moment $t_0 = 0$ the capacitor is charged with electric charge q_0, or,

$$\begin{cases} q(t_0 = 0) = 0, \\ \dot{q}(t_0 = 0)\big(\equiv i(t_0 = 0)\big) = 0, \end{cases} \tag{4.6.123}$$

if before closing the switch K the capacitor was uncharged.

Therefore, in this second case, in agreement with relation (4.6.101) of the Lagrange's method of variation of parameters, the general solution of the non-homogeneous differential equation (4.6.103) is

$$q(t) = C_1 t\, e^{-\frac{t}{\tau}} + C_2 e^{-\frac{t}{\tau}} + \frac{t\tau E}{L} + \tau^2 \frac{E}{L}\left(1 - \frac{t}{\tau}\right)$$

$$= C_1 t\, e^{-\frac{t}{\tau}} + C_2 e^{-\frac{t}{\tau}} + \tau^2 \frac{E}{L}. \tag{4.6.124}$$

The initial conditions expressed by Eq. (4.6.122) lead in this case to

$$\begin{cases} C_2 + \dfrac{\tau^2 E}{L} = q_0, \\ C_1 - \dfrac{C_2}{\tau} = 0. \end{cases}$$

The first equation of this system easily leads to

$$C_2 = q_0 - \frac{\tau^2 E}{L}, \tag{4.6.125}$$

while the second equation gives

$$C_1 = \frac{C_2}{\tau} = \frac{q_0}{\tau} - \frac{\tau E}{L}. \tag{4.6.126}$$

Introducing C_1 and C_2 determined above into Eq. (4.6.124), one obtains the final solution of the problem for this case (the critically damped regime of operation of the circuit):

$$q(t) = \left(q_0 - \frac{\tau^2 E}{L}\right)\frac{t}{\tau}e^{-\frac{t}{\tau}} + \left(q_0 - \frac{\tau^2 E}{L}\right)e^{-\frac{t}{\tau}} + \frac{\tau^2 E}{L}$$

$$= \left(q_0 - \frac{\tau^2 E}{L}\right)\left(1 + \frac{t}{\tau}\right)e^{-\frac{t}{\tau}} + \frac{\tau^2 E}{L},$$

or, since according to Eq. (4.6.125) the expression $\frac{\tau^2 E}{L}$ has dimensions of an electric charge, so that it can be denoted by Q_0 (indeed,

$$\frac{4L^2}{R^2}\frac{E}{L} \xrightarrow{R=R_{cr}} LC\frac{E}{L} = EC \equiv Q_0),$$

we finally have

$$q(t) = Q_0 + (q_0 - Q_0)\left(1 + \frac{t}{\tau}\right)e^{-\frac{t}{\tau}}. \tag{4.6.127}$$

If, in particular, $q_0 = 0$ (initially, the capacitor is not charged), the solution of the problem for the critically damped regime of operation of the circuit is

$$q(t) = Q_0\left[1 - \left(1 + \frac{t}{\tau}\right)e^{-\frac{t}{\tau}}\right], \tag{4.6.128}$$

whose graphic representation is given in Fig. 4.24.

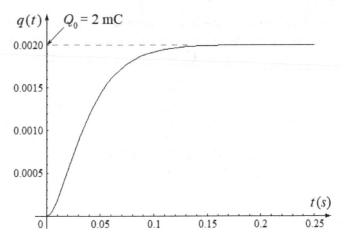

FIGURE 4.24

Graphical representation of time variation of the electric charge $q = q(t)$ on the plates of a capacitor in a DC powered series RLC circuit, for the critically damped regime of operation, in the particular case when the capacitor is not initially charged.

Observation. As one can observe, the solution to the problem in the case of critically damped regime results from that corresponding to the overdamped regime, when the quantity ωt takes such values that we can use the approximations $\cosh \omega t \simeq 1$ and $\sinh \omega t \simeq \omega t$. Indeed, using these approximation relations in Eqs. (4.6.116) and (4.6.117), the relations (4.6.127) and (4.6.128) corresponding to the critically damped regime are straightly obtained.

Below are given the command lines written using Mathematica 5.0 software to plot the time variation $q = q(t)$, in the particular case $q_0 = 0$, for the critically damped regime.

$R = 10^4$;

$L = 100$;

$CC = 10 * 10^{-6}$;

$EE = 200$;

$Q_0 = CC * EE$;

$\tau = \dfrac{2 * L}{R}$;

$\omega = \sqrt{\dfrac{R^2}{4 * L^2} - \dfrac{1}{L * CC}}$;

$q[t_-] := Q_0 - Q_0 * (1 + t/\tau)e^{-\frac{t}{\tau}}$;

$\text{Plot}[q[t], (t, 0, 0.25)]$

For $q_0 \neq 0$ the graphic representation of $q = q(t)$ corresponding to the critically damped regime is given in Fig. 4.25.

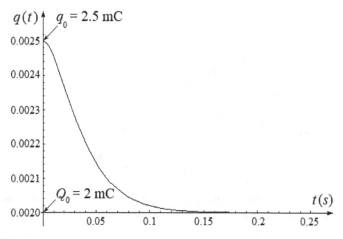

FIGURE 4.25
Graphical representation of time variation of the electric charge $q = q(t)$ on the plates of a capacitor in a DC powered series RLC circuit, for the critically damped regime of operation, in the particular case when the capacitor is initially charged with electric charge q_0.

As can be observed to a direct calitative comparison of the dependencies $q = q(t)$ corresponding to the two regimes of operation of the circuit studied so far, namely the overdamped and critically damped regimes, there are no distinguishable differences between the two cases (the curve in Fig. 4.22 is very similar in shape to the corresponding curve in Fig. 4.24, and the curve in Fig. 4.23 is very similar to the curve in Fig. 4.25, with which it corresponds).

Case c)

In this case, the system (4.6.100) writes as follows:

$$\begin{cases} C_1'(t)e^{-\frac{t}{\tau}}\cos\omega t + C_2'(t)e^{-\frac{t}{\tau}}\sin\omega t = 0, \\ \left(-\omega\sin\omega t - \frac{1}{\tau}\cos\omega t\right)e^{-\frac{t}{\tau}}C_1'(t) \\ + \left(\omega\cos\omega t - \frac{1}{\tau}\sin\omega t\right)e^{-\frac{t}{\tau}}C_2'(t) = \frac{E}{L}, \end{cases} \quad (4.6.129)$$

or, simplifying the first equation by $e^{-\frac{t}{\tau}} \neq 0$, and amplifying the second equation by $\tau e^{\frac{t}{\tau}}$,

$$\begin{cases} C_1'(t)\cos\omega t + C_2'\sin\omega t = 0, \\ -(\omega\tau\sin\omega t + \cos\omega t)C_1'(t) \\ +(\omega\tau\cos\omega t - \sin\omega t)C_2'(t) = \frac{E\tau}{L}e^{\frac{t}{\tau}}. \end{cases} \quad (4.6.130)$$

Multiplying the first equation of the above system by $(\omega\tau\sin\omega t + \cos\omega t)$

and the second equation by $\cos \omega t$, then adding member by member the resulting equations, we obtain

$$\omega\tau\, C_2'(t) = \frac{\tau E}{L} e^{\frac{t}{\tau}} \cos \omega t,$$

that is

$$C_2'(t) = \frac{E}{\omega L} e^{\frac{t}{\tau}} \cos \omega t, \qquad (4.6.131)$$

so that

$$\begin{aligned}
C_2(t) = \int C_2'(t)dt &= \frac{E}{\omega L} \int e^{\frac{t}{\tau}} \cos \omega t\, dt \\
&= \frac{\tau E}{\omega L} \frac{\cos \omega t + \omega\tau \sin \omega t}{1 + \omega^2\tau^2} e^{\frac{t}{\tau}} + \mathcal{C}_2, \qquad (4.6.132)
\end{aligned}$$

where $\mathcal{C}_2$ is a true arbitrary integration constant, which can be determined by means of the initial conditions. Indeed, we have

$$\begin{aligned}
I_2 \equiv \int e^{\frac{t}{\tau}} \cos \omega t\, dt &= \int \cos \omega t\, d\left(\tau e^{\frac{t}{\tau}}\right) = \tau e^{\frac{t}{\tau}} \cos \omega t \\
&\quad - \int \tau e^{\frac{t}{\tau}} d(\cos \omega t) = \tau e^{\frac{t}{\tau}} \cos \omega t + \omega\tau \int e^{\frac{t}{\tau}} \sin \omega t\, dt \\
&= \tau e^{\frac{t}{\tau}} \cos \omega t + \omega\tau \int \sin \omega t\, d\left(\tau e^{\frac{t}{\tau}}\right) = \tau e^{\frac{t}{\tau}} \cos \omega t
\end{aligned}$$

$$\begin{aligned}
&+ \omega\tau \int \sin \omega t\, dt \left(\tau e^{\frac{t}{\tau}}\right) = \tau e^{\frac{t}{\tau}} \cos \omega t \\
&+ \omega\tau \left[\tau e^{\frac{t}{\tau}} \sin \omega t - \int \tau e^{\frac{t}{\tau}} d(\sin \omega t)\right] \\
&= \tau e^{\frac{t}{\tau}} \cos \omega t + \omega\tau \left(\tau e^{\frac{t}{\tau}} \sin \omega t - \omega\tau \int e^{\frac{t}{\tau}} \cos \omega t\, dt\right) \\
&= \tau e^{\frac{t}{\tau}} \cos \omega t + \omega\tau^2 e^{\frac{t}{\tau}} \sin \omega t - \omega^2\tau^2 I_2,
\end{aligned}$$

that is

$$I_2\left(1 + \omega^2\tau^2\right) = \tau e^{\frac{t}{\tau}}\left(\cos \omega t + \omega\tau \sin \omega t\right),$$

which yields

$$I_2 = \int e^{\frac{t}{\tau}} \cos \omega t\, dt = \tau e^{\frac{t}{\tau}} \frac{\cos \omega t + \omega\tau \sin \omega t}{1 + \omega^2\tau^2}.$$

Since the integral is indefinite, we must add an arbitrary integration constant, therefore

$$I_2 = \int e^{\frac{t}{\tau}} \cos \omega t\, dt = \tau e^{\frac{t}{\tau}} \frac{\cos \omega t + \omega\tau \sin \omega t}{1 + \omega^2\tau^2} + \mathcal{C}_2,$$

which is just the result we were looking for, *i.e*, the formula (4.6.132).

We also observe that the first equation of the system (4.6.130) gives

$$C_1'(t) = -C_2'(t)\,\frac{\sin\omega t}{\cos\omega t}\,,$$

or, by means of Eq. (4.6.131),

$$C_1'(t) = -\frac{E}{\omega L}\,e^{\frac{t}{\tau}}\sin\omega t,$$

which leads to

$$C_1(t) = \int C_1'(t)\,dt = -\frac{E}{\omega L}\int e^{\frac{t}{\tau}}\sin\omega t\,dt$$

$$= -\frac{\tau E}{\omega L}\,\frac{\sin\omega t - \omega\tau\cos\omega t}{1+\omega^2\tau^2}\,e^{\frac{t}{\tau}} + C_1,\quad (4.6.133)$$

where C_1 is also a true arbitrary integration constant, which, like C_2, can be determined by means of the initial conditions:

$$\begin{cases} q(t_0 = 0) = q_0, \\ \dot{q}(t_0 = 0)\big(\equiv i(t_0 = 0)\big) = 0, \end{cases} \quad (4.6.134)$$

if at the initial moment $t_0 = 0$ the capacitor was charged with the electric charge q_0, or,

$$\begin{cases} q(t_0 = 0) = 0, \\ \dot{q}(t_0 = 0)\big(\equiv i(t_0 = 0)\big) = 0, \end{cases} \quad (4.6.135)$$

if the capacitor was uncharged before closing the switch K.

Indeed, we have

$$I_1 \equiv \int e^{\frac{t}{\tau}}\sin\omega t\,dt = \int \sin\omega t\,d\left(\tau e^{\frac{t}{\tau}}\right) = \tau e^{\frac{t}{\tau}}\sin\omega t$$

$$-\int \tau e^{\frac{t}{\tau}}\,d(\sin\omega t) = \tau e^{\frac{t}{\tau}}\sin\omega t - \omega\tau\int e^{\frac{t}{\tau}}\cos\omega t\,dt$$

$$= \tau e^{\frac{t}{\tau}}\sin\omega t - \omega\tau\int \cos\omega t\,d\left(\tau e^{\frac{t}{\tau}}\right) = \tau e^{\frac{t}{\tau}}\sin\omega t$$

$$-\omega\tau\left[\tau e^{\frac{t}{\tau}}\cos\omega t - \int \tau e^{\frac{t}{\tau}}d(\cos\omega t)\right]$$

$$= \tau e^{\frac{t}{\tau}}\sin\omega t - \omega\tau\left(\tau e^{\frac{t}{\tau}}\cos\omega t + \omega\tau\int e^{\frac{t}{\tau}}\sin\omega t\,dt\right)$$

$$= \tau e^{\frac{t}{\tau}}\sin\omega t - \omega\tau^2 e^{\frac{t}{\tau}}\cos\omega t - \omega^2\tau^2 I_1,$$

that is

$$I_1\left(1+\omega^2\tau^2\right) = \tau e^{\frac{t}{\tau}}\left(\sin\omega t - \omega\tau\cos\omega t\right),$$

which gives

$$I_1 = \int e^{\frac{t}{\tau}}\sin\omega t\,dt = \tau e^{\frac{t}{\tau}}\,\frac{\sin\omega t - \omega\tau\cos\omega t}{1+\omega^2\tau^2}\,,$$

and, since I_1 is an indefinite integral, we must add an arbitrary integration constant, therefore

$$I_1 = \int e^{\frac{t}{\tau}} \sin \omega t \, dt = \tau e^{\frac{t}{\tau}} \frac{\sin \omega t - \omega \tau \cos \omega t}{1 + \omega^2 \tau^2} + C_1,$$

which is exaclly what we had to prove, *i.e.*, the formula (4.6.133).

Consequently, in this third case, in agreement with relation (4.6.101) related to the Lagrange's method of variation of parameters, the general solution of the non-homogeneous differential equation (4.6.103) is

$$q(t) = C_1 e^{-\frac{t}{\tau}} \cos \omega t + C_2 e^{-\frac{t}{\tau}} \sin \omega t$$
$$+ \Big[\sin \omega t \big(\cos \omega t + \omega \tau \sin \omega t \big) - \cos \omega t \big(\sin \omega t - \omega \tau \cos \omega t \big) \Big]$$
$$\times \frac{\tau E}{\omega L} \frac{1}{1 + \omega^2 \tau^2} = C_1 e^{-\frac{t}{\tau}} \cos \omega t + C_2 e^{-\frac{t}{\tau}} \sin \omega t + \frac{E}{L} \frac{\tau^2}{1 + \omega^2 \tau^2}. \quad (4.6.136)$$

The initial conditions expressed by Eq. (4.6.134) lead in this case to

$$\begin{cases} C_1 + \dfrac{\tau^2}{1 + \omega^2 \tau^2} \dfrac{E}{L} = q_0, \\[3mm] -\dfrac{C_1}{\tau} + \omega C_2 = 0. \end{cases}$$

The first equation of the above system gives

$$C_1 = q_0 - \frac{\tau^2}{1 + \omega^2 \tau^2} \frac{E}{L} = q_0 - Q_0,$$

since dimension of the expression

$$\frac{\tau^2}{1 + \omega^2 \tau^2} \frac{E}{L}$$

is that of an electric charge. This fact allowed us to denote

$$\frac{\tau^2}{1 + \omega^2 \tau^2} \frac{E}{L} \equiv Q_0.$$

Indeed, we have

$$\frac{\tau^2}{1 + \omega^2 \tau^2} \frac{E}{L} = \frac{4L^2}{R^2} \frac{1}{1 + \left(\frac{1}{LC} - \frac{R^2}{4L^2} \right) \frac{4L^2}{R^2}} \frac{E}{L}$$
$$= \frac{4L^2}{R^2} \frac{R^2 LC}{4L^2} \frac{E}{L} = EC \equiv Q_0.$$

Next, using the second equation from the same above system, one obtains

$$C_2 = \frac{C_1}{\omega \tau} = \frac{1}{\omega \tau} \left(q_0 - \frac{\tau^2}{1 + \omega^2 \tau^2} \frac{E}{L} \right) = \frac{1}{\omega \tau} (q_0 - Q_0).$$

Introducing C_1 and C_2 this way determined into Eq. (4.6.136), we obtain the final solution of the problem in this case (underdamped regime of operation of the circuit):

$$q(t) = \left(q_0 - Q_0\right)e^{-\frac{t}{\tau}}\cos\omega t + \frac{1}{\omega\tau}\left(q_0 - Q_0\right)e^{-\frac{t}{\tau}}\sin\omega t$$

$$+ \frac{E}{L}\frac{\tau^2}{1+\omega^2\tau^2} = Q_0 + \left(q_0 - Q_0\right)\left(\cos\omega t + \frac{1}{\omega\tau}\sin\omega t\right)e^{-\frac{t}{\tau}}. \quad (4.6.137)$$

It is easily noticed that the only difference between relation (4.6.137) and relation (4.6.116) − corresponding to the overdamped regime of operation of the circuit − is the fact that instead of hyperbolic (aperiodic) trigonometric functions, this time appear the common trigonometric functions (which are periodic).

In the particular case $q_0 = 0$ (the capacitor is not initially charged), the solution to the problem for the underdamped ("periodic") regime of operation of the circuit is

$$q(t) = Q_0 - Q_0 e^{-\frac{t}{\tau}}\left(\cos\omega t + \frac{1}{\omega\tau}\sin\omega t\right), \quad (4.6.138)$$

whose graphical representation is given in Fig. 4.26. The shape of the curve represented in this figure justifies the name of "periodic" given to this regime.

Below are given the command lines written using Mathematica 5.0 software to plot the time variation $q = q(t)$, in the particular case $q_0 = 0$, for the underdamped ("periodic") damped regime.

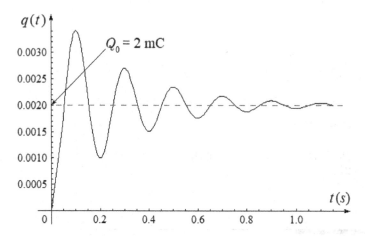

FIGURE 4.26
Graphical representation of time variation of the electric charge $q = q(t)$ on the plates of a capacitor in a DC powered series RLC circuit, for the underdamped/"periodic" regime of operation, in the particular case when the capacitor is not initially charged.

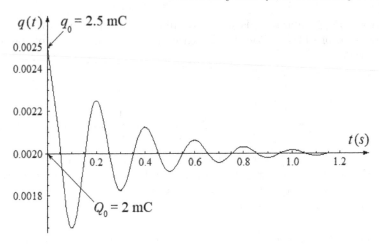

FIGURE 4.27

Graphical representation of time variation of the electric charge $q = q(t)$ on the plates of a capacitor in a DC powered series RLC circuit, for the underdamped/"periodic" regime of operation, in the particular case when the capacitor is initially charged with electric charge q_0.

$R = 700;$

$L = 100;$

$CC = 10 * 10^{-6};$

$EE = 200;$

$Q_0 = CC * EE;$

$\tau = \dfrac{2 * L}{R};$

$\omega = \sqrt{\dfrac{-R^2}{4 * L^2} + \dfrac{1}{L * CC}};$

$q[t\_] := Q_0 - Q_0 * e^{-\frac{t}{\tau}}(\text{Cos}[\omega * t] + \dfrac{1}{\omega * \tau} * \text{Sin}[\omega * t]);$

$\text{Plot}[q[t], (t, 0, 1.15)]$

For $q_0 \neq 0$, the graphical representation of the dependence $q = q(t)$ is given in Fig. 4.27.

4.7   Problem No. 33

Determine the time variation of the electric charge on the plates of a capacitor in a series RLC circuit, powered by an AC source with $e(t) = E \cos(\omega_0 t + \alpha)$.

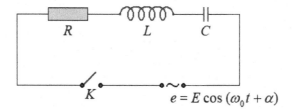

$$e = E \cos(\omega_0 t + \alpha)$$

FIGURE 4.28
A series RLC circuit powered by an AC source with $e(t) = E \cos(\omega_0 t + \alpha)$.

The circuit elements are considered ideal and $R = \text{const.}$, $L = \text{const.}$, $C = \text{const.}$

Solution

This time the series RLC circuit is powered by an AC source (see Fig. 4.28). Given what was discussed at the beginning of the **Problem No. 32** solving, in this case we have to solve the equation

$$\frac{d^2q}{dt^2} + \frac{R}{L}\frac{dq}{dt} + \frac{1}{LC}q = \frac{e}{L} = \frac{E}{L}\cos(\omega_0 t + \alpha). \qquad (4.7.139)$$

To this end, given the type of this differential equation, we will use the same Lagrange's method of variation of parameters.

The solving of the homogeneous differential equation attached to the non-homogeneous differential equation (4.7.139) and finding of the fundamental systems of solutions corresponding to the three cases (the "aperiodic", "critically aperiodic" and, respectively, "periodic" regimes of operation of the circuit) is performed exactly as in the case of **Problem No. 32**. Using the same notations, we have:

Case a)

In this case the system of Eqs. (4.6.100) takes the form

$$\begin{cases} C_1'(t)e^{\left(-\frac{1}{\tau}+\omega\right)t} + C_2'(t)e^{-\left(\frac{1}{\tau}+\omega\right)t} = 0, \\ \left(-\frac{1}{\tau}+\omega\right)C_1'(t)e^{\left(-\frac{1}{\tau}+\omega\right)t} - \left(\frac{1}{\tau}+\omega\right)C_2'(t)e^{-\left(\frac{1}{\tau}+\omega\right)t} \\ = \frac{E}{L}\cos(\omega_0 t + \alpha). \end{cases} \qquad (4.7.140)$$

Multiplying the first equation of the above system with $\left(\frac{1}{\tau}+\omega\right)$ and adding member by member the obtained equation to the second equation of the system, we have

$$2\omega C_1'(t)e^{\left(-\frac{1}{\tau}+\omega\right)t} = \frac{E}{L}\cos(\omega_0 t + \alpha),$$

which gives

$$C_1'(t) = \frac{E}{2\omega L}e^{\left(\frac{1}{\tau}-\omega\right)t}\cos(\omega_0 t + \alpha), \qquad (4.7.141)$$

so that

$$C_1(t) = \int C_1'(t)\,dt = \frac{E}{2\omega L}\int e^{\left(\frac{1}{\tau}-\omega\right)t}\cos(\omega_0 t + \alpha)\,dt$$

$$= \frac{E}{2\omega L}\frac{\tau e^{\left(\frac{1}{\tau}-\omega\right)t}}{\omega_0^2\tau^2 + (1-\omega\tau)^2}$$

$$\times \left[(1-\omega\tau)\cos(\omega_0 t + \alpha) + \omega_0\tau\sin(\omega_0 t + \alpha)\right] + \mathcal{C}_1, \qquad (4.7.142)$$

where $\mathcal{C}_1$ is an arbitrary, true, integration constant. Next, the first equation of the system (4.7.140) gives

$$C_2'(t) = -C_1'(t)e^{2\omega t},$$

or, by means of Eq. (4.7.141),

$$C_2'(t) = -\frac{E}{2\omega L}e^{\left(\frac{1}{\tau}-\omega\right)t}e^{2\omega t}\cos(\omega_0 t + \alpha)$$

$$= -\frac{E}{2\omega L}e^{\left(\frac{1}{\tau}+\omega\right)t}\cos(\omega_0 t + \alpha),$$

which gives

$$C_2(t) = \int C_2'(t)\,dt = -\frac{E}{2\omega L}\int e^{\left(\frac{1}{\tau}+\omega\right)t}\cos(\omega_0 t + \alpha)\,dt$$

$$= -\frac{E}{2\omega L}\frac{\tau e^{\left(\frac{1}{\tau}+\omega\right)t}}{\omega_0^2\tau^2 + (1+\omega\tau)^2}$$

$$\times \left[(1+\omega\tau)\cos(\omega_0 t + \alpha) + \omega_0\tau\sin(\omega_0 t + \alpha)\right] + \mathcal{C}_2, \qquad (4.7.143)$$

where $\mathcal{C}_2$ is also a true, arbitrary integration constant which, as well as $\mathcal{C}_1$, can be determined by means of the following initial conditions:

$$\begin{cases} q(t_0 = 0) = q_0, \\ \dot{q}(t_0 = 0)\big(\equiv i(t_0 = 0)\big) = 0, \end{cases} \qquad (4.7.144)$$

if at the initial moment $t_0 = 0$ the capacitor was charged with the electric charge q_0, or,

$$\begin{cases} q(t_0 = 0) = 0, \\ \dot{q}(t_0 = 0)\big(\equiv i(t_0 = 0)\big) = 0, \end{cases} \qquad (4.7.145)$$

if before closing the switch K the capacitor was uncharged.

Therefore, in this (first) case, in agreement with relation (4.6.101) from the theory related to Lagrange's method of variation of parameters, the general solution of the non-homogeneous equation (4.7.139) is

$$q(t) = C_1 e^{-\left(\frac{1}{\tau} - \omega\right)t} + C_2 e^{-\left(\frac{1}{\tau} + \omega\right)t}$$

$$+ \frac{E}{2\omega L} \left\{ \frac{\tau}{\omega_0^2 \tau^2 + (1 - \omega\tau)^2} \left[(1 - \omega\tau) \cos(\omega_0 t + \alpha) \right. \right.$$

$$\left. + \omega_0 \tau \sin(\omega_0 t + \alpha) \right] - \left[(1 + \omega\tau) \cos(\omega_0 t + \alpha) + \omega_0 \tau \sin(\omega_0 t + \alpha) \right]$$

$$\times \left. \frac{\tau}{\omega_0^2 \tau^2 + (1 + \omega\tau)^2} \right\} = C_1 e^{-\left(\frac{1}{\tau} - \omega\right)t} + C_2 e^{-\left(\frac{1}{\tau} + \omega\right)t}$$

$$+ \tau^2 \frac{E}{L} \frac{\left[1 - \tau^2(\omega^2 + \omega_0^2)\right] \cos(\omega_0 t + \alpha) + 2\omega_0 \tau \sin(\omega_0 t + \alpha)}{\left[\omega_0^2 \tau^2 + (1 - \omega\tau)^2\right]\left[\omega_0^2 \tau^2 + (1 + \omega\tau)^2\right]}. \quad (4.7.146)$$

The initial conditions expressed by Eq. (4.7.144) then lead to

$$\begin{cases} C_1 + C_2 + \dfrac{E\tau^2}{L} \dfrac{\left[1 - \tau^2(\omega^2 + \omega_0^2)\right]\cos\alpha + \omega_0\tau\sin\alpha}{\left[\omega_0^2\tau^2 + (1-\omega\tau)^2\right]\left[\omega_0^2\tau^2 + (1+\omega\tau)^2\right]} = q_0, \\[3mm] C_1 + C_2 + \omega\tau(C_2 - C_1) \\[2mm] + \omega_0\tau \dfrac{E\tau^2}{L} \dfrac{\left[1 - \tau^2(\omega^2 + \omega_0^2)\right]\sin\alpha - 2\omega_0\tau\cos\alpha}{\left[\omega_0^2\tau^2 + (1-\omega\tau)^2\right]\left[\omega_0^2\tau^2 + (1+\omega\tau)^2\right]} = 0, \end{cases}$$

with the solutions

$$\begin{cases} C_1 = q_0 \dfrac{1 + \omega\tau}{2\omega\tau} - \dfrac{E\tau}{2\omega L} \dfrac{\omega_0\tau\sin\alpha + (1 - \omega\tau)\cos\alpha}{\omega^2\tau^2 + (1 - \omega\tau)^2}, \\[3mm] C_2 = -q_0 \dfrac{1 - \omega\tau}{2\omega\tau} + \dfrac{E\tau}{2\omega L} \dfrac{\omega_0\tau\sin\alpha + (1 + \omega\tau)\cos\alpha}{\omega^2\tau^2 + (1 + \omega\tau)^2}, \end{cases}$$

Introducing these results into Eq. (4.7.146), we obtain the final solution to the problem in this case (the "aperiodic regime" of operation of the circuit):

$$q(t) = C_1 e^{-\left(\frac{1}{\tau} - \omega\right)t} + C_2 e^{-\left(\frac{1}{\tau} + \omega\right)t}$$

$$+ \tau^2 \frac{E}{L} \frac{\left[1 - \tau^2(\omega^2 + \omega_0^2)\right]\cos(\omega_0 t + \alpha) + 2\omega_0\tau\sin(\omega_0 t + \alpha)}{\left[\omega_0^2\tau^2 + (1 - \omega)^2\right]\left[\omega_0^2\tau^2 + (1 + \omega)^2\right]}$$

$$= q_0 e^{-\frac{t}{\tau}} \left(\cosh\omega t + \frac{1}{\omega\tau}\sinh\omega t \right) - \frac{E\tau}{2\omega L} e^{-\frac{t}{\tau}}$$

$$\times \left[\frac{\omega_0\tau\sin\alpha + (1 - \omega\tau)\cos\alpha}{\omega_0^2\tau^2 + (1 - \omega\tau)^2} e^{\omega t} \right.$$

$$\left. - \frac{\omega_0\tau\sin\alpha + (1 + \omega\tau)\cos\alpha}{\omega_0^2\tau^2 + (1 + \omega\tau)^2} e^{-\omega t} \right]$$

$$+ \tau^2 \frac{E}{L} \frac{\left[1 - \tau^2(\omega^2 + \omega_0^2)\right]\cos(\omega_0 t + \alpha) + 2\omega_0\tau\sin(\omega_0 t + \alpha)}{\left[\omega_0^2\tau^2 + (1 - \omega\tau)^2\right]\left[\omega_0^2\tau^2 + (1 + \omega\tau)^2\right]}. \quad (4.7.147)$$

In particular, if $q_0 = 0$ (the capacitor is not initially charged), the solution to the problem for the "aperiodic regime" of operation of the circuit is

$$q(t) = \frac{E\tau}{2\omega L} e^{-\frac{t}{\tau}} \left[\frac{\omega_0 \tau \sin\alpha + (1 + \omega\tau)\cos\alpha}{\omega_0^2 \tau^2 + (1 + \omega\tau)^2} e^{-\omega t} \right.$$

$$\left. - \frac{\omega_0 \tau \sin\alpha + (1 - \omega\tau)\cos\alpha}{\omega_0^2 \tau^2 + (1 - \omega\tau)^2} e^{\omega t} \right]$$

$$+ \tau^2 \frac{E}{L} \frac{\left[1 - \tau^2\left(\omega^2 + \omega_0^2\right)\right] \cos(\omega_0 t + \alpha) + 2\omega_0 \tau \sin(\omega_0 t + \alpha)}{\left[\omega_0^2 \tau^2 + (1 - \omega\tau)^2\right]\left[\omega_0^2 \tau^2 + (1 + \omega\tau)^2\right]}, \quad (4.7.148)$$

whose graphic representation is given in Fig. 4.29. We give below the command lines written using Mathematica 5.0 software to plot the time variation of the electric charge $q = q(t)$, in the particular case $q_0 \neq 0$, for the "aperiodic regime" of operation of the circuit.

$R = 300;$

$L = 2 * 10^{-2};$

$CC = 10 * 10^{-6};$

$EE = 10;$

$q_0 = 2.5 * 10^{-4};$

$\tau = \dfrac{2 * L}{R};$

$\omega = \sqrt{\dfrac{R^2}{4*L^2} - \dfrac{1}{L*CC}};$

$\omega_0 = 10 * \pi;$

$\alpha = \frac{\pi}{6};$

$q[t\_] := q_0 * e^{-\frac{t}{\tau}} * (\text{Cosh}[\omega * t] + \frac{1}{\omega * \tau} * \text{Sinh}[\omega * t])$

$- \dfrac{EE * \tau}{2 * \omega * L} * e^{-\frac{t}{\tau}} * (e^{\omega * t} * \dfrac{\omega_0 * \tau * \text{Sin}[\alpha] + (1 - \omega * \tau) * \text{Cos}[\alpha]}{\omega_0^2 * \tau^2 + (1 - \omega * \tau)^2}$

$- e^{-\omega * t} * \dfrac{\omega_0 * \tau * \text{Sin}[\alpha] + (1 + \omega * \tau) * \text{Cos}[\alpha]}{\omega_0^2 * \tau^2 + (1 + \omega * \tau)^2}) + \dfrac{EE * \tau^2}{L}$

$* \dfrac{(1 - \tau^2 * (\omega^2 + \omega_0^2)) * \text{Cos}[\omega_0 * t + \alpha] + 2 * \omega_0 * \tau * \text{Sin}[\omega_0 * t + \alpha]}{(\omega_0^2 * \tau^2 + (1 - \omega * \tau)^2) * (\omega_0^2 * \tau^2 + (1 + \omega * \tau)^2)};$

$\text{Plot}[q[t], \{t, 0, 0.3\}]$

For $q_0 \neq 0$, the graphic representation of the dependency $q = q(t)$ is given in Fig. 4.30.

Case b)

In this case, the system (4.6.100) is written as follows:

$$\begin{cases} C_1'(t)te^{-\frac{t}{\tau}} + C_2'(t)e^{-\frac{t}{\tau}} = 0, \\ \left(1 - \frac{t}{\tau}\right) C_1'(t) - \frac{1}{\tau}C_2'(t)e^{-\frac{t}{\tau}} = \frac{E}{L}\cos(\omega_0 t + \alpha). \end{cases} \quad (4.7.149)$$

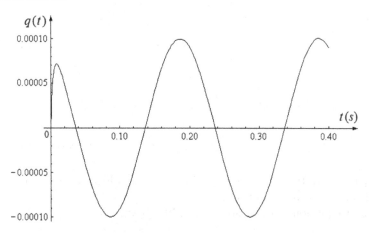

FIGURE 4.29
Graphical representation of time variation of the electric charge $q = q(t)$ on the plates of a capacitor in an AC powered series RLC circuit, for the aperiodic regime of operation, in the particular case when the capacitor is not initially charged.

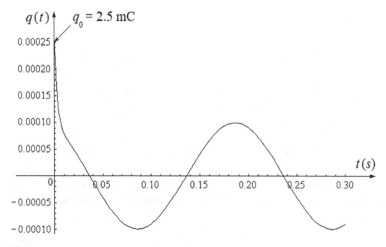

FIGURE 4.30
Graphical representation of time variation of the electric charge $q = q(t)$ on the plates of a capacitor in an AC powered series RLC circuit, for the aperiodic regime of operation, in the particular case when the capacitor is initially charged with electric charge q_0.

Multiplying the first equation of the above system by $\frac{1}{\tau}$ and adding member by member to the second equation of the system, one obtains

$$C_1'(t)e^{-\frac{t}{\tau}} = \frac{E}{L}\cos(\omega_0 t + \alpha),$$

which yields

$$C_1'(t) = \frac{E}{L} e^{\frac{t}{\tau}} \cos(\omega_0 t + \alpha), \tag{4.7.150}$$

so that

$$C_1(t) = \int C_1'(t) \, dt = \frac{E}{L} \int e^{\frac{t}{\tau}} \cos(\omega_0 t + \alpha) \, dt + \mathcal{C}_1$$

$$= \frac{E}{L} \frac{\tau e^{\frac{t}{\tau}} \left[\cos(\omega_0 t + \alpha) + \omega_0 \tau \sin(\omega_0 t + \alpha) \right]}{1 + \omega_0^2 \tau^2} + \mathcal{C}_1, \tag{4.7.151}$$

where $\mathcal{C}_1$ is an arbitrary, true, integration constant which can be determined by means of the initial conditions.

The first equation of the system (4.7.149) gives

$$C_2'(t) = -t C_1'(t),$$

or, by using Eq. (4.7.150),

$$C_2'(t) = -\frac{E}{L} t e^{\frac{t}{\tau}} \cos(\omega_0 t + \alpha),$$

so that we have

$$C_2(t) = \int C_2'(t) \, dt = -\frac{E}{L} \int t e^{\frac{t}{\tau}} \cos(\omega_0 t + \alpha) \, dt + \mathcal{C}_2$$

$$= \mathcal{C}_2 - \frac{E}{L} \frac{\tau e^{\frac{t}{\tau}}}{\left(1 + \omega_0^2 \tau^2 \right)^2} \Big[(t - \tau) \cos(\omega_0 t + \alpha)$$

$$+ (t - 2\tau) \omega_0 \tau \sin(\omega_0 t + \alpha)$$

$$+ (t + \tau) \omega_0^2 \tau^2 \cos(\omega_0 t + \alpha) + t \omega_0^3 \tau^3 \sin(\omega_0 t + \alpha) \Big], \tag{4.7.152}$$

where $\mathcal{C}_2$ is a new, arbitrary, true integration constant which, as well as $\mathcal{C}_1$, can be determined by means of the following initial conditions:

$$\begin{cases} q(t_0 = 0) = q_0, \\ \dot{q}(t_0 = 0) \big(\equiv i(t_0 = 0) \big) = 0, \end{cases} \tag{4.7.153}$$

if at the initial moment $t_0 = 0$ the capacitor is charged with electric charge q_0, or,

$$\begin{cases} q(t_0 = 0) = 0, \\ \dot{q}(t_0 = 0) \big(\equiv i(t_0 = 0) \big) = 0, \end{cases} \tag{4.7.154}$$

if before closing the switch K the capacitor was uncharged.

Therefore, in this (second) case, in agreement with relation (4.6.101) met in the theory related to Lagrange's method of variation of parameters, the general solution of the non-homogeneous equation (4.7.139) is

$$q(t) = C_1 t e^{-\frac{t}{\tau}} + C_2 t e^{-\frac{t}{\tau}} + \frac{E}{L} \frac{t\tau \left[\cos(\omega_0 t + \alpha) + \omega_0 \tau \sin(\omega_0 t + \alpha) \right]}{1 + \omega_0^2 \tau^2}$$

$$- \frac{E}{L} \frac{\tau}{(1 + \omega_0^2 \tau^2)^2} \left\{ \left[(t - \tau) + (t + \tau)\omega_0^2 \tau^2 \right] \cos(\omega_0 t + \alpha) \right.$$

$$\left. + \left[(t - 2\tau)\omega_0 \tau + t\omega_0^3 \tau^3 \right] \sin(\omega_0 t + \alpha) \right\}. \tag{4.7.155}$$

The initial conditions lead in this case to

$$\begin{cases} C_2 + \dfrac{E}{L} \dfrac{\tau^2}{\left(1 + \omega_0^2 \tau^2\right)^2} \left[\left(1 - \omega_0^2 \tau^2\right) \cos\alpha + 2\omega_0 \tau \sin\alpha \right] = q_0, \\[4mm] C_1 - \dfrac{C_2}{\tau} + \dfrac{E\tau(\cos\alpha + \omega_0 \tau \sin\alpha)}{L\left(1 + \omega_0^2 \tau^2\right)} \\[4mm] \quad - \dfrac{E\tau \left[\left(1 - \omega_0^2 \tau^2\right) \cos\alpha + 2\omega_0 \tau \sin\alpha \right]}{L\left(1 + \omega_0^2 \tau^2\right)} = 0. \end{cases}$$

The first equation of this system easily gives

$$C_2 = q_0 - \frac{E}{L} \frac{\tau^2}{\left(1 + \omega_0^2 \tau^2\right)^2} \left[\left(1 - \omega_0^2 \tau^2\right) \cos\alpha + 2\omega_0 \tau \sin\alpha \right], \tag{4.7.156}$$

and the second equation gives

$$C_1 = \frac{q_0}{\tau} - \frac{E}{L} \frac{\tau}{1 + \omega_0^2 \tau^2} (\cos\alpha + \omega_0 \tau \sin\alpha). \tag{4.7.157}$$

Using C_1 and C_2 determined this way, into Eq. (4.7.155), we obtain the final solution of the problem in this case ("critically aperiodic regime" of operation of the circuit):

$$q(t) = q_0 \left(1 + \frac{t}{\tau}\right) e^{-\frac{t}{\tau}} + \frac{E}{L} \frac{\tau t}{1 + \omega_0^2 \tau^2}$$

$$\times \left[\cos(\omega_0 t + \alpha) + \omega_0 \tau \sin(\omega_0 t + \alpha) - (\cos\alpha + \omega_0 \tau \sin\alpha) \right.$$

$$\left. \times e^{-\frac{t}{\tau}} \right] - \frac{E}{L} \frac{\tau}{\left(1 + \omega_0^2 \tau^2\right)^2} \left\{ (t - \tau) \cos(\omega_0 t + \alpha) + (t - 2\tau)\omega_0 \tau \right.$$

$$\times \sin(\omega_0 t + \alpha) + (t + \tau)\omega_0^2 \tau^2 \cos(\omega_0 t + \alpha) + t\tau^3 \omega_0^3$$

$$\left. \times \sin(\omega_0 t + \alpha) + \tau \left[\left(1 - \omega_0^2 \tau^2\right) \cos\alpha + 2\omega_0 \tau \sin\alpha \right] e^{-\frac{t}{\tau}} \right\}. \tag{4.7.158}$$

In particular, if $q_0 = 0$ (the capacitor is not initially charged), the solution to the problem for the "critically aperiodic regime" of operation of the circuit is

$$q(t) = \frac{E}{L} \frac{\tau t}{1 + \omega_0^2 \tau^2} \Big[\cos(\omega_0 t + \alpha) + \omega_0 \tau \sin(\omega_0 t + \alpha)$$

$$- (\cos \alpha + \omega_0 \tau \sin \alpha) e^{-\frac{t}{\tau}} \Big] - \frac{E}{L} \frac{\tau}{\left(1 + \omega_0^2 \tau^2\right)^2}$$

$$\times \Big\{ (t - \tau) \cos(\omega_0 t + \alpha) + (t - 2\tau) \omega_0 \tau \sin(\omega_0 t + \alpha)$$

$$+ (t + \tau) \omega_0^2 \tau^2 \cos(\omega_0 t + \alpha) + t \tau^3 \omega_0^3 \sin(\omega_0 t + \alpha)$$

$$+ \tau \Big[\left(1 - \omega_0^2 \tau^2\right) \cos \alpha + 2\omega_0 \tau \sin \alpha \Big] e^{-\frac{t}{\tau}} \Big\}, \qquad (4.7.159)$$

whose graphical representation is given in Fig. 4.31.

We give below the command lines written using Mathematica 5.0 software to plot the time variation of the electric charge $q = q(t)$, in the particular case $q_0 \neq 0$, for the "critically aperiodic regime" of operation of the circuit.

$R = 40 * \text{Sqrt}[5];$

$L = 2 * 10^{-2};$

$CC = 10 * 10^{-6};$

$EE = 10;$

$q_0 = 2.5 * 10^{-4};$

$\tau = \dfrac{2 * L}{R};$

$\omega = \sqrt{\dfrac{R^2}{4 * L^2} - \dfrac{1}{L * CC}};$

$\omega_0 = 10 * \pi$

$\alpha = \frac{\pi}{6};$

$q[t\_] := q_0 * (1 + \dfrac{t}{\tau}) * e^{-\frac{t}{\tau}} + \dfrac{EE * \tau * t}{L * (1 + \omega_0^2 * \tau^2)} *$

$(\text{Cos}[\omega_0 * \tau + \alpha] + \omega_0 * \tau * \text{Sin}[\omega_0 * t + \alpha] - (\text{Cos}[\alpha] + \omega_0 * \tau * \text{Sin}[\alpha] * e^{-\frac{t}{\tau}})$

$-\dfrac{EE}{L} * \dfrac{\tau}{(1 + \omega_0^2 * \tau^2)^2} * ((t - \tau) * \text{Cos}[\omega_0 * t + \alpha] + (t - 2 * \tau) * \omega_0 * \tau$

$*\text{Sin}[\omega_0 * t + \alpha] + (t + \tau) * \omega_0^2 * \tau^2 * \text{Cos}[\omega_0 * t + \alpha] + t * \omega_0^3 * \tau^3 * \text{Sin}[\omega_0 * t + \alpha]$

$+((1 - \omega_0^2 * \tau^2) * \text{Cos}[\alpha] + 2 * \omega_0 * \tau * \text{Sin}[\alpha]) * \tau * e^{-\frac{t}{\tau}});$

$\text{Plot}[q[t], (t, 0, 0.2)]$

For $q_0 \neq 0$, the graphical representation of the dependence $q = q(t)$ corresponding to the "critically aperiodic regime" is given in Fig. 4.32.

Case c)

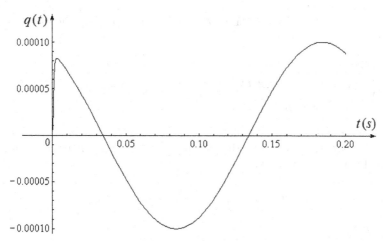

FIGURE 4.31
Graphical representation of time variation of the electric charge $q = q(t)$ on the plates of a capacitor in an AC powered series RLC circuit, for the critically aperiodic regime of operation, in the particular case when the capacitor is not initially charged.

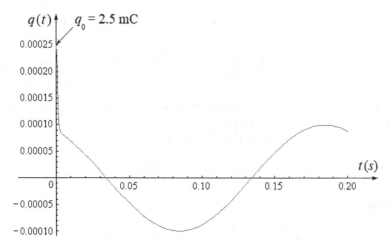

FIGURE 4.32
Graphical representation of time variation of the electric charge $q = q(t)$ on the plates of a capacitor in an AC powered series RLC circuit, for the critically aperiodic regime of operation, in the particular case when the capacitor is initially charged with electric charge q_0.

In this case, the system (4.6.100) will be written as

$$
\begin{cases}
C_1'(t)e^{-\frac{t}{\tau}}\cos\omega t + C_2'(t)e^{-\frac{t}{\tau}}\sin\omega t = 0, \\
\left(-\omega\sin\omega t - \frac{1}{\tau}\cos\omega t\right)e^{-\frac{t}{\tau}}C_1'(t) \\
+\left(\omega\cos\omega t - \frac{1}{\tau}\sin\omega t\right)e^{-\frac{t}{\tau}}C_2'(t) = \frac{E}{L}\cos(\omega_0 t + \alpha),
\end{cases}
\qquad (4.7.160)
$$

or, simplifying the first equation by $e^{-\frac{t}{\tau}} \neq 0$ and amplifying the second equation by $\tau e^{\frac{t}{\tau}}$,

$$
\begin{cases}
C_1'(t) \cos \omega t + C_2'(t) \sin \omega t = 0, \\
-(\omega\tau \sin \omega t + \cos \omega t) C_1'(t) + (\omega\tau \cos \omega t - \sin \omega t) C_2'(t) \\
= \frac{E\tau}{L} e^{\frac{t}{\tau}} \cos(\omega_0 t + \alpha).
\end{cases} \tag{4.7.161}
$$

Multiplying now the first equation of this system by $(\omega\tau \sin \omega t + \cos \omega t)$ and the second equation by $\cos \omega t$, then adding member by member the equations obtained this way, one finds

$$
\omega\tau C_2'(t) = \frac{\tau E}{L} e^{\frac{t}{\tau}} \cos \omega t \cos(\omega_0 t + \alpha),
$$

which gives

$$
C_2'(t) = \frac{E}{\omega L} e^{\frac{t}{\tau}} \cos \omega t \cos(\omega_0 t + \alpha), \tag{4.7.162}
$$

so that

$$
C_2(t) = \int C_2'(t)\, dt = \frac{E}{\omega L} \int e^{\frac{t}{\tau}} \cos \omega t \cos(\omega_0 t + \alpha)\, dt = \frac{E}{2\omega L}
$$
$$
\times \tau e^{\frac{t}{\tau}} \left\{ \frac{\cos\left[\alpha - (\omega - \omega_0)t\right] - \tau(\omega - \omega_0)\sin\left[\alpha - (\omega - \omega_0)t\right]}{1 + \tau^2(\omega - \omega_0)^2} \right.
$$
$$
\left. + \frac{\cos\left[\alpha + (\omega + \omega_0)t\right] + \tau(\omega + \omega_0)\sin\left[\alpha + (\omega + \omega_0)t\right]}{1 + \tau^2(\omega + \omega_0)^2} \right\} + C_2, \tag{4.7.163}
$$

where C_2 is an arbitrary, true, integration constant, which can be determined by means of the initial conditions.

Then, the first equation of the system (4.7.160) leads to

$$
C_1'(t) = -C_2'(t) \frac{\sin \omega t}{\cos \omega t},
$$

or, using Eq. (4.7.162),

$$
C_1'(t) = -\frac{E}{\omega L} e^{\frac{t}{\tau}} \sin \omega t \cos(\omega_0 t + \alpha),
$$

so that we get

$$
C_1(t) = \int C_1'(t)\, dt = -\frac{E}{\omega L} \int e^{\frac{t}{\tau}} \sin \omega t \cos(\omega_0 t + \alpha)\, dt = \frac{E}{2\omega L}
$$
$$
\times \tau e^{\frac{t}{\tau}} \left\{ \frac{\sin\left[\alpha - (\omega - \omega_0)t\right] + \tau(\omega - \omega_0)\cos\left[\alpha - (\omega - \omega_0)t\right]}{1 + \tau^2(\omega - \omega_0)^2} \right.
$$
$$
\left. - \frac{\sin\left[\alpha + (\omega - \omega_0)t\right] + \tau(\omega + \omega_0)\cos\left[\alpha + (\omega + \omega_0)t\right]}{1 + \tau^2(\omega + \omega_0)^2} \right\} + C_1, \tag{4.7.164}
$$

where C_1 is also an arbitrary, true, integration constant which, as well as C_2, can be determined by means of the following initial conditions:

$$\begin{cases} q(t_0 = 0) = q_0, \\ \dot{q}(t_0 = 0)\big(\equiv i(t_0 = 0)\big) = 0, \end{cases} \tag{4.7.165}$$

if at the initial moment $t_0 = 0$ the capacitor was charged with the electric charge q_0, or,

$$\begin{cases} q(t_0 = 0) = 0, \\ \dot{q}(t_0 = 0)\big(\equiv i(t_0 = 0)\big) = 0, \end{cases} \tag{4.7.166}$$

if before closing the switch K the capacitor was uncharged.

Therefore, in this (third) case, in agreement with relation (4.6.101) from the theory of Lagrange's method of variation of parameters, the general solution of the non-homogeneous equation (4.7.139) is

$$q(t) = \frac{\tau E \cos \omega t}{2\omega L}$$

$$\times \left\{ \frac{\sin\big[\alpha - (\omega - \omega_0)t\big] + \tau(\omega - \omega_0)\cos\big[\alpha - (\omega - \omega_0)t\big]}{1 + \tau^2(\omega - \omega_0)^2} \right.$$

$$\left. - \frac{\sin\big[\alpha + (\omega + \omega_0)t\big] - \tau(\omega + \omega_0)\cos\big[\alpha + (\omega + \omega_0)t\big]}{1 + \tau^2(\omega + \omega_0)^2} \right\}$$

$$+ C_1 e^{-\frac{t}{\tau}} \cos \omega t + \frac{\tau E \sin \omega t}{2\omega L}$$

$$\times \left\{ \frac{\cos\big[\alpha - (\omega - \omega_0)t\big] - \tau(\omega - \omega_0)\sin\big[\alpha - (\omega - \omega_0)t\big]}{1 + \tau^2(\omega - \omega_0)^2} \right.$$

$$\left. + \frac{\cos\big[\alpha + (\omega + \omega_0)t\big] + \tau(\omega + \omega_0)\sin\big[\alpha + (\omega + \omega_0)t\big]}{1 + \tau^2(\omega + \omega_0)^2} \right\}$$

$$+ C_2 e^{-\frac{t}{\tau}} \sin \omega t. \tag{4.7.167}$$

The initial conditions expressed by Eq. (4.7.165) lead in this case to

$$\begin{cases} C_1 + \dfrac{\tau E}{2\omega L}\left[\dfrac{\sin\alpha + \tau(\omega - \omega_0)\cos\alpha}{1 + \tau^2(\omega - \omega_0)^2} - \dfrac{\sin\alpha - \tau(\omega + \omega_0)\cos\alpha}{1 + \tau^2(\omega + \omega_0)^2}\right] = q_0, \\[4mm] -\dfrac{C_1}{\tau} + \omega C_2 + \dfrac{\tau E}{2\omega L}\left\{\dfrac{\tau(\omega - \omega_0)^2 \sin\alpha}{1 + \tau^2(\omega - \omega_0)^2}\right. \\[4mm] \quad - \dfrac{(\omega + \omega_0)\cos\alpha + \tau(\omega + \omega_0)^2 \sin\alpha}{1 + \tau^2(\omega + \omega_0)^2} \\[4mm] \left. + \omega\left[\dfrac{\cos\alpha - \tau(\omega - \omega_0)\sin\alpha}{1 + \tau^2(\omega - \omega_0)^2} + \dfrac{\cos\alpha + \tau(\omega + \omega_0)\sin\alpha}{1 + \tau^2(\omega + \omega_0)^2}\right]\right\} = 0, \end{cases}$$

with the solution

$$\begin{cases} C_1 = q_0 - \dfrac{\tau E}{2\omega L}\left[\dfrac{\sin\alpha + \tau(\omega - \omega_0)\cos\alpha}{1 + \tau^2(\omega - \omega_0)^2} - \dfrac{\sin\alpha - \tau(\omega + \omega_0)\cos\alpha}{1 + \tau^2(\omega + \omega_0)^2}\right], \\[2mm] C_2 = \dfrac{q_0}{\omega\tau} - \dfrac{E}{2\omega^2 L}\left[\dfrac{\sin\alpha + \tau(\omega - \omega_0)\cos\alpha}{1 + \tau^2(\omega - \omega_0)^2}\right. \\[2mm] \qquad \left. - \dfrac{\sin\alpha - \tau(\omega + \omega_0)\cos\alpha}{1 + \tau^2(\omega + \omega_0)^2}\right] - \dfrac{E}{2\omega^2 L}\left\{\dfrac{\tau^2(\omega - \omega_0)^2\sin\alpha}{1 + \tau^2(\omega - \omega_0)^2}\right. \\[2mm] \qquad - \dfrac{\tau(\omega + \omega_0)\cos\alpha + \tau^2(\omega + \omega_0)^2\sin\alpha}{1 + \tau^2(\omega + \omega_0)^2} \\[2mm] \qquad \left. + \omega\tau\left[\dfrac{\cos\alpha - \tau(\omega - \omega_0)\sin\alpha}{1 + \tau^2(\omega - \omega_0)^2} + \dfrac{\cos\alpha + \tau(\omega + \omega_0)\sin\alpha}{1 + \tau^2(\omega + \omega_0)^2}\right]\right\}. \end{cases}$$

If C_1 and C_2 just determined are introduced into Eq. (4.7.167), one follows the final solution to the problem in this case (the "periodic regime" of operation of the circuit):

$$\begin{aligned} q(t) &= q_0 e^{-\frac{t}{\tau}}\left(\cos\omega t + \frac{\sin\omega t}{\omega\tau}\right) - \frac{\tau E e^{-\frac{t}{\tau}}\cos\omega t}{2\omega L} \\[2mm] &\times \left[\frac{\sin\alpha + \tau(\omega - \omega_0)\cos\alpha}{1 + \tau^2(\omega - \omega_0)^2} - \frac{\sin\alpha - \tau(\omega + \omega_0)\cos\alpha}{1 + \tau^2(\omega + \omega_0)^2}\right] \\[2mm] &- e^{-\frac{t}{\tau}}\sin\omega t \frac{E}{2\omega^2 L}\left\{\frac{\sin\alpha + \tau(\omega - \omega_0)\cos\alpha}{1 + \tau^2(\omega - \omega_0)^2}\right. \\[2mm] &- \frac{\sin\alpha - \tau(\omega + \omega_0)\cos\alpha}{1 + \tau^2(\omega + \omega_0)^2} + \frac{\tau^2(\omega - \omega_0)^2\sin\alpha}{1 + \tau^2(\omega - \omega_0)^2} \\[2mm] &- \frac{\tau(\omega + \omega_0)\cos\alpha + \tau^2(\omega + \omega_0)^2\sin\alpha}{1 + \tau^2(\omega + \omega_0)^2} + \omega\tau \\[2mm] &\times \left.\left[\frac{\cos\alpha - \tau(\omega - \omega_0)\sin\alpha}{1 + \tau^2(\omega - \omega_0)^2} + \frac{\cos\alpha + \tau(\omega + \omega_0)\sin\alpha}{1 + \tau^2(\omega + \omega_0)^2}\right]\right\} + \cos\omega t \\[2mm] &\times \frac{\tau E}{2\omega L}\left\{\frac{\sin\left[\alpha - (\omega - \omega_0)t\right] + \tau(\omega - \omega_0)\cos\left[\alpha - (\omega - \omega_0)t\right]}{1 + \tau^2(\omega - \omega_0)^2}\right. \\[2mm] &\left. - \frac{\sin\left[\alpha + (\omega + \omega_0)t\right] - \tau(\omega + \omega_0)\cos\left[\alpha + (\omega + \omega_0)t\right]}{1 + \tau^2(\omega + \omega_0)^2}\right\} + \sin\omega t \\[2mm] &\times \frac{\tau E}{2\omega L}\left\{\frac{\cos\left[\alpha - (\omega - \omega_0)t\right] - \tau(\omega - \omega_0)\sin\left[\alpha - (\omega - \omega_0)t\right]}{1 + \tau^2(\omega - \omega_0)^2}\right. \\[2mm] &\left. + \frac{\cos\left[\alpha + (\omega + \omega_0)t\right] + \tau(\omega + \omega_0)\sin\left[\alpha + (\omega + \omega_0)t\right]}{1 + \tau^2(\omega + \omega_0)^2}\right\}. \end{aligned} \qquad (4.7.168)$$

If, in particular, $q_0 = 0$ (the capacitor is not initially charged), the solution to the problem for the "periodic regime" of operation of the circuit is given by

$$q(t) = \frac{\tau E e^{-\frac{t}{\tau}} \cos \omega t}{2\omega L} \left[\frac{\sin \alpha - \tau(\omega + \omega_0) \cos \alpha}{1 + \tau^2(\omega + \omega_0)^2} \right.$$

$$\left. - \frac{\sin \alpha + \tau(\omega - \omega_0) \cos \alpha}{1 + \tau^2(\omega - \omega_0)^2} \right] - e^{-\frac{t}{\tau}} \sin(\omega t) \frac{E}{2\omega^2 L}$$

$$\times \left\{ \frac{\sin \alpha + \tau(\omega - \omega_0) \cos \alpha}{1 + \tau^2(\omega - \omega_0)^2} - \frac{\sin \alpha - \tau(\omega + \omega_0) \cos \alpha}{1 + \tau^2(\omega + \omega_0)^2} \right.$$

$$+ \frac{\tau^2(\omega - \omega_0)^2 \sin \alpha}{1 + \tau^2(\omega - \omega_0)^2} - \frac{\tau(\omega + \omega_0) \cos \alpha + \tau^2(\omega + \omega_0)^2 \sin \alpha}{1 + \tau^2(\omega + \omega_0)^2} + \omega\tau$$

$$\left. \times \left[\frac{\cos \alpha - \tau(\omega - \omega_0) \sin \alpha}{1 + \tau^2(\omega - \omega_0)^2} + \frac{\cos \alpha + \tau(\omega + \omega_0) \sin \alpha}{1 + \tau^2(\omega + \omega_0)^2} \right] \right\} + \cos \omega t$$

$$\times \frac{\tau E}{2\omega L} \left\{ \frac{\sin \left[\alpha - (\omega - \omega_0)t \right] + \tau(\omega - \omega_0) \cos \left[\alpha - (\omega - \omega_0)t \right]}{1 + \tau^2(\omega - \omega_0)^2} \right.$$

$$\left. - \frac{\sin \left[\alpha + (\omega + \omega_0)t \right] - \tau(\omega + \omega_0) \cos \left[\alpha + (\omega + \omega_0)t \right]}{1 + \tau^2(\omega + \omega_0)^2} \right\} + \sin \omega t$$

$$\times \frac{\tau E}{2\omega L} \left\{ \frac{\cos \left[\alpha - (\omega - \omega_0)t \right] - \tau(\omega - \omega_0) \sin \left[\alpha - (\omega - \omega_0)t \right]}{1 + \tau^2(\omega - \omega_0)^2} \right.$$

$$\left. + \frac{\cos \left[\alpha + (\omega + \omega_0)t \right] + \tau(\omega + \omega_0) \sin \left[\alpha + (\omega + \omega_0)t \right]}{1 + \tau^2(\omega + \omega_0)^2} \right\}, \quad (4.7.169)$$

and its graphical representation looks like in Fig. 4.33.

Below are given the command lines written using Mathematica 5.0 software to plot the time variation of the electric charge $q = q(t)$, in the particular case $q_0 \neq 0$, for the "periodic regime" of operation of the circuit.

$R = 2;$

$L = 2 * 10^{-6};$

$CC = 10 * 10^{-6};$

$EE = 10;$

$q_0 = 2.5 * 10^{-4};$

$\tau = \dfrac{2 * L}{R};$

$\omega = \sqrt{\text{Abs}[\dfrac{R^2}{4 * L^2} - \dfrac{1}{L * CC}]};$

$\omega_0 = 10 * \pi;$

$\alpha = \dfrac{\pi}{6};$

$$q[t_-] := q_0 * e^{-\frac{t}{\tau}} * (\text{Cos}[\omega * t] + \frac{1}{\omega * \tau} * \text{Sin}[\omega * t])$$

$$-e^{-\frac{t}{\tau}} * \text{Cos}[\omega * t] * \frac{\tau * EE}{2 * \omega * L} * (\frac{\text{Sin}[\alpha] + \tau * (\omega - \omega_0) * \text{Cos}[\alpha]}{1 + \tau^2 * (\omega - \omega_0)^2}$$

$$-\frac{\text{Sin}[\alpha] - \tau * (\omega + \omega_0) * \text{Cos}[\alpha]}{1 + \tau^2 * (\omega + \omega_0)^2}) - e^{-\frac{t}{\tau}} * \text{Sin}[\omega * t] * \frac{EE}{2 * \omega^2 * L}$$

$$*(\frac{\text{Sin}[\alpha] + \tau * (\omega - \omega_0) * \text{Cos}[\alpha]}{1 + \tau^2 * (\omega - \omega_0)^2} - \frac{\text{Sin}[\alpha] - \tau * (\omega + \omega_0) * \text{Cos}[\alpha]}{1 + \tau^2 * (\omega + \omega_0)^2})$$

$$-e^{-\frac{t}{\tau}} * \text{Sin}[\omega * t] * \frac{EE}{2 * \omega^2 * L} * (\frac{\tau^2 * (\omega - \omega_0)^2 * \text{Sin}[\alpha]}{1 + \tau^2 * (\omega - \omega_0)^2}$$

$$-\frac{\tau * (\omega + \omega_0) * \text{Cos}[\alpha] + \tau^2 * (\omega + \omega_0)^2 * \text{Sin}[\alpha]}{1 + \tau^2 * (\omega + \omega_0)^2}$$

$$+\omega * \tau * (\frac{\text{Cos}[\alpha] - \tau * (\omega - \omega_0) * \text{Sin}[\alpha]}{1 + \tau^2 * (\omega - \omega_0)^2} + \frac{\text{Cos}[\alpha]}{1 + \tau^2 * (\omega + \omega_0)^2}$$

$$+\frac{\tau * (\omega + \omega_0) * \text{Sin}[\alpha]}{1 + \tau^2 * (\omega + \omega_0)^2})) + \frac{\tau * EE}{2 * \omega * L} * (\text{Cos}[\omega * t]$$

$$*(\frac{\text{Sin}[\alpha - (\omega - \omega_0) * t] + \tau * (\omega - \omega_0) * \text{Cos}[\alpha - (\omega - \omega_0) * t]}{1 + \tau^2 * (\omega - \omega_0)^2}$$

$$-\frac{\text{Sin}[\alpha + (\omega + \omega_0) * t] + \tau * (\omega + \omega_0) * \text{Cos}[\alpha + (\omega + \omega_0) * t]}{1 + \tau^2 * (\omega + \omega_0)^2})$$

$$+\text{Sin}[\omega * t] * (\text{Cos}[\alpha - (\omega - \omega_0) * t] * \frac{1}{1 + \tau^2 * (\omega - \omega_0)^2}$$

$$+\frac{\tau * (\omega - \omega_0)}{1 + \tau^2 * (\omega - \omega_0)^2} * \text{Sin}[\alpha - (\omega - \omega_0) * t]$$

$$+\frac{\text{Cos}[\alpha + (\omega + \omega_0) * t] + \tau * (\omega + \omega_0) * \text{Sin}[\alpha + (\omega + \omega_0) * t]}{1 + \tau^2 * (\omega + \omega_0)^2}));$$

$\text{Plot}[q[t], (t, 0, 0.25)]$

For $q_0 \neq 0$, the dependence $q = q(t)$ is graphically represented in Fig. 4.34.

4.8   Problem No. 34

Using the phasor method, find the electric current through a series *RLC* circuit, operating in a permanent sinusoidal regime.

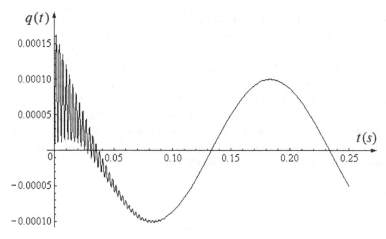

FIGURE 4.33
Graphical representation of time variation of the electric charge $q = q(t)$ on the plates of a capacitor in an AC powered series RLC circuit, for the periodic regime of operation, in the particular case when the capacitor is not initially charged.

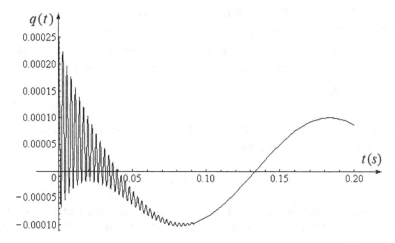

FIGURE 4.34
Graphical representation of time variation of the electric charge $q = q(t)$ on the plates of a capacitor in an AC powered series RLC circuit, for the periodic regime of operation, in the particular case when the capacitor is initially charged with electric charge q_0.

Complements

A physical quantity $y(t)$ of the form

$$y(t) = A \cos\left(\omega t + \varphi\right), \tag{4.8.170}$$

is called a *sinusoidal quantity*. Terminology:

→ A is called the *amplitude* of the sinusoidal quantity;

→ $\phi(t) = \omega t + \varphi$ is called the *phase* of the sinusoidal quantity;

→ ω is called the *pulsation* of the sinusoidal quantity;

→ φ is called the *initial phase* of the sinusoidal quantity;

→ $T = 2\pi/\omega$ is called the *period* of the sinusoidal quantity;

→ $\nu = 1/T$ is called the *frequency* of the sinusoidal quantity;

For the convenience of calculation, to a sinusoidal physical quantity $y(t)$, one associates a complex function (denoted by $\underline{y}$) of real variable t, called *complex image* (or *complex phasor* or simply *phasor*) of the considered function, through

$$\underline{y} = A\, e^{i(\omega t + \varphi)}. \tag{4.8.171}$$

We also have,

$$y(t) = \operatorname{Re}\underline{y}.$$

The product of a sinusoidal quantity with a real number is also a sinusoidal quantity (with the same pulsation) and also the sum of two sinusoidal quantities having the same pulsation ω is also a sinusoidal quantity with ω pulsation. It is verified that the set of ω-pulsation sinusoidal quantities forms a *vector space*. On the other hand, the set of complex functions of the form of those in Eq. (4.8.171) with fixed ω also forms a vector space, and the correspondence between the two sets of elements is an *isomorphism* of linear spaces.

Operations of derivation and integration of sinusoidal quantities

The derivative of the sinusoidal quantity defined by relation (4.8.170), *i.e.*,

$$\frac{dy(t)}{dt} = -\omega A \sin\left(\omega t + \varphi\right) = \omega A \cos\left(\omega t + \varphi + \frac{\pi}{2}\right),$$

is also a sinusoidal quantity, but of amplitude ωA and initial phase $\varphi + \pi/2$. Its complex image is the phasor

$$\omega A\, e^{i\left(\omega t + \varphi + \frac{\pi}{2}\right)} = i\,\omega A\, e^{i(\omega t + \varphi)} = \frac{d\underline{y}}{dt},$$

and corresponds to the derivative with respect to t (time) of the complex image (phasor) given by Eq. (4.8.171). This phasor can be obtained simply from the relation

$$\frac{d\underline{y}}{dt} = \left(i\,\omega\right)\underline{y}. \tag{4.8.172}$$

Of all the primitives of the sinusoidal quantity $y(t)$, only one is also sinusoidal, namely

$$\int A \cos\left(\omega t + \varphi\right) dt = \frac{1}{\omega}A \sin\left(\omega t + \varphi\right) = \frac{1}{\omega}A \cos\left(\omega t + \varphi - \frac{\pi}{2}\right).$$

The complex image of this primitive is the phasor

$$\frac{1}{\omega}A\, e^{i\left(\omega t + \varphi - \frac{\pi}{2}\right)} = \frac{1}{i\,\omega}A\, e^{i(\omega t + \varphi)} = \frac{1}{i\,\omega}\underline{y}. \tag{4.8.173}$$

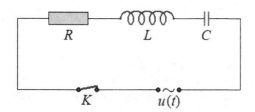

FIGURE 4.35
A series RLC circuit powered by an AC source with $u(t) = U \cos(\omega_0 t + \alpha)$.

In the same way, higher order derivatives and primitives can be calculated. This gives the following

Theorem Let $y(t)$ be a sinusoidal quantity having the pulsation ω and let $\underline{y}$ be its corresponding phasor. Then, to the k−order derivative $y^{(k)}(t)$ it corresponds the phasor

$$\left(i\,\omega\right)^{k} \underline{y}, \qquad (4.8.174)$$

and to the periodic k−order primitive of the complex function $y(t)$ it corresponds the phasor

$$\frac{1}{\left(i\,\omega\right)^{k}} \underline{y} = \left(i\,\omega\right)^{-k} \underline{y}. \qquad (4.8.175)$$

The properties of the phasors make them useful in the study of the sinusoidal steady states in physical systems modelled by equations or systems of integro-differential equations with constant coefficients and sinusoidal input functions.

Solution

The series RLC circuit (containing a resistor R, an inductor L and a capacitor C) in Fig. 4.35 is characterized by the integro-differential equation

$$L\,\frac{di(t)}{dt} + R\,i(t) + \frac{1}{C}\int i(t)\,dt = u(t), \qquad (4.8.176)$$

resulting from the application of Kirchhoff's mesh law.

If the input function $u(t)$ is a sinusoidal quantity of pulsation ω, we determine the current $i(t)$ using the complex image of the integro-differential equation (4.8.176). According to the theorem stated above, we can write that

$$i\,\omega\,L\,\underline{i} + R\,\underline{i} + \frac{1}{i\,\omega}\,\frac{1}{C}\,\underline{i} = \underline{u}, \qquad (4.8.177)$$

or

$$\underline{i} = \frac{\underline{u}}{\underline{Z}}, \qquad (4.8.178)$$

where

$$\underline{Z} = R + i\,\omega\,L + \frac{1}{i\,\omega\,C}, \tag{4.8.179}$$

is the *complex impedance* of the circuit.

One observes that the complex impedance $\underline{Z}$ contains the complex impedances of the constitutive elements of the circuit, namely

$$\begin{cases} \underline{Z}_R = R, \\ \underline{Z}_L = i\,\omega\,L, \\ \underline{Z}_C = \dfrac{1}{i\,\omega\,C}, \end{cases} \tag{4.8.180}$$

so that we can write

$$\underline{Z} = \underline{Z}_R + \underline{Z}_L + \underline{Z}_C. \tag{4.8.181}$$

Then relation (4.8.177) can also be written as

$$\underline{u} = \underline{u}_R + \underline{u}_L + \underline{u}_C, \tag{4.8.182}$$

where

$$\begin{cases} \underline{u}_R = \underline{Z}_R\,\underline{i}_R, \\ \underline{u}_L = \underline{Z}_L\,\underline{i}_L, \\ \underline{u}_C = \underline{Z}_C\,\underline{i}_C, \end{cases} \tag{4.8.183}$$

and

$$\underline{i}_R = \underline{i}_L = \underline{i}_C = \underline{i}. \tag{4.8.184}$$

It is like having a DC circuit containing only resistors connected in series, except that the "resistances" of the three "resistors" can also have complex values (more precisely, purely dissipative elements, *i.e.*, ideal resistors have real "resistances", while ideal reactive circuit elements, *i.e.*, ideal coils and ideal capacitors have only complex "resistances".

Actually, a real resistor has also a certain inductance, and a real capacitor has also a certain resistance. Besides, an ideal dissipative element (an ideal resistor) do not introduce any phase difference between the voltage and the current through it, while an ideal reactive element does; as we know, an ideal coil introduces an out of phase between the current and voltage of $\pi/2$ rads, in the sense that the phase of the current is 90 degrees behind that of the voltage, and an ideal capacitor introduces an out of phase of the same value but in the opposite sense (the phase of the current is $\pi/2$ rads ahead that of the voltage). In other words, in a purely inductive AC circuit the current "lags" the applied voltage by 90°, or $\pi/2$ rads, while in a purely capacitive circuit the exact opposite is true, the current "leads" the voltage by 90°).

In this way, the AC problem has been reduced to a DC problem, except that the quantities we work with are complex.

The phasor diagram of the series RLC circuit was graphically represented in Fig. 4.36.

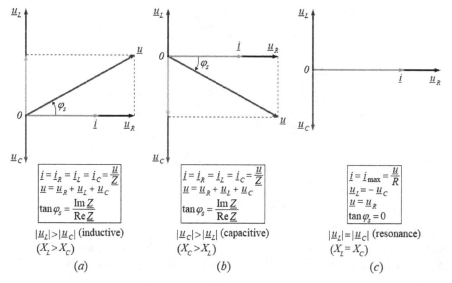

FIGURE 4.36
Phasor diagram of an AC series RLC circuit: (a) inductive character, (b) capacitive character, and (c) at resonance.

Moreover, we can write that

$$\underline{Z} = Z e^{i\varphi_s},$$ (4.8.185)

where

$$Z = \sqrt{R^2 + \left(\omega L - \frac{1}{\omega C}\right)^2},$$ (4.8.186)

and

$$\varphi_s = \arctan \frac{\operatorname{Im} \underline{Z}}{\operatorname{Re} \underline{Z}} = \arctan \frac{\omega L - \frac{1}{\omega C}}{R},$$ (4.8.187)

where, from Eq. (4.8.179) we have

$$\underline{Z} = R + i\left(\omega L - \frac{1}{\omega C}\right).$$ (4.8.188)

Finally, according to Eq. (4.8.178) the complex current (the phasor of the current) will be

$$\underline{i} = \frac{\underline{u}}{\underline{Z}} = \frac{U\, e^{i(\omega t + \alpha)}}{Z\, e^{i\varphi_s}} = \frac{U}{Z} e^{i(\omega t + \alpha - \varphi_s)},$$ (4.8.189)

provided the input voltage is

$$u(t) = U \cos\left(\omega t + \alpha\right),$$ (4.8.190)

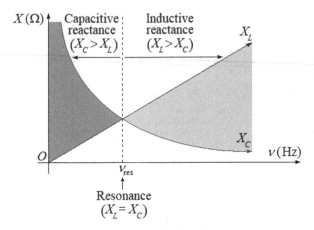

FIGURE 4.37
Graphical representation of the inductive and capacitive reactances variation
as functions of frequency, $X_L = X_L(\nu)$ and $X_C = X_C(\nu)$, respectively.

i.e.,

$$\underline{u} = U e^{i(\omega t + \alpha)}. \tag{4.8.191}$$

By taking the real part of the current phasor we obtain the alternative
current through the circuit,

$$i(t) = I \cos\left(\omega t + \alpha - \varphi_s\right) = \frac{U}{Z} \cos\left(\omega t + \alpha - \varphi_s\right). \tag{4.8.192}$$

If the supply voltage of a series RLC circuit has a fixed amplitude but
different frequencies and the frequency can be continuously varied, there becomes a frequency value at which the inductive reactance $X_L = \omega L$ of the
coil becomes equal in value to the capacitive reactance $X_C = \left(\omega C\right)^{-1}$ of the
capacitor (see Fig. 4.37). The point at which this occurs is called the *resonant frequency point* of the circuit, and as we are analysing a series RLC
circuit this resonance frequency produces a *series resonance* (see Fig. 4.38).
The resonance frequency of the series RLC circuit is therefore

$$\nu_{\text{res}} = \frac{1}{2\pi}\sqrt{\frac{1}{LC}}, \tag{4.8.193}$$

or, equivalently,

$$\omega_{\text{res}} = \sqrt{\frac{1}{LC}}. \tag{4.8.194}$$

Since we are dealing with a series circuit, if at resonance the two reac-
tances are equal and cancelling, the two corresponding voltages, $\underline{u}_L$ and $\underline{u}_C$
must also be opposite and equal in value, thereby cancelling each other out,
because with pure components the phasor voltages are drawn at $+90°$ and
$-90°$, respectively.

Out of resonance   At resonance

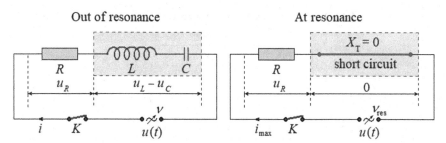

FIGURE 4.38
The series RLC circuit powered by an AC source, represented in its two fundamental characteristic poses: out of resonance and at resonance.

Then in a series resonance circuit as $\underline{u}_L = -\underline{u}_C$ the resulting reactive voltage is zero and all the supply voltage is dropped across the resistor. Therefore, $\underline{u}_R = \underline{u}$ and it is for this reason that series resonance circuits are known as *voltage resonance* circuits, (as opposed to parallel resonance circuits which are *current resonance* circuits).

Because at resonance the impedance $\underline{Z}$ is at its minimum value, $\underline{Z}_{min} = R$ (see Fig. 4.39), and the current flowing through a series resonance circuit is given by the voltage to impedance ratio, the circuit current at this frequency will be at its maximum value of $\underline{i}_{max} = \underline{u}/R$ (see Fig. 4.40).

Let's assume that the series RLC circuit is driven by a variable frequency at a constant voltage. At voltage resonance the power absorbed by the circuit is at its maximum value, because in this situation the current reaches its maximum value. If now the frequency is reduced or increased until the average power absorbed by the resistor in the series resonance circuit is half that of its maximum value, then two frequency points called the *half-power points* are defined, which are $-3\,\mathrm{dB}$ down from maximum, taking $0\,\mathrm{dB}$ as the maximum current reference (see Fig. 4.41).

The point corresponding to the lower frequency at half the power is called the *lower cut-off frequency* and is denoted by ν_L, while the point corresponding to the upper frequency at half power is called the *upper cut-off frequency* and is labelled by ν_H. The "distance" between these two points, *i.e.*, $\Delta\nu_{(-3\,\mathrm{dB})} = \nu_H - \nu_L$ is called the *bandwidth* and is denoted by BW. It represents the range of frequencies over which at least half of the maximum power and current is provided as we have shown.

The current value corresponding to these $-3\,\mathrm{dB}$ points is $70.71\,\%$ of its maximum resonant value which is defined as

$$i_{(-3\,\mathrm{dB})} = i_{\left(P=\frac{P_{\max}}{2}\right)} = \frac{i_{\max}}{\sqrt{2}},$$

and $\frac{1}{\sqrt{2}} = 0.7071$.

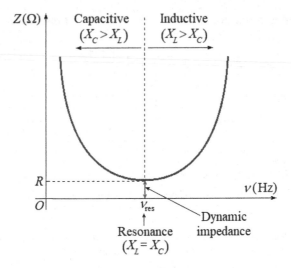

FIGURE 4.39
Graphical representation of the impedance variation of an AC series RLC circuit as a function of frequency, $Z = Z(\nu)$.

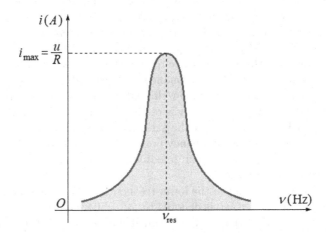

FIGURE 4.40
Graphical representation of the electric current intensity variation in an AC series RLC circuit as a function of frequency, $i = i(\nu)$.

The sharpness of the power/current peak is measured quantitatively and is called the *quality factor*, Q, of the circuit. The quality factor relates the maximum or peak energy stored in the circuit (the total reactance being responsible for this) to the energy dissipated (by the resistance) during each cycle of oscillation, meaning that it is a ratio of resonant frequency to

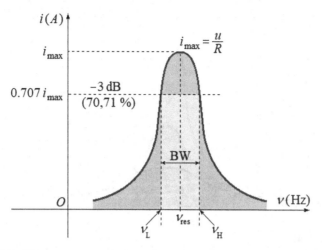

FIGURE 4.41
Same dependence as in Fig. 4.40, but the graphical representation has been redesigned (completed with appropriate elements) to highlight the BW of an AC powered series RLC circuit.

bandwidth, and, the higher the circuit Q, the smaller the bandwidth:

$$Q = \frac{\nu_{\text{res}}}{\Delta\nu_{(-3\,\text{dB})}} = \frac{\nu_{\text{res}}}{\text{BW}}.$$

As the bandwidth is defined by the two $-3\,\text{dB}$ points, the selectivity of the circuit is a measure of its ability to "reject" any frequencies either side of these two points; the wider the BW, the less selective the circuit will be and vice-versa. Since

$$Q = \frac{(X_L)_{\text{res}}}{R} = \frac{(X_C)_{\text{res}}}{R},$$

the selectivity of a series RLC resonance circuit can be adjusted by modifying the value of the resistance only, the other two components (L and C) being kept unchanged (see Fig. 4.42).

As in practice a series resonance circuit usually functions on resonant frequency (the series resonance circuits are one of the most important circuits used electrical and even electronic circuits; they can be found in a very wide range of forms such as in radio and TV tuning circuits, noise filters and more), this type of circuit is also known as an *acceptor circuit* because at resonance, the impedance of the circuit is at its minimum so easily accepts the current whose frequency is equal or very close to its ν_{res}.

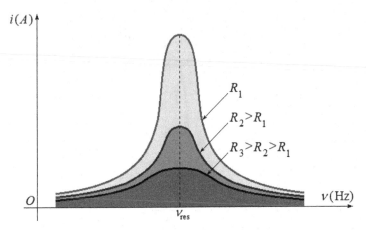

FIGURE 4.42
Same dependence as in Fig. 4.40, but the graphical representation has been redesigned for highlighting the selectivity of an AC powered series RLC resonance circuit; the wider the BW, the less selective the circuit is and vice-versa. The figure highlights the decisive role of the circuit resistance, which alone determines the selectivity of the circuit: the larger the resistance, the less selective the circuit is and vice-versa.

4.9   Problem No. 35

Using the phasor method, study the parallel RLC circuit operating in a permanent sinusoidal regime.

Solution

Applying the two Kirchhoff's laws for the parallel RLC circuit in Fig. 4.43, which is powered by an AC source of voltage

$$u(t) = U \cos\left(\omega t + \alpha\right), \quad \left(\underline{u} = U e^{i(\omega t + \alpha)}\right), \qquad (4.9.195)$$

one can write

$$\begin{cases} i_R(t) + i_L(t) + i_C(t) = i_T(t), \\ u_R(t) = u_L(t) = u_C(t) = u(t). \end{cases} \qquad (4.9.196)$$

Using complex quantities (phasors), relations (4.9.196) re-write as follows:

$$\begin{cases} \underline{i}_R + \underline{i}_L + \underline{i}_C = \underline{i}_T, \\ \underline{u}_R = \underline{u}_L = \underline{u}_C = \underline{u}. \end{cases} \qquad (4.9.197)$$

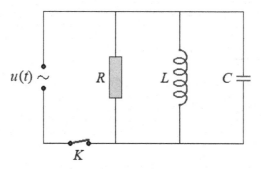

FIGURE 4.43

A parallel RLC circuit powered by an AC source with $u(t) = U\cos(\omega_0 t + \alpha)$.

Introducing the complex admittances

$$
\begin{cases}
\underline{Y}_R \equiv \dfrac{1}{\underline{Z}_R} = \dfrac{1}{R}, \\[2mm]
\underline{Y}_L \equiv \dfrac{1}{\underline{Z}_L} = \dfrac{1}{i\omega L}, \\[2mm]
\underline{Y}_C \equiv \dfrac{1}{\underline{Z}_C} = i\omega C,
\end{cases}
\tag{4.9.198}
$$

the Ohm's law for a portion of a circuit is written as follows (for each of the three arms of the parallel RLC circuit):

$$
\begin{cases}
\underline{i}_R = \underline{Y}_R \underline{u}_R, \\
\underline{i}_L = \underline{Y}_L \underline{u}_L, \\
\underline{i}_C = \underline{Y}_C \underline{u}_C.
\end{cases}
\tag{4.9.199}
$$

From Eqs. (4.9.197), (4.9.198) and (4.9.199) it results that

$$
\underline{i}_T \equiv \underline{i} = \underline{Y}\,\underline{u},
\tag{4.9.200}
$$

where

$$
\underline{Y} = \underline{Y}_R + \underline{Y}_L + \underline{Y}_C = \frac{1}{R} + \frac{1}{i\omega L} + i\omega C,
\tag{4.9.201}
$$

is the total admittance of the parallel RLC circuit.

The phasor diagram of the parallel RLC circuit is graphically represented in Fig. 4.44.

The phasor of the total admittance of the parallel RLC circuit is written as

$$
\underline{Y} = Y\,e^{i\varphi_p},
\tag{4.9.202}
$$

where φ_p is the current-voltage phase mismatch of the parallel RLC circuit. According to Eqs. (4.9.202) and (4.9.195), relation (4.9.200) is re-written as

$$
\underline{i} = \underline{Y}\,\underline{u} = YU\,e^{i(\omega t + \alpha + \varphi_p)},
\tag{4.9.203}
$$

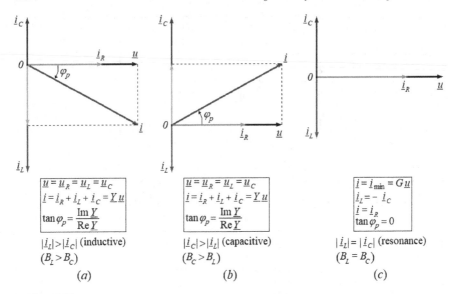

FIGURE 4.44
Phasor diagram of an AC parallel RLC circuit: (a) inductive character, (b) capacitive character and (c) at resonance.

and then the total current $i_T(t)$ will be given by

$$i_T(t) \equiv i(t) = \operatorname{Re}\underline{i} = YU \cos\left(\omega t + \alpha + \varphi_p\right) \equiv I \cos\left(\omega t + \alpha + \varphi_p\right). \quad (4.9.204)$$

Besides we have

$$Y = \operatorname{Re}\underline{Y} = \sqrt{\left(\operatorname{Re}\underline{Y}\right)^2 + \left(\operatorname{Im}\underline{Y}\right)^2} = \sqrt{\frac{1}{R^2} + \left(\omega C - \frac{1}{\omega L}\right)^2}, \quad (4.9.205)$$

and

$$\tan\varphi_p = \frac{\operatorname{Im}\underline{Y}}{\operatorname{Re}\underline{Y}} = \frac{\omega C - \dfrac{1}{\omega L}}{\dfrac{1}{R}}. \quad (4.9.206)$$

Regarding the resonance phenomenon, in many ways, a parallel RLC resonance circuit is exactly the same as the series RLC resonance circuit. As in the case of the series RLC circuit, the two reactive components make a second-order circuit also here. Moreover, both series and parallel RLC circuits are influenced by variations in the supply frequency at constant voltage supply and both have a frequency point where their two reactive components cancel each other out producing the resonance phenomenon.

However, there is a clear difference here in the occurrence of the resonance phenomenon: while in the case of the series circuit the resulting reactive voltage is zero and all the supply voltage is dropped across the resistor, talking

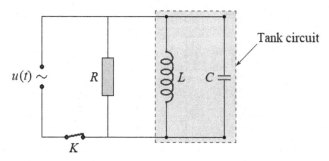

FIGURE 4.45
Same circuit as in Fig. 4.43 redesigned for highlighting the parallel LC tank circuit which plays a very important role at resonance.

about the *voltage resonance*, in the case of the parallel RLC circuit the resonance phenomenon is produced by the particular currents flowing through each parallel branch within the parallel reactive circuit which is known as the *LC tank circuit* (a tank circuit is a parallel combination of a coil and a capacitor that is used in filter networks to either select or reject AC frequencies – see Fig. 4.45).

The resonance phenomenon that appears in a parallel RLC circuit is also called *anti-resonance* and occurs when the resultant current through the LC tank circuit is in phase with the supply voltage. Thus, at resonance there appears a large current that circulates between the coil and the capacitor due to the energy of the oscillations, and this is why in this case we talk about *current resonance*. The parallel resonance is also called anti-resonance because the characteristics and graphs drawn for a parallel circuit are exactly opposite to that of a series circuit (the minimum and maximum impedance, current, etc. are reversed from those of the series circuit).

The mathematics behind the parallel RLC circuit resonance is very simple. In this case too, the resonance occurs when $X_L = X_C$, even if the reason is different (this equality comes from that of corresponding susceptances and not directly from the equality of reactances, as in the case of series resonance). So, in the case of parallel resonance the mathematical resonance condition writes

$$\mathrm{Im}\,\underline{Y} = 0,$$

i.e., the imaginary part of the total admittance phasor must equal zero. Of course, the resonance frequency will be the same as that for the series resonance, *i.e.*,

$$\nu_{\mathrm{res}} = \frac{1}{2\pi}\frac{1}{\sqrt{LC}}.$$

Thus, because at resonance the parallel RLC circuit produces the same equation as for the series RLC resonance circuit, it makes no difference – from this point of view – if the coil and capacitor are connected in parallel or series.

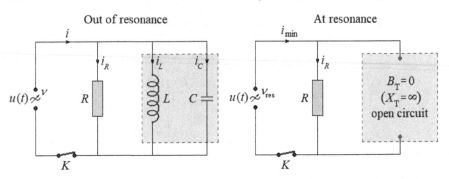

FIGURE 4.46
The parallel RLC circuit powered by an AC source, represented in its two fundamental characteristic poses: out of resonance and at resonance. As can be observed, at resonance the parallel LC tank circuit acts like an open circuit.

Besides, at resonance the parallel LC tank circuit acts like an open circuit with the circuit current being determined by the resistor only (see Fig. 4.46). So the total impedance of a parallel resonance circuit at resonance becomes just the value of R.

The circuit's frequency response can be changed by changing the value of this resistance. Changing the value of R affects the amount of current that flows through the circuit at resonance, if both L and C remain unchanged. Then the impedance of the circuit at parallel resonance $Z_{res} = R$ is called the *dynamic impedance* of the circuit (see Fig. 4.47). But if the parallel circuit's impedance is at its maximum at resonance, then the circuit's admittance must be at its minimum and one of the characteristics of a parallel resonance circuit is that admittance is very low limiting the circuit's current. Consequently, unlike the series resonance circuit, the resistor in a parallel resonance circuit has a damping effect on the circuit's BW, making the circuit less selective.

As the total susceptance phasor $\underline{B}_T = \underline{B}_L + \underline{B}_C$ is zero for $\nu = \nu_{res}$, the total admittance is at its minimum and is equal to the conductance $G = 1/R$ of the circuit. Therefore at parallel resonance the current flowing through the circuit must also be at its minimum as the inductive and capacitive branch current phasors are equal in modulus but 180° out of phase: $\underline{i}_L = -\underline{i}_C$. Thus, the net reactive current is equal to zero and so, at resonance the total current will be

$$i_T = \sqrt{i_R^2 + 0^2} = i_R,$$

and because at resonance the admittance Y is at its minimum value $(= G)$, the circuit current will also be at its minimum value of $Gu = u/R$ and the graph of current vs frequency for a parallel resonance circuit will look like in Fig. 4.48.

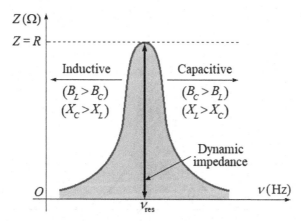

FIGURE 4.47
Graphical representation of the impedance variation of an AC parallel RLC circuit as a function of frequency, $Z = Z(\nu)$. The figure highlights the dynamic impedance of the circuit (the impedance of the circuit at parallel resonance, $Z_{\text{res}} = R$.

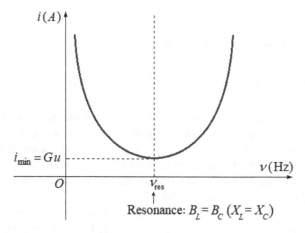

FIGURE 4.48
Graphical representation of the electric current intensity variation in an AC parallel RLC circuit as a function of frequency, $i = i(\nu)$.

The bandwidth of a parallel resonance circuit is defined in exactly the same manner as for the series resonance circuit (see Fig. 4.49). Thus we have

$$\text{BW} = \nu_H - \nu_L = \frac{\nu_{\text{res}}}{Q},$$

where Q is the quality factor (or the selectivity) of the parallel resonance circuit, which in this case is defined as the ratio of the circulating branch

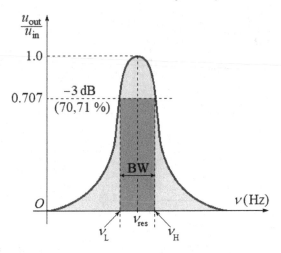

FIGURE 4.49
Graphical representation of the variation of ratio $u_{\text{out}}/u_{\text{in}}$ in an AC parallel
RLC circuit as a function of frequency, which highlights the BW of this kind
of circuit.

currents to the supply current, being given by

$$Q = \frac{\left(B_L\right)_{\text{res}}}{G} = \frac{\left(B_C\right)_{\text{res}}}{G} = \frac{R}{\omega_{\text{res}}L} = \omega_{\text{res}}R = R\sqrt{\frac{C}{L}}.$$

As can be observed, the Q–factor of a parallel resonance circuit is the inverse of the expression for its series circuit correspondent. Besides, in series
resonance circuits the Q–factor gives the voltage magnification of the circuit,
while in a parallel circuit it gives the current magnification.

Note that the bandwidth of a parallel circuit can also be defined in terms
of the impedance curve (see Fig. 4.50).

Whereas a series RLC resonant circuit is also called an *acceptor circuit*
(and, in general, because in practice a series RLC circuit is mostly used in
the resonant regime, it is generally called an acceptor circuit), in the case
of a parallel RLC resonant circuit we talk, on the contrary, about a *rejecter
circuit*. Since in practice a parallel circuit also usually functions on resonant
frequency, this type of circuit is also known as a *rejecter circuit* because at
resonance the impedance of the circuit is at its maximum, thereby suppressing
or rejecting the current whose frequency is equal or very close to its ν_{res}.

In the case of more complex linear AC circuits, theoretically the problem
can be reduced to a linear integro-differential equation with constant coefficients (which can be obtained by applying Kirchhoff's laws), which relates the
quantity to be determined, $y(t)$ to the sinusoidal input signal $x(t)$. By making
a corresponding/needed number of successive derivatives all integrals can be
eliminated, so that the connection between the input function and the output

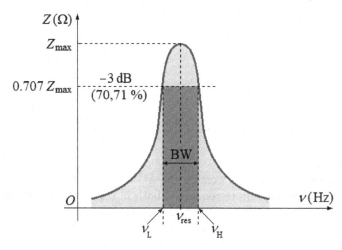

FIGURE 4.50
Another way to highlight the BW of the AC parallel *RLC* circuit, which appeals to the variation of circuit impedance as a function of frequency, $Z = Z(\nu)$.

function can be put in the form

$$D_n y(t) = D_m x(t), \qquad (4.9.207)$$

where D_n and D_m are the following linear differential operators:

$$
\begin{cases}
D_n = b_n \dfrac{d^n}{dt^n} + b_{n-1} \dfrac{d^{n-1}}{dt^{n-1}} + \cdots + b_1 \dfrac{d}{dt} + b_0, \\[2mm]
D_m = a_m \dfrac{d^m}{dt^m} + a_{m-1} \dfrac{d^{m-1}}{dt^{m-1}} + \cdots + a_1 \dfrac{d}{dt} + a_0,
\end{cases}
\qquad (4.9.208)
$$

a_i and b_j, $i = \overline{0, m}$, $j = \overline{0, n}$, being constant coefficients. Switching to the phasors using relation

$$\frac{d^k}{dt^k}\xi(t) \;\rightarrow\; \left(i\omega\right)^k \underline{\xi}, \quad k = 1, 2, 3, \ldots,$$

the relation (4.9.207) becomes

$$\left[\sum_{k=0}^{n} (i\omega)^k b_k\right] \underline{y} = \left[\sum_{k=0}^{m} (i\omega)^k a_k\right] \underline{x}. \qquad (4.9.209)$$

The ratio

$$\frac{\displaystyle\sum_{k=0}^{m} (i\omega)^k a_k}{\displaystyle\sum_{k=0}^{n} (i\omega)^k b_k} \equiv F(i\omega), \qquad (4.9.210)$$

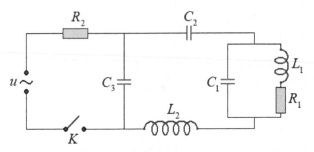

FIGURE 4.51
(a) A more complex circuit powered by an AC source, containing two resistors, two coils and three capacitors connected both in series and parallel as shown in the picture.

is called *complex network function* and
  − if $\underline{x}$ is a current, and $\underline{y}$ is a voltage, then $\underline{F}(i\omega)$ is a *complex impedance*;
  − if $\underline{x}$ is a voltage, and $\underline{y}$ is a current, then $\underline{F}(i\omega)$ is a *complex admittance*;
From Eq. (4.9.210) then it comes that

$$\underline{y} = \underline{F}(i\omega)\,\underline{x}, \tag{4.9.211}$$

and passing to sinusoidal quantities, the solution of the problem in question can be obtained.

The mathematical apparatus presented is useful not only in this case, but also in the study of linear systems with constant parameters, which are excited by sinusoidal signals. For example, this formalism is easily applied in the case of mathematically modeled physical systems by systems of linear integro-differential equations with constant coefficients, provided that the sinusoidal excitation has constant pulsation.

4.10   Problem No. 36

a) Draw the phasor diagram of the circuit in Fig. 4.51.a which operates in a permanent sinusoidal regime, knowing that all circuit elements are ideal;
b) For the circuit in Fig. 4.51.b find the resonance frequencies.

Solution

a) From the very beginning we specify that, in fact, we can imagine a phasor as a "rotating vector", which has a fixed origin (a fixed point of application) and rotates directly (trigonometrically) at a constant angular velocity ω (which is always imposed by the AC power source).

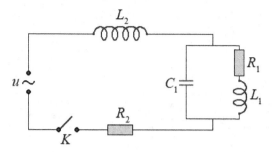

FIGURE 4.51
(b) An AC circuit containing two resistors, two coils and one capacitor connected both in series and parallel as shown in the picture.

Under these conditions, the *phasor diagram* of a circuit operating in a permanent sinusoidal regime is nothing but the "captured" image at a certain moment in time (at an instant of time), as if we were taking a picture/photograph of the entire set of corresponding phasors of all the currents and voltages related to the circuit elements at that moment (instant) of time.

Because all phasors rotate at the same angular velocity, their relative position will always be the same, so if we take that "snapshot" at any time, the phasor diagram will look exactly the same all the time, and, at most could be rotated *as a whole* by a certain angle (which is irrelevant to solving the problem).

Therefore, we can only draw the phasor diagram for a circuit that operates in a permanent sinusoidal regime, and which is powered by one or more alternating voltage sources that *all* provide signals of the same pulsation (frequency) and in no case of different pulsations (frequencies); otherwise the relative position of all phasors will not remain unchanged over time, and the phasor diagram will change its "shape" according to the relations between the (different) pulsations of the two or more AC sources, which is clearly unacceptable.

In order to draw a phasor diagram of a circuit operating in a permanent sinusoidal regime, the following results must be taken into account (all the circuit elements are supposed to be ideal):

1) an ideal resistor do not introduce any phase difference between the voltage on it and current that flows through it; so, the two corresponding phasors, *i.e.*, graphically the current phasor i_R and the voltage phasor u_R will be parallel (they have the same direction, *i.e.*, they will be collinear "vectors");

2) a coil will always introduce a phase difference between the current and voltage; if the coil is ideal, then the phase of the current through the coil is $\pi/2$ radians behind that of the voltage across coil terminals;

3) a capacitor will always introduce a phase difference between the current and voltage; if the capacitor is ideal, then the phase of the current through the capacitor is $\pi/2$ radians ahead that of the voltage across capacitor terminals.

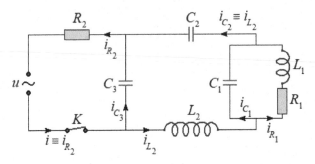

FIGURE 4.52
Same circuit as in Fig. 4.51.a redesigned for highlighting the currents which
flow through the circuit elements.

First of all we draw the two axes of the complex plane; the horizontal
one usually corresponds to the real part of the complex quantities, while the
vertical one is associated with their imaginary part. Thus, any phasor of the
form

$$\underline{P} = P_0\, e^{i(\omega t + \varphi)},$$

will be graphically depicted as an arrow of length P_0 leaving the origin and
making an angle of φ radians with the horizontal axis in direct sense (counter-
clockwise, or trigonometric sense).

It is not necessary to be very rigorous in terms of respecting the length
of the phasors, but it is still good to at least respect the proportions. Indeed,
for example, if a current of $I = 1A$ flows through a resistor, and it has a
resistance of $R = 10\,\mathrm{k\Omega}$, then the length of the current phasor should be, let's
say, one unit, and that of the voltage phasor should be 10,000 units! Similarly,
if a current of $1A$ flows through a coil, the phasor of the voltage on the coil
should be about 377 units long, if the AC frequency is $\nu = 60\,\mathrm{Hz}$.

To effectively draw a phasor diagram, it usually starts from the most "particular" portion/grouping of the circuit, which obviously in our case is the
grouping consisting of resistor R_1 and coil L_1, which are connected in parallel
with capacitor C_1 (see Fig. 4.52). For this grouping we first figure the phasor
of the current through resistor R_1. Since in reality the phasor diagram rotates
counter-clockwise as a whole with angular velocity ω, it is not so important
what the actual initial phase φ of this phasor is, because the subsequent construction of the phase diagram will respect the relative positions of the other
phasors, these positions being determined by the phase differences introduced
by the different circuit elements. Therefore, we can very well figure this first
phasor right on the horizontal axis.

If a particular circuit has more than one of these "particular" groupings,
then starting to draw the phasor diagram depends more on the solver's flair,
intuition and experience in solving this kind of problem.

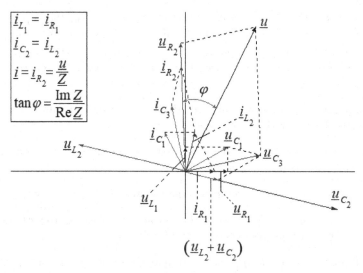

FIGURE 4.53
Phasor diagram of the circuit in Fig. 4.51.a powered by an AC source.

As soon as the phasor diagram is constructed, the AC problem actually becomes a plane geometry problem.

That said, let's get to work! For the proposed example, the order of drawing the phasors in Fig. 4.53 is as follows:

i) the phasor $i_{R_1} = i_{L_1}$ of the current through the resistor R_1 and coil L_1 (the resistor R_1 and the coil L_1 are connected in series);

ii) the phasor u_{R_1} of the voltage on the resistor R_1 (collinear with i_{R_1});

iii) the phasor u_{L_1} of the voltage on the coil L_1 (counter-clockwise rotated by $\pi/2$ rads with respect to i_{L_1});

iv) the phasor $u_{C_1} = u_{R_1} + u_{L_1}$ of the voltage on the capacitor C_1 (the same with the sum of the voltages on the resistor R_1 and the coil L_1);

v) the phasor i_{C_1} of the current through the capacitor C_1 (counter-clockwise rotated by $\pi/2$ rads with respect to u_{C_1});

vi) the phasor $i_{L_2} = i_{C_2} = i_{R_1} + i_{C_1}$ of the current through the coil L_2 and capacitor C_2 (they are connected in series);

vii) the phasor u_{L_2} of the voltage on the coil L_2 (counter-clockwise rotated by $\pi/2$ rads with respect to i_{L_2});

viii) the phasor u_{C_2} of the voltage on the capacitor C_2 (clockwise rotated by $\pi/2$ rads with respect to $i_{C_2} = i_{L_2}$));

ix) the phasor $\underline{u}_{C_3} = \underline{u}_{C_1} + \underline{u}_{C_2} + \underline{u}_{L_2}$ of the voltage on the capacitor C_3 (with respect to the AC voltage source, the capacitor C_3 is connected in parallel with the group consisting of coil L_2, capacitor C_2 and grouping made of C_1, R_1 and L_1; a second reason is provided by Kirchhoff's mesh law);

x) the phasor $\underline{i}_{C_3}$ of the current through the capacitor C_3 (counter-clockwise rotated by $\pi/2$ rads with respect to $\underline{u}_{C_3}$);

xi) the phasor $\underline{i}_{R_2} = \underline{i}_{C_2} + \underline{i}_{C_3}$ of the current through the coil L_2 and capacitor C_2 (they are connected in series);

xii) the phasor $\underline{u}_{R_2}$ of the voltage on the resistor R_3 (collinear with $\underline{i}_{R_3}$);

xiii) the phasor $\underline{u} = \underline{u}_{R_3} + \underline{u}_{C_3}$ of the voltage supplied by the AC voltage source (according to the Kirchhoff's mesh law);

b) Of course, the current i_{R_2} also represents the "total" current (that flows through the AC source) and thus the angle between the phasors $\underline{i}_{R_2} \equiv \underline{i}$ and $\underline{u}$ will represent the "total" phase difference $\Delta\varphi$ between the "total" current and "total" voltage.

Whenever there are reactive elements in a circuit, and the current or voltage resonance conditions are not met, there is a non-zero (net) phase difference $\Delta\varphi \neq 0$ between the "total" current i and the "total" voltage u. In other words, in these circumstances the phasors of "total" current and "total" voltage are out of phase.

In general, the "total" phase difference $\Delta\varphi$ can be:

1) *positive*; in this case, the "total" current phasor is offset behind the "total" voltage phasor by the angle $\Delta\varphi$, and we say that the circuit has an *inductive* global character;

2) *negative*; in this case, the "total" current phasor is offset ahead the "total" voltage phasor by the angle $\Delta\varphi$, and we say that the circuit has a *capacitive* global character;

3) *zero*; in this case, the "total" current phasor is in phase with the "total" voltage phasor (the two phasors are collinear) and we say that the circuit works under resonance conditions. In general, the resonance condition can easily be found.

By working with phasors, an AC problem basically becomes a DC problem. By attaching to a resistor R the "complex impedance" $\underline{Z}_R \equiv R$, to a coil the complex impedance $\underline{Z}_L = i\omega L$, and to a capacitor the complex impedance $\underline{Z}_C = \frac{1}{i\omega C}$, then any AC network can be treated as a DC network. This means that all circuit elements (whether resistors, coils or capacitors) can be regarded as *sui-generis* "resistors" (having the above complex impedances) and we can then relatively easily calculate the equivalent complex impedance of the whole network, as this will be nothing but the "equivalent resistance" of the whole "DC network". And the whole network is nothing but a "combination" of "resistors" connected in series and/or parallel (obviously, we work with the formulas from the series and parallel groups in DC, but with the complex values of the "resistances").

No matter how complicated the AC network may be, after calculations one finds for the equivalent complex impedance of the whole network an expression of the form

$$\underline{Z}_{equiv} = Z_{\text{real}} + iZ_{\text{img}} = Z\,e^{i\Delta\varphi_s}, \qquad (4.10.212)$$

where

$$Z = \sqrt{Z_{\text{real}}^2 + Z_{\text{img}}^2}, \qquad (4.10.213)$$

and

$$\tan\Delta\varphi_s = \frac{Z_{\text{img}}}{Z_{\text{real}}}. \qquad (4.10.214)$$

We can also determine the equivalent complex admittance of the whole network as being

$$\underline{Y}_{equiv} = Y_{\text{real}} + iY_{\text{img}} = Y\,e^{i\Delta\varphi_p}, \qquad (4.10.215)$$

where

$$Y = \sqrt{Y_{\text{real}}^2 + Y_{\text{img}}^2}, \qquad (4.10.216)$$

and

$$\tan\Delta\varphi_p = \frac{Y_{\text{img}}}{Y_{\text{real}}}. \qquad (4.10.217)$$

In the case of simple series and parallel networks, the equivalent/"total" impedance $\underline{Z}_{equiv}$, and equivalent/"total" admittance $\underline{Y}_{equiv}$, respectively, are determined.

As we have a series or parallel network (as we determine an *equivalent impedance* or an *equivalent admittance*, respectively), we talk about *voltage resonance* and *current resonance*, respectively. In the first case the current is the same through the whole circuit (through every circuit element), but the voltages are different on each element of the circuit, while in the second case the voltage is the same at the terminals of each circuit element, but the currents are different through each element of the circuit.

The voltage resonance condition simply writes $\Delta\varphi_s = 0$, that is the two "total" phasors ("total" current and "total" voltage) must be in phase (collinear). According to Eq. (4.10.214) this means we must have

$$Z_{\text{img}} = 0. \qquad (4.10.218)$$

The current resonance condition will be written as $\Delta\varphi_p = 0$, that is (according to Eq. (4.10.217)):

$$Y_{\text{img}} = 0. \qquad (4.10.219)$$

For more complicated networks (which contain elements or groups of elements connected both in series and in parallel) one cannot speak separately about voltage resonance and/or current resonance, but the unique resonance condition. Thus, in this case we are simply talking about the *resonance condition* of the circuit, *i.e.*, without specifying that it is the resonance of voltages or currents. For such more complicated circuit we can write

$$\underline{Z}_{equiv} = Z_{\text{real}} + iZ_{\text{img}} = Z\,e^{i\Delta\varphi}, \tag{4.10.220}$$

where

$$Z = \sqrt{Z_{\text{real}}^2 + Z_{\text{img}}^2}, \tag{4.10.221}$$

and

$$\tan\Delta\varphi = \frac{Z_{\text{img}}}{Z_{\text{real}}}. \tag{4.10.222}$$

Of course, the unique resonance condition now can be written as $\Delta\varphi = 0$, *i.e.*,

$$Z_{\text{img}} = 0. \tag{4.10.223}$$

For the circuit in Fig. 4.51.b, the phasor of the equivalent/"total" impedance is

$$\underline{Z}_{equiv} = Z_{\text{real}} + iZ_{\text{img}} \equiv Z_r + iZ_i, \tag{4.10.224}$$

where

$$Z_r = \frac{\omega^4 C_1^2 L_1^2 R_2 + \omega^2 C_1 R_2\left(C_1 R_1^2 - 2L_1\right) + R_1 + R_2}{\omega^4 C_1^2 L_1^2 + \omega^2 C_1\left(C_1 R_1^2 - 2L_1\right) + 1}, \tag{4.10.225}$$

and

$$Z_i = \omega$$
$$\times\ \frac{\omega^4 C_1^2 L_1^2 L_2 + \omega^2 C_1\left(C_1 R_1^2 L_2 - 2L_1 L_2 - L_1^2\right) + L_1 + L_2 - C_1 R_1^2}{\omega^4 C_1^2 L_1^2 + \omega^2 C_1\left(C_1 R_1^2 - 2L_1\right) + 1}. \tag{4.10.226}$$

The resonance condition $Z_i = 0$ then writes

$$\omega\left[\omega^4 C_1^2 L_1^2 L_2 + \omega^2 C_1\left(C_1 R_1^2 L_2 - 2L_1 L_2 - L_1^2\right) + L_1 + L_2 - C_1 R_1^2\right] = 0, \tag{4.10.227}$$

with supplementary condition

$$\omega^4 C_1^2 L_1^2 + \omega^2 C_1\left(C_1 R_1^2 - 2L_1\right) + 1 \neq 0, \tag{4.10.228}$$

which must be satisfied by the solutions of Eq. (4.10.227). With the notation

$$\alpha \equiv \sqrt{L_1^4 + 2C_1 L_1^2 L_2 R_1^2 - 4C_1 L_1 L_2^2 R_1^2 + C_1^2 L_2^2 R_1^4},$$

these solutions are given by

$$
\begin{cases}
\omega_1 = 0, \\[2mm]
\omega_2 = \dfrac{1}{\sqrt{2}} \sqrt{\dfrac{2}{C_1 L_1} + \dfrac{1}{C_1 L_2} - \dfrac{R_1^2}{L_1^2} - \dfrac{\alpha}{C_1 L_1^2 L_2}}, \\[3mm]
\omega_3 = -\omega_2, \\[2mm]
\omega_4 = \dfrac{1}{\sqrt{2}} \sqrt{\dfrac{2}{C_1 L_1} + \dfrac{1}{C_1 L_2} - \dfrac{R_1^2}{L_1^2} + \dfrac{\alpha}{C_1 L_1^2 L_2}}, \\[3mm]
\omega_5 = -\omega_4.
\end{cases}
\tag{4.10.229}
$$

As can be easily verified, all the solutions in Eq. (4.10.229) of Eq. (4.10.227) satisfy the condition expressed by Eq. (4.10.228), so, from this point of view they are valid solutions. However, there are other supplementary conditions that must be satisfied by the solutions in Eq. (4.10.229) for these to be resonant frequencies of the circuit in Fig. 4.51.b.

Obviously, the solution $\omega_1 = 0$ is not acceptable. Having negative values, the solutions ω_3 and ω_5 are also non acceptable. This leaves only two resonant frequencies for the circuit in Fig. 4.51.b, namely ω_2 and ω_4.

Because the resonance frequencies must be not only positive but also real quantities, the following inequalities should be analysed:

$$
L_1^4 + 2C_1 L_1^2 L_2 R_1^2 + C_1^2 L_2^2 R_1^4 \geq 4C_1 L_1 L_2^2 R_1^2,
\tag{4.10.230}
$$

$$
\frac{2}{C_1 L_1} + \frac{1}{C_1 L_2} > \frac{R_1^2}{L_1^2} + \frac{\alpha}{C_1 L_1^2 L_2},
\tag{4.10.231}
$$

$$
\frac{2}{C_1 L_1} + \frac{1}{C_1 L_2} + \frac{\alpha}{C_1 L_1^2 L_2} > \frac{R_1^2}{L_1^2}.
\tag{4.10.232}
$$

So, for the solution ω_2 to be a resonance frequency of the circuit in Fig. 4.51.b, inequalities (4.10.230) and (4.10.231) must be satisfied, and for ω_4 to be a resonance frequency, inequalities (4.10.230) and (4.10.232) must be satisfied.

4.11   Problem No. 37

Consider the circuit given in Fig. 4.54. Determine the variation range of the real, positive and finite parameter g, so that the circuit works within the stability domain.

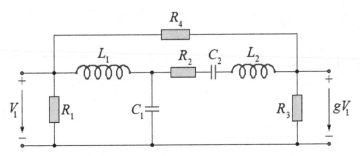

FIGURE 4.54
A DC circuit containing four resistors, two coils and two capacitors, viewed as a dynamic system.

Solution

Before solving the problem in concrete terms, we will briefly present the essential elements concerning the concept of *stability of dynamic systems*. The following discussion concerns the linear dynamic systems, invariant with respect to time, with a single input and a single output. Using the superposition principle, the generalization to systems with many inputs and outputs is not difficult to perform.

A dynamic system is stable if to any finite excitation corresponds a finite response. Let's suppose that the connection between the excitation $x(t)$ and the response $y(t)$ is mathematically modelled by an ordinary, linear, non-homogeneous differential equation with real, constant coefficients, of the form

$$\alpha_0 \frac{d^n y}{dt^n} + \alpha_1 \frac{d^{n-1} y}{dt^{n-1}} + \alpha_2 \frac{d^{n-2} y}{dt^{n-2}} + \dots$$

$$+ \alpha_{n-1} \frac{dy}{dt} + \alpha_n y = x(t), \ \alpha_i \in \mathbb{R}, \ \forall i = 0, 1, 2, \dots . \ (4.11.233)$$

The general solution of this equation is

$$y(t) = y_p(t) + c_1 e^{r_1 t} + c_2 e^{r_2 t} + \dots + c_{n-1} e^{r_{n-1} t} + c_n e^{r_n t}, \qquad (4.11.234)$$

where $y_p(t)$ is a particular solution of the non-homogeneous differential equation (which can be determined, for example, by the Lagrange's method of variation of parameters), $c_1, c_2, \dots, c_n$ are arbitrary constants which are going to be determined by means of initial conditions of the problem, and $r_1, r_2, \dots, r_n$ are the solutions of the characteristic equation

$$P(r) = \alpha_0 r^n + \alpha_1 r^{n-1} + \alpha_2 r^{n-2} + \dots + \alpha_{n-1} r + \alpha_n = 0, \qquad (4.11.235)$$

attached to the differential equation (4.11.233). In the relation (4.11.234), the solutions $r_1, r_2, .., r_n$ were supposed to be simple (*i.e.*, of multiplicity 1).

In general, the characteristic equation (4.11.235) has complex solutions of the form $r = \rho + i\omega, i = \sqrt{-1}$, $\rho, \omega \in \mathbb{R}$, and to a complex root $r_j = \rho_j + i\omega_j$ corresponds – in the solution given by Eq. (4.11.234) – a term of the form

$$e^{r_j t} = e^{\rho_j t}(\cos\omega_j t + i\sin\omega_j t), \tag{4.11.236}$$

representing an oscillation of pulsation ω_j and exponentially time-variable amplitude.

If $\rho_j < 0$, the oscillation is damped, for $\rho_j = 0$ the oscillation is harmonic, and if $\rho_j > 0$ the oscillation is unbounded (its amplitude grows endlessly over time: $y(t) \to \infty$ for $t \to \infty$).

From what has been discussed so far it follows that if $\rho_j < 0$, $\forall j = 1, 2, ..., n$, then the system in question is *stable* (it can be demonstrated that, in this case, for finite excitations there is at least one bounded particular solution $y_p(t)$).

For $\rho_j > 0$ the amplitude of the corresponding term in the solution given by Eq. (4.11.234) increases boundlessly in time and, consequently, the studied system is *unstable*.

If $\rho_j = 0$, then the contribution of the term expressed by Eq. (4.11.236) in the solution given by Eq. (4.11.234) is *bounded*, but the solution $y(t)$ *can be unstable* in the case of multiple roots (*i.e.*, roots having multiplicity greater than or equal to 2) because of the secular terms (i.e., the terms of the form $t\cos\omega_j t, t\sin\omega_j t, t^2\cos\omega_j t, ...$), or due to excitation $x(t)$, in the case of resonance phenomenon. In this case ($\rho_j = 0$), the studied system is *marginal stable* if in Eq. (4.11.234) do not appear secular terms and there is no resonance, and *unstable* otherwise.

This way, the study of the stability of dynamic system described by Eq. (4.11.233) has been reduced to the localization in the complex plane of the roots of the characteristic equation (4.11.235):

– if all roots of Eq. (4.11.235) are situated in the $\text{Re}(r) < 0$ half-plane, then the system is stable;

– if the characteristic equation has roots with positive real part, then the system is unstable;

– if the characteristic equation has simple roots with null real part, then the study of stability of the dynamical system also requires the consideration of the excitation $x(t)$.

There are several criteria used in the study of the stability of dynamical systems. We will present here four of them.

1) The *Stodola's condition*: A necessary condition for all the roots of Eq. (4.11.235) to have negative real part is that all coefficients have the same sign.

Indeed, if $\rho_j + i\omega_j$ is a root of the equation $P(r) = 0$, then $\rho_j - i\omega_j$ will be also a root of the same equation, since the coefficients of the polynomial $P(r)$ are real. This means that the polynomial contains the quadratic factor

$$(r - \rho_j - i\omega_j)(r - \rho_j + i\omega_j) = r^2 - 2r\rho_j + \rho_j^2 + \omega_j^2, \tag{4.11.237}$$

which, if $\rho_j < 0$, has all the positive coefficients. The polynomial $P(r)$ will be then the product of some squared and linear factors with positive coefficients and, therefore, all its coefficients will have the same sign.

2) The *Hurwitz criterion* or the *Routh-Hurwitz stability criterion*: In order that all roots of the equation $P(r) = 0$ with real coefficients α_j, $j = 1, 2, 3, \dots$, $\alpha_0 > 0$ have negative real part, it is necessary and sufficient to be simultaneously fulfilled all the following inequalities:

$$D_1 = \alpha_1 > 0, \quad D_2 = \begin{vmatrix} \alpha_1 & \alpha_0 \\ \alpha_3 & \alpha_2 \end{vmatrix} > 0, \quad D_3 = \begin{vmatrix} \alpha_1 & \alpha_0 & 0 \\ \alpha_3 & \alpha_2 & \alpha_1 \\ \alpha_5 & \alpha_4 & \alpha_3 \end{vmatrix} > 0,$$

$$D_4 = \begin{vmatrix} \alpha_1 & \alpha_0 & 0 & 0 \\ \alpha_3 & \alpha_2 & \alpha_1 & \alpha_0 \\ \alpha_5 & \alpha_4 & \alpha_3 & \alpha_2 \\ \alpha_7 & \alpha_6 & \alpha_5 & \alpha_4 \end{vmatrix} > 0, \quad D_5 = \begin{vmatrix} \alpha_1 & \alpha_0 & 0 & 0 & 0 \\ \alpha_3 & \alpha_2 & \alpha_1 & \alpha_0 & 0 \\ \alpha_5 & \alpha_4 & \alpha_3 & \alpha_2 & \alpha_1 \\ \alpha_7 & \alpha_6 & \alpha_5 & \alpha_4 & \alpha_3 \\ \alpha_9 & \alpha_8 & \alpha_7 & \alpha_6 & \alpha_5 \end{vmatrix} > 0,$$

$$D_n = \begin{vmatrix} \alpha_1 & \alpha_0 & 0 & 0 & 0 & \dots & 0 \\ \alpha_3 & \alpha_2 & \alpha_1 & \alpha_0 & 0 & \dots & 0 \\ \cdot & \cdot & \cdot & \cdot & \cdot & \dots & \cdot \\ \cdot & \cdot & \cdot & \cdot & \cdot & \dots & \cdot \\ \cdot & \cdot & \cdot & \cdot & \cdot & \dots & \cdot \\ \alpha_{2n-3} & \alpha_{2n-4} & \alpha_{2n-5} & \alpha_{2n-6} & \alpha_{2n-7} & \dots & \alpha_{n-2} \\ \alpha_{2n-1} & \alpha_{2n-2} & \alpha_{2n-3} & \alpha_{2n-4} & \alpha_{2n-5} & \dots & \alpha_n \end{vmatrix} > 0, \quad (4.11.238)$$

where $\alpha_j = 0$ for $j > n$. The proof of the theorem that constitutes this criterion can be found, for example, in the reference [50].

The Hurwitz stability criterion is a relatively simple procedure for the study of stability of dynamical systems, and requires the exact knowledge of all the coefficients of the characteristic equation, which is not always possible. For situations when, for example, the function $P(r)$ is experimentally determined (is given "through points"), more useful proves to be the Nyquist's procedure, based on the *principle of variation of the argument*.

Let, more general, $P(r)$ be a rational function without zeros and poles[1] on the imaginary axis. In order to determine the number of the roots of the

---

[1]It is called the *pole* of the complex function $w(z) = u(x, y) + i\,v(x, y)$, of the complex variable $z = x + iy$, a singular isolated point z_0 which satisfies the relation $\lim\limits_{z \to z_0} w(z) = \infty$. It is said that the function $w(z)$ has at the point z_0 an *isolated singular point* if there exists a real positive number ρ, so that the function is holomorphic for $0 < |z - z_0| < \rho$, but it is not monogenic at the point z_0. A function $w(z)$ is called *monogenic* or *derivable* at the point z_0 if the quantity $w'(z_0) = \lim\limits_{z \to z_0} \frac{w(z) - w(z_0)}{z - z_0}$ exists and it is finite. The function $w(z)$ is called *holomorphic* at the point z_0 (in which case z_0 is also called a *regular point* of the function), if it is monogenic in a neighbourhood of this point.

equation $P(r) = 0$ in the $\rho > 0$ half-plane (the half-plane on the right side in Fig. 4.55), we will apply the theorem[2] of variation of the argument for function $P(\gamma)$ on the contour γ from Fig. 4.55, composed by the segment $[-iR, \, iR]$ and the semicircle $|r| = R$, in the $r = \rho + i\omega$ complex plane. For large enough R, it results

$$\Delta_\gamma \operatorname{Arg} P(r) = \Delta_\gamma \operatorname{Arg} w = 2\pi(\zeta - p), \qquad (4.11.239)$$

where ζ is the number of zeros of the function $w = P(r)$ from the $\operatorname{Re}(r) > 0$ half-plane, and p is the number of poles of the function, in the same domain. This way is obtained

3) The *Nyquist criteria* or the *Nyquist stability criterion*: For a dynamical system described by a meromorphic[3] function $P(r)$, which has p poles in the $\operatorname{Re}(r) > 0$ half-plane, to be stable, it is necessary and sufficient that, when r goes along the contour γ, its image $w = P(r)$ encircles p times the origin of the coordinate axes clockwise.

In the case of an automatic control system (*i.e.*, a system containing a feedback loop, and so, being capable of automatic adjustment; most often, such systems are called *feedback control systems*), the transfer function has the form

$$H(r) = \frac{kG(r)}{1 + kG(r)}, \qquad (4.11.240)$$

where $G(r)$ is a rational function with p zeros in the domain $\operatorname{Re}(r) > 0$, and the positive parameter k designates the amplification/gain factor. The study of the stability of such a system leads to the condition that the function

$$P(r) \equiv \frac{1}{H(r)} = 1 + \frac{1}{kG(r)} \qquad (4.11.241)$$

does not have roots with positive real part. The relation (4.11.239) then

---

[2]This theorem states that if $w = w(z)$ is a meromorphic function in the simple-connected domain $\mathcal{D}$ and γ is a closed simple curve contained in $\mathcal{D}$ which does not pass by any of the poles or zeros of the function $w(z)$, then is valid the relation $(2\pi)^{-1}\Delta_\gamma \operatorname{Arg} w(z) = \zeta - p$, where ζ and p represent the number of zeros and poles, respectively, of the function $w(z)$, zeros and poles that are situated inside the closed curve γ (each zero or multiple pole is considered as many times as its order of multiplicity), and the symbol Δ_γ signifies the variation of the function that follows that symbol, when going along the curve γ in direct sense (the trigonometric sense). Here is the geometric interpretation: if $\gamma' = w(\gamma)$, where γ' is the curve shown in Fig. 4.56, then the number $(2\pi)^{-1}\Delta_\gamma \operatorname{Arg} w$ represents the number of complete rotations around the origin, performed by the vector w, when the point w describes the curve γ'.

[3]A function that in a domain $\mathcal{D}$ has no singularities other than poles and eliminable singularities is called a *meromorphic function* in $\mathcal{D}$. A rational function (the ratio of two polynomials) is an example of meromorphic function in the whole complex plane. An isolated singular point z_0 of the function $w(z)$ is called *eliminable singularity* if and only if $\lim_{z \to z_0} (z - z_0)w(z) = 0$.

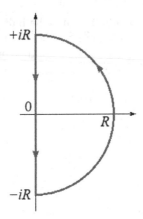

FIGURE 4.55
The contour γ in the $r = \rho + i\omega$ complex plane, composed by the segment $[-iR, iR]$ and the semicircle $|r| = R$.

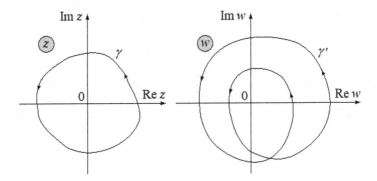

FIGURE 4.56
An auxiliary geometrical construction showing the z and w complex planes containing the two curves, γ and $\gamma' = w(\gamma)$, respectively, referred to in the theorem of variation of the argument from mathematical analysis of complex functions of complex variable.

becomes

$$\Delta_\gamma \mathrm{Arg} P(r) = \Delta_\gamma \mathrm{Arg} \left\{ 1 + \frac{1}{kG(r)} \right\}$$

$$= \Delta_\gamma \mathrm{Arg} \left\{ k + \frac{1}{G(r)} \right\} = \Delta_\gamma \mathrm{Arg}\{W - (-k)\}, \quad (4.11.242)$$

where $W = 1/G(r)$. This way one obtains

4) The *Nyquist stability criterion for feedback control systems*: For a feedback control system which has the transfer function given by Eq. (4.11.240) to be

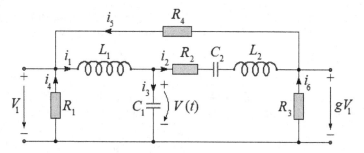

FIGURE 4.57
Same circuit as in Fig. 4.54 redesigned for highlighting the currents which flow through the circuit elements and voltage drop on the capacitor C_1.

stable, it is necessary and sufficient that when the point r goes along the contour γ in direct sense, the vector $W = 1/G(r)$ should bypass p times in reverse direction the point $W = -k$.

These being specified, let's move on to solving the problem now. In this respect, we will appeal to the Hurwitz stability criterion; therefore, it is necessary to determine the coefficients (real numbers) α_j, $j = 1, 2, 3, ...$, of the characteristic equation attached to the differential equation of a specific variable of the studied physical system. We will consider, to this purpose, the situation shown in Fig. 4.57.

From Kirchhoff's first and second laws one obtains the following three sets of equations:

$$\begin{cases} i_3(t) = C_1 \dfrac{dV(t)}{dt}, \\ i_4(t)R_1 = -V_1(t), \\ i_6(t)R_3 = -gV_1(t), \end{cases} \qquad (4.11.243)$$

$$\begin{cases} i_1(t) = i_4(t) + i_5(t), \\ i_3(t) = i_1(t) - i_2(t), \\ i_5(t) = i_2(t) + i_6(t), \end{cases} \qquad (4.11.244)$$

$$\begin{cases} i_5(t)R_4 + V_1(t) = gV_1(t), \\ L_1 \dfrac{di_1(t)}{dt} + V(t) = V_1(t), \\ -L_2 \dfrac{di_2(t)}{dt} - \dfrac{1}{C_2} \displaystyle\int i_2(t)dt, \\ -i_2(t)R_2 + V(t) = gV_1(t), \end{cases} \qquad (4.11.245)$$

where $V(t)$ is the electric voltage on the capacitor of capacitance C_1.

Let us determine the differential equation satisfied by the current $i_2(t)$. To this end we will start with Eq. $(4.11.245)_3$. Taking the first derivative with respect to time of this equation and omitting (for writing simplicity) the functional time dependence of the time-variable quantities, we have

$$-L_2\frac{d^2i_2}{dt^2} - \frac{i_2}{C_2} - R_2\frac{di_2}{dt} + \frac{dV}{dt} = g\frac{dV_1}{dt}. \qquad (4.11.246)$$

In this equation we will replace $\frac{dV}{dt}$ with the corresponding quantity obtained from Eqs. $(4.11.243)_1$ and $(4.11.244)_2$. The result is

$$-L_2\frac{d^2i_2}{dt^2} - \frac{i_2}{C_2} - R_2\frac{di_2}{dt} + \frac{i_1}{C_1} - \frac{i_2}{C_1} = g\frac{dV_1}{dt}. \qquad (4.11.247)$$

From Eq. $(4.11.245)_1$ it follows that

$$i_5 = \frac{(g-1)V_1}{R_4},$$

and Eq. $(4.11.243)_2$ gives $i_4 = -\frac{V_1}{R_1}$. Introducing these values into Eq. $(4.11.244)_1$ we obtain the following expression for i_1:

$$i_1 = \left(\frac{g-1}{R_4} - \frac{1}{R_1}\right)V_1 = \beta V_1, \qquad (4.11.248)$$

where we used the obvious notation

$$\beta \equiv \frac{g-1}{R_4} - \frac{1}{R_1}.$$

By introducing Eq. (4.11.248) into Eq. (4.11.247), it results the following differential equation for the current $i_2(t)$:

$$-L_2\frac{d^2i_2}{dt^2} - R_2\frac{di_2}{dt} - \left(\frac{1}{C_1} + \frac{1}{C_2}\right)i_2 + \beta\frac{V_1}{C_1} = g\frac{dV_1}{dt}. \qquad (4.11.249)$$

Taking the derivatives with respect to time of both sides of Eq. (4.11.249) we obtain the following third-order differential equation for $i_2(t)$:

$$L_2\frac{d^3i_2}{dt^3} + R_2\frac{d^2i_2}{dt^2} + \left(\frac{1}{C_1} + \frac{1}{C_2}\right)\frac{di_2}{dt} - \frac{\beta}{C_1}\frac{dV_1}{dt} + g\frac{d^2V_1}{dt^2} = 0. \qquad (4.11.250)$$

Let us now determine $\frac{d^2V_1}{dt^2}$ in terms of i_2, V_1 and $\frac{dV_1}{dt}$. To this end, we will use the relations $(4.11.243)_1$, $(4.11.244)_2$, $(4.11.245)_2$ and $(4.11.248)$. By eliminating of i_1 and V from all these already mentioned relations, we have

$$\frac{d^2V_1}{dt^2} = \frac{1}{\beta L_1 C_1}i_2 - \frac{1}{L_1 C_1}V_1 + \frac{1}{\beta L_1}\frac{dV_1}{dt}.$$

By introducing this result into Eq. (4.11.250), we get

$$L_2\frac{d^3i_2}{dt^3} + R_2\frac{d^2i_2}{dt^2} + \left(\frac{1}{C_1} + \frac{1}{C_2}\right)\frac{di_2}{dt}$$

$$+ \left(\frac{g}{\beta L_1} - \frac{\beta}{C_1}\right)\frac{dV_1}{dt} + \frac{g}{\beta L_1 C_1}i_2 - \frac{g}{L_1 C_1}V_1 = 0.$$

In a manner similar to the deduction of relation (4.11.248), from Eq. (4.11.243)$_3$ we have $i_6 = -\frac{gV_1}{R_3}$. By introducing this result together with $i_5 = \frac{(g-1)V_1}{R_4}$ (that was determined above) into Eq. (4.11.244)$_3$, we obtain for i_2 the following expression:

$$i_2 = \left(\frac{g-1}{R4} + \frac{g}{R_3} \right) = \sigma V_1, \qquad (4.11.251)$$

where the new notation

$$\sigma \equiv \frac{g-1}{R_4} + \frac{g}{R_3}$$

is also obvious.

Now, in the last equation containing the third-order derivative with respect to time of i_2 we will introduce V_1 and $\frac{dV_1}{dt}$ from Eq. (4.11.251) as being $\frac{i_2}{\sigma}$ and, respectively, $\frac{1}{\sigma}\frac{di_2}{dt}$ and, finally, we have

$$L_2\frac{d^3i_2}{dt^3} + R_2\frac{d^2i_2}{dt^2} + \left[\left(\frac{1}{C_1} + \frac{1}{C_2} \right) + \frac{1}{\sigma} \left(\frac{g}{\beta L_1} - \frac{\beta}{C_1} \right) \right] \frac{di_2}{dt}$$

$$+ \frac{g}{L_1C_1} \left(\frac{1}{\beta} - \frac{1}{\sigma} \right) i_2 = 0. \qquad (4.11.252)$$

This is an ordinary, homogeneous, third-order differential equation with real, constant coefficients. The attached characteristic equation writes

$$L_2r^3 + R_2r^2 + \left[\left(\frac{1}{C_1} + \frac{1}{C_2} \right) + \frac{1}{\sigma} \left(\frac{g}{\beta L_1} - \frac{\beta}{C_1} \right) \right] r$$

$$+ \frac{g}{L_1C_1} \left(\frac{1}{\beta} - \frac{1}{\sigma} \right) = 0. \qquad (4.11.253)$$

Using the notations

$$\alpha_0 \equiv L_2, \ \alpha_1 \equiv R_2, \ \alpha_2 \equiv \left(\frac{1}{C_1} + \frac{1}{C_2} \right) + \frac{1}{\sigma} \left(\frac{g}{\beta L_1} - \frac{\beta}{C_1} \right),$$

$$\alpha_3 \equiv \frac{g}{L_1C_1} \left(\frac{1}{\beta} - \frac{1}{\sigma} \right),$$

Eq. (4.11.253) takes a form that is similar to Eq. (4.11.235), that is

$$P(r) \equiv \alpha_0r^3 + \alpha_1r^2 + \alpha_2r + \alpha_3 = 0. \qquad (4.11.254)$$

Therefore, the stability of the studied circuit can be most easily investigated by means of the Hurwitz stability criterion. In agreement with this criterion, the system stability implies the simultaneous validity of the following inequalities:

$$D_1 = \alpha_1 > 0, \ D_2 = \begin{vmatrix} \alpha_1 & \alpha_0 \\ \alpha_3 & \alpha_2 \end{vmatrix} > 0,$$

$$D_3 = \begin{vmatrix} \alpha_1 & \alpha_0 & 0 \\ \alpha_3 & \alpha_2 & \alpha_1 \\ \alpha_5 & \alpha_4 & \alpha_3 \end{vmatrix} = \begin{vmatrix} \alpha_1 & \alpha_0 & 0 \\ \alpha_3 & \alpha_2 & \alpha_1 \\ 0 & 0 & \alpha_3 \end{vmatrix} > 0,$$

that is

$$\begin{cases} \alpha_1 > 0, \\ \alpha_1\alpha_2 - \alpha_0\alpha_3 > 0, \\ \alpha_1\alpha_2\alpha_3 - \alpha_0\alpha_3^2 = \alpha_3(\alpha_1\alpha_2 - \alpha_0\alpha_3) > 0, \end{cases}$$

which implies the simultaneous fulfillment of the following new inequalities:

$$\alpha_1 > 0, \quad \alpha_3 > 0, \quad \alpha_1\alpha_2 - \alpha_0\alpha_3 > 0.$$

In the concrete case of our problem, that means:

1) $\alpha_1 > 0 \ \Leftrightarrow \ R_2 > 0$ (which is obvious);

2) $\alpha_3 > 0 \ \Leftrightarrow \ \dfrac{g}{L_1 C_1}\left(\dfrac{1}{\beta} - \dfrac{1}{\sigma}\right) > 0$. Since $L_1 > 0$ and $C_1 > 0$, the fulfillment

of this inequality requires $g > 0$ and $\dfrac{1}{\beta} > \dfrac{1}{\sigma} \ \Leftrightarrow \ g > -\dfrac{R_3}{R_1}$;

3) $\alpha_1\alpha_2 - \alpha_0\alpha_3 > 0 \quad \Leftrightarrow$

$$R_2\left[\left(\frac{1}{C_1} + \frac{1}{C_2}\right) + \frac{1}{\sigma}\left(\frac{g}{\beta L_1} - \frac{\beta}{C_1}\right)\right] > \frac{g}{C_1}\frac{L_2}{L_1}\left(\frac{1}{\beta} - \frac{1}{\sigma}\right).$$

Given the expressions of β and σ, the last condition can be written as

$$Ag^2 + Bg + C > 0, \tag{4.11.255}$$

where

$$A \equiv R_1^2\left(R_2R_3L_1C_1 + R_2R_4L_1C_1 + R_2R_4L_1C_2 - R_4^2L_2C_2\right), \tag{4.11.256}$$

$$\begin{aligned} B \equiv R_1\big(&R_1R_2R_3R_4^2C_1C_2 + R_2R_3R_4L_1C_2 - 2R_1R_2R_3L_1C_1 \\ &-R_1R_2R_4L_1C_1 - R_2R_3R_4L_1C_1 - R_2R_4^2L_1C_1 \\ &-R_1R_2R_4L_1C_2 - R_2R_4^2L_1C_2 - R_3R_4^2L_2C_2\big), \end{aligned} \tag{4.11.257}$$

and

$$C \equiv R_2R_3L_1\left(R_1^2C_1 + R_1R_4C_1 - R_4^2C_2 - R_1R_4C_2\right). \tag{4.11.258}$$

Considering the fact that the parameter g (which, according to the conditions of the problem, is a real, positive and finite number) must satisfy the inequality (4.11.255), which implies a quadratic inequation in g, in order to exist values of g that are able to verify this inequality — under the given conditions — it is necessary that:

a) $A < 0$, that is the function $f(g) = Ag^2 + Bg + C$ must have a maximum point, and not a minimum one (the branches of the parable must be facing down; otherwise, the positive variable g would diverge);

b) The equation $f(g) = Ag^2 + Bg + C = 0$ must have real roots, therefore its discriminant must be positive: $\Delta = B^2 - 4AC > 0$;

c) At least one solution of the equation $f(g) = 0$ must be positive, otherwise the searched values of g (those situated between the two solutions of the equation $f(g) = 0$) could not be positive. Since $A < 0$, for this requirement to be satisfied, it is necessary that:

    i) $C > 0$ — to have different sign solutions for the equation $f(g) = 0$. In this case, the allowed range for the values of g is $g \in (0, g_p)$, where g_p is the positive solution;

    ii) $C < 0$ — to have solutions of the same sign for the equation $f(g) = 0$. In this case, for the solutions to be positive (which is mandatory), it is necessary that $B > 0$ (otherwise — *i.e.*, if $B < 0$ — both solutions would be negative).

Since the solution of the problem depends on a large number of "variables" (there are no less than eight quantities playing the role of real positive parameters, namely $R_1, R_2, R_3, R_4, L_1, L_2, C_1, C_2$), its finding generally involves a laborious analysis. We will finish our investigation by considering a single particular case, by taking, for instance, $R_1 = 1.2 \times 10^5\,\Omega, R_2 = 10\,\Omega, R_3 = 10^5\,\Omega, R_4 = 1.6 \times 10^5\,\Omega, L_1 = 4.5 \times 10^{-3}\,\text{H}, L_2 = 2 \times 10^{-3}\,\text{H}, C_1 = 1.2 \times 10^{-8}\,\text{F}$ and $C_2 = 2 \times 10^{-4}\,\text{F}$. In this case the equation $f(g) = 0$ writes

$$f(g) = -1.47435 \times 10^{14}\, g^2 + 7.61825 \times 10^{14}\, g - 4.03182 \times 10^{10} = 0,$$

and has the solutions

$$g_1 = 0.0000529237 \simeq 0; \quad g_2 = 5.16713 \simeq 5.$$

Corroborating this result to those from the points 1) — which is automatically satisfied — and 2) — which imposes $g > -R_3/R_1 = -0,8(3)$ — as well as with the requirement that g be real, positive, and finite, it results the following interval for the acceptable values of the parameter g (the value range of g for which the circuit can be stable — of course, only in the particular considered case):

$$g \in (g_1, g_2) \quad \Leftrightarrow \quad g \in (0, 5).$$

5

Liénard-Wiechert Potentials. Electromagnetic Waves

5.1 Problem No. 38

Determine the electrodynamic potentials $\vec{A}(\vec{r},t)$ and $V(\vec{r},t)$ associated to the electromagnetic field created by an electric point charge, moving in vacuum, its movement being given.

Solution

Let e $(e > 0)$ be the charge of the particle, $\vec{x} = \vec{x}(t)$ its trajectory, written in a parametric form (see Fig. 5.1), and c the speed of light in vacuum.

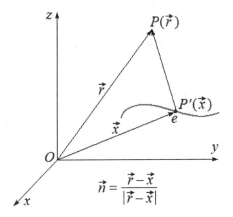

FIGURE 5.1
Some point $P(\vec{r})$ in three-dimensional space where the field generated by a moving charged particle is calculated/determined and the trajectory $\vec{x} = \vec{x}(t)$ of the relativistic charged particle.

It is convenient to write the velocity of the particle as

$$\frac{d\vec{x}}{dt} \equiv c\vec{\beta}(t), \qquad (5.1.1)$$

DOI: 10.1201/9781003402602-5

while the charge and current densities, $\rho(\vec{r}, t)$ and $\vec{j}(\vec{r}, t)$, respectively, will be expressed by means of Dirac *delta* distribution as

$$\rho(\vec{r}, t) = e\,\delta\left[\vec{r} - \vec{x}(t)\right], \tag{5.1.2}$$

$$\vec{j}(\vec{r}, t) = ec\vec{\beta}(t)\delta\left[\vec{r} - \vec{x}(t)\right]. \tag{5.1.3}$$

To meet the requirements of the problem, we will use the well-known expressions of the *retarded (causal) potentials*,

$$V(\vec{r}, t) = \frac{1}{4\pi\varepsilon_0} \int\limits_{(V')} \frac{\rho\left(\vec{r}', t - \frac{|\vec{r} - \vec{r}'|}{c}\right)}{|\vec{r} - \vec{r}'|} d\vec{r}', \tag{5.1.4}$$

$$\vec{A}(\vec{r}, t) = \frac{\mu_0}{4\pi} \int\limits_{(V')} \frac{\vec{j}\left(\vec{r}', t - \frac{|\vec{r} - \vec{r}'|}{c}\right)}{|\vec{r} - \vec{r}'|} d\vec{r}', \tag{5.1.5}$$

where V' is the volume of the domain D' in which the sources $\rho(\vec{r}', t')$ and $\vec{j}(\vec{r}', t')$ lie, while $\vec{r}'$ is the position vector of an arbitrary point $P'(\vec{r}') \in D'$ in this domain. Let us first calculate the scalar potential $V(\vec{r}, t)$. According to Eqs. (5.1.2) and (5.1.4), we have

$$V(\vec{r}, t) = \frac{e}{4\pi\varepsilon_0} \int\limits_{(V')} \frac{\delta\left[\vec{r}' - \vec{x}\left(t - \frac{|\vec{r} - \vec{r}'|}{c}\right)\right]}{|\vec{r} - \vec{r}'|} d\vec{r}'. \tag{5.1.6}$$

Observing that the *delta* distribution interfering in the integral can be written as

$$\delta\left[\vec{r}' - \vec{x}\left(t - \frac{|\vec{r} - \vec{r}'|}{c}\right)\right]$$

$$= \int \delta[\vec{r}' - \vec{x}(t')]\delta\left(t' - t + \frac{|\vec{r} - \vec{r}'|}{c}\right) dt',$$

we still have

$$V(\vec{r}, t) = \frac{e}{4\pi\varepsilon_0} \int dt' \int \frac{\delta[\vec{r}' - \vec{x}(t')]\delta\left(t' - t + \frac{|\vec{r} - \vec{r}'|}{c}\right)}{|\vec{r} - \vec{r}'|} d\vec{r}'$$

$$= \frac{e}{4\pi\varepsilon_0} \int dt' \frac{1}{|\vec{r} - \vec{x}(t')|}\delta\left(t' - t + \frac{|\vec{r} - \vec{x}(t')|}{c}\right). \tag{5.1.7}$$

On the other hand, the *delta* distribution theory says that (see **Appendix E**):

$$\int g(t')\delta[f(t') - t]dt' = \left[\frac{g(t')}{\frac{df}{dt'}}\right]_{f(t')=t}. \tag{5.1.8}$$

Since in our case

$$g(t') = \frac{1}{|\vec{r} - \vec{x}(t')|}; \; f(t') = t' + \frac{|\vec{r} - \vec{x}(t')|}{c},$$

and

$$\frac{df}{dt'} = 1 + \frac{1}{c}\frac{d}{dt'}\left\{\left[r_i - x_i(t')\right]\left[r_i - x_i(t')\right]\right\}^{1/2}$$

$$= 1 + \frac{1}{c}\frac{r_i - x_i(t')}{|\vec{r} - \vec{x}(t')|}\left(-\frac{dx_i}{dt'}\right) = 1 - \frac{r_i - x_i(t')}{|\vec{r} - \vec{x}(t')|}\beta_i = 1 - \vec{n}\cdot\vec{\beta},$$

where

$$\vec{n} = \frac{\vec{r} - \vec{x}(t')}{|\vec{r} - \vec{x}(t')|}$$

is the unit vector of the direction $\vec{r} - \vec{x}(t')$, we finally have

$$V(\vec{r}, t) = \left[\frac{e}{4\pi\varepsilon_0 |\vec{r} - \vec{x}|\left(1 - \vec{n}\cdot\vec{\beta}\right)}\right]_{t=t'+\frac{|\vec{r}-\vec{x}(t')|}{c}}. \qquad (5.1.9)$$

Here $\vec{n}, \vec{\beta}$ and $\vec{x}$ are functions of t'.

In a similar way can be determined the vector potential

$$\vec{A}(\vec{r}, t) = \left[\frac{\mu_0 ec\vec{\beta}}{4\pi |\vec{r} - \vec{x}|\left(1 - \vec{n}\cdot\vec{\beta}\right)}\right]_{t=t'+\frac{|\vec{r}-\vec{x}(t')|}{c}}. \qquad (5.1.10)$$

The potentials determined by relations (5.1.9) and (5.1.10) are called *Liénard-Wiechert potentials*. They describe the classical electromagnetic effect of a moving electric point charge in terms of a scalar potential and a vector potential in the Lorenz gauge. Stemming directly from Maxwell's equations, these potentials describe the complete, relativistically correct, time-varying electromagnetic field for a point charge in arbitrary motion, but are not corrected for quantum effects. Electromagnetic radiation in the form of waves can be obtained from these potentials. These expressions were developed in part by Alfred-Marie Liénard in 1898 and independently by Emil Wiechert in 1900. Of course, if we want to find the two components $\vec{E}(\vec{r}, t)$ and $\vec{B}(\vec{r}, t)$ of the electromagnetic field generated by a moving electric point charge in arbitrary motion, we should use the well-known relationships between the fields and potentials, namely

$$\begin{cases} \vec{E}(\vec{r}, t) = -\nabla V(\vec{r}, t) - \dfrac{\partial \vec{A}(\vec{r}, t)}{\partial t}, \\ \vec{B}(\vec{r}, t) = \nabla \times \vec{A}(\vec{r}, t). \end{cases}$$

In view of the above considerations, it is useful to mention that it is not correct to express, for example, the scalar potential of the field produced by a number of moving electrons (in vacuum) by means of the formula

$$V(\vec{r}, t) = \frac{1}{4\pi\varepsilon_0} \int\limits_{(V')} \frac{\rho(\vec{r}', t')}{|\vec{r} - \vec{r}'|} \, d\vec{r}' = \frac{1}{4\pi\varepsilon_0} \frac{q}{R},$$

because the microscopic charge density depends on t' (both directly and indirectly, through $\vec{r}'(t')$), which, in its turn, depends on $|\vec{r} - \vec{r}'|$. Therefore, the correct result is that expressed by Eq. (5.1.9).

5.2   Problem No. 39

Interpret the *Lorenz's gauge condition*

$$\nabla \cdot \vec{A} + \varepsilon\mu \frac{\partial V}{\partial t} = 0,$$

in the light of the results obtained by solving the non-homogeneous, second order partial differential equations satisfied by the electrodynamic potentials $V(\vec{r}, t)$ and $\vec{A}(\vec{r}, t)$.

Solution

In electrodynamics, the retarded potentials are the electromagnetic/electrodynamic potentials for the electromagnetic field generated by time-varying charge and/or electric current distributions in the past. The fields propagate at the speed of light c (if the propagation takes place in vacuum), so the delay of the fields connecting cause and effect at earlier and later times is an important factor: the signal takes a finite time to propagate from a point in the domain D' of charge and/or current distributions (the point $P'(\vec{r}') \in D'$ of cause) to another point in space (the point $P(\vec{r}) \in D$, where the effect is measured – see Fig. 5.2).

These potentials are also called *causal potentials* or *delayed potentials* and they are solutions (most elegantly obtained by using the Green's function method of solving linear differential equations – see **Appendix G**), of the following non-homogeneous, second-order partial differential equations:

$$\begin{cases} \Delta V(\vec{r}, t) - \varepsilon\mu \dfrac{\partial^2 V(\vec{r}, t)}{\partial t^2} = -\dfrac{\rho(\vec{r}, t)}{\varepsilon}, \\[4mm] \Delta \vec{A}(\vec{r}, t) - \varepsilon\mu \dfrac{\partial^2 \vec{A}(\vec{r}, t)}{\partial t^2} = -\mu \vec{j}(\vec{r}, t), \end{cases}$$

where ε and μ are the electric permittivity and magnetic permeability, respectively, of the medium through which the electromagnetic perturbation propagates.

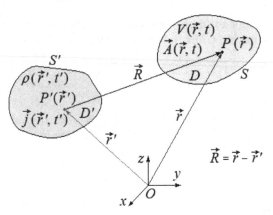

FIGURE 5.2

The domain D' of the sources $\rho(\vec{r}', t')$ and $\vec{j}(\vec{r}', t')$ (the cause) and the domain D of the fields $V(\vec{r}, t)$ and $\vec{A}(\vec{r}, t)$ (the effect).

Let us denote by x, y, z, t the space-time coordinates of the observation point, $P(\vec{r}) \in D$, and by $x', y', z', t'\left(= t - \frac{R}{u} \right)$ the space-time coordinates of the current point $P'(\vec{r}') \in D'$ from the domain occupied by sources, where $R = |\vec{r} - \vec{r}'|$ is the distance between the two points. The relation known as *Lorenz's condition*,

$$\nabla \cdot \vec{A} + \varepsilon\mu \frac{\partial V}{\partial t} = 0, \qquad (5.2.11)$$

must be satisfied in the observation point (the point of effect), while the sources ρ and $\vec{j}$ must depend on the space-time coordinates of the "cause point".

As one knows, the retarded electrodynamic potentials are given by

$$V(\vec{r}, t) = \frac{1}{4\pi\varepsilon_0} \int\limits_{(V')} \frac{\rho\left(\vec{r}', t - \frac{|\vec{r} - \vec{r}'|}{u} \right)}{|\vec{r} - \vec{r}'|} d\vec{r}' \equiv \frac{1}{4\pi\varepsilon_0} \int\limits_{(V')} \frac{\rho(\vec{r}', t')}{R} d\vec{r}', \quad (5.2.12)$$

$$\vec{A}(\vec{r}, t) = \frac{\mu_0}{4\pi} \int\limits_{(V')} \frac{\vec{j}\left(\vec{r}', t - \frac{|\vec{r} - \vec{r}'|}{u} \right)}{|\vec{r} - \vec{r}'|} d\vec{r}' \equiv \frac{\mu_0}{4\pi} \int\limits_{(V')} \frac{\vec{j}(\vec{r}', t')}{R} d\vec{r}'. \quad (5.2.13)$$

Thus we have

$$\frac{\partial V(\vec{r}, t)}{\partial t} = \frac{1}{4\pi\varepsilon} \int\limits_{(V')} \frac{\partial}{\partial t} \left[\frac{\rho\left(\vec{r}', t - \frac{|\vec{r} - \vec{r}'|}{u} \right)}{|\vec{r} - \vec{r}'|} \right] d\vec{r}'$$

$$= \frac{1}{4\pi\varepsilon} \int\limits_{(V')} \frac{1}{R} \frac{\partial \rho\left(\vec{r}', t - \frac{R}{u} \right)}{\partial \left(t - \frac{R}{u} \right)} \frac{\partial \left(t - \frac{R}{u} \right)}{\partial t} d\vec{r}'$$

$$= \frac{1}{4\pi\varepsilon} \int\limits_{(V')} \frac{1}{R} \frac{\partial\rho\left(\vec{r}',t-\frac{R}{u}\right)}{\partial\left(t-\frac{R}{u}\right)} d\vec{r}' = \frac{1}{4\pi\varepsilon} \int\limits_{(V')} \frac{1}{R} \frac{\partial\rho\left(\vec{r}',t'\right)}{\partial t'} d\vec{r}'.$$

To determine $\nabla \cdot \vec{A}$, let's do some preliminary calculations. Since ∇ acts only on coordinates without prime (the coordinates of the observation/"effect" point), while ∇' acts only on coordinates with prime (the coordinates of the sources/"cause" point), we have

$$\nabla \cdot \vec{j} \equiv \nabla \cdot \vec{j}(\vec{r}',t') = \nabla \cdot \vec{j}\left(\vec{r}',t-\frac{R}{u}\right) = \frac{\partial\vec{j}}{\partial\left(t-\frac{R}{u}\right)} \cdot \nabla\left(t-\frac{R}{u}\right)$$

$$= -\frac{1}{u} \frac{\partial\vec{j}}{\partial\left(t-\frac{R}{u}\right)} \cdot \nabla R = \frac{1}{u} \frac{\partial\vec{j}}{\partial\left(t-\frac{R}{u}\right)} \cdot \nabla'R = \frac{1}{u} \frac{\partial\vec{j}}{\partial t'} \cdot \nabla'R,$$

and

$$\nabla' \cdot \vec{j} \equiv \nabla' \cdot \vec{j}(\vec{r}',t') = \nabla' \cdot \vec{j}\left(\vec{r}',t-\frac{R}{u}\right)$$

$$= \left(\nabla' \cdot \vec{j}\right)_{t-\frac{R}{u}=const.} - \frac{1}{u} \frac{\partial\vec{j}}{\partial\left(t-\frac{R}{u}\right)} \cdot \nabla'R$$

$$= \left(\nabla' \cdot \vec{j}\right)_{t-\frac{R}{u}=const.} - \frac{1}{u} \frac{\partial\vec{j}}{\partial t'} \cdot \nabla'R,$$

leading to

$$\nabla \cdot \vec{j} = \left(\nabla' \cdot \vec{j}\right)_{t-\frac{R}{u}=const.} - \nabla' \cdot \vec{j}.$$

Using the above results, we can write

$$\nabla \cdot \vec{A} \equiv \nabla \cdot \vec{A}(\vec{r},t) = \frac{\mu}{4\pi} \nabla \cdot \left(\int\limits_{(V')} \frac{\vec{j}(\vec{r}',t')}{|\vec{r}-\vec{r}'|} d\vec{r}'\right) = \frac{\mu}{4\pi}$$

$$\times \int\limits_{(V')} \nabla \cdot \left(\frac{\vec{j}(\vec{r}',t')}{R}\right) d\vec{r}' = \frac{\mu}{4\pi} \int\limits_{(V')} \left[\frac{1}{R} \nabla \cdot \vec{j}(\vec{r}',t')\right.$$

$$\left. + \vec{j}(\vec{r}',t') \cdot \nabla\left(\frac{1}{R}\right)\right] d\vec{r}' = \frac{\mu}{4\pi} \int\limits_{(V')} \left\{\frac{1}{R}\left[\left(\nabla' \cdot \vec{j}\right)_{t-\frac{R}{u}=const.}\right.\right.$$

$$\left.\left. - \nabla' \cdot \vec{j}\right] - \vec{j} \cdot \nabla'\left(\frac{1}{R}\right)\right\} d\vec{r}' = \frac{\mu}{4\pi} \int\limits_{(V')} \frac{1}{R}\left(\nabla' \cdot \vec{j}\right)_{t-\frac{R}{u}=const.}$$

$$\times d\vec{r}' - \frac{\mu}{4\pi} \int\limits_{(V')} \nabla' \cdot \left(\frac{\vec{j}}{R}\right) d\vec{r}' = \frac{\mu}{4\pi} \int\limits_{(V')} \frac{1}{R}\left(\nabla' \cdot \vec{j}\right)_{t-\frac{R}{u}=const.}$$

$$\times d\vec{r}' - \frac{\mu}{4\pi} \oint\limits_{(S')} \frac{1}{R}\vec{j} \cdot d\vec{S}' = \frac{\mu}{4\pi} \int\limits_{(V')} \frac{1}{R}\left(\nabla' \cdot \vec{j}\right)_{t-\frac{R}{u}=const.} d\vec{r}',$$

where we took into account that $\vec{j}\big|_{S'} = 0$ (the currents are distributed inside the volume V' and not on the surface S').

So we found that

$$\frac{\partial V(\vec{r}, t)}{\partial t} = \frac{1}{4\pi\varepsilon} \int\limits_{(V')} \frac{1}{R} \frac{\partial \rho(\vec{r}', t')}{\partial t'} d\vec{r}'. \tag{5.2.14}$$

and

$$\nabla \cdot \vec{A} = \frac{\mu}{4\pi} \int\limits_{(V')} \frac{1}{R} \left(\nabla' \cdot \vec{j}\right)_{t - \frac{R}{u} = const.} d\vec{r}'. \tag{5.2.15}$$

Introducing these results into Lorenz's condition (5.2.11), we have

$$\nabla \cdot \vec{A} + \varepsilon\mu \frac{\partial V}{\partial t}$$

$$= \frac{\mu}{4\pi} \int_V \frac{1}{R} \left[\frac{\partial \rho}{\partial \left(t - \frac{R}{u}\right)} + \left(\nabla' \cdot \vec{j}\right)_{t - \frac{R}{u} = const.} \right] d\vec{r}' = 0, \tag{5.2.16}$$

resulting in

$$\frac{\partial \rho}{\partial \left(t - \frac{R}{u}\right)} + \left(\nabla' \cdot \vec{j}\right)_{t - \frac{R}{u} = const.} = 0, \tag{5.2.17}$$

which is nothing else but the *continuity equation* in electromagnetism, valid at the moment $t' = t - \frac{R}{u}$ (the time of the sources), at an arbitrary point $P'(\vec{r}')$ from the domain of the sources D', as it should be.

We draw the reader's attention once again that while the Lorenz's gauge condition must be satisfied at the "effect point" and at the "effect time" (*i.e.*, *where* and *when* the field is measured), the continuity equation must be satisfied in the "cause domain" and at the "cause time" (*i.e.*, *where* the sources are located and *when* they vary to produce the electromagnetic perturbation/wave), as it is natural.

Thus, the Lorenz gauge condition is satisfied by the retarded electrodynamic potentials given by Eqs. (5.1.4) and (5.1.5), as it translates into the equation of continuity for the electric charge. However, the Lorenz gauge condition *is not equivalent* to the equation of continuity. For example, if the solutions of Maxwell's equations are found in another gauge, say the Coulomb gauge, and we plug the solutions into the Lorenz gauge condition, the latter will not be satisfied. However, the equation of continuity for electric charge will always be valid, irrespective of which gauge condition we are using. The equation of continuity expresses a physical law, which is the conservation of electric charge, while gauge fixing conditions do not have any physical significance, being just some supplementary relations by which we pick up a certain form for the potentials out of an infinity of physically equivalent possibilities.

5.3   Problem No. 40

Using the Hertz vector/potential $\vec{Z}(\vec{r}, t)$, determine the electromagnetic field generated by an oscillating electric dipole, located at the coordinates origin and oriented along the z-axis.

Complements

The Lorenz's condition written for vacuum $(\vec{j} = 0,\ \rho = 0)$:

$$\mathrm{div}\,\vec{A} + \varepsilon_0 \mu_0 \frac{\partial V}{\partial t} = 0 \tag{5.3.18}$$

is satisfied if we choose

$$V = -\,\mathrm{div}\,\vec{Z}, \quad \vec{A} = \varepsilon_0 \mu_0 \frac{\partial \vec{Z}}{\partial t}, \tag{5.3.19}$$

where the vector $\vec{Z}$ is called *Hertz vector/potential*. The electromagnetic field is then expressed by the relations

$$\begin{cases} \vec{E} = -\nabla V - \dfrac{\partial \vec{A}}{\partial t} = \nabla\big(\nabla \cdot \vec{Z}\big) - \varepsilon_0 \mu_0 \dfrac{\partial^2 \vec{Z}}{\partial t^2}, \\[2mm] \vec{B} = \nabla \times \vec{A} = \varepsilon_0 \mu_0 \nabla \times \left(\dfrac{\partial \vec{Z}}{\partial t} \right). \end{cases} \tag{5.3.20}$$

Just like the electrodynamic potentials V and $\vec{A}$, the Hertz potential is submitted to a gauge transformation of the form

$$\vec{Z}' = \vec{Z} + \nabla \times \vec{F}, \tag{5.3.21}$$

where $\vec{F}$ is a vector function which, in agreement with relations that define V and $\vec{A}$, does not explicitly depend on time. If the electric displacement field (electric induction) $\vec{D}$ is expressed in terms of the electric polarization vector $\vec{P}$,

$$\vec{D} = \varepsilon_0 \vec{E} + \vec{P},$$

and Eq. (5.3.20) is introduced into Maxwell's source equation

$$\nabla \times \vec{B} = \varepsilon_o \mu_0 \frac{\partial \vec{E}}{\partial t} + \mu_0 \frac{\partial \vec{P}}{\partial t},$$

one finds

$$\Delta \vec{Z} - \varepsilon_0 \mu_0 \frac{\partial^2 \vec{Z}}{\partial t^2} = -\frac{1}{\varepsilon_0} \vec{P}. \tag{5.3.22}$$

This is a second-order partial differential equation of hyperbolic type, non-homogeneous, perfectly similar to equations satisfied by the electrodynamic potentials V and $\vec{A}$, the polarization density $\vec{P}$ being the source of the vector field $\vec{Z}$. The causal solution of Eq. (5.3.22) is – as one knows – the retarded potential

$$\vec{Z}(\vec{r},t) = \frac{1}{4\pi\varepsilon_0} \int\limits_{(V')} \frac{\vec{P}\left(\vec{r}',t-\frac{R}{c}\right)}{R} d\vec{r}', \quad R = |\vec{r} - \vec{r}'|. \tag{5.3.23}$$

It is not difficult to show that if one considers a time-periodical variation of the source of the form

$$\vec{P}\left(\vec{r}', t - \frac{R}{c}\right) = \vec{P}(\vec{r}')e^{-i\omega t}e^{ikR},$$

then, in the approximation $R \simeq r$, the Hertz vector/potential becomes

$$\vec{Z} \simeq \frac{1}{4\pi\varepsilon_0} \frac{\vec{p}}{r}, \tag{5.3.24}$$

where

$$\vec{p} = \vec{p_0}e^{i(kr-\omega t)}$$

is the *oscillating electric dipole moment*, with

$$\vec{p_0} \equiv \int\limits_{(V')} \vec{P}(\vec{r}')d\vec{r}'.$$

In the same approximation, in accordance with Eq. (5.3.19), the vector potential $\vec{A}$ is given – in this case – by

$$\vec{A} = \frac{\mu_0}{4\pi} \frac{\dot{\vec{p}}}{r}.$$

Solution

According to the problem conditions, $|\vec{p}| = p_z = p$, and then

$$Z_z \equiv Z = \frac{1}{4\pi\varepsilon_o} \frac{p_0 e^{i(kr-\omega t)}}{r}, \quad A_z \equiv A = -\frac{i\omega\mu_0}{4\pi} \frac{p}{r}, \tag{5.3.25}$$

leading to

$$A = \frac{-ik}{4\pi} \sqrt{\frac{\mu_0}{\varepsilon_0}} \frac{p}{r}. \tag{5.3.26}$$

Also,

$$V = -\frac{1}{4\pi\varepsilon_0} \vec{p_0} \cdot \nabla \left(\frac{e^{i(kr-\omega t)}}{r} \right),$$

or, if we neglect the terms containing r^{-n} for $n \geq 2$,

$$V = \frac{p\cos\theta}{4\pi\varepsilon_0 r}\left(\frac{1}{r} - ik\right). \tag{5.3.27}$$

Due to the spherical symmetry of the problem, we also have to know the corresponding components of the vector $\vec{A}$, namely

$$A_r = A\cos\theta, \quad A_\theta = -A\sin\theta, \quad A_\varphi = 0. \tag{5.3.28}$$

Now, we can calculate the field at some point M, distant enough so that we can use the approximation $\mathcal{O}(r^{-n}, \ n \geq 2) = 0$. Using only the real part of the above results, we then have

$$\begin{cases} E_r = -\dfrac{\partial V}{\partial r} - \dfrac{\partial A_r}{\partial t} = -\dfrac{p\cos\theta}{4\pi\varepsilon_0}\left(\dfrac{ik}{r^2} - \dfrac{2}{r^3}\right) \\[2mm] \qquad -\dfrac{ikp\cos\theta}{4\pi r}\left(\dfrac{1}{r} - ik\right) \simeq 0, \\[3mm] E_\theta = -\dfrac{1}{r}\dfrac{\partial V}{\partial t} - \dfrac{\partial A_\theta}{\partial t} = \dfrac{p\sin\theta}{4\pi\varepsilon_0 r^2}\left(\dfrac{1}{r} - ik\right) - \dfrac{\omega kp\sin\theta}{4\pi r}\sqrt{\dfrac{\mu_0}{\varepsilon_0}} \\[3mm] \qquad \simeq -\dfrac{\omega^2\sin\theta}{4\pi\varepsilon_0 c^2 r}p_0\cos(kr - \omega t), \\[3mm] E_\varphi = -\dfrac{1}{r\sin\theta}\dfrac{\partial V}{\partial\varphi} = 0. \end{cases} \tag{5.3.29}$$

The components of the field $\vec{B}$ are calculated in the same way. The result is the following:

$$\begin{cases} B_r = \dfrac{1}{r^2\sin\theta}\left[\dfrac{\partial}{\partial\theta}\left(\sin\theta A_\varphi\right) - \dfrac{\partial}{\partial\varphi}\left(rA_\theta\right)\right] = 0, \\[3mm] B_\theta = \dfrac{1}{r\sin\theta}\left[\dfrac{\partial}{\partial\varphi}\left(A_r\right) - \dfrac{\partial}{\partial r}\left(r\sin\theta A_\varphi\right)\right] = 0, \\[3mm] B_\varphi = \dfrac{1}{r}\left[\dfrac{\partial}{\partial r}\left(rA_\theta\right) - \dfrac{\partial}{\partial\theta}\left(A_r\right)\right] \simeq -\dfrac{\omega^2\mu_0\sin\theta}{4\pi rc}p_0\cos(kr - \omega t). \end{cases} \tag{5.3.30}$$

As was to be expected, we have

$$|E_\theta| = c|B_\varphi|. \tag{5.3.31}$$

Observation. Using Eqs. (5.3.19) and (5.3.25), one easily finds

$$\begin{cases} \vec{E} = -k^2 Z\cos\theta\,\vec{s} + k^2 Z\,\vec{u}_z, \\[2mm] \vec{B} = -\dfrac{k^2}{c}Z\sin\theta\,\vec{u}_\varphi, \end{cases}$$

where $\vec{s}$ is the unit vector of direction of the wave propagation. The relations resulting by expressing Z in terms of p, projected on the spherical coordinates directions, lead to Eqs. (5.3.29) and (5.3.30).

5.4   Problem No. 41

Determine the angular distribution of the temporal average of the power radiated per unit solid angle by a thin, linear antenna of length d, which is excited at its middle, supposing that the current distribution along the antenna is sinusoidal.

Solution

Suppose that the antenna is oriented along z-axis, as shown in Fig. 5.3. If

$$\vec{\tilde{\Pi}} = \frac{1}{2}\vec{E} \times \vec{H}^*$$

is the *complex Poynting vector*, then the *density of the average flow density of the electromagnetic energy released per unit time* (or, differently saying, the *time average of the radiated power per unit surface*) is the real part of the normal component of the $\vec{\tilde{\Pi}}$ vector,

$$\langle \varphi_{em} \rangle \equiv \frac{dP}{dS} = \text{Re}\,\tilde{\Pi}_n = \frac{1}{2}\text{Re}\big[\vec{n} \cdot (\vec{E} \times \vec{H}^*)\big], \qquad (5.4.32)$$

where $\langle\,\rangle$ means the average over a period, and $\vec{n}$ is the unit vector in the direction of $\vec{r}$ (which direction is perpendicular on the surface S that determines − by means of r − the solid angle Ω).

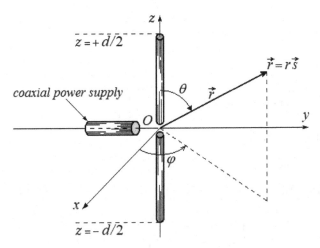

FIGURE 5.3
Orientation of the linear, centrally-fed antenna with respect to a Cartesian reference frame.

This follows immediately if one takes into account the relationship

$$\langle \vec{\Pi} \rangle \equiv \langle \vec{E} \times \vec{H} \rangle = \langle \tilde{\vec{\Pi}} \rangle = \frac{1}{2}\mathrm{Re}\big(\vec{E} \times \vec{H}^{*}\big), \qquad (5.4.33)$$

where $\vec{\Pi} = \vec{E} \times \vec{H} = EH\vec{s}$ is the (real) Poynting vector, which has the significance of the radiant *energy flux* (the energy transfer per unit area per unit time) or *power flow* of an electromagnetic field, provided the electromagnetic energy flows through a surface that is perpendicular to the vector $\vec{\Pi}$. The direction and sense of the vector $\vec{\Pi}$ are given by the vector $\vec{s}$, which is the versor (the unit vector) of the direction of propagation of the electromagnetic wave.

The time-averaged power per unit solid angle radiated by the antenna then is

$$\frac{dP}{d\Omega} = \mathrm{Re}\big(r^{2}\tilde{\Pi}_{n}\big) = \frac{1}{2}\mathrm{Re}\Big[r^{2}\vec{n} \cdot \big(\vec{E} \times \vec{H}^{*}\big)\Big]. \qquad (5.4.34)$$

Therefore, to calculate $dP/d\Omega$ one must first determine the fields $\vec{E}$ and $\vec{H}$. Supposing that antenna is in vacuum (or even in the air, for which $\mu_r \simeq 1$), the magnetic component of the field can be determined by using the relation

$$\vec{H} = \frac{1}{\mu_0}\vec{B} = \frac{1}{\mu_0}\nabla \times \vec{A},$$

while the component $\vec{E}$ can be obtained by using the following Maxwell's equation – assuming that there are no conduction currents in the area where the electromagnetic field is determined:

$$\nabla \times \vec{B}\big(= \mu_0 \nabla \times \vec{H}\big) = \frac{1}{c^2}\frac{\partial \vec{E}}{\partial t}. \qquad (5.4.35)$$

If outside the source we consider for the component $\vec{E}$ a periodical time variation of the form

$$\vec{E}(\vec{r},t) = \vec{E}_0(\vec{r})e^{-i\omega t},$$

then Eq. (5.4.35) gives

$$\vec{E} = \frac{ic^2}{\omega}\nabla \times \vec{B} = \frac{ic^2\mu_0}{\omega}\nabla \times \vec{H} = \frac{i}{\omega\varepsilon_0}\nabla \times \vec{H} = \frac{i}{k}\sqrt{\frac{\mu_0}{\varepsilon_0}}\,\nabla \times \vec{H}. \qquad (5.4.36)$$

To conclude, in order to determine $dP/d\Omega$ it is enough to find the magnetic vector potential $\vec{A}$, since

$$\frac{dP}{d\Omega} = \mathrm{Re}\big(r^{2}\tilde{\Pi}_{n}\big) = \frac{1}{2}\mathrm{Re}\Big[r^{2}\vec{n} \cdot \big(\vec{E} \times \vec{H}^{*}\big)\Big]$$

$$= \frac{1}{2}\mathrm{Re}\left\{ \frac{ir^{2}}{k}\sqrt{\frac{\mu_0}{\varepsilon_0}}\Big[\big(\nabla \times \vec{H}\big) \times \vec{H}^{*}\Big] \cdot \vec{n} \right\}, \qquad (5.4.37)$$

and the magnetic field $\vec{H}$ (and, implicitly, $\vec{H}^{*}$) is directly expressed through the magnetic vector potential $\vec{A}$: $\vec{H} = \mu_0^{-1}\nabla \times \vec{A}$.

The magnetic vector potential $\vec{A}$ is determined as the solution of the equation

$$\Delta \vec{A}(\vec{r},t) - \varepsilon_0 \mu_0 \frac{\partial^2 \vec{A}(\vec{r},t)}{\partial t^2} = -\mu_0 \vec{J}(\vec{r},t). \qquad (5.4.38)$$

This last equation can be easily deduced by using Maxwell's equations and Lorenz's gauge condition. The most convenient way to determine the solution of Eq. (5.4.38) is to use the Green's function method, which yields the following expression for $\vec{A}(\vec{r},t)$:

$$\vec{A}(\vec{r},t) = \frac{\mu_0}{4\pi} \int d^3 \vec{r}' \int dt' \frac{J(\vec{r}',t')}{|\vec{r}-\vec{r}'|} \delta\left(t' + \frac{|\vec{r}-\vec{r}'|}{c} - t\right), \qquad (5.4.39)$$

if the boundary surfaces are absent and the electromagnetic perturbation propagates in vacuum. The Dirac's delta function assures the causal behaviour of the fields (meaning that the cause − the current $\vec{J}(\vec{r}',t')$ − must precede the effect which is the "field" $\vec{A}(\vec{r},t)$).

Since for a system of time-variable electric charges and currents can be performed a Fourier analysis of time-dependence of these quantities, and each component can be "manipulated" separately, there is no loss in generality if one considers the potentials, fields and radiation of a localized system of charges and currents (as in our case) as having a sinusoidal variation in time of the form

$$\vec{J}(\vec{r},t) = \vec{J}(\vec{r})e^{-i\omega t}. \qquad (5.4.40)$$

Using this time-dependence of $\vec{J}$ in Eq. (5.4.39), the solution for $\vec{A}(\vec{r},t)$ becomes

$$\vec{A}(\vec{r},t) = \frac{\mu_0}{4\pi} e^{-i\omega t} \int\limits_{(D')} \vec{J}(\vec{r}') \frac{e^{ik|\vec{r}-\vec{r}'|}}{|\vec{r}-\vec{r}'|} d^3 \vec{r}', \qquad (5.4.41)$$

where the sifting property (or sampling property − as sometimes this property is called) of Dirac's delta function has been used (see **Appendix E**, formula (E.1.6)), while D' represents the space domain where the sources are distributed/localized (in the case of our application, D' is represented by the closed interval $[-d/2, d/2]$, *i.e.*, in our particular case D' is a one-dimensional domain).

In the wave zone ($kr \gg 1$) the exponential $e^{ik|\vec{r}-\vec{r}'|}$ oscillates rapidly and determine the behaviour of the magnetic vector potential $\vec{A}(\vec{r},t)$. In this region one can approximate

$$|\vec{r}-\vec{r}'| \simeq r - \vec{n} \cdot \vec{r}'. \qquad (5.4.42)$$

This relation keeps its validity even for $r \gg d$ (where the distance d is of the order of source dimensions), independently of the value of the product kr. Therefore, this approximation is adequate even in the static neighbouring area, characterized by $d \ll r \ll \lambda$ and, even more appropriate in the intermediate (induction) zone, $d \ll r \sim \lambda$, where $\lambda = 2\pi c/\omega$ is the wave length of the electromagnetic radiation. More than that, if the principal term is the only required, the inverse of distance in Eq. (5.4.41) can be simply replaced by r and then the vector potential becomes

$$\vec{A}(\vec{r},t) = \frac{\mu_0}{4\pi} \frac{e^{i(kr-\omega t)}}{r} \int\limits_{(D')} \vec{J}(\vec{r}')e^{-ik(\vec{n}\cdot\vec{r}')}d^3\vec{r}'. \tag{5.4.43}$$

This investigation shows that in the far zone (wave zone) the magnetic vector potential behaves as an emerging spherical wave, with an angle-dependent coefficient. After all, it is easy to show that the relationships

$$\vec{H} = \frac{\vec{B}}{\mu_0} = \frac{1}{\mu_0}\nabla \times \vec{A}$$

and

$$\vec{E} = \frac{i}{k}\sqrt{\frac{\mu_0}{\varepsilon_0}}\nabla \times \vec{H}$$

lead to fields which are orthogonal to the position vector $\vec{r}$ and decrease as r^{-1} (radiative fields).

If the amortization due to emission of radiation is neglected and the antenna is thin enough, then one can consider with a good enough precision that along the antenna the current is sinusoidal in time and space, with the wave number $k = \omega/c$. In addition, the current is symmetrical on the two antenna arms and vanishes at its ends. These characteristics are mathematically modelled by

$$\vec{J}(\vec{r}) = I\sin\left(\frac{kd}{2} - k|z|\right)\delta(x)\delta(y)\vec{u}_3, \tag{5.4.44}$$

where $\vec{u}_3 \equiv \hat{z}$ is the unit vector of z-axis. The delta functions ensure the fact that the current "drain" takes place only along the z-axis. If $kd \geq \pi$, then I is the maximum value of the current, and at the coordinates origin (which is the central excitation point of the antenna) the current has the constant value $I\sin(kd/2)$.

By using Eq. (5.4.44), it follows from Eq. (5.4.43) that in the radiative zone (where relation (5.4.43) is valid) the magnetic vector potential $\vec{A}(\vec{r},t)$ is oriented in the direction of z-axis, and

$$\vec{A}(\vec{r},t) = \frac{\mu_0}{4\pi}\frac{e^{i(kr-\omega t)}}{r}\int\limits_{(D')} \vec{J}(\vec{r}')e^{-ik(\vec{n}\cdot\vec{r}')}d^3\vec{r}' = \vec{u}_3\frac{\mu_0 I}{4\pi}\frac{e^{i(kr-\omega t)}}{r}$$

$$\times \int\limits_{(D')}\sin\left(\frac{kd}{2} - k|z'|\right)e^{-ik(\vec{n}\cdot\vec{r}')}\delta(x')\delta(y')\,dx'dy'dz'$$

$$= \vec{u}_3\frac{\mu_0 I}{4\pi r}e^{ikr}e^{-i\omega t}\int_{-\infty}^{+\infty}\delta(x')dx'\int_{-\infty}^{+\infty}\delta(y')dy'$$

$$\times \int_{-d/2}^{+d/2}\sin\left(\frac{kd}{2} - k|z'|\right)e^{-ikz'\cos\theta}dz' = \vec{u}_3\frac{\mu_0 I}{4\pi r}e^{ikr}e^{-i\omega t}$$

$$\times \int_{-d/2}^{+d/2} \sin\left(\frac{kd}{2} - k|z'|\right) e^{-ikz'\cos\theta} dz', \tag{5.4.45}$$

where, without affecting the result, the integration domain over x' and y' has been extended from $-\infty$ to $+\infty$ and the property expressed by Eq. (E.1.5) has been used (see **Appendix E**).

Therefore, at least for the moment, our problem is to calculate the integral

$$J = \int_{-d/2}^{+d/2} \sin\left(\frac{kd}{2} - k|x|\right) e^{-ikx\cos\theta} dx. \tag{5.4.46}$$

To do this, following the usual procedure, one considers the auxiliary integral

$$K = \int_{-d/2}^{+d/2} \cos\left(\frac{kd}{2} - k|x|\right) e^{-ikx\cos\theta} dx, \tag{5.4.47}$$

and two combinations are formed, $C_1 = K - iJ$ and $C_2 = K + iJ$. These combinations are easier calculated and then the searched integral is given by

$$J = \frac{C_2 - C_1}{2i}.$$

This way, using Euler's formula, one finds

$$C_1 = K - iJ = \int_{-d/2}^{+d/2} e^{-i\left(\frac{kd}{2} - k|x|\right)} e^{-ikx\cos\theta} dx$$

$$= e^{-i\frac{kd}{2}} \int_{-d/2}^{+d/2} e^{-ik(x\cos\theta - |x|)} dx$$

$$= e^{-i\frac{kd}{2}} \left[\int_{-d/2}^{0} e^{-ik(x\cos\theta + x)} dx + \int_{0}^{+d/2} e^{-ik(x\cos\theta - x)} dx\right]$$

$$= e^{-i\frac{kd}{2}} \left[\int_{-d/2}^{0} e^{-ikx(1+\cos\theta)} dx + \int_{0}^{+d/2} e^{ikx(1-\cos\theta)} dx\right]$$

$$= e^{-i\frac{kd}{2}} \left\{\left[\frac{e^{-ikx(1+\cos\theta)}}{-ik(1+\cos\theta)}\right]_{-d/2}^{0} + \left[\frac{e^{ikx(1-\cos\theta)}}{ik(1-\cos\theta)}\right]_{0}^{+d/2}\right\}$$

$$= \frac{i}{k} e^{-i\frac{kd}{2}} \left[\frac{1 - e^{i\frac{kd}{2}(1+\cos\theta)}}{1+\cos\theta} + \frac{1 - e^{i\frac{kd}{2}(1-\cos\theta)}}{1-\cos\theta}\right]$$

$$= \frac{i}{k} \frac{e^{-i\frac{kd}{2}}}{\sin^2\theta} \left[2 - 2e^{i\frac{kd}{2}} \cos\left(\frac{kd}{2}\cos\theta\right)\right.$$

$$\left. + 2ie^{i\frac{kd}{2}} \cos\theta \sin\left(\frac{kd}{2}\cos\theta\right)\right], \tag{5.4.48}$$

and, similarly

$$
\begin{aligned}
C_2 = K + iJ &= \int_{-d/2}^{+d/2} e^{i\left(\frac{kd}{2} - k|x|\right)} e^{-ikx\cos\theta} dx \\
&= e^{i\frac{kd}{2}} \int_{-d/2}^{+d/2} e^{-ik(x\cos\theta+|x|)} dx \\
&= e^{i\frac{kd}{2}} \left[\int_{-d/2}^{0} e^{-ik(x\cos\theta-x)} dx + \int_{0}^{+d/2} e^{-ik(x\cos\theta+x)} dx \right] \\
&= e^{i\frac{kd}{2}} \left[\int_{-d/2}^{0} e^{ikx(1-\cos\theta)} dx + \int_{0}^{+d/2} e^{-ikx(1+\cos\theta)} dx \right] \\
&= e^{i\frac{kd}{2}} \left\{ \left[\frac{e^{ikx(1-\cos\theta)}}{ik(1-\cos\theta)} \right]_{-d/2}^{0} + \left[\frac{e^{-ikx(1+\cos\theta)}}{-ik(1+\cos\theta)} \right]_{0}^{+d/2} \right\} \\
&= -\frac{i}{k} e^{i\frac{kd}{2}} \left[\frac{1 - e^{-i\frac{kd}{2}(1-\cos\theta)}}{1-\cos\theta} + \frac{1 - e^{-i\frac{kd}{2}(1+\cos\theta)}}{1+\cos\theta} \right] \\
&= -\frac{i}{k} \frac{e^{i\frac{kd}{2}}}{\sin^2\theta} \left[2 - 2e^{-i\frac{kd}{2}} \cos\left(\frac{kd}{2}\cos\theta\right) \right. \\
&\quad \left. - 2ie^{-i\frac{kd}{2}} \cos\theta \sin\left(\frac{kd}{2}\cos\theta\right) \right].
\end{aligned}
\tag{5.4.49}
$$

Then,

$$
\begin{aligned}
J &= \int_{-d/2}^{+d/2} \sin\left(\frac{kd}{2} - k|x|\right) e^{-ikx\cos\theta} dx = \frac{C_2 - C_1}{2i} \\
&= \frac{1}{2i} \left\{ \frac{-i}{k} \frac{e^{i\frac{kd}{2}}}{\sin^2\theta} \left[2 - 2e^{-i\frac{kd}{2}} \cos\left(\frac{kd}{2}\cos\theta\right) \right. \right. \\
&\quad \left. - 2ie^{-i\frac{kd}{2}} \cos\theta \sin\left(\frac{kd}{2}\cos\theta\right) \right] \\
&\quad - \frac{i}{k} \frac{e^{-i\frac{kd}{2}}}{\sin^2\theta} \left[2 - 2e^{i\frac{kd}{2}} \cos\left(\frac{kd}{2}\cos\theta\right) \right. \\
&\quad \left. \left. + 2ie^{i\frac{kd}{2}} \cos\theta \sin\left(\frac{kd}{2}\cos\theta\right) \right] \right\} \\
&= -\frac{1}{k\sin^2\theta} \left[e^{i\frac{kd}{2}} + e^{-i\frac{kd}{2}} - 2\cos\left(\frac{kd}{2}\cos\theta\right) \right] \\
&= \frac{2}{k} \frac{\cos\left(\frac{kd}{2}\cos\theta\right) - \cos\frac{kd}{2}}{\sin^2\theta}.
\end{aligned}
\tag{5.4.50}
$$

By using this result in Eq. (5.4.45), the magnetic vector potential $\vec{A}(\vec{r}, t)$ becomes

$$
\begin{aligned}
\vec{A}(\vec{r}, t) &= \vec{u}_3 \frac{\mu_0 I e^{ikr} e^{-i\omega t}}{4\pi r} \int_{-d/2}^{+d/2} \sin\left(\frac{kd}{2} - k|z'|\right) e^{-ikz'\cos\theta} dz' \\
&= \vec{u}_3 \frac{\mu_0 I e^{ikr} e^{-i\omega t}}{2\pi kr} \frac{\cos\left(\frac{kd}{2}\cos\theta\right) - \cos\frac{kd}{2}}{\sin^2\theta} \equiv A_0 \frac{e^{ikr}}{r} \vec{u}_3 ,
\end{aligned}
\tag{5.4.51}
$$

where

$$
A_0 \equiv \frac{\mu_0}{4\pi} \frac{2I e^{-i\omega t}}{k} \frac{\cos\left(\frac{kd}{2}\cos\theta\right) - \cos\frac{kd}{2}}{\sin^2\theta}.
$$

To obtain the angular distribution of the power radiated by the antenna, $dP/d\Omega$, in agreement with the relation (5.4.37) we must calculate the quantity $\left[(\nabla \times \vec{H}) \times \vec{H}^*\right] \cdot \vec{n}$, where $\vec{H} = \mu_0^{-1}\nabla \times \vec{A}$. Therefore, we have

$$
\begin{aligned}
\vec{H} &= \mu_0^{-1}\nabla \times \vec{A} = \mu_0^{-1} A_0 \nabla \times \left(\frac{e^{ikr}}{r} \vec{u}_3\right) \\
&= \mu_0^{-1} A_0 \left[\nabla\left(\frac{e^{ikr}}{r}\right) \times \vec{u}_3 + \frac{e^{ikr}}{r} \nabla \times \vec{u}_3\right] \\
&= \frac{A_0}{\mu_0} \nabla\left(\frac{e^{ikr}}{r}\right) \times \vec{u}_3 = \frac{A_0}{\mu_0}\left[\frac{1}{r}\nabla e^{ikr} + e^{ikr}\nabla\left(\frac{1}{r}\right)\right] \times \vec{u}_3 \\
&= \mu_0^{-1} A_0 \left(\frac{ik\vec{r}}{r^2} e^{ikr} - e^{ikr}\frac{\vec{r}}{r^3}\right) \times \vec{u}_3 \\
&= \mu_0^{-1} A_0 e^{ikr}\left(\frac{ik}{r} - \frac{1}{r^2}\right)\frac{\vec{r}}{r} \times \vec{u}_3.
\end{aligned}
\tag{5.4.52}
$$

In the radiative zone (wave zone), this relation receives the asymptotic form

$$
\begin{aligned}
\vec{H} &= \mu_0^{-1} A_0 e^{ikr}\frac{ik\vec{r}}{r^2} \times \vec{u}_3 \\
&= \mu_0^{-1}\frac{\mu_0 I e^{-i\omega t}}{2\pi k}\frac{\cos\left(\frac{kd}{2}\cos\theta\right) - \cos\frac{kd}{2}}{\sin^2\theta} e^{ikr}\frac{ik}{r} \vec{n} \times \vec{u}_3 \\
&= \frac{iI e^{-i(\omega t - kr)}}{2\pi r}\frac{\cos\left(\frac{kd}{2}\cos\theta\right) - \cos\frac{kd}{2}}{\sin^2\theta} \vec{n} \times \vec{u}_3,
\end{aligned}
\tag{5.4.53}
$$

or, simpler,

$$
\vec{H} = \frac{ik}{\mu_0} \vec{n} \times \vec{A},
\tag{5.4.54}
$$

and then,

$$
|\vec{H}| = \sqrt{\vec{H} \cdot \vec{H}^*} = \frac{I}{2\pi r}\frac{\cos\left(\frac{kd}{2}\cos\theta\right) - \cos\frac{kd}{2}}{\sin\theta}.
\tag{5.4.55}
$$

For $\nabla \times \vec{H}$, we have

$$\nabla \times \vec{H} = \nabla \times \left[\mu_0^{-1} A_0 e^{ikr} \left(\frac{ik}{r} - \frac{1}{r^2} \right) \frac{\vec{r}}{r} \times \vec{u}_3 \right]$$

$$= \frac{Ie^{-i\omega t}}{2\pi k} \frac{\cos \left(\frac{kd}{2} \cos \theta \right) - \cos \frac{kd}{2}}{\sin^2 \theta} \nabla \times \left[e^{ikr} \left(\frac{ik}{r} - \frac{1}{r^2} \right) \vec{n} \times \vec{u}_3 \right]$$

$$\equiv B_0 \nabla \times \left[e^{ikr} \left(\frac{ik}{r} - \frac{1}{r^2} \right) \vec{n} \times \vec{u}_3 \right] = B_0 \nabla \times \left[\left(\frac{ike^{ikr}}{r} \right. \right.$$

$$\left. \left. - \frac{e^{ikr}}{r^2} \right) \vec{n} \times \vec{u}_3 \right] = B_0 \left(\frac{ike^{ikr}}{r} - \frac{e^{ikr}}{r^2} \right) \underbrace{\nabla \times (\vec{n} \times \vec{u}_3)}_{=0}$$

$$+ B_0 \nabla \left(\frac{e^{ikr}}{r} - \frac{e^{ikr}}{r^2} \right) \times (\vec{n} \times \vec{u}_3)$$

$$= ik B_0 \nabla \left(\frac{ike^{ikr}}{r} \right) \times (\vec{n} \times \vec{u}_3) - B_0 \nabla \left(\frac{e^{ikr}}{r^2} \right) \times (\vec{n} \times \vec{u}_3)$$

$$= -ik B_0 e^{ikr} \left(\frac{ik}{r} - \frac{1}{r^2} \right) (\vec{n} \times \vec{u}_3) \times \vec{n} + B_0 e^{ikr} \left(\frac{ik}{r^2} - \frac{2}{r^3} \right)$$

$$\times (\vec{n} \times \vec{u}_3) \times \vec{n} = k^2 B_0 \frac{e^{ikr}}{r} (\vec{n} \times \vec{u}_3) \times \vec{n}$$

$$+ 2 B_0 e^{ikr} \left(\frac{ik}{r^2} - \frac{1}{r^3} \right) (\vec{n} \times \vec{u}_3) \times \vec{n} = \frac{kIe^{-i(\omega t - kr)}}{2\pi r}$$

$$\times \frac{\cos \left(\frac{kd}{2} \cos \theta \right) - \cos \frac{kd}{2}}{\sin^2 \theta} (\vec{n} \times \vec{u}_3) \times \vec{n} + \frac{Ie^{-i(\omega t - kr)}}{\pi k}$$

$$\times \frac{\cos \left(\frac{kd}{2} \cos \theta \right) - \cos \frac{kd}{2}}{\sin^2 \theta} \left[\frac{ik}{r^2} - \frac{1}{r^3} \right] (\vec{n} \times \vec{u}_3) \times \vec{n}. \qquad (5.4.56)$$

Taking into account this result, it follows that in the radiative zone the electric field takes the following asymptotic form:

$$\vec{E} = \frac{I}{k} \sqrt{\frac{\mu_0}{\varepsilon_0}} \nabla \times \vec{H} = \frac{i}{k} \sqrt{\frac{\mu_0}{\varepsilon_0}} \frac{kIe^{-i(\omega t - kr)}}{2\pi r}$$

$$\times \frac{\cos \left(\frac{kd}{2} \cos \theta \right) - \cos \frac{kd}{2}}{\sin^2 \theta} (\vec{n} \times \vec{u}_3) \times \vec{n}$$

$$= \frac{iIe^{-i(\omega t - kr)}}{2\pi r} \frac{\cos \left(\frac{kd}{2} \cos \theta \right) - \cos \frac{kd}{2}}{\sin^2 \theta} \sqrt{\frac{\mu_0}{\varepsilon_0}} (\vec{n} \times \vec{u}_3) \times \vec{n}$$

$$= \sqrt{\frac{\mu_0}{\varepsilon_0}} \vec{H} \times \vec{n}. \qquad (5.4.57)$$

In the wave zone, we therefore can write

$$
\left[(\nabla \times \vec{H}) \times \vec{H}^* \right] \cdot \vec{n} = \left\{ \left[\frac{kIe^{-i(\omega t - kr)}}{2\pi r} \right. \right.
$$

$$
\times \frac{\cos\left(\frac{kd}{2}\cos\theta\right) - \cos\frac{kd}{2}}{\sin^2\theta} (\vec{n} \times \vec{u}_3) \times \vec{n} \left. \right] \times
$$

$$
\times \left[\frac{iIe^{-i(\omega t - kr)}}{2\pi r} \frac{\cos\left(\frac{kd}{2}\cos\theta\right) - \cos\frac{kd}{2}}{\sin^2\theta} \vec{n} \times \vec{u}_3 \right]^* \left. \right\} \cdot \vec{n}
$$

$$
= \frac{-ikI^2}{4\pi^2 r^2} \left| \frac{\cos\left(\frac{kd}{2}\cos\theta\right) - \cos\frac{kd}{2}}{\sin^2\theta} \right|^2 \left\{ \left[(\vec{n} \times \vec{u}_3) \times \vec{n} \right] \times \right.
$$

$$
\times (\vec{n} \times \vec{u}_3) \left. \right\} \cdot \vec{n} = \frac{-ikI^2}{4\pi^2 r^2} \left| \frac{\cos\left(\frac{kd}{2}\cos\theta\right) - \cos\frac{kd}{2}}{\sin^2\theta} \right|^2
$$

$$
\times \left\{ (\vec{n} \times \vec{u}_3) \times \left[\vec{n} \times (\vec{n} \times \vec{u}_3) \right] \right\} \cdot \vec{n}
$$

$$
= \frac{-ikI^2}{4\pi^2 r^2} \left| \frac{\cos\left(\frac{kd}{2}\cos\theta\right) - \cos\frac{kd}{2}}{\sin^2\theta} \right|^2 \left\{ (\vec{n} \times \vec{u}_3) \times \right.
$$

$$
\times \left[\vec{n}(\vec{n} \cdot \vec{u}_3) - \vec{u}_3(\vec{n} \cdot \vec{n}) \right] \left. \right\} \cdot \vec{n}
$$

$$
= \frac{-ikI^2}{4\pi^2 r^2} \left| \frac{\cos\left(\frac{kd}{2}\cos\theta\right) - \cos\frac{kd}{2}}{\sin^2\theta} \right|^2 \left[(\vec{n} \times \vec{u}_3) \times \right.
$$

$$
\times (\vec{n}\cos\theta - \vec{u}_3) \left. \right] \cdot \vec{n} = \frac{-ikI^2}{4\pi^2 r^2} \left| \frac{\cos\left(\frac{kd}{2}\cos\theta\right) - \cos\frac{kd}{2}}{\sin^2\theta} \right|^2
$$

$$
\times \left[\vec{n}(\vec{u}_3 \cdot \vec{u}_3) - \vec{u}_3(\vec{u}_3 \cdot \vec{n}) - \vec{n}(\vec{n} \cdot \vec{u}_3)\cos\theta \right.
$$

$$
+ \vec{u}_3(\vec{n} \cdot \vec{n})\cos\theta \left. \right] \cdot \vec{n} = \frac{-ikI^2}{4\pi^2 r^2} \left| \frac{\cos\left(\frac{kd}{2}\cos\theta\right) - \cos\frac{kd}{2}}{\sin^2\theta} \right|^2
$$

$$
\times \left(\vec{n} - \vec{u}_3\cos\theta - \vec{n}\cos^2\theta + \vec{u}_3\cos\theta \right) \cdot \vec{n}
$$

$$
= \frac{-ikI^2}{4\pi^2 r^2} \left| \frac{\cos\left(\frac{kd}{2}\cos\theta\right) - \cos\frac{kd}{2}}{\sin^2\theta} \right|^2 \sin^2\theta. \qquad (5.4.58)
$$

If this expression is introduced into Eq. (5.4.37), one obtains

$$
\frac{dP}{d\Omega} = \frac{1}{2}\mathrm{Re}\left\{ \frac{ir^2}{k}\sqrt{\frac{\mu_0}{\varepsilon_0}} \left[(\nabla \times \vec{H} \times \vec{H}^*) \cdot \vec{n} \right] \right\}
$$

$$= \frac{1}{2}\mathrm{Re}\left[\frac{ir^2}{k}\sqrt{\frac{\mu_0}{\varepsilon_0}}\left(\frac{-ikI^2}{4\pi^2 r^2}\left|\frac{\cos(\frac{kd}{2}\cos\theta) - \cos\frac{kd}{2}}{\sin^2\theta}\right|^2\sin^2\theta\right)\right]$$

$$= \frac{I^2}{8\pi^2}\sqrt{\frac{\mu_0}{\varepsilon_0}}\left|\frac{\cos(\frac{kd}{2}\cos\theta) - \cos\frac{kd}{2}}{\sin\theta}\right|^2. \tag{5.4.59}$$

The electric vector is along the direction of the component of $\vec{A}$ which is perpendicular to $\vec{n}$ (i.e., on direction of the vector $(\vec{n} \times \vec{u}_3) \times \vec{n}$, or, equivalently, on the direction of the vector $\vec{H} \times \vec{n}$). Therefore, the polarization of radiation is located in the plane containing antenna and the radius vector at the observation point.

The angular distribution given by Eq. (5.4.59) depends, obviously, on the value of the product kd. Within the limit of big wavelengths ($kd \ll 1$) one can easily show that it reduces to the result corresponding to the electric dipole,

$$\frac{dP}{d\Omega} = \frac{I^2}{32\pi^2}\sqrt{\frac{\mu_0}{\varepsilon_0}}\left(\frac{kd}{2}\right)^4\sin^2\theta = \frac{\pi^2 I^2 d^4}{32}\frac{1}{\lambda^4}\sqrt{\frac{\mu_0}{\varepsilon_0}}\sin^2\theta, \tag{5.4.60}$$

which is the Lord Rayleigh's "blue sky law" ($dP/d\Omega \sim \lambda^{-4}$). Indeed, for $kd \ll 1$, we have

$$\cos\left(\frac{kd}{2}\right) \simeq 1 - \frac{k^2 d^2}{8}, \quad \cos\left(\frac{kd}{2}\cos\theta\right) \simeq 1 - \frac{k^2 d^2 \cos^2\theta}{8},$$

and then

$$\left|\frac{\cos\left(\frac{kd}{2}\cos\theta\right) - \cos\frac{kd}{2}}{\sin\theta}\right|^2 \simeq \left|\frac{\frac{k^2 d^2}{8}(1 - \cos^2\theta)}{\sin\theta}\right|^2 = \frac{k^4 d^4}{64}\sin^2\theta.$$

As can be easily verified, if this relation is introduced into Eq. (5.4.60), then the relation (5.4.60) is obtained.

For the special values $kd = \pi$ and $kd = 2\pi$, corresponding to the length of an antenna of one half-wavelength ($d = \lambda/2$) and one wavelength ($d = \lambda$), respectively, of the current oscillating along the antenna, the angular distributions are

$$\frac{dP}{d\Omega} = \frac{I^2}{8\pi^2}\sqrt{\frac{\mu_0}{\varepsilon_0}}\left|\frac{\cos\left(\frac{kd}{2}\cos\theta\right) - \cos\frac{kd}{2}}{\sin^2\theta}\right|^2$$

$$= \frac{I^2}{8\pi^2}\sqrt{\frac{\mu_0}{\varepsilon_0}}\begin{cases} \dfrac{\cos^2\left(\frac{\pi}{2}\cos\theta\right)}{\sin^2\theta}, & kd = \pi, \quad \left(d = \frac{\lambda}{2}\right), \\[3mm] \dfrac{4\cos^4\left(\frac{\pi}{2}\cos\theta\right)}{\sin^2\theta}, & kd = 2\pi, \quad (d = \lambda). \end{cases} \tag{5.4.61}$$

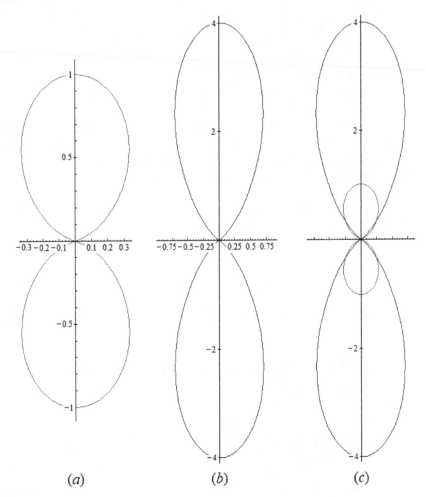

(a)　　　　　　　　　　(b)　　　　　　　　　　(c)

FIGURE 5.4

Dipole angular distributions for a thin, linear, centrally-fed antenna (a) dipole angular distribution of the power radiated by a half-wave antenna; (b) dipole angular distribution of the power radiated by a full-wave antenna; (c) the two distributions from cases (a) and (b) plotted on the same graph.

These angular distributions are graphically represented in Fig. 5.4: Fig. 5.4.a for the half-wave antenna, Fig. 5.4.b for the full-wave antenna, and Fig. 5.4.c for both distributions on the same graphic, for comparison. The angular distributions of the power radiated by antenna are graphically represented as diagrams of polar intensities, by means of the following command lines written using Mathematica 5.0 analytical and numerical software:

ParametricPlot[{$\dfrac{\text{Cos}[(\frac{\pi}{2})*\text{Cos}[t]]^\wedge 2}{\text{Sin}[t]^\wedge 2}*\text{Cos}[t]$,

$\dfrac{\text{Cos}[(\frac{\pi}{2})*\text{Cos}[t]]^\wedge 2}{\text{Sin}[t]^\wedge 2}*\text{Sin}[t]\}, \{t, -\pi, \pi\}$,

AspectRatio → Automatic];

ParametricPlot[{$\dfrac{4*\text{Cos}[(\frac{\pi}{2})*\text{Cos}[t]]^\wedge 4}{\text{Sin}[t]^\wedge 2}*\text{Cos}[t]$,

$\dfrac{4*\text{Cos}[(\frac{\pi}{2})*\text{Cos}[t]]^\wedge 4}{\text{Sin}[t]^\wedge 2}*\text{Sin}[t]\}, \{t, -\pi, \pi\}$,

AspectRatio → Automatic];

ParametricPlot[

{{$\dfrac{\text{Cos}[(\frac{\pi}{2})*\text{Cos}[t]]^\wedge 2}{\text{Sin}[t]^\wedge 2}*\text{Cos}[t], \dfrac{\text{Cos}[(\frac{\pi}{2})*\text{Cos}[t]]^\wedge 2}{\text{Sin}[t]^\wedge 2}*\text{Sin}[t]\}$,

{$\dfrac{4*\text{Cos}[(\frac{\pi}{2})*\text{Cos}[t]]^\wedge 4}{\text{Sin}[t]^\wedge 2}*\text{Cos}[t]$,

{$\dfrac{4*\text{Cos}[(\frac{\pi}{2})*\text{Cos}[t]]^\wedge 4}{\text{Sin}[t]^\wedge 2}*\text{Sin}[t]\}\}, \{t, -\pi, \pi\}$,

AspectRatio → Automatic];

or, equivalently,

<< Graphics'Graphics'

grhalfwave = PolarPlot[{$\dfrac{\text{Cos}[(\frac{\pi}{2})*\text{Cos}[t]]^\wedge 2}{\text{Sin}[t]^\wedge 2}, \dfrac{\text{Cos}[(\frac{\pi}{2})*\text{Cos}[t]]^\wedge 2}{\text{Sin}[t]^\wedge 2}\}$,

$\{t, 0, 2\pi\}]$

grwave = PolarPlot[{$\dfrac{4*\text{Cos}[(\frac{\pi}{2})*\text{Cos}[t]]^\wedge 4}{\text{Sin}[t]^\wedge 2}, \dfrac{4*\text{Cos}[(\frac{\pi}{2})*\text{Cos}[t]]^\wedge 4}{\text{Sin}[t]^\wedge 2}$,

$\{t, 0, 2\pi\}]$

Show[grhalfwave, grwave]

The distribution of the half-wave antenna is very similar to a dipolar simple figure, while the (full)-wave antenna has a more sharp distribution. The distribution of the (full)-wave antenna can be imagined as a coherent superposition of the fields of two half-wave antennas, one over the other, excited in the phase. The intensity at $\theta = \pi/2$, where the waves are algebraically added, is four times bigger than that of a half-wave antenna. At angles far from $\theta = \pi/2$ the amplitudes tend to interfere, giving a narrower graphical figure. By means of an appropriate arrangement of system of basic antennas, like the half-wave antennas, with current phases suitable chosen, it can be obtained arbitrary radiative figures by means of coherent superposition.

5.5   Problem No. 42

Consider a centre-fed thin linear antenna of length d, which is oriented along the z-axis. Using the sinusoidal current approximation, *i.e.*, considering that the antenna is excited at its centre and the current is sinusoidally distributed along the antenna according to law

$$\vec{j}(\vec{r}) = I \sin\left(\frac{kd}{2} - k|z|\right) \delta(x)\,\delta(y)\,\hat{z},$$

($k = \omega/c$ is the wave number), it can be shown that the angular distribution of the time average of the power radiated by antenna per unit solid angle is given by formula (see **Problem No. 41**):

$$\frac{dP_{rad}}{d\Omega} = \frac{I^2 Z_0}{8\pi^2} \left| \frac{\cos\left(\frac{kd}{2}\cos\theta\right) - \cos\frac{kd}{2}}{\sin\theta} \right|^2,$$

where

$$Z_0 = \sqrt{\frac{\mu_0}{\varepsilon_0}}$$

is the impedance of the free space. Calculate:
a) the total power emitted by the antenna;
b) the radiative resistance of the antenna.

Solution

a) The total power emitted by the antenna can be simply obtained as follows:

$$P_{\text{rad}} = \int dP_{\text{rad}} = \iint \frac{dP_{\text{rad}}}{d\Omega}\,d\Omega$$

$$= \frac{I^2 Z_0}{8\pi^2} \int_0^\pi \left| \frac{\cos\left(\frac{kd}{2}\cos\theta\right) - \cos\frac{kd}{2}}{\sin\theta} \right|^2 \sin\theta\,d\theta \int_0^{2\pi} d\varphi$$

$$= \frac{I^2 Z_0}{4\pi} \left\{ \gamma - 2\cos^2\left(\frac{kd}{2}\right) \operatorname{Ci}(kd) + \ln(kd) \right.$$

$$+ \frac{1}{2}\left[\gamma + \operatorname{Ci}(2kd) + \ln\left(\frac{kd}{2}\right)\right] \cos(kd)$$

$$+ \left. \frac{1}{2}\left[\operatorname{Si}(2kd) - 2\operatorname{Si}(kd)\right] \sin(kd) \right\}, \qquad (5.5.62)$$

where

$$\gamma = \lim_{n\to\infty}\left[\left(\sum_{k=1}^{\infty}\frac{1}{k}\right) - \ln n\right] \cong 0.577216$$

is the *Euler-Mascheroni constant* (sometimes simply called the *Euler constant*), while

$$\mathrm{Si}\, z = \int\limits_0^z \frac{\sin t}{t}\, dt\,,$$

and

$$\mathrm{Ci}\, z = -\int\limits_z^\infty \frac{\cos t}{t}\, dt$$

are the special functions *sine integral* and *cosine integral*, respectively. As can be observed, for a given excitation current, the total power radiated by the antenna depends on frequency through the function

$$f(k) = f\left(\frac{2\pi\nu}{c}\right) = \gamma - 2\cos^2\left(\frac{kd}{2}\right)\mathrm{Ci}(kd) + \ln(kd)$$

$$+ \frac{1}{2}\left[\gamma + \mathrm{Ci}(2kd) + \ln\left(\frac{kd}{2}\right)\right]\cos(kd)$$

$$+ \frac{1}{2}\left[\mathrm{Si}(2kd) - 2\,\mathrm{Si}(kd)\right]\sin(kd). \tag{5.5.63}$$

As one knows, the relative extremum points of the function $f(k)$ can be determined by equating with zero its first derivative with respect to k:

$$0 = \frac{df(k)}{dk} = \frac{1}{k} - \frac{2}{k}\cos(kd)\cos^2\left(\frac{kd}{2}\right) + \frac{\cos(kd)}{2k}\left[1 + \cos(2kd)\right]$$

$$+ 2d\cos\left(\frac{kd}{2}\right)\mathrm{Ci}(kd)\sin\left(\frac{kd}{2}\right) - \frac{d\sin(kd)}{2}\left[\gamma + \mathrm{Ci}(2kd)\right.$$

$$+ \ln\left(\frac{kd}{2}\right)\right] + \frac{\sin(kd)}{2k}\left[\sin(2kd) - 2\sin(kd)\right] + \frac{d\cos(kd)}{2}$$

$$\times \left[\mathrm{Si}(2kd) - 2\,\mathrm{Si}(kd)\right]. \tag{5.5.64}$$

For a typical antenna with $d = 2\times 10^{-1}$m, the first solution of the above equation (the first relative maximum) is $k_{M_1} \cong 28.1617\,\mathrm{m}^{-1}$, and it corresponds to the frequency $\nu_{M_1} \cong 1.345\,\mathrm{GHz}$ and to the wavelength $\lambda_{M_1} \cong 22.31\,\mathrm{cm}$.

In Fig. 5.5 the dependence of the total power radiated by the antenna on the wave number $k = 2\pi\nu/c$, for the value $I = 10\,\mathrm{mA}$ of the excitation current is depicted. This graphical representation corresponds to the interval $k \in [(2\pi/3)\times 10^{-6}, 10\pi]$ for the wave number, or, equivalently, frequency interval $\nu \in [0.1\,\mathrm{kHz}, 1.5\,\mathrm{GHz}]$, or, still, the wavelength interval $\lambda \in [20\,\mathrm{cm}, 3000\,\mathrm{km}]$.

Using the power series expansion of the sine and cosine functions, as well as of the special functions sine integral

$$\mathrm{Si}\, x = x - \frac{x^3}{18} + \frac{x^5}{600} - \frac{x^7}{35280} + \mathcal{O}(x^9),$$

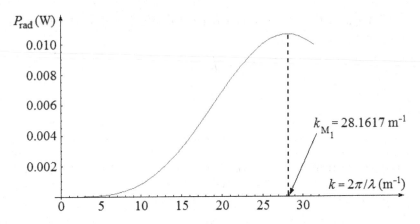

FIGURE 5.5
Dependence of the total power radiated by a thin, linear, centrally-fed antenna on the wave number $k = |\vec{k}|$, for $k \in [(2\pi/3) \times 10^{-6}, 10\pi]$.

and cosine integral,

$$\mathrm{Ci}\, x = \gamma + \ln x - \frac{x^2}{4} + \frac{x^4}{96} - \frac{x^6}{4320} + \mathcal{O}\left(x^8\right),$$

for a fixed excitation current and small frequency values (corresponding to large wavelengths, $kd \ll 1$), the radiated power increases with frequency to approximately the fourth power. Indeed, we have

$$F(k) = \frac{k^4 d^4}{48} + \mathcal{O}\left[(kd)^6\right] \cong \frac{k^4 d^4}{48},$$

showing that in the range of large wavelengths this type of antenna behaves like an electric dipole radiator.

Due to the contribution of the multipoles of higher order, the approximation $F(k) \cong k^4 d^4/48$ (in other words, $\mathcal{O}[(kd)^6] = 0$) is valid only in the limit of large wavelengths. This fact is displayed in Fig. 5.6, showing the dependencies

$$P_{\mathrm{rad}}^{\mathrm{exact}} = P_{\mathrm{rad}}^{\mathrm{exact}}(k) = \frac{I^2 Z_0}{4\pi}\, F(k),$$

and

$$P_{\mathrm{rad}}^{\mathrm{dipole\, type}} \equiv P_{\mathrm{rad}}^{\mathrm{dtype}}(k) = \frac{I^2 Z_0}{4\pi}\frac{k^4 d^4}{48}.$$

As can be observed (see Fig. 5.6), the two curves overlap with a good approximation only for frequencies not higher than $477\,\mathrm{MHz}$, or, equivalently, wavelengths not lower than $\lambda \simeq 62.8\,\mathrm{cm}$. The percentage "error" in calculating the total power radiated by the antenna when taking into consideration only the dipole-type component (*i.e.*, if the contribution of the higher-order dipoles

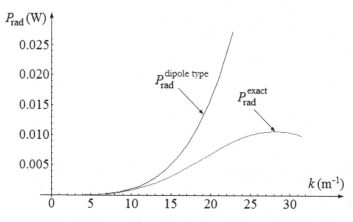

FIGURE 5.6
Contribution of higher-order multipoles to the total power radiated by a thin, linear, centrally-fed antenna.

is neglected) is graphically represented in Fig. 5.7 as a function of the wave vector modulus $k = |\vec{k}|$:

$$\frac{P_{\text{rad}}^{\text{dipole type}} - P_{\text{rad}}^{\text{exact}}}{P_{\text{rad}}^{\text{dipole type}}} \times 100\% \equiv \frac{P_{\text{rad}}^{\text{dtype}} - P_{\text{rad}}}{P_{\text{rad}}^{\text{dtype}}} \times 100\%$$

$$= \left\{ 1 - \frac{48}{k^4 d^4} \left(\left[\text{Si}(2kd) - 2\,\text{Si}(kd) \right] \frac{\sin(kd)}{2} \right. \right.$$

$$- 2\cos^2\left(\frac{kd}{2}\right) \text{Ci}(kd) + \gamma + \ln(kd)$$

$$\left. \left. + \frac{1}{2} \left[\gamma + \ln\left(\frac{kd}{2}\right) + \text{Ci}(2kd) \right] \cos(kd) \right) \right\} \times 100\%. \tag{5.5.65}$$

As Fig. 5.7 shows, at least within the frequency domain corresponding to the interval $k \in [0, 40]\,\text{m}^{-1}$, the larger the wavelength, the smaller this relative "error" is.

As the range for frequencies increases more and more, one remarks an "oscillatory" behaviour of the total power radiated by the antenna in terms of the wave number/frequency (see Fig. 5.8), representing the dependence $P_{\text{rad}} = P_{\text{rad}}(k)$, for $I = 10\,\text{mA}$ and $d = 20\,\text{cm}$ in the frequency interval $\nu \in [0.1\,\text{kHz}, 55\,\text{GHz}]$. The values of k at the points of relative minimum and maximum can be determined using the numeric calculus.

b) As can be easily verified, the unit of measurement of the coefficient of $I^2/2$ appearing in Eq. (5.5.62) is that of an electric resistance; it is called the *radiative resistance*, R_{rad}, of the antenna.

Normally, electrical resistance is the "proportionality factor" of I^2 in the expression of an electrical power, but here the supplementary factor $1/2$

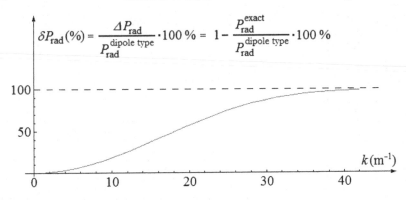

$$\delta P_{\rm rad}(\%) = \frac{\Delta P_{\rm rad}}{P_{\rm rad}^{\rm dipole\ type}} \cdot 100\,\% = 1 - \frac{P_{\rm rad}^{\rm exact}}{P_{\rm rad}^{\rm dipole\ type}} \cdot 100\,\%$$

FIGURE 5.7
Dependence of the percentage relative error of the total power radiated by a thin, linear, centrally-excited antenna on the wave number modulus $k = |\vec{k}|$.

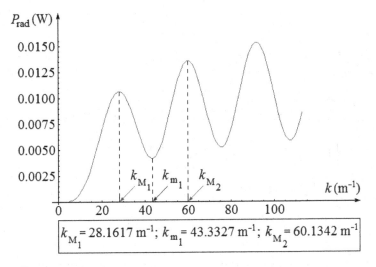

FIGURE 5.8
The total power radiated by a thin, linear, centrally-fed antenna, as a function of the wave number modulus, $k = |\vec{k}|$, for an extended frequency interval.

appears. This is because the involved harmonic fields (*i.e.*, fields with the time dependence of the form $e^{-i\omega t}$, that is

$$\vec{E}(\vec{r},t) = \operatorname{Re}\left[\vec{E}(\vec{r})e^{i\omega t}\right] = \frac{1}{2}\left[\vec{E}(\vec{r})e^{i\omega t} + \vec{E}^{*}(\vec{r})e^{i\omega t}\right].)$$

have a complex (and not a real) character. When calculating the average flux density of the electromagnetic energy emitted per unit time (differently speaking, the time average of the power radiated per unit surface) the complex Poynting vector is used rather than its real counterpart. In the "same spirit", the real mechanical work performed in unit time by the field $\vec{E}$ on currents $\vec{j}$, must be replaced by its complex correspondent:

$$\int_V \vec{j} \cdot \vec{E} \, d\vec{r} \; \rightarrow \; \frac{1}{2} \int_V \vec{j}^* \cdot \vec{E} \, d\vec{r}$$

$$\left(\vec{\Pi} = \vec{E} \times \vec{H} \; \rightarrow \; \tilde{\vec{\Pi}} = \frac{1}{2} \vec{E} \times \vec{H}^* \right).$$

The use of complex quantities is often preferred because it offers a significant advantage: it allows an easier identification and, at the same time, tracking of the active from reactive quantities.

In view of all the above, we can write

$$R_{\text{rad}} = \frac{Z_0}{2\pi} \left\{ \gamma - 2\cos^2\left(\frac{kd}{2}\right) \text{Ci}(kd) + \ln(kd) \right.$$

$$+ \frac{1}{2}\left[\gamma + \text{Ci}(2kd) + \ln\left(\frac{kd}{2}\right)\right] \cos(kd)$$

$$\left. + \frac{1}{2}\Big[\text{Si}(2kd) - 2\,\text{Si}(kd)\Big] \sin(kd) \right\}. \tag{5.5.66}$$

Let study the two cases of interest, namely: the half-wave antenna, for which $d = \frac{\lambda}{2}$ ($kd = \pi$) and the full-wave antenna, for which $d = \lambda$ ($kd = 2\pi$).

Thus, we have

$$R_{\text{rad}}^{\text{half}-\text{wave}} = \frac{Z_0}{2\pi} \left\{ \gamma + \ln\pi - \frac{1}{2}\left[\gamma + \text{Ci}(2\pi) + \ln\left(\frac{\pi}{2}\right)\right] \right\}$$

$$\cong 1.22\,\frac{Z_0}{2\pi} \cong 73.14\,\Omega, \tag{5.5.67}$$

and, respectively,

$$R_{\text{rad}}^{\text{full}-\text{wave}} = \frac{Z_0}{2\pi} \left\{ \gamma - 2\,\text{Ci}(2\pi) + \ln(2\pi) + \frac{1}{2}\left[\gamma + \text{Ci}(4\pi) + \ln\pi\right] \right\}$$

$$\cong 3.32\,\frac{Z_0}{2\pi} \cong 199,05\,\Omega. \tag{5.5.68}$$

As can easily be observed, in these two particular cases, the radiative resistance of the antenna does not depend on frequency.

5.6 Problem No. 43

Consider a centre-fed thin linear antenna of length d, which is oriented along the z-axis. Supposing that the antenna is excited in such a way that a sinusoidal current performs a complete wave-length oscillation as in Fig. 5.9, the

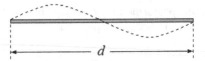

FIGURE 5.9
Excitation of a thin, linear, centrally-fed antenna with a sinusoidal current performing a complete wavelength oscillation.

current distribution along the antenna can be written as

$$\vec{j}(\vec{r}) = I_0 \sin(kz)\delta(x)\delta(y)\hat{z}.$$

Starting with a vector potential that behaves like an emergent spherical wave with an angle-dependent coefficient, that is

$$\vec{A}(\vec{r},t) = \frac{\mu_0}{4\pi}\frac{e^{i(kr-\omega t)}}{r}\int\limits_{(V')}\vec{j}(\vec{r}')\,e^{-ik(\vec{s}\cdot\vec{r}')}d\vec{r}', \qquad (5.6.69)$$

(where $\vec{s} = \vec{r}/r$) and using the method of multi-polar expansion,
a) show that both the electric and magnetic dipole contributions/components vanishes;
b) calculate the temporal average of the power radiated by the antenna per unit solid angle, corresponding to the electric quadrupole radiative component;
c) calculate the total power radiated by the antenna under the conditions of point **b)**;
d) calculate the radiative resistance of the antenna under the conditions of point **b)**;

Solution

a) If the dimensions d of the domain D' in which the sources are located are much smaller than the wavelength of the electromagnetic radiation ($d \ll \lambda \ll r$), then it is recommended to expand the integral in Eq. (5.6.69) in powers of k:

$$\lim_{kr\to\infty}\vec{A}(\vec{r}) = \frac{\mu_0}{4\pi}\frac{e^{i(kr-\omega t)}}{r}\int\limits_{(V')}\vec{j}(\vec{r}')\,e^{-ik(\vec{s}\cdot\vec{r}')}\,d\vec{r}'$$

$$= \frac{\mu_0}{4\pi}\frac{e^{i(kr-\omega t)}}{r}\sum_{n=0}^{\infty}\frac{(-ik)^n}{n!}\int\limits_{(V')}\vec{j}(\vec{r}')\,(\vec{s}\cdot\vec{r}')^n\,d\vec{r}'. \quad (5.6.70)$$

If in Eq. (5.6.70) one takes $n = 0$, then the *electric dipole* contribution is obtained. Thus we have

$$\vec{A}_{\text{el dipole}} = \frac{\mu_0}{4\pi} \frac{e^{i(kr-\omega t)}}{r} \int \vec{j}(\vec{r}')\, d\vec{r}' = \frac{\mu_0 I_0}{4\pi} \frac{e^{i(kr-\omega t)}}{r} \hat{z}$$

$$\times \int \delta(x')\, dx' \int \delta(y')\, dy' \int_{-d/2}^{+d/2} \sin(kz')\, dz'$$

$$= \frac{\mu_0 I_0}{4\pi} \frac{e^{i(kr-\omega t)}}{r} \hat{z} \int_{-d/2}^{+d/2} \sin(kz')\, dz' = 0, \tag{5.6.71}$$

that is, the electric dipole contribution vanishes.

For $n = 1$, if the quantity $\vec{j}(\vec{r}')(\vec{s} \cdot \vec{r}')$ is transcribed in the form

$$\vec{j}(\vec{r}')(\vec{s} \cdot \vec{r}') \equiv \vec{j}(\vec{s} \cdot \vec{r}')$$

$$= \frac{1}{2}\left[(\vec{s} \cdot \vec{r}')\vec{j} - (\vec{s} \cdot \vec{j})\vec{r}'\right] + \frac{1}{2}\left[(\vec{s} \cdot \vec{r}')\vec{j} + (\vec{s} \cdot \vec{j})\vec{r}'\right],$$

i.e., it is written as the sum of an antisymmetric and a symmetric part, then the *magnetic dipole* and *electric quadrupole* contributions, respectively, are obtained.

The magnetic dipole contribution is given by

$$\vec{A}_{\text{mag dipole}} = -\frac{\mu_0}{4\pi} \frac{e^{i(kr-\omega t)}}{r} \frac{ik}{2} \int \left[(\vec{s} \cdot \vec{r}')\vec{j} - (\vec{s} \cdot \vec{j})\vec{r}'\right] d\vec{r}'$$

$$= -\frac{\mu_0}{4\pi} \frac{e^{i(kr-\omega t)}}{r} \frac{ik}{2} \int \left(\vec{r}' \times \vec{j}\right) \times \vec{s}\, d\vec{r}'$$

$$= \frac{\mu_0 I_0}{4\pi} \frac{e^{i(kr-\omega t)}}{r} \frac{ik}{2} \vec{s} \times \int \delta(x')\delta(y')\sin(kz')\left(\vec{r}' \times \hat{z}'\right)$$

$$\times dx'dy'dz' = \frac{\mu_0 I_0}{4\pi} \frac{e^{i(kr-\omega t)}}{r} \frac{ik}{2}\left(\vec{s} \times \hat{x}\right) \int y'\delta(y')dy'$$

$$\times \int_{-d/2}^{+d/2} \sin(kz')\, dz' - \frac{\mu_0 I_0}{4\pi} \frac{e^{i(kr-\omega t)}}{r} \frac{ik}{2}\left(\vec{s} \times \hat{y}\right)$$

$$\times \int x'\delta(x')\, dx' \int_{-d/2}^{+d/2} \sin(kz')\, dz' = 0, \tag{5.6.72}$$

since the integrals over x' and y' can be extended over the whole real axis, in which case both integrals vanish. Consequently, the magnetic dipole contribution also vanishes.

b) Next, the electric quadrupole contribution is given by

$$\vec{A}_{\text{el q-pole}} = -\frac{\mu_0}{4\pi} \frac{e^{i(kr-\omega t)}}{r} \frac{ik}{2} \int \left[(\vec{s}\cdot\vec{r}')\vec{j} + (\vec{s}\cdot\vec{j})\vec{r}' \right] d\vec{r}'$$

$$= -\hat{z}\frac{\mu_0 I_0}{4\pi} \frac{e^{i(kr-\omega t)}}{r} \frac{ik}{2} \int \left[z'\cos\theta\sin(kz') \right.$$

$$\left. + z'\cos\theta\sin(kz') \right] \delta(x')\delta(y')dx'dy'dz'$$

$$= -\hat{z}\frac{\mu_0 I_0}{4\pi} \frac{e^{i(kr-\omega t)}}{r} ik\cos\theta \int \delta(x')dx' \int \delta(y')dy'$$

$$\times \int_{-d/2}^{+d/2} z'\sin(kz')dz' = e^{i(kr-\omega t)}\frac{\mu_0 I_0}{4\pi}\frac{ik\hat{z}\cos\theta}{r}$$

$$\times \int_{-d/2}^{+d/2} \frac{\partial}{\partial k}\left[\cos(kz')\right] dz' = e^{i(kr-\omega t)}\frac{\mu_0 I_0}{4\pi}\frac{ik\hat{z}\cos\theta}{r}$$

$$\times \frac{\partial}{\partial k}\left[\int_{-d/2}^{+d/2} \cos(kz')dz' \right] = I\hat{z}\frac{\mu_0 I_0}{4\pi}\frac{e^{i(kr-\omega t)}}{kr}$$

$$\times \left[kd\cos\left(\frac{kd}{2}\right) - 2\sin\left(\frac{kd}{2}\right) \right]\cos\theta. \tag{5.6.73}$$

According to the general definition, the temporal average of the power radiated by the antenna per unit solid angle is

$$\frac{dP_{\text{rad}}}{d\Omega} = \text{Re}\left(r^2\tilde{\Pi}_n\right) = \frac{1}{2}\text{Re}\left[r^2\vec{s}\cdot\left(\vec{E}\times\vec{H}^*\right)\right] = \frac{r^2}{2\mu_0}\text{Re}\left[\left(\vec{E}\times\vec{B}^*\right)\cdot\vec{s}\right].$$

Considering the wave-zone ($\mathcal{O}(r^{-2}) = 0$), the magnetic component of the field is

$$\vec{B} = \nabla \times \vec{A}$$

$$= \nabla \times \left\{ i\hat{z}\frac{\mu_0 I_0}{4\pi}\frac{e^{i(kr-\omega t)}}{kr}\left[kd\cos\left(\frac{kd}{2}\right) - 2\sin\left(\frac{kd}{2}\right) \right]\cos\theta \right\}$$

$$= i\frac{\mu_0 I_0}{4\pi k}\left[kd\cos\left(\frac{kd}{2}\right) - 2\sin\left(\frac{kd}{2}\right) \right]\cos\theta \,\nabla\left(\frac{e^{i(kr-\omega t)}}{r}\right)\times\hat{z}$$

$$= i\frac{\mu_0 I_0}{4\pi k}\cos\theta\left[kd\cos\left(\frac{kd}{2}\right) - 2\sin\left(\frac{kd}{2}\right) \right]\frac{e^{i(krt-\omega t)}}{r}$$

$$\times \left(ik - \frac{1}{r}\right)(\vec{s}\times\hat{z}) \cong -\vec{u}_\varphi\frac{\mu_0 I_0}{4\pi}\left[kd\cos\left(\frac{kd}{2}\right) - 2\sin\left(\frac{kd}{2}\right) \right]$$

$$\times \frac{e^{i(kr-\omega t)}}{r}\sin\theta\cos\theta, \tag{5.6.74}$$

while the electric component is given by.

$$\vec{E} = \frac{i}{k}\sqrt{\frac{\mu_0}{\varepsilon_0}}\,\nabla\times\vec{H}.$$

As a result of some calculations similar to those leading to relation (5.4.57), within the same approximation $\mathcal{O}(r^{-2}) = 0$, the above relation for $\vec{E}$ becomes

$$\vec{E} = \sqrt{\frac{\mu_0}{\varepsilon_0}}\, \vec{H} \times \vec{s} = c\, \vec{B} \times \vec{s}.$$

Then, using the relation (5.6.74), we can write

$$\mathrm{Re}\left[\left(\vec{E} \times \vec{B}^*\right) \cdot \vec{s}\right] = c\,\mathrm{Re}\left\{\left[\left(\vec{B} \times \vec{s}\right) \times \vec{B}^*\right] \cdot \vec{s}\right\} = c\left(|\vec{B}|^2 - |\vec{s} \cdot \vec{B}|^2\right)$$

$$= c\,|\vec{B}|^2 = \frac{c\mu_0^2 I_0^2}{16\pi^2 r^2}\left[kd\cos\left(\frac{kd}{2}\right) - 2\sin\left(\frac{kd}{2}\right)\right]^2$$

$$\times \sin^2\theta\cos^2\theta.$$

Now we are prepared to calculate, in this particular case (for the electric quadrupole component), the angular distribution of the power radiated by the antenna; it is given by

$$\frac{dP_{\mathrm{rad}}}{d\Omega} = \mathrm{Re}\left(r^2\tilde{\Pi}_n\right) = \frac{1}{2}\mathrm{Re}\left[r^2\vec{s} \cdot \left(\vec{E} \times \vec{H}^*\right)\right]$$

$$= \frac{c\mu_0 I_0^2}{32\pi^2}\left[kd\cos\left(\frac{kd}{2}\right) - 2\sin\left(\frac{kd}{2}\right)\right]^2 \sin^2\theta\cos^2\theta. \quad (5.6.75)$$

It can be easily shown that, in the limit of large wavelengths ($kd \ll 1$), this result leads to an oscillatory, spheroidal distribution of electric charge, representing one of the simplest examples of quadrupolar radiative source:

$$\frac{dP_{\mathrm{rad}}}{d\Omega} = \frac{c\mu_0 I_0^2}{32\pi^2}\frac{k^6 d^6}{144}\sin^2\theta\cos^2\theta = \frac{ck^6 Q_0^2}{512\varepsilon_0\pi^2}\sin^2\theta\cos^2\theta. \quad (5.6.76)$$

The quadrupolar momentum tensor corresponding to this symmetric charge distribution is diagonal; the elements of the tensor are

$$Q_{33} = Q_0, \quad Q_{11} = Q_{22} = -\frac{1}{2}Q_0,$$

where

$$Q_0 = \frac{I_0 d^3}{3c}.$$

Indeed, for $kd \ll 1$ we have

$$\cos\left(\frac{kd}{2}\right) \cong 1 - \frac{k^2 d^2}{8},$$

as well as

$$\sin\left(\frac{kd}{2}\right) \cong \frac{kd}{2} - \frac{k^3 d^3}{48},$$

therefore

$$\left[kd\cos\left(\frac{kd}{2}\right)-2\sin\left(\frac{kd}{2}\right)\right]^2 \cong \left[kd\left(1-\frac{k^2d^2}{8}\right)\right.$$

$$\left.-2\left(\frac{kd}{2}-\frac{k^3d^3}{48}\right)\right]^2 = \frac{k^6d^6}{144}.$$

Introducing this relation into Eq. (5.6.75), we arrive at Eq. (5.6.76).

For the particular values $kd = \pi$ and $kd = 2\pi$, corresponding to one half-wave $(d = \lambda/2)$ and two half-waves $(d = \lambda)$ of the current oscillating along antenna, respectively, the corresponding angular distributions are given by

$$\frac{dP_{\rm rad}}{d\Omega} = \frac{c\mu_0 I_0^2}{32\pi^2}\sin^2 2\theta \times \begin{cases} 1, & kd = \pi, \quad d = \dfrac{\lambda}{2}, \\ \pi^2, & kd = 2\pi, \quad d = \lambda. \end{cases} \tag{5.6.77}$$

The above relations show that both the half-wave and the full-wave antennas have the same angular distribution of radiated power. This distribution is graphically represented in Fig. 5.10. The figure shows the four characteristic lobes (hence the name *quadrupole radiation*), with maxima for $\theta = \pi/4$ and $\theta = 3\pi/4$.

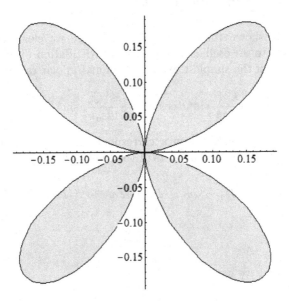

FIGURE 5.10
Quadrupole angular distribution of radiated power for a thin, linear, centrally-fed antenna.

c) The total power radiated by such an electric quadrupole is given by

$$P_{rad} = \int dP_{rad} = \iint \frac{dP_{rad}}{d\Omega}\, d\Omega = \frac{c\mu_0 I_0^2}{32\pi^2} \left[kd\, \cos\left(\frac{kd}{2}\right) \right.$$
$$\left. - 2\sin\left(\frac{kd}{2}\right) \right]^2 \int_0^\pi \sin^2\theta \cos^2\theta \sin\theta\, d\theta \int_0^{2\pi} d\varphi$$
$$= \frac{c\mu_0 I_0^2}{30\pi} \left[\frac{kd}{2}\cos\left(\frac{kd}{2}\right) - \sin\left(\frac{kd}{2}\right) \right]^2. \tag{5.6.78}$$

As we can easily see, in the general case, the total radiated power depends on the frequency.

d) As one knows, the radiative resistance of the antenna is the coefficient of $I_0^2/2$ appearing in the expression given by Eq. (5.6.78) of the total power radiated by the antenna, that is

$$R_{\text{rad}} = R_{\text{rad}}(k) = \frac{c\mu_0}{15\pi} \left[\frac{kd}{2}\cos\left(\frac{kd}{2}\right) - \sin\left(\frac{kd}{2}\right) \right]^2. \tag{5.6.79}$$

For the particular values $kd = \pi$ and $kd = 2\pi$, the total radiated power is

$$P_{\text{rad}} = \frac{c\mu_0 I_0^2}{30\pi} \times \begin{cases} 1, & kd = \pi \quad \left(d = \frac{\lambda}{2}\right), \\ \pi^2, & kd = 2\pi \quad (d = \lambda), \end{cases} \tag{5.6.80}$$

and depends only on the current I_0 traveling through the antenna.

The corresponding radiative resistance of the antenna is

$$R_{rad} = \frac{c\mu_0}{15\pi} \times \begin{cases} 1, & kd = \pi, \quad d = \frac{\lambda}{2}, \\ \pi^2, & kd = 2\pi, \quad d = \lambda \end{cases}$$
$$= \frac{Z_0}{15\pi} \times \begin{cases} 1, & kd = \pi, \quad d = \frac{\lambda}{2}, \\ \pi^2, & kd = 2\pi, \quad d = \lambda \end{cases}$$
$$\cong \begin{cases} 8\,\Omega, & kd = \pi, \quad d = \frac{\lambda}{2}, \\ 78.96\,\Omega, & kd = 2\pi, \quad d = \lambda. \end{cases} \tag{5.6.81}$$

As can be seen from relation (5.6.81), in the particular cases of half-wave and full-wave antennas, the radiative resistance of the antenna depends neither on the frequency of the current flowing through the antenna nor on its length.

5.7 Problem No. 44

Determine the spatial limit of applicability of the Classical Electrodynamics in the case of an electron.

Solution

The spatial limit of applicability of the Classical Electrodynamics for the electron is given by the electron "radius", considered as an elementary particle. At distances smaller than the electron radius, we expect that classical electrodynamics will no longer provide accurate results, and therefore become inapplicable.

In fact, as we will see, the quantum effects are making known their presence even beginning with distances over one thousand times larger than the electron "classical" radius, which implies the consideration of quantum effects even at such a distance.

Since, as is well known, in the Special Theory of Relativity the mass of any particle vary with its velocity, in the following we will determine the electron radius in two important cases:

a) in the static limit case (supposing the electron at rest), on the basis of some relativistic considerations, and

b) in the case of an electron moving with a non-relativistic speed (its velocity, $\vec{v}$, satisfies the condition $|\vec{v}|/c \ll 1$).

a) In this case, the electron radius can be evaluated by considering the equivalence between the electron proper electromagnetic energy and its rest energy.

As is well known, the electrostatic energy of an assembly of electrically charged particles, with charges q_k, $k = 1, 2, 3, ...$, and assimilable to some material points is given by

$$W = \frac{1}{2} \sum_k q_k V_k, \qquad (5.7.82)$$

where V_k is the electrostatic potential created by all the particles at the point where the particle with order number k is situated. This result can be deduced both directly, on the basis of equating the mechanical work required to bring an electric charge from infinity to a given point, with its electrostatic potential energy at that point, and starting from the more general case, namely that which make use of notion of electromagnetic (hence not electrostatic) field. Let us start with the second alternative. Since the energy density of the electromagnetic field is

$$w = \frac{1}{2} \left(\varepsilon E^2 + \mu H^2 \right),$$

in the static case, when, obviously, the magnetic field vanishes in the proper reference system, this becomes

$$w_{static} = \frac{1}{2} \varepsilon E^2,$$

and the energy of the system (containing only electric charges – but no electric currents) writes

$$W = \frac{1}{2} \int \varepsilon E^2 d\tau,$$

where the integral extends over the whole space. Since $\vec{E} = -\nabla V$, the last relation can also be written as

$$W = -\frac{1}{2} \int \varepsilon \vec{E} \cdot \nabla V \, d\tau = -\frac{1}{2}\varepsilon \int \nabla \cdot \left(V\vec{E} \right) d\tau + \frac{1}{2}\varepsilon \int V\nabla \cdot \vec{E} \, d\tau, \quad (5.7.83)$$

where the following relation has been used: $\nabla \cdot \left(V\vec{E} \right) = \vec{E} \cdot \nabla V + V\nabla \cdot \vec{E}$. But, according to Green-Gauss-Ostrogradski theorem (divergence theorem),

$$\int\limits_{(D)} \nabla \cdot \left(V\vec{E} \right) d\tau = \oint\limits_{(S)} V\vec{E} \cdot d\vec{S} = 0,$$

because the integration domain D is the whole space, meaning that the closing surface S is at infinity, where both V and $\vec{E}$ vanish. In addition, using the Gauss flux theorem (written in the local/differential form),

$$\nabla \cdot \vec{D} = \varepsilon \nabla \cdot \vec{E} = \rho_e,$$

where ρ_e is the volume charge density of the system, the relation (5.7.83) becomes

$$W = \frac{1}{2} \int \rho_e V \, d\tau.$$

In the case of a discret system, the integral is replaced by the corresponding sum expressed in Eq. (5.7.82).

Observation: Even if relation (5.7.82) can be obtained by a direct procedure, starting with analysis of a discrete system of electric point charges, we have preferred to show its deductibility from the general expression of energy of the electromagnetic field, in order to put into evidence the general frame, that which make allowance of the notion of field, in the argumentation of case a).

Let us particularise the result given by Eq. (5.7.82) for a single particle, namely the electron, with the electric charge $e = -1,6 \times 10^{-19}$C and the rest mass $m_0 = 9,1 \times 10^{-31}$kg. The electrostatic potential energy of a single electron at rest will be

$$W_{e-} = \frac{1}{2}eV_{e-} = \frac{e^2}{8\pi\varepsilon_0 R}, \quad (5.7.84)$$

where $V_{e-} = \dfrac{e}{4\pi\varepsilon_0 R}$ is the electrostatic potential of the electrostatic field produced by the electron at rest.

If the electron is considered a material point, then $R = 0$ and – obviously – both V_{e-} and W_{e-} are infinite. Then, in view of the connection between mass and energy, since the electrostatic potential energy of the electron is infinite, it follows that its mass should be also infinite. Classical Electrodynamics cannot tell us if the proper mass of the electron is only electromagnetic in nature.

In Quantum Field Theory, specifically the procedure of renormalization, the following formula holds:

$$m = m_0 + \delta m,$$

where m is the experimentally measurable mass of the particle, m_0 is the so-called *bare mass* of the particle, and δm is the increase in mass which is due to the interaction of the particle with the field (in our case, with the electromagnetic field of the electron itself). To avoid any confusion, we draw the reader's attention that in this paragraph the quantities denoted by m and m_0 have different meanings from those used in the rest of the problem solving.

The bare mass of a particle is the limit of its mass as the scale of distance approaches zero or, equivalently, as the energy of the particle collision approaches infinity. Obviously, the bare mass differs from the *invariant mass* as usually understood because the invariant mass includes the "garment" of the particle by particle-antiparticle pairs of virtual quanta which are temporarily created by the force-fields (the so-called *gauge fields*) around the particle. In some versions of QFT, the bare mass of some particles may even be minus or plus infinity. In the Standard Model, where the Higgs boson plays a very important role, all particles have a zero bare mass. The *invariant mass* (or *intrinsic mass*, or *rest mass*, or even *proper mass*) is "the portion" of the total mass of a particle that is independent of the overall motion of the particle. More exactly, it is a characteristic of the particle's total energy and momentum that is the same in all inertial frames of reference (frames which are related by Lorentz transformations).

Formally, the renormalization of the electron mass could be accomplished considering an infinite, negative, non-electromagnetic mass, which could "compensate" the infinite value of the electromagnetic mass.

In the frame of Quantum Electrodynamics (QED – which is a particular chapter of QFT), the quantum image "of the day" is that mass renormalization involves the emission and reabsorption of a virtual photon by a moving electron (or any other charged particle). This is a quantum-mechanical interaction between the electron and the electron's own electromagnetic field. However, the intuitive idea of mass renormalization can be explained without resorting to any quantum theory. Being a charge carrier, any electron simply produces an electromagnetic field around itself, which brings contributions to the energy-momentum tensor, so it possesses energy and mass. If we want to accelerate the electron, we also have to accelerate the electromagnetic field attached to it, and this changes the effective, measurable mass of the electron. This is the essence of mass renormalization, with the electromagnetic field adding to mass of any charged particle (adding an infinite amount, if the calculation is not regularized).

At present, within the QED there are a lot of models of electron mass renormalization, some of them very recent, proving the topicality of this subject. For example, the researchers Volker Bach, Thomas Chen, Juerg Froehlich and Israel Michael Sigal solved this problem (in the frame of non-relativistic quantum electrodynamics), in a paper published in the *Journal of Functional Analysis* [*J. Func. Anal.*, 243(2), 426–535 (2007)] by means of the smooth isospectral Feshbach map within the operational theory of the renormalization group.

The conclusion is that, for small distances, the Classical Electrodynamics (CED) becomes contradictory: on the one hand it admits the experimentally determined value for the electron mass, $m_0 = 9, 1 \times 10^{-31}$kg (which is a very small value), and on the other hand – as we have shown above – it predicts an infinite mass.

Considering for the electron at rest the "equivalence" between its electromagnetic energy, given by Eq. (5.7.84), and its rest energy, $W_0 = m_0 c^2$, it follows for the electron a "proper radius" a little bigger than $1\,\text{fm} = 10^{-15}$m:

$$\frac{e^2}{8\pi\varepsilon_0 R_0} = m_0 c^2 \Rightarrow R_0 = \frac{e^2}{8\pi\varepsilon_0 m_0 c^2} \simeq 1, 4 \times 10^{-15}\text{m}. \qquad (5.7.85)$$

This distance gives the applicability limit of Classical Electrodynamics for the electron (as one observes, this distance depends on the particle characteristics: the electric charge and the rest mass). This limit of applicability of Classical Electrodynamics is given by the basic principles of CED themselves. These principles lose their validity in the spatial domain beginning with distances even bigger than 10^{-15}m, such as

$$\lambda = \frac{h}{m_0 c} = 2, 4 \times 10^{-12}\text{m},$$

when the quantum effects become notable.

b) In this case (of an electron moving with a speed $\vec{v}$, so that $|\vec{v}|/c \ll 1$), the electromagnetic field created by the electron is given by

$$\begin{cases} \vec{E} = -\nabla V - \dfrac{\partial \vec{A}}{\partial t} = -\dfrac{\partial V}{\partial \vec{r}} - \dfrac{\partial \vec{A}}{\partial t}, \\[2mm] \vec{B} = \nabla \times \vec{A}, \end{cases} \qquad (5.7.86)$$

where $V(\vec{r}, t)$ and $\vec{A}(\vec{r}, t)$ are the electrodynamic potentials of this field – the Liénard-Wiechert potentials (see **Problem No. 38**):

$$V(\vec{r}, t) = \frac{1}{4\pi\varepsilon_0}\left[\frac{e}{|\vec{r} - \vec{x}|\left(1 - \vec{n} \cdot \vec{\beta}\right)}\right]_{t=t' + \frac{|\vec{r} - \vec{x}(t')|}{c}}, \qquad (5.7.87)$$

$$\vec{A}(\vec{r}, t) = \frac{\mu_0}{4\pi}\left[\frac{ec\vec{\beta}}{|\vec{r} - \vec{x}|\left(1 - \vec{n} \cdot \vec{\beta}\right)}\right]_{t=t' + \frac{|\vec{r} - \vec{x}(t')|}{c}}, \qquad (5.7.88)$$

where $\vec{x} = \vec{x}(t')$ is the equation of the electron trajectory, $\vec{r}$ is the position vector of the point where the potentials are determined (the "observation"

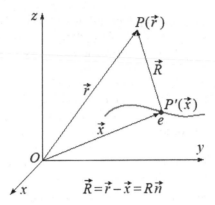

FIGURE 5.11
the relative position $P'(\vec{x})$ of the electron in motion, with respect to the observation point $P(\vec{r})$.

point), $\vec{n}$ is the unit vector of direction $\vec{r}-\vec{x}$, and $c\vec{\beta} = d\vec{x}/dt'$ (with $\vec{v} = d\vec{x}/dt'$) is the electron velocity on its trajectory. We mention the fact that in relations (5.7.86) the derivatives are performed with respect to the coordinates x, y, z and the time t, corresponding to the observer.

To simplify the writing, let us denote by $\vec{R}$ the vector $\vec{r} - \vec{x}(= \vec{R})$, which is the position vector of the observation point with respect to the current/actual position of the electron (see Fig. 5.11).

If $P(\vec{r})$ is the observation point, and $P'(\vec{x})$ is the current position of the electron, the electromagnetic signal emitted by the electron at the moment t' will arrive at the observation point later, at the moment

$$t = t' + \frac{|\vec{r} - \vec{x}(t')|}{c} = t' + \frac{R(t')}{c}, \qquad (5.7.89)$$

where $\vec{r} = \vec{x} + \vec{R}$. If we consider the relation (5.7.89) as an equation in t', then the second postulate of Einstein's Special Theory of Relativity (STR) assure the fact that this equation has a unique solution. Indeed, since the electron velocity may not exceed the speed of light in vacuum, and only the events coming from the absolute past can affect the observer, one follows that only the signal emitted in $P'(\vec{x})$ will arrive at $P(\vec{r})$.

The second point of intersection of the electron worldline with the light cone cannot lead to a causal solution, since the upper light cone corresponds to the absolute future (this solution of the equation violates the causality principle) – see Fig. 5.12.

Using the new notations, the expressions of the two electromagnetic potentials become

$$V(\vec{r}, t) = \frac{1}{4\pi\varepsilon_0} \left[\frac{e}{\left(R - \frac{\vec{R}\cdot\vec{v}}{c}\right)} \right]_{t=t'+\frac{R}{c}} \equiv \frac{1}{4\pi\varepsilon_0} \frac{e}{R - \frac{\vec{R}\cdot\vec{v}}{c}}, \qquad (5.7.90)$$

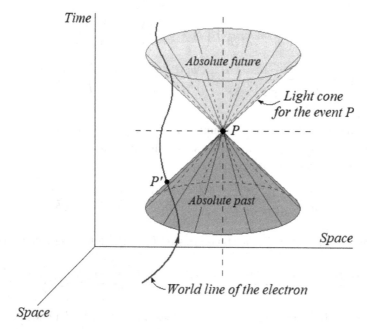

FIGURE 5.12
Schematically representation of the light (hyper)cone (only two of the three spatial axes are depicted) for the event P (the observation "point"), and the worldline of the moving electron. P' is the current/actual position of the electron, being the only point of intersection of the electron worldline with the light cone which lead to a causal solution.

and

$$\vec{A}(\vec{r}, t) = \frac{\mu_0}{4\pi} \left[\frac{e\vec{v}}{\left(R - \frac{\vec{R}\cdot\vec{v}}{c}\right)} \right]_{t=t'+\frac{R}{c}} \equiv \frac{\mu_0}{4\pi} \frac{e\vec{v}}{R - \frac{\vec{R}\cdot\vec{v}}{c}}, \qquad (5.7.91)$$

where, in order to simplify the writing, we have omitted the index of square brackets, being understood that the quantities on the right side must be considered at the time $t' = t - \dfrac{R}{c}$.

Analysing the relations (5.7.90) and (5.7.91), one observes that $\vec{A} = \dfrac{\vec{v}}{c^2}V$, so that[1]:

$$\vec{E} = -\frac{\partial V}{\partial \vec{r}} - \frac{\partial \vec{A}}{\partial t} = -\frac{\partial V}{\partial \vec{r}} - \frac{1}{c^2}\frac{\partial}{\partial t}(\vec{v}V) = -\left(\frac{\partial}{\partial \vec{r}} + \frac{\vec{v}}{c^2}\frac{\partial}{\partial t}\right)V - \frac{V}{c^2}\frac{\partial \vec{v}}{\partial t}$$

---

[1]In the following, in solving this problem we will use the formal writing of the gradient as $\frac{\partial}{\partial \vec{r}}$; for example, the gradient of V will be written as $\frac{\partial V}{\partial \vec{r}}$.

$$= -e \frac{1}{4\pi\varepsilon_0} \left(\frac{\partial}{\partial \vec{r}} + \frac{\vec{v}}{c^2} \frac{\partial}{\partial t} \right) \left(R - \frac{\vec{R} \cdot \vec{v}}{c} \right)^{-1} - \frac{V}{c^2} \frac{\partial \vec{v}}{\partial t'} \frac{\partial t'}{\partial t}$$

$$= \frac{1}{4\pi\varepsilon_0} \frac{e}{\left(R - \frac{\vec{R} \cdot \vec{v}}{c} \right)^2} \left(\frac{\partial}{\partial \vec{r}} + \frac{\vec{v}}{c^2} \frac{\partial}{\partial t} \right) \left(R - \frac{\vec{R} \cdot \vec{v}}{c} \right) - \frac{\vec{a} V}{c^2} \frac{\partial t'}{\partial t}, \quad (5.7.92)$$

where it has been taken into account that $\vec{a} = \frac{\partial \vec{v}}{\partial t'}$ is the electron acceleration. The derivative $\frac{\partial t'}{\partial t}$ can be found by using the obvious relation $R = c(t - t')$ and taking the partial time derivative

$$\frac{\partial R}{\partial t} \left(= \frac{\partial R}{\partial t'} \frac{\partial t'}{\partial t} \right) = c \left(1 - \frac{\partial t'}{\partial t} \right) \Rightarrow \frac{\partial t'}{\partial t} = \frac{c}{c + \frac{\partial R}{\partial t'}}.$$

The partial derivative $\frac{\partial R}{\partial t'}$ can be expressed from the following calculations:

$$\frac{\partial R^2}{\partial t'} = \frac{\partial (\vec{R} \cdot \vec{R})}{\partial t'} = 2\vec{R} \cdot \frac{\partial \vec{R}}{\partial t'} = 2\vec{R} \cdot \frac{\partial (\vec{r} - \vec{x})}{\partial t'} = -2\vec{R} \cdot \vec{v},$$

where it has been used the fact that $\vec{r}$ does not explicitly depend on t', and $\vec{v} = \frac{d\vec{x}}{dt'} \equiv \frac{\partial \vec{x}}{\partial t'}$, because $\vec{x} = \vec{x}(t')$. But the derivative of the squared modulus of the vector $\vec{R}$ can be also written as

$$\frac{\partial R^2}{\partial t'} = 2R \frac{\partial R}{\partial t'},$$

so that we have

$$2R \frac{\partial R}{\partial t'} = -2\vec{R} \cdot \vec{v} \Rightarrow \frac{\partial R}{\partial t'} = -\frac{\vec{R} \cdot \vec{v}}{R},$$

and so,

$$\frac{\partial t'}{\partial t} = \frac{c}{c + \frac{\partial R}{\partial t'}} = \frac{c}{c - \frac{\vec{R} \cdot \vec{v}}{R}} = \frac{R}{R - \frac{\vec{R} \cdot \vec{v}}{c}}.$$

By using this result in the formula (5.7.92), we have

$$\vec{E} = \frac{1}{4\pi\varepsilon_0} \frac{e}{\left(R - \frac{\vec{R} \cdot \vec{v}}{c} \right)^2} \left(\frac{\partial}{\partial \vec{r}} + \frac{\vec{v}}{c^2} \frac{\partial}{\partial t} \right)$$

$$\times \left(R - \frac{\vec{R} \cdot \vec{v}}{c} \right) - \frac{1}{4\pi\varepsilon_0} \frac{e R \vec{a}}{c^2 \left(R - \frac{\vec{R} \cdot \vec{v}}{c} \right)^2}. \quad (5.7.93)$$

There are still to be calculated the derivatives $\frac{\partial R}{\partial \vec{r}}, \frac{\partial R}{\partial t}, \frac{\partial}{\partial \vec{r}}(\vec{R} \cdot \vec{v})$ and $\frac{\partial}{\partial t}(\vec{R} \cdot \vec{v})$. We have (we point out once again that both here and in the calculations that will follow in solving this problem, the partial derivative with respect to $\vec{r}$ actually represents the formal writing of the gradient):

$$\nabla t' = \operatorname{grad} t' \equiv \frac{\partial t'}{\partial \vec{r}} = \frac{\partial}{\partial \vec{r}} \left(t - \frac{R}{c} \right) = -\frac{1}{c}\frac{\partial R}{\partial \vec{r}},$$

because x, y, z and t are independent variables. Thus, we can write

$$\frac{\partial R}{\partial \vec{r}} = -c\frac{\partial t'}{\partial \vec{r}},$$

and so, determining $\frac{\partial R}{\partial \vec{r}}$ requires knowing $\frac{\partial t'}{\partial \vec{r}}$. Let's now determine $\frac{\partial t'}{\partial \vec{r}}$. To this end, note that the quantity R can be represented either in the observation point coordinates, $R = R(t, x, y, z)$, or in the sources point coordinates (the electron, in our case), $R = R(t', x', y', z')$, where x', y' and z' are the coordinates of the point defined by vector[2] $\vec{x}(t')$. Choosing the second option, it follows that the derivatives of R with respect to $\vec{r}$ have to be expressed by means of variables t', x', y' and z' as

$$\frac{\partial R}{\partial \vec{r}} = \frac{\partial R}{\partial t'}\frac{\partial t'}{\partial \vec{r}} + \frac{\partial R}{\partial x'}\frac{\partial x'}{\partial \vec{r}} + \frac{\partial R}{\partial y'}\frac{\partial y'}{\partial \vec{r}} + \frac{\partial R}{\partial z'}\frac{\partial z'}{\partial \vec{r}}.$$

But

$$\frac{\partial R}{\partial x'}\frac{\partial x'}{\partial \vec{r}} = \frac{\partial R}{\partial x}\frac{\partial x}{\partial x'}\frac{\partial x'}{\partial x}\frac{\partial x}{\partial \vec{r}} = \frac{\partial R}{\partial x}\frac{\partial x}{\partial \vec{r}}.$$

Similarly, we can write

$$\frac{\partial R}{\partial y'}\frac{\partial y'}{\partial \vec{r}} = \frac{\partial R}{\partial y}\frac{\partial y}{\partial \vec{r}},$$

and

$$\frac{\partial R}{\partial z'}\frac{\partial z'}{\partial \vec{r}} = \frac{\partial R}{\partial z}\frac{\partial z}{\partial \vec{r}}.$$

Also,

$$\frac{\partial R}{\partial x} = \frac{\partial}{\partial x}\sqrt{(x - x')^2 + (y - y')^2 + (z - z')^2} = \frac{x - x'}{R},$$

and, analogously,

$$\frac{\partial R}{\partial y} = \frac{y - y'}{r} \quad \text{and} \quad \frac{\partial R}{\partial z} = \frac{z - z'}{r}.$$

Since

$$\frac{\partial x}{\partial \vec{r}} = \operatorname{grad} x = \vec{i}\frac{\partial x}{\partial x} + \vec{j}\frac{\partial x}{\partial y} + \vec{k}\frac{\partial x}{\partial z} = \vec{i},$$

---

[2]Note that x (the modulus of the vector $\vec{x}(t')$) has nothing to do with the x-component of the vector $\vec{r}$, the fact that both quantities are denoted by the same letter being — shall we say — an "uninspired notation" for the vector $\vec{x}(t')$. Actually, as the reader can easily see, in solving this problem the components of the 3-vector $\vec{x}(t')$ are denoted by x', y' and z'.

and, also, $\frac{\partial y}{\partial \vec{r}} = \vec{j}$, and $\frac{\partial z}{\partial \vec{r}} = \vec{k}$, where $\vec{i}, \vec{j}$ and $\vec{k}$ are the versors of axes Ox, Oy and Oz, respectively, it follows that

$$\frac{\partial R}{\partial x'}\frac{\partial x'}{\partial \vec{r}} + \frac{\partial R}{\partial y'}\frac{\partial y'}{\partial \vec{r}} + \frac{\partial R}{\partial z'}\frac{\partial z'}{\partial \vec{r}} = \frac{\partial R}{\partial x}\frac{\partial x}{\partial \vec{r}} + \frac{\partial R}{\partial y}\frac{\partial y}{\partial \vec{r}} + \frac{\partial R}{\partial z}\frac{\partial z}{\partial \vec{r}}$$

$$= \vec{i}\frac{x - x'}{R} + \vec{j}\frac{y - y'}{R} + \vec{k}\frac{z - z'}{R} = \frac{\vec{R}}{R}.$$

Then

$$\frac{\partial R}{\partial \vec{r}} = \frac{\partial R}{\partial t'}\frac{\partial t'}{\partial \vec{r}} + \frac{\partial R}{\partial x'}\frac{\partial x'}{\partial \vec{r}} + \frac{\partial R}{\partial y'}\frac{\partial y'}{\partial \vec{r}} + \frac{\partial R}{\partial z'}\frac{\partial z'}{\partial \vec{r}}$$

$$= \frac{\partial R}{\partial t'}\frac{\partial t'}{\partial \vec{r}} + \frac{\vec{R}}{R} = \frac{\vec{R}}{R} - \frac{\vec{R}\cdot\vec{v}}{R}\frac{\partial t'}{\partial \vec{r}}.$$

Therefore

$$\nabla t' \equiv \frac{\partial t'}{\partial \vec{r}} \left[= \frac{\partial}{\partial \vec{r}}\left(t - \frac{R}{c}\right)\right] = -\frac{1}{c}\left(\frac{\vec{R}}{R} - \frac{\vec{R}\cdot\vec{v}}{R}\frac{\partial t'}{\partial \vec{r}}\right) = \frac{\vec{R}\cdot\vec{v}}{cR}\frac{\partial t'}{\partial \vec{r}} - \frac{\vec{R}}{cR},$$

which gives

$$\frac{\partial t'}{\partial \vec{r}} = -\frac{\vec{R}}{cR}\frac{1}{1 - \frac{\vec{R}\cdot\vec{v}}{cR}} = -\frac{\vec{R}}{c\left(R - \frac{\vec{R}\cdot\vec{v}}{c}\right)},$$

and therefore

$$\frac{\partial R}{\partial \vec{r}} = -c\frac{\partial t'}{\partial \vec{r}} = c\frac{\vec{R}}{c\left(R - \frac{\vec{R}\cdot\vec{v}}{c}\right)} = \frac{\vec{R}}{R - \frac{\vec{R}\cdot\vec{v}}{c}}. \tag{5.7.94}$$

The derivative $\frac{\partial R}{\partial t}$ is immediately obtained from the following two relations already proved:

$$\frac{\partial R}{\partial t} = c\left(1 - \frac{\partial t'}{\partial t}\right),$$

and

$$\frac{\partial t'}{\partial t} = \frac{R}{R - \frac{\vec{R}\cdot\vec{v}}{c}}.$$

It follows immediately that

$$\frac{\partial R}{\partial t} = c\left(1 - \frac{\partial t'}{\partial t}\right) = c\left(1 - \frac{R}{R - \frac{\vec{R}\cdot\vec{v}}{c}}\right) = -\frac{\vec{R}\cdot\vec{v}}{R - \frac{\vec{R}\cdot\vec{v}}{c}}. \tag{5.7.95}$$

Let us now calculate the gradient of $(\vec{R}\cdot\vec{v})$, *i.e.*, formally, the derivative $\frac{\partial}{\partial \vec{r}}(\vec{R}\cdot\vec{v})$. In view of B.3.55, we can write

$$\frac{\partial}{\partial \vec{r}}(\vec{R}\cdot\vec{v}) \equiv \nabla(\vec{R}\cdot\vec{v}) = \vec{R}\times(\nabla\times\vec{v}) + \vec{v}\times(\nabla\times\vec{R}) + (\vec{R}\cdot\nabla)\vec{v} + (\vec{v}\cdot\nabla)\vec{R}.$$

Let's calculate all the terms of this expression one by one. We have

$$\nabla \times \vec{v} = \varepsilon_{ijk} \frac{\partial v_k}{\partial x_j} \vec{u}_i = \varepsilon_{ijk} \frac{\partial v_k}{\partial t'} \frac{\partial t'}{\partial x_j} \vec{u}_i = \varepsilon_{ijk} a_k \frac{\partial t'}{\partial x_j} \vec{u}_i = \varepsilon_{ijk}$$

$$\times a_k \left[\frac{1}{c} \frac{-R_j}{R - \frac{\vec{R} \cdot \vec{v}}{c}} \right] \vec{u}_i = -\frac{1}{c} \frac{1}{R - \frac{\vec{R} \cdot \vec{v}}{c}} \varepsilon_{ijk} R_j a_k \vec{u}_i$$

$$= -\frac{1}{c} \frac{1}{R - \frac{\vec{R} \cdot \vec{v}}{c}} (\vec{R} \times \vec{a}) = -\frac{1}{c} \frac{\vec{R} \times \vec{a}}{R - \frac{\vec{R} \cdot \vec{v}}{c}},$$

and then, for the first term in the r.h.s. of the relation giving the gradient of $(\vec{R} \cdot \vec{v})$ we get the following result:

$$\vec{R} \times (\nabla \times \vec{v}) = -\frac{1}{c} \frac{1}{R - \frac{\vec{R} \cdot \vec{v}}{c}} \vec{R} \times (\vec{R} \times \vec{a}) = -\frac{1}{c} \frac{1}{R - \frac{\vec{R} \cdot \vec{v}}{c}} \left[(\vec{R} \cdot \vec{a})\vec{R} - R^2 \vec{a} \right].$$

Let us now deal with the second term of the same relationship. Thus, we have

$$\nabla \times \vec{R} = \varepsilon_{ijk} \frac{\partial R_k}{\partial x_j} \vec{u}_i = \varepsilon_{ijk} \left(\frac{\partial x_k}{\partial x_j} - \frac{\partial x'_k}{\partial x_j} \right) \vec{u}_i = \varepsilon_{ijk} \frac{\partial x_k}{\partial x_j} \vec{u}_i - \varepsilon_{ijk} \frac{\partial x'_k}{\partial x_j} \vec{u}_i$$

$$= \varepsilon_{ijk} \delta_{kj} \vec{u}_i - \varepsilon_{ijk} \frac{\partial x'_k}{\partial t'} \frac{\partial t'}{\partial x_j} \vec{u}_i = -\varepsilon_{ijk} v_k \frac{\partial t'}{\partial x_j} \vec{u}_i$$

$$= -\varepsilon_{ijk} v_k \left[\frac{1}{c} \frac{-R_j}{R - \frac{\vec{R} \cdot \vec{v}}{c}} \right] \vec{u}_i = \frac{1}{c} \frac{1}{R - \frac{\vec{R} \cdot \vec{v}}{c}} \varepsilon_{ijk} R_j v_k \vec{u}_i$$

$$= \frac{1}{c} \frac{1}{R - \frac{\vec{R} \cdot \vec{v}}{c}} (\vec{R} \times \vec{v}) = \frac{1}{c} \frac{\vec{R} \times \vec{v}}{R - \frac{\vec{R} \cdot \vec{v}}{c}},$$

where we have taken into account that always the product of an antisymmetric tensor and a symmetric tensor, both tensors having the symmetry property *in the same index pair*, is zero. Indeed, if A_{ij} is an antisymmetric tensor, *i.e.*, $A_{ij} = -A_{ji}$, and S_{ij} is a symmetric tensor, *i.e.*, $S_{ij} = S_{ji}$ (note that the index pair is the same for the two tensors), then

$$A_{ij} S_{ij} = \frac{1}{2}(A_{ij} S_{ij} + A_{ij} S_{ij}) = \frac{1}{2}(A_{ij} S_{ij} - A_{ji} S_{ji}) = \frac{1}{2}(A_{ij} S_{ij} - A_{ij} S_{ij}) = 0.$$

Then, for the second term in the r.h.s. of the relation giving the gradient of $(\vec{R} \cdot \vec{v})$ we get

$$\vec{v} \times (\nabla \times \vec{R}) = \frac{1}{c} \frac{1}{R - \frac{\vec{R} \cdot \vec{v}}{c}} \vec{v} \times (\vec{R} \times \vec{v}) = \frac{1}{c} \frac{1}{R - \frac{\vec{R} \cdot \vec{v}}{c}} \left[v^2 \vec{R} - (\vec{R} \cdot \vec{v})\vec{v} \right].$$

Next, for the third term of the same relationship we have

$$(\vec{R} \cdot \nabla)\vec{v} = R_j \frac{\partial \vec{v}}{\partial x_j} = R_j \frac{\partial \vec{v}}{\partial t'} \frac{\partial t'}{\partial x_j} = R_j \vec{a} \left[\frac{1}{c} \frac{-R_j}{R - \frac{\vec{R} \cdot \vec{v}}{c}} \right] = -\frac{1}{c} \frac{R^2 \vec{a}}{R - \frac{\vec{R} \cdot \vec{v}}{c}}.$$

Finally, for the fourth (and last) term in the r.h.s. of the same relation we can write

$$(\vec{v} \cdot \nabla)\vec{R} = v_j \frac{\partial \vec{R}}{\partial x_j} = v_j \frac{\partial}{\partial x_j}(\vec{r} - \vec{x}) = v_j \frac{\partial \vec{r}}{\partial x_j} - v_j \frac{\partial \vec{x}}{\partial x_j} = v_j \vec{u}_j - v_j \frac{\partial \vec{x}}{\partial t'} \frac{\partial t'}{\partial x_j}$$

$$= \vec{v} - v_j \vec{v} \left[\frac{1}{c} \frac{-R_j}{R - \frac{\vec{R} \cdot \vec{v}}{c}} \right] = \vec{v} + \frac{1}{c} \frac{(\vec{R} \cdot \vec{v})\vec{v}}{R - \frac{\vec{R} \cdot \vec{v}}{c}}.$$

Now adding all these four terms together, the gradient of $(\vec{R} \cdot \vec{v})$ takes the following form:

$$\frac{\partial}{\partial \vec{r}}(\vec{R} \cdot \vec{v}) \equiv \nabla(\vec{R} \cdot \vec{v}) = -\frac{1}{c} \frac{1}{R - \frac{\vec{R} \cdot \vec{v}}{c}} \left[(\vec{R} \cdot \vec{a})\vec{R} - R^2 \vec{a} \right]$$

$$+ \frac{1}{c} \frac{1}{R - \frac{\vec{R} \cdot \vec{v}}{c}} \left[v^2 \vec{R} - (\vec{R} \cdot \vec{v})\vec{v} \right] - \frac{1}{c} \frac{R^2 \vec{a}}{R - \frac{\vec{R} \cdot \vec{v}}{c}} + \vec{v}$$

$$+ \frac{1}{c} \frac{(\vec{R} \cdot \vec{v})\vec{v}}{R - \frac{\vec{R} \cdot \vec{v}}{c}} = \vec{v} + \frac{1}{c} \frac{v^2 - \vec{R} \cdot \vec{a}}{R - \frac{\vec{R} \cdot \vec{v}}{c}} \vec{R}. \qquad (5.7.96)$$

In this way it only remained to calculate the derivative $\frac{\partial}{\partial t}(\vec{R} \cdot \vec{v})$. We have

$$\frac{\partial}{\partial t}(\vec{R} \cdot \vec{v}) = \frac{\partial \vec{R}}{\partial t} \cdot \vec{v} + \vec{R} \cdot \frac{\partial \vec{v}}{\partial t} = \vec{v} \cdot \frac{\partial \vec{R}}{\partial t} + \vec{R} \cdot \frac{\partial \vec{v}}{\partial t} = \vec{v} \cdot \frac{\partial \vec{R}}{\partial t'} \frac{\partial t'}{\partial t}$$

$$+ \vec{R} \cdot \frac{\partial \vec{v}}{\partial t'} \frac{\partial t'}{\partial t} = \frac{\partial t'}{\partial t} \left(\vec{v} \cdot \frac{\partial \vec{R}}{\partial t'} + \vec{R} \cdot \frac{\partial \vec{v}}{\partial t'} \right) = \frac{R}{R - \frac{\vec{R} \cdot \vec{v}}{c}}$$

$$\times \left[\vec{v} \cdot (-\vec{v}) + \vec{R} \cdot \vec{a} \right] = \frac{\vec{R} \cdot \vec{a} - v^2}{R - \frac{\vec{R} \cdot \vec{v}}{c}} R. \qquad (5.7.97)$$

Introducing now Eqs. (5.7.94)–(5.7.97) into Eq. (5.7.93), we find

$$\vec{E} = \frac{e}{4\pi\varepsilon_0 \left(R - \frac{\vec{R} \cdot \vec{v}}{c} \right)^2} \left(\frac{\partial}{\partial \vec{r}} + \frac{\vec{v}}{c^2} \frac{\partial}{\partial t} \right) \left(R - \frac{\vec{R} \cdot \vec{v}}{c} \right)$$

$$- \frac{eR\vec{a}}{4\pi\varepsilon_0 c^2 \left(R - \frac{\vec{R} \cdot \vec{v}}{c} \right)^2} = \frac{e}{4\pi\varepsilon_0 \left(R - \frac{\vec{R} \cdot \vec{v}}{c} \right)^2} \left[\frac{\partial R}{\partial \vec{r}} + \frac{\vec{v}}{c^2} \frac{\partial R}{\partial t} \right.$$

$$\left. - \frac{1}{c} \frac{\partial}{\partial \vec{r}}(\vec{R} \cdot \vec{v}) - \frac{\vec{v}}{c^3} \frac{\partial}{\partial t}(\vec{R} \cdot \vec{v}) \right] - \frac{eR\vec{a}}{4\pi\varepsilon_0 c^2 \left(R - \frac{\vec{R} \cdot \vec{v}}{c} \right)^2}$$

$$= \frac{e}{4\pi\varepsilon_0 \left(R - \frac{\vec{R}\cdot\vec{v}}{c}\right)^2} \left[\frac{\partial R}{\partial \vec{r}} + \frac{\vec{v}}{c^2}\frac{\partial R}{\partial t} - \frac{1}{c}\frac{\partial}{\partial \vec{r}}(\vec{R}\cdot\vec{v}) - \frac{\vec{v}}{c^3}\frac{\partial}{\partial t}(\vec{R}\cdot\vec{v})\right]$$

$$- \frac{eR\vec{a}}{4\pi\varepsilon_0 c^2 \left(R - \frac{\vec{R}\cdot\vec{v}}{c}\right)^2} = \frac{e(4\pi\varepsilon_0)^{-1}}{\left(R - \frac{\vec{R}\cdot\vec{v}}{c}\right)^2}\left[\frac{\vec{R}}{R - \frac{\vec{R}\cdot\vec{v}}{c}} - \frac{\vec{v}}{c^2}\frac{\vec{R}\cdot\vec{v}}{R - \frac{\vec{R}\cdot\vec{v}}{c}}\right.$$

$$\left. - \frac{1}{c}\left(\vec{v} + \frac{v^2 - \vec{R}\cdot\vec{a}}{cR - \vec{R}\cdot\vec{v}}\vec{R}\right) - \frac{\vec{v}}{c^3}\frac{\vec{R}\cdot\vec{a} - v^2}{R - \frac{\vec{R}\cdot\vec{v}}{c}}R\right] - \frac{eR\vec{a}(4\pi\varepsilon_0)^{-1}}{c^2\left(R - \frac{\vec{R}\cdot\vec{v}}{c}\right)^2}$$

$$= \frac{e}{4\pi\varepsilon_0 \left(R - \frac{\vec{R}\cdot\vec{v}}{c}\right)^3}\left[\vec{R} - \frac{\vec{v}(\vec{R}\cdot\vec{v})}{c^2} - \frac{\vec{v}}{c}\left(R - \frac{\vec{R}\cdot\vec{v}}{c}\right) - \frac{v^2\vec{R}}{c^2}\right.$$

$$\left. + \frac{\vec{R}(\vec{R}\cdot\vec{a})}{c^2} - \frac{\vec{v}R(\vec{R}\cdot\vec{a})}{c^3} + \frac{v^2}{c^2}\frac{\vec{v}R}{c} - \frac{R\vec{a}}{c^2}\left(R - \frac{\vec{R}\cdot\vec{v}}{c}\right)\right]$$

$$= \frac{e}{4\pi\varepsilon_0 \left(R - \frac{\vec{R}\cdot\vec{v}}{c}\right)^3}\left[\left(1 - \frac{v^2}{c^2} + \frac{\vec{R}\cdot\vec{a}}{c^2}\right)\left(\vec{R} - \frac{R\vec{v}}{c}\right)\right.$$

$$\left. - \frac{R\vec{a}}{c^2}\left(R - \frac{\vec{R}\cdot\vec{v}}{c}\right)\right],$$

or, if we use the formula $\vec{A} \times (\vec{B} \times \vec{C}) = \vec{B}(\vec{A}\cdot\vec{C}) - \vec{C}(\vec{A}\cdot\vec{B})$ for the vectors $\vec{A} \equiv \vec{R}, \vec{B} \equiv \vec{R} - \dfrac{R\vec{v}}{c}$ and $\vec{C} \equiv \vec{a}$,

$$\vec{E} = \frac{e}{4\pi\varepsilon_0}\frac{\left(1 - \frac{v^2}{c^2}\right)\left(\vec{R} - \frac{R\vec{v}}{c}\right) + \frac{\vec{R}}{c^2}\times\left[\left(\vec{R} - \frac{R\vec{v}}{c}\right)\times\vec{a}\right]}{\left(R - \frac{\vec{R}\cdot\vec{v}}{c}\right)^3} = \vec{E}_v + \vec{E}_a, \quad (5.7.98)$$

where

$$\vec{E}_v = \frac{e}{4\pi\varepsilon_0}\frac{\left(1 - \frac{v^2}{c^2}\right)\left(\vec{R} - \frac{R\vec{v}}{c}\right)}{\left(R - \frac{\vec{R}\cdot\vec{v}}{c}\right)^3} \quad (5.7.99)$$

is the electric component of the electromagnetic field created by the moving electron, which depends on the electron velocity only, while the component

$$\vec{E}_a = \frac{e}{4\pi\varepsilon_0 c^2}\frac{\vec{R}\times\left[\left(\vec{R} - \frac{R\vec{v}}{c}\right)\times\vec{a}\right]}{\left(R - \frac{\vec{R}\cdot\vec{v}}{c}\right)^3} = \frac{\mu_0}{4\pi}\frac{e\vec{R}\times\left[\left(\vec{R} - \frac{R\vec{v}}{c}\right)\times\vec{a}\right]}{\left(R - \frac{\vec{R}\cdot\vec{v}}{c}\right)^3} \quad (5.7.100)$$

depends on both the speed and acceleration of the electron.

The induction of the magnetic component of the electromagnetic field is

$$\vec{B} = \nabla \times \vec{A} = \nabla \times \left(\frac{\vec{v}V}{c^2}\right) = \frac{1}{c^2}\nabla V \times \vec{v} + \frac{V}{c^2}\nabla \times \vec{v}, \quad (5.7.101)$$

where

$$\nabla V \equiv \frac{\partial V}{\partial \vec{r}} = \frac{\partial}{\partial \vec{r}} \left(\frac{e}{4\pi\varepsilon_0} \frac{1}{R - \frac{\vec{R}\cdot\vec{v}}{c}} \right) = \frac{e}{4\pi\varepsilon_0} \frac{\partial}{\partial \vec{r}} \left(R - \frac{\vec{R}\cdot\vec{v}}{c} \right)^{-1}$$

$$= -\frac{e}{4\pi\varepsilon_0} \frac{1}{\left(R - \frac{\vec{R}\cdot\vec{v}}{c} \right)^2} \frac{\partial}{\partial \vec{r}} \left(R - \frac{\vec{R}\cdot\vec{v}}{c} \right) = -\frac{e}{4\pi\varepsilon_0} \frac{1}{\left(R - \frac{\vec{R}\cdot\vec{v}}{c} \right)^2}$$

$$\times \left[\frac{\partial R}{\partial \vec{r}} - \frac{1}{c} \frac{\partial}{\partial \vec{r}} (\vec{R}\cdot\vec{v}) \right] = -\frac{e}{4\pi\varepsilon_0} \frac{1}{\left(R - \frac{\vec{R}\cdot\vec{v}}{c} \right)^2} \left[\frac{\vec{R}}{R - \frac{\vec{R}\cdot\vec{v}}{c}} \right.$$

$$\left. - \frac{1}{c} \left(\vec{v} + \frac{v^2 - \vec{R}\cdot\vec{a}}{cR - \vec{R}\cdot\vec{v}} \vec{R} \right) \right]$$

$$= -\frac{e}{4\pi\varepsilon_0} \frac{\vec{R} - \frac{\vec{v}}{c}\left(R - \frac{\vec{R}\cdot\vec{v}}{c} \right) - \frac{v^2}{c^2}\vec{R} + \frac{\vec{R}\cdot\vec{a}}{c^2}\vec{R}}{\left(R - \frac{\vec{R}\cdot\vec{v}}{c} \right)^3}$$

$$= -\frac{e}{4\pi\varepsilon_0} \frac{\vec{R}\left(1 - \frac{v^2}{c^2} + \frac{\vec{R}\cdot\vec{a}}{c^2} \right) - \frac{\vec{v}}{c}\left(R - \frac{\vec{R}\cdot\vec{v}}{c} \right)}{\left(R - \frac{\vec{R}\cdot\vec{v}}{c} \right)^3}, \qquad (5.7.102)$$

and, as we already calculated,

$$\nabla \times \vec{v} = -\frac{1}{c} \frac{\vec{R}\times\vec{a}}{R - \frac{\vec{R}\cdot\vec{v}}{c}}. \qquad (5.7.103)$$

By means of Eqs. (5.7.102) and (5.7.103), Eq. (5.7.101) becomes

$$\vec{B} = \nabla \times \vec{A} = \nabla \times \left(\frac{\vec{v}V}{c^2} \right) = \frac{1}{c^2} \nabla V \times \vec{v} + \frac{V}{c^2} \nabla \times \vec{v}$$

$$= -\frac{1}{c^2} \frac{e}{4\pi\varepsilon_0} \frac{\vec{R}\left(1 - \frac{v^2}{c^2} + \frac{\vec{R}\cdot\vec{a}}{c^2} \right) - \frac{\vec{v}}{c}\left(R - \frac{\vec{R}\cdot\vec{v}}{c} \right)}{\left(R - \frac{\vec{R}\cdot\vec{v}}{c} \right)^3} \times \vec{v}$$

$$+ \frac{V}{c^2} \left(-\frac{1}{c} \frac{\vec{R}\times\vec{a}}{R - \frac{\vec{R}\cdot\vec{v}}{c}} \right) = -\frac{e}{4\pi\varepsilon_0 c^2} \frac{\left(1 - \frac{v^2}{c^2} + \frac{\vec{R}\cdot\vec{a}}{c^2} \right)(\vec{R}\times\vec{v})}{\left(R - \frac{\vec{R}\cdot\vec{v}}{c} \right)^3}$$

$$- \frac{e}{4\pi\varepsilon_0 c^2} \frac{\frac{\vec{R}\times\vec{a}}{c}}{\left(R - \frac{\vec{R}\cdot\vec{v}}{c} \right)^2} = -\frac{e}{4\pi\varepsilon_0 c^2}$$

$$\times \frac{\left(1 - \frac{v^2}{c^2} + \frac{\vec{R}\cdot\vec{a}}{c^2} \right)(\vec{R}\times\vec{v}) + \left(R - \frac{\vec{R}\cdot\vec{v}}{c} \right)\frac{\vec{R}\times\vec{a}}{c}}{\left(R - \frac{\vec{R}\cdot\vec{v}}{c} \right)^3}.$$

Therefore, the final expressions of the two components of the electromagnetic field created by an electron moving with the velocity $\vec{v}$, in the approximation $|\vec{v}| \ll c$, are

$$\vec{E} = \frac{e}{4\pi\varepsilon_0} \frac{\left(1 - \frac{v^2}{c^2}\right)\left(\vec{R} - \frac{R\vec{v}}{c}\right) + \frac{\vec{R}}{c^2} \times \left[\left(\vec{R} - \frac{R\vec{v}}{c}\right) \times \vec{a}\right]}{\left(R - \frac{\vec{R}\cdot\vec{v}}{c}\right)^3}, \tag{5.7.104}$$

and

$$\vec{B} = \frac{\mu_0}{4\pi} \frac{\left(1 - \frac{v^2}{c^2} + \frac{\vec{R}\cdot\vec{a}}{c^2}\right)\left(e\vec{v} \times \vec{R}\right) + \left(R - \frac{\vec{R}\cdot\vec{v}}{c}\right)\frac{e\vec{a}\times\vec{R}}{c}}{\left(R - \frac{\vec{R}\cdot\vec{v}}{c}\right)^3}. \tag{5.7.105}$$

In the particular case when the electron motion is rectilinear and uniform (there is no acceleration) and $|\vec{v}| \ll c$, one can omit not only the ratio v^2/c^2, but also the ratio $|\vec{v}|/c$, and the the two components given by Eqs. (5.7.104) and (5.7.105) of the electromagnetic field generated by the moving electron become

$$\vec{E} = \frac{e}{4\pi\varepsilon_0}\frac{\vec{R}}{R^3}, \tag{5.7.106}$$

and

$$\vec{B} = \frac{\mu_0}{4\pi}\frac{e\vec{v} \times \vec{R}}{R^3}, \tag{5.7.107}$$

respectively. If we adopt for the electron the model of being a sphere of radius R_0, uniformly charged with electric charge of superficial density $\sigma_e = e/4\pi R_0^2$, and use the approximation above — which involves the use of expressions in Eqs. (5.7.106) and (5.7.107) — its energy will be

$$W = \frac{1}{2}\varepsilon_0 \int\limits_{(D)} E^2 dV + \frac{1}{2\mu_0}\int\limits_{(D)} B^2 dV$$

$$= \frac{e^2}{32\pi^2\varepsilon_0}\int_{R_0}^{\infty}\int_0^{\pi}\int_0^{2\pi}\frac{1}{r^2}\sin\theta dr d\theta d\varphi$$

$$+ \frac{\mu_0 e^2 v^2}{32\pi^2}\int_{R_0}^{\infty}\int_0^{\pi}\int_0^{2\pi}\frac{1}{r^2}\sin^3\theta dr d\theta d\varphi$$

$$= \frac{e^2}{32\pi^2\varepsilon_0}\int_{R_0}^{\infty}\frac{dr}{r^2}\int_0^{\pi}\sin\theta d\theta\int_0^{2\pi}d\varphi$$

$$+ \frac{\mu_0 e^2 v^2}{32\pi^2}\int_{R_0}^{\infty}\frac{dr}{r^2}\int_0^{\pi}\sin^3\theta d\theta\int_0^{2\pi}d\varphi = \frac{e^2}{32\pi^2\varepsilon_0}\left(-\frac{1}{r}\right)_{R_0}^{\infty}\cdot 4\pi$$

$$+ \frac{\mu_0 e^2 v^2}{32\pi^2}\left(-\frac{1}{r}\right)_{R_0}^{\infty}\left(\frac{1}{12}\cos 3\theta - \frac{3}{4}\cos\theta\right)_0^{\pi}\cdot 2\pi$$

$$= \frac{e^2}{8\pi\varepsilon_0 R_0} + \frac{\mu_0 e^2 v^2}{12\pi R_0} = \frac{e^2}{8\pi\varepsilon_0 R_0}\left(1 + \frac{2}{3}\frac{v^2}{c^2}\right) \simeq \frac{e^2}{8\pi\varepsilon_0 R_0},$$

that is, approximately the same value as that obtained for the electron at rest (see formula (5.7.84)), leading to approximately the same value for the electron radius as that given by Eq. (5.7.85) for the electron at rest. So, the fact that the electron has a non-relativistic motion (compared to when it is at rest) has almost no influence on its energy or radius. In other words, the non-relativistic motion of the electron does not significantly affect either its energy or radius.

On the other side, the energy of the magnetic field of the electron moving with velocity $v \ll c$ is

$$W_{mag} = \frac{1}{2\mu_o} \int_{(D)} B^2 dV$$

$$= \frac{\mu_0 e^2 v^2}{32\pi^2} \int_{R_0}^{\infty} \int_0^{\pi} \int_0^{2\pi} \frac{1}{r^2} \sin^3 \theta dr d\theta d\varphi = \frac{\mu_0 e^2 v^2}{12\pi R_0}. \quad (5.7.108)$$

Since the electron motion is supposed to be non-relativistic, its kinetic energy is given by the classical formula $E_{kin} = mv^2/2$, which gives $v^2 = 2E_{kin}/m$ and, if this last relation is introduced into Eq. (5.7.108), one finds the energy of the magnetic field in the form

$$W_{mag} = \frac{\mu_0 e^2 v^2}{12\pi R_0} = \frac{e^2 E_{kin}}{6\pi \varepsilon_0 mc^2 R_0} = \frac{2}{3} \frac{e^2}{4\pi\varepsilon_0 R_0} \frac{E_{kin}}{mc^2}. \quad (5.7.109)$$

But the energy of the magnetic field "disappears" when the electron is at rest or, in other words, when its kinetic energy vanishes. This means that $W_{mag} = E_{kin}$, and formula (5.7.109) becomes

$$mc^2 = \frac{2}{3} \frac{e^2}{4\pi\varepsilon_0 R_0}. \quad (5.7.109')$$

Adopting for the "classical" radius of the electron the value

$$R_0^{classic} = \frac{e^2}{4\pi\varepsilon_0 mc^2} = 2,8 \times 10^{-15} m, \quad (5.7.110)$$

where $m = 9,1 \times 10^{-31}$kg is the electron rest mass, relation (5.7.109') can be considered as being a *sui-generis* "definition" of the *electromagnetic mass* of the electron. This was settled by J.J. Thomson in 1881, and can be considered as one of the electron radius "definitions":

$$R_0 = \frac{2e^2}{12\pi\varepsilon_0 mc^2} = \frac{2}{3} \frac{e^2}{4\pi\varepsilon_0 mc^2} = \frac{2}{3} R_0^{classic} = 0,9(3) \times 10^{-15} m. \quad (5.7.111)$$

As one can be observed, regardless of how the electron radius is defined/determined, it has a value of the order 10^{-15}m $= 1$ fm. This value also expresses the spatial "limit" of applicability of Classical Electrodynamics to the electron.

5.8   Problem No. 45

Decompose in plane waves the electromagnetic field generated by an electron that moves uniformly and rectilinearly in vacuum.

Solution

Let $\vec{v}$ be the electron velocity, e its electric charge (in modulus), ρ the charge density, and $V(\vec{r}, t)$ and $\vec{A}(\vec{r}, t)$ the electrodynamic potentials associated to the electron electromagnetic field. If the electron position at the moment $t = 0$ is chosen as the origin of its movement, the charge density can be written by means of Dirac's delta distribution as

$$\rho(\vec{r}, t) = e\,\delta(\vec{r} - \vec{r}_0) = e\,\delta(\vec{r} - \vec{v}t). \qquad (5.8.112)$$

The d'Alembert type non-homogeneous second-order differential equations satisfied by the electromagnetic potentials $\vec{A}(\vec{r}, t)$ and $V(\vec{r}, t)$ then are

$$\Delta\vec{A} - \frac{1}{c^2}\frac{\partial^2\vec{A}}{\partial t^2} = -\mu_0 e\,\vec{v}\,\delta(\vec{r} - \vec{v}t), \qquad (5.8.113)$$

and respectively,

$$\Delta V - \frac{1}{c^2}\frac{\partial^2 V}{\partial t^2} = -\frac{e}{c}\,\delta(\vec{r} - \vec{v}t). \qquad (5.8.114)$$

As can be seen,

$$\vec{A} = \frac{V}{c^2}\vec{v}, \qquad (5.8.115)$$

therefore it's enough to deal with just one potential, say V. As usual, the spatial variable $\vec{r}$ is associated through the Fourier transform with the new variable $\vec{k}$, while the temporal variable t with the new variable ω. Then, keeping unchanged the time t and denoting by $V_{\vec{k}}$ the Fourier transform of $V(\vec{r}, t)$ with respect to the variable pair $(\vec{r}, \vec{k})$, we have

$$V_{\vec{k}}(t) \equiv V(\vec{k}, t) = \frac{1}{(2\pi)^{3/2}}\int V(\vec{r}, t)e^{-i\vec{k}\cdot\vec{r}}d\vec{r},$$

which is usually called the *direct Fourier transform*, while

$$V(\vec{r}, t) = \frac{1}{(2\pi)^{3/2}}\int V(\vec{k}, t)e^{i\vec{k}\cdot\vec{r}}d\vec{k} \equiv \frac{1}{(2\pi)^{3/2}}\int V_{\vec{k}}(t)e^{i\vec{k}\cdot\vec{r}}d\vec{k}, \qquad (5.8.116)$$

is called the *inverse Fourier transform*.

Let us now introduce Eq. (5.8.116) into Eq. (5.8.114) and use the Fourier transform of $\delta(\vec{r} - \vec{v}t)$. This gives

$$\frac{1}{(2\pi)^{3/2}}\int\left[-k^2 V_{\vec{k}}(t) - \frac{1}{c^2}\frac{\partial^2 V_{\vec{k}}(t)}{\partial t^2}\right]e^{i\vec{k}\cdot\vec{r}}d\vec{k} = -\frac{e}{(2\pi)^3\varepsilon_0}\int e^{i\vec{k}\cdot(\vec{r} - \vec{v}t)}d\vec{k}.$$

By equalizing the integrands in the relation above, one obtains

$$\frac{\partial^2 V_{\vec{k}}(t)}{\partial t^2} + c^2 k^2 V_{\vec{k}}(t) = \frac{c^2 e}{(2\pi)^{3/2}\varepsilon_0} e^{-i\vec{k}\cdot\vec{v}t},$$

or, equivalently,

$$\frac{d^2 V_{\vec{k}}(t)}{dt^2} + c^2 k^2 V_{\vec{k}}(t) = \frac{c^2 e}{(2\pi)^{3/2}\varepsilon_0} e^{-i\vec{k}\cdot\vec{v}t}. \qquad (5.8.117)$$

We will look for a solution of this equation of the form

$$V_{\vec{k}}(t) = V_{0\vec{k}}(t) e^{-i\vec{k}\cdot\vec{v}t}. \qquad (5.8.118)$$

Asking that $V_{\vec{k}}(t)$ given by Eq. (5.8.118) verify Eq. (5.8.117), one obtains

$$-(\vec{k}\cdot\vec{v})^2 V_{0\vec{k}}(t) e^{-i\vec{k}\cdot\vec{v}t} + c^2 k^2 V_{0\vec{k}}(t) e^{-i\vec{k}\cdot\vec{v}t} = \frac{c^2 e}{(2\pi)^{3/2}\varepsilon_0} e^{-i\vec{k}\cdot\vec{v}t},$$

whence

$$V_{0\vec{k}}(t) = \frac{e}{(2\pi)^{3/2}\varepsilon_0} \frac{1}{k^2 - \left(\dfrac{\vec{k}\cdot\vec{v}}{c}\right)^2},$$

so that, Eq. (5.8.118) becomes

$$V_{\vec{k}}(t) = \frac{e}{(2\pi)^{3/2}\varepsilon_0} \frac{e^{-i\vec{k}\cdot\vec{v}t}}{k^2 - \left(\dfrac{\vec{k}\cdot\vec{v}}{c}\right)^2}. \qquad (5.8.119)$$

With this result, Eq. (5.8.116) reads as follows:

$$V(\vec{r},t) = \frac{e}{(2\pi)^3\varepsilon_0} \int \frac{e^{i\vec{k}\cdot(\vec{r}-\vec{v}t)}}{k^2 - \left(\dfrac{\vec{k}\cdot\vec{v}}{c}\right)^2} d\vec{k}, \qquad (5.8.120)$$

and, also, Eq. (5.8.115) gives

$$\vec{A}(\vec{r},t) = \frac{e\vec{v}}{(2\pi)^3 c^2\varepsilon_0} \int \frac{e^{i\vec{k}\cdot(\vec{r}-\vec{v}t)}}{k^2 - \left(\dfrac{\vec{k}\cdot\vec{v}}{c}\right)^2} d\vec{k}. \qquad (5.8.121)$$

Defining $V_{\vec{k}}(\vec{r},t)$ through

$$V(\vec{r},t) = \int V_{\vec{k}}(\vec{r},t)\, d\vec{k},$$

and $\vec{A}_{\vec{k}}(\vec{r}, t)$ through a similar relation. *i.e.*,

$$\vec{A}(\vec{r}, t) = \int \vec{A}_{\vec{k}}(\vec{r}, t)\, d\vec{k},$$

the searched Fourier transforms, $V_{\vec{k}}(\vec{r}, t)$ and $\vec{A}_{\vec{k}}(\vec{r}, t)$ then are

$$V_{\vec{k}}(\vec{r}, t) = \frac{e}{(2\pi)^3 \varepsilon_0} \frac{e^{i\vec{k}\cdot(\vec{r} - \vec{v}t)}}{k^2 - \left(\dfrac{\vec{k}\cdot\vec{v}}{c}\right)^2}, \qquad (5.8.122)$$

and, respectively,

$$\vec{A}_{\vec{k}}(\vec{r}, t) = \frac{\vec{v}}{c^2} V_{\vec{k}}(\vec{r}, t) = \frac{e\vec{v}}{(2\pi)^3 \varepsilon_0 c^2} \frac{e^{i\vec{k}\cdot(\vec{r} - \vec{v}t)}}{k^2 - \left(\dfrac{\vec{k}\cdot\vec{v}}{c}\right)^2}. \qquad (5.8.123)$$

Therefore, in this representation, the electromagnetic field associated with the electron is given by the following two relations:

$$\vec{e}_{\vec{k}} = -\nabla V_{\vec{k}} - \frac{\partial \vec{A}_{\vec{k}}}{\partial t} = -\nabla V_{\vec{k}} - \frac{\vec{v}}{c^2} \frac{\partial V_{\vec{k}}}{\partial t}, \qquad (5.8.124)$$

and

$$\vec{b}_{\vec{k}} = \nabla \times \vec{A}_{\vec{k}} = -\frac{1}{c^2} \vec{v} \times \nabla V_{\vec{k}}, \qquad (5.8.125)$$

where we took into account that $\vec{v} = \text{const.}$, and we also used the relation (B.3.53) from **Appendix B**. By using Eqs. (5.8.122) and (5.8.123) in Eqs. (5.8.124) and (5.8.125) and performing the calculations, one obtains

$$\vec{e}_{\vec{k}} = -\frac{ie}{2\pi^2} \frac{1}{4\pi\varepsilon_0} \frac{\vec{k} - \dfrac{\vec{v}}{c}\left(\dfrac{\vec{k}\cdot\vec{v}}{c}\right)}{k^2 - \left(\dfrac{\vec{k}\cdot\vec{v}}{c}\right)^2} e^{i\vec{k}\cdot(\vec{r} - \vec{v}t)}, \qquad (5.8.126)$$

and, respectively,

$$\vec{b}_{\vec{k}} = \frac{ie}{2\pi^2} \frac{\mu_0}{4\pi} \frac{\vec{k} \times \vec{v}}{k^2 - \left(\dfrac{\vec{k}\cdot\vec{v}}{c}\right)^2} e^{i\vec{k}\cdot(\vec{r} - \vec{v}t)}. \qquad (5.8.127)$$

It is observed that, unlike the free plane waves which are purely transversal, the electric field $\vec{e}_{\vec{k}}$ has also a longitudinal component, whose modulus is

$$\left| \vec{e}_{\vec{k}}^{\text{long}} \right| = A \frac{\vec{k} \cdot \vec{s} - \dfrac{\vec{s} \cdot \vec{v}}{c} \left(\dfrac{\vec{k} \cdot \vec{v}}{c} \right)}{k^2 - \left(\dfrac{\vec{k} \cdot \vec{v}}{c} \right)^2} = \frac{A}{k}, \tag{5.8.128}$$

where A is a constant, and $\vec{s}$ is the versor of the wave vector $\vec{k}$. We therefore have

$$\vec{e}_{\vec{k}}^{\text{long}} = \frac{A}{k} \vec{s} = -\frac{ie}{2\pi^2} \frac{1}{4\pi\varepsilon_0} \frac{\vec{k}}{k^2} e^{i\vec{k} \cdot (\vec{r} - \vec{v}t)}. \tag{5.8.129}$$

One also observes that

$$\vec{b}_{\vec{k}}^{\text{long}} = 0. \tag{5.8.130}$$

If $\vec{v} \to 0$, $\vec{e}_{\vec{k}}^{\text{trans}} \to 0$, and the (electro)static case is re-found.

5.9   Problem No. 46

An atom radiates electromagnetic waves and remains in an excited state a time τ. The time dependence of the electric component of the electromagnetic field emitted by the atom is of the form

$$E(t) = E_0 e^{-\frac{t}{\tau} + i\omega_0 t}.$$

Determine the width of the spectral line emitted by the atom.

Solution

Before emitting the electromagnetic wave, both electric $\vec{E}$ and magnetic $\vec{B}$ components of the wave are zero. Since the magnitude of the magnetic component of the electromagnetic field emitted by the atom is negligible as compared to the electrical component, in what follows we will consider only the last one.

In view of the above observations, we can write the time dependence of the electric field emitted by the atom in the form

$$E(t) = \begin{cases} 0, & \text{for } t < 0, \\ E_0 \, e^{-\frac{t}{\tau} + i\omega_0 t}, & \text{for } t > 0. \end{cases} \tag{5.9.131}$$

Due to the continuous dependence on time and on pulsation ω of the electric component of the electromagnetic field emitted by the atom, we must use the integral form of the Fourier transform of the field. The inverse Fourier transform is given by

$$E(t) = \frac{1}{\sqrt{2\pi}} \int_{-\infty}^{+\infty} E(\omega)e^{i\omega t}d\omega, \qquad (5.9.132)$$

while the direct Fourier transform − which gives the spectrum $E(\omega)$ − is

$$
\begin{aligned}
E(\omega) &= \frac{1}{\sqrt{2\pi}} \int_{-\infty}^{+\infty} E(t)e^{-i\omega t}dt = \frac{1}{\sqrt{2\pi}} \int_{-\infty}^{0} 0 \cdot e^{-i\omega t}dt \\
&+ \frac{1}{\sqrt{2\pi}} \int_{0}^{+\infty} E_0 e^{-\frac{t}{\tau}+i\omega_0 t}e^{-i\omega t}dt = \frac{E_0}{\sqrt{2\pi}} \int_{0}^{+\infty} e^{-\frac{t}{\tau}-i(\omega-\omega_0)t}dt \\
&= \frac{E_0}{\sqrt{2\pi}} \int_{0}^{+\infty} e^{-t\left[\frac{1}{\tau}+i(\omega-\omega_0)\right]}dt = \frac{E_0}{\sqrt{2\pi}} \left[\frac{-1}{\frac{1}{\tau}+i(\omega-\omega_0)} \right. \\
&\left. \times e^{-t\left[\frac{1}{\tau}+i(\omega_0-\omega)\right]} \right]_{t=0}^{t=+\infty} = \frac{E_0}{\sqrt{2\pi}\left[\frac{1}{\tau}+i(\omega_0-\omega)\right]}.
\end{aligned}
$$

The intensity of the radiation emitted by the atom is

$$
\begin{aligned}
I(\omega) \propto |E(\omega)|^2 &= \frac{E_0^2}{2\pi} \frac{1}{\left|\frac{1}{\tau}+i(\omega-\omega_0)\right|^2} \\
&= \frac{E_0^2}{2\pi} \frac{1}{\left[\frac{1}{\tau}+i(\omega-\omega_0)\right]\left[\frac{1}{\tau}-i(\omega-\omega_0)\right]} \\
&= \frac{E_0^2}{2\pi} \frac{1}{\frac{1}{\tau^2}+(\omega-\omega_0)^2},
\end{aligned}
$$

or,

$$I(\omega) = \frac{I_0}{\frac{1}{\tau^2}+(\omega-\omega_0)^2}, \qquad (5.9.133)$$

where the exact expression of I_0 (which is not what we are interested in here) is obtained from

$$I_0 = \int_{-\infty}^{+\infty} I(\omega)d\omega. \qquad (5.9.134)$$

As one knows from the radiation theory, the half-width of a spectral line emitted by an atom is determined by the relation (see Fig. 5.13):

$$I\left(\omega_0 \pm \frac{\Delta\omega}{2}\right) = \frac{1}{2}I(\omega_0). \qquad (5.9.135)$$

In view of Eq. (5.9.133), this relation can also be written as

$$\frac{I_0}{\frac{1}{\tau^2}+\left[\left(\omega_0 \pm \frac{\Delta\omega}{2}\right)-\omega_0\right]^2} = \frac{I_0\tau^2}{2},$$

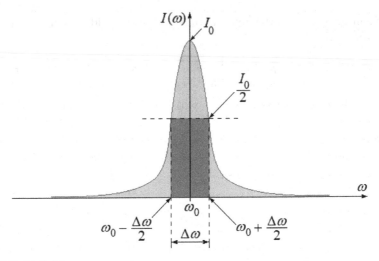

FIGURE 5.13
The spectral line emitted by an atom that radiates electromagnetic waves.
The figure shows how to determine the spectral line width $\Delta\omega$.

or

$$\frac{1}{\tau^2} + \left[\left(\omega_0 \pm \frac{\Delta\omega}{2}\right) - \omega_0\right]^2 = \frac{2}{\tau^2},$$

and, finally

$$\left(\pm\frac{\Delta\omega}{2}\right)^2 = \frac{1}{\tau^2} \Leftrightarrow \frac{\Delta\omega}{2} = \frac{1}{\tau} \Rightarrow \Delta\omega = \frac{2}{\tau}. \tag{5.9.136}$$

6

Motion of Charged Particles in the
Electromagnetic Field – Non-relativistic
Approach

6.1   Problem No. 47

Determine the motion of an electron in a static and uniform electromagnetic
field, for which the vectors $\vec{E}$ and $\vec{B}$ are parallel, supposing that its initial
velocity $\vec{v}_0$ is perpendicular to the common direction of the two fields.

Solution

The equation of motion of an electron in the static and uniform electromagnetic field $(\vec{E}, \vec{B})$ is

$$m\ddot{\vec{r}} = -e(\vec{E} + \vec{v} \times \vec{B}), \qquad (6.1.1)$$

where m is the electron mass, and $-e$ ($e > 0$) its electric charge.

Without restricting the generality of the problem, we suppose that the
(parallel) vectors $\vec{E}$ and $\vec{B}$ are oriented in the direction and sense of the z-
axis, while the initial velocity $\vec{v}_0$ of the electron (orthogonal to the common
direction of the fields) is oriented along the x-axis. Then, since

$$\vec{v} \times \vec{B} = \begin{vmatrix} \vec{i} & \vec{j} & \vec{k} \\ \dot{x} & \dot{y} & \dot{z} \\ 0 & 0 & B \end{vmatrix} = \dot{y}B\vec{i} - \dot{x}B\vec{j},$$

the projections of Eq. (6.1.1) on coordinate axes write

$$\begin{cases} Ox: & m\ddot{x} = -eB\dot{y}, \\ Oy: & m\ddot{y} = +eB\dot{x}, \\ Oz: & m\ddot{z} = -eE. \end{cases} \qquad (6.1.2)$$

The last equation of system (6.1.2) shows that the motion of the electron
along the z-axis is uniformly accelerated ($E = \text{const.}$):

$$z(t) = -\frac{eE}{2m}t^2 + \frac{C_1}{m}t + \frac{C_2}{m},$$

DOI: 10.1201/9781003402602-6

where C_1 and C_2 are two arbitrary integration constants, determinable from the initial conditions. Supposing that the initial position of the electron at the initial moment $t_0 = 0$ is the origin of the coordinate system, we have $z(0) = 0$ and, according to the previous supposition regarding the initial velocity, we also have $\dot{z}(0) = 0$. These two conditions lead to null values for the two constants, so that

$$z(t) = -\frac{eE}{2m}t^2. \tag{6.1.3}$$

It has remained to integrate the system of second-order coupled ordinary differential equations

$$\begin{cases} m\ddot{x} = -eB\dot{y}, \\ m\ddot{y} = +eB\dot{x}. \end{cases}$$

To integrate this system, we will use the complex variable $\xi = x + iy$. Multiplying the second equation by $i = \sqrt{-1}$ and adding the result to the first equation, we obtain

$$\ddot{\xi} - i\frac{eB}{m}\dot{\xi} = 0. \tag{6.1.4}$$

This equation can be easily integrated; its solution is

$$\xi = ae^{i\omega t} + b, \tag{6.1.5}$$

where $\omega = \frac{eB}{m}$, while $a = a_x + ia_y$ and $b = b_x + ib_y$ are two arbitrary complex integration constants. Separating in Eq. (6.1.5) the real and imaginary parts, we will come back to variables $x(t)$ and $y(t)$, which are written as

$$\begin{cases} x(t) = a_x \cos \omega t - a_y \sin \omega t + b_x, \\ y(t) = a_x \sin \omega t + a_y \cos \omega t + b_y. \end{cases} \tag{6.1.6}$$

The four real, arbitrary integration constants a_x, a_y, b_x and b_y can be uniquely determined by means of the following four initial conditions:

$$\begin{cases} x(0) = 0, \\ y(0) = 0, \\ \dot{x}(0) = v_0, \\ \dot{y}(0) = 0. \end{cases} \tag{6.1.7}$$

As a result of some simple calculations, one obtains

$$x(t) = \frac{v_0}{\omega} \sin \omega t, \tag{6.1.8}$$

and

$$y(t) = \frac{v_0}{\omega}(1 - \cos \omega t). \tag{6.1.9}$$

It is noted that the relation

$$x^2(t) + \left[y(t) - \frac{v_0}{\omega}\right]^2 = \frac{v_0^2}{\omega^2} \tag{6.1.10}$$

is valid at any moment t. This means that the projection of the electron trajectory in the xy-plane (this plane being perpendicular to the common direction of the two fields) is a circle. This circle has a constant radius, $R = v_0/\omega$ and its center is located at the point $C(x_0, y_0) = C(0, v_0/\omega)$.

We also have

$$\dot{x}^2(t) + \dot{y}^2(t) = v_0^2,$$

meaning that the motion of rotation around the field is uniform. Its angular velocity is

$$\frac{v_0}{R} = \frac{v_0}{v_0/\omega} = \omega = \frac{eB}{m}.$$

The full motion of the electron results by composing the uniform rotation given by relations (6.1.8) and (6.1.9), with the uniformly accelerated translation given by Eq. (6.1.3). The trajectory is therefore a *helix with variable pitch*, wrapped on the circular cylinder of radius R.

6.2 Problem No. 48

Determine the motion of an electron in a static, uniform electromagnetic field, with non-parallel vectors $\vec{E}$ and $\vec{B}$, supposing that the initial velocity of the electron, $\vec{v}_0$, is perpendicular to the plane determined by the two vectors.

Solution

The equation of motion of an electron with electric charge $-e$ ($e > 0$) in the static and uniform electromagnetic field $(\vec{E}, \vec{B})$ is

$$m\ddot{\vec{r}} = -e(\vec{E} + \vec{v} \times \vec{B}). \qquad (6.2.11)$$

Without restricting the generality of the problem, let us consider the z-axis along the magnetic field, and the yz-plane being determined by the vectors $\vec{E}$ and $\vec{B}$ – that is $\vec{E} = (0, E_y, E_z)$ and $\vec{B} = (0, 0, B)$. In this case, the statement of the problem requires that $\vec{v}_0 = (v_0, 0, 0)$. Since

$$\vec{v} \times \vec{B} = \begin{vmatrix} \vec{i} & \vec{j} & \vec{k} \\ \dot{x} & \dot{y} & \dot{z} \\ 0 & 0 & B \end{vmatrix} = \dot{y}B\vec{i} - \dot{x}B\vec{j},$$

the components of the equation of motion (6.2.11) are

$$\begin{cases} m\ddot{x} = -eB\dot{y}, \\ m\ddot{y} = -eE_y + eB\dot{x}, \\ m\ddot{z} = -eE_z. \end{cases} \qquad (6.2.12)$$

Supposing that at the initial moment the electron is at the origin of the reference frame and its initial velocity is $\vec{v}_0 = (v_0, 0, 0)$, the last equation of the system (6.2.12) leads to

$$z(t) = -\frac{eE_z}{2m}t^2. \tag{6.2.13}$$

Using the complex variable $\xi = x + iy$, the first two equations of the system (6.2.12) can be replaced by a single equation, namely

$$\ddot{\xi} - i\frac{eB}{m}\dot{\xi} = -i\frac{e}{m}E_y, \tag{6.2.14}$$

which is an ordinary, non-homogeneous, second-order differential equation, with constant coefficients. The general solution of this equation is given by adding the general solution of the attached homogeneous equation to a particular solution of the non-homogeneous equation. Since the attached homogeneous equation has exactly the same form as that met in the previous problem, its solution is

$$\xi_0(t) = ae^{i\omega t} + b,$$

where a and b are two complex constants, and $\omega = eB/m$. The particular solution of the non-homogeneous equation cannot be searched as having the form of the term that gives the non-homogeneity, because the equation does not also contain the underived function which is going to be determined. Because of this, in order to find the particular solution of the non-homogeneous equation we will use the Lagrange's method of variation of parameters. Since the fundamental system of solutions of the homogeneous equation is

$$\begin{cases} \xi_1(t) = e^{i\omega t}, \\ \xi_2(t) = 1, \end{cases}$$

according to the Lagrange's method of variation of parameters, one follows that the general solution of the problem is given by

$$\xi(t) = e^{i\omega t}\int a'(t)dt + \int b'(t)dt,$$

where $a'(t)$ and $b'(t)$ are the solutions of the system of first-order differential equations

$$\begin{cases} \xi_1(t)a'(t) + \xi_2(t)b'(t) = 0, \\ \xi_1'(t)a'(t) + \xi_2'(t)b'(t) = -i\frac{e}{m}E_y, \end{cases}$$

that is, of the system

$$\begin{cases} e^{i\omega t}a'(t) + b'(t) = 0, \\ i\omega e^{i\omega t}a'(t) = -i\frac{e}{m}E_y. \end{cases}$$

The second equation easily yields

$$a(t) = -\frac{ieE_y}{m\omega^2}e^{-i\omega t} + c,$$

where c is an arbitrary, veritable, complex integration constant, while the first equation of the same system gives

$$b(t) = \frac{eE_y t}{m\omega} + d,$$

where d is also an arbitrary, veritable, complex integration constant. The general solution of Eq. (6.2.14) therefore is

$$\xi(t) = c\,e^{i\omega t} - \frac{ieE_y}{m\omega^2} + \frac{eE_y}{m\omega}t + d$$

$$= c\,e^{i\omega t} + \frac{eE_y}{m\omega}t + g = ce^{i\omega t} + \frac{E_y}{B}t + g, \qquad (6.2.15)$$

where g is a new, veritable, complex constant, whose expression can be easily identified.

Separating in Eq. (6.2.15) the real and imaginary parts, we find

$$\begin{cases} x(t) = c_x \cos\omega t - c_y \sin\omega t + \dfrac{E_y}{B}t + g_x, \\ y(t) = c_x \sin\omega t + c_y \cos\omega t + g_y. \end{cases} \qquad (6.2.16)$$

The initial conditions

$$\begin{cases} x(0) = 0, \\ y(0) = 0, \\ \dot{x}(0) = v_0, \\ \dot{y}(0) = 0, \end{cases} \qquad (6.2.17)$$

lead to

$$\begin{cases} c_x + g_x = 0, \\ c_y + g_y = 0, \\ c_y\omega - \dfrac{E_y}{B} = -v_0, \\ c_x\omega = 0, \end{cases}$$

so that

$$\begin{cases} c_x = 0, \\ g_x = 0, \\ c_y = \dfrac{E_y}{\omega B} - \dfrac{v_0}{\omega}, \\ g_y = \dfrac{v_0}{\omega} - \dfrac{E_y}{\omega B}, \end{cases}$$

so that the relations (6.2.16) become

$$\begin{cases} x(t) = \alpha \sin \omega t + \dfrac{E_y}{B} t, \\ y(t) = \alpha(1 - \cos \omega t), \end{cases} \qquad (6.2.18)$$

where

$$\alpha \equiv \frac{v_0}{\omega} - \frac{E_y}{\omega B}.$$

It is convenient to introduce a new notation, namely $\alpha \equiv \alpha_0 \frac{E_y}{\omega B}$, with

$$\alpha_0 \equiv \frac{B}{E_y}\left(y_0 - \frac{E_y}{B}\right),$$

in which case the relations (6.2.18) take the form

$$\begin{cases} x(t) = \dfrac{E_y}{\omega B}(\omega t + \alpha_0 \sin \omega t), \\ y(t) = \alpha_0 \dfrac{E_y}{\omega B}(1 - \cos \omega t). \end{cases} \qquad (6.2.19)$$

The parametric equations of the trajectory therefore are

$$\begin{cases} x(t) = \dfrac{E_y}{\omega B}(\omega t + \alpha_0 \sin \omega t), \\ y(t) = \alpha_0 \dfrac{E_y}{\omega B}(1 - \cos \omega t). \\ z(t) = -\dfrac{eE_z}{2m} t^2. \end{cases} \qquad (6.2.20)$$

Fig. 6.1 represents an example of trajectory determined by the parametric equations (6.2.20), for the numerical values (all quantities are expressed in SI units) given below, along with the command lines necessary to draw the trajectory, written using Mathematica 5.0 analytical and numerical software:

$v_0 = 10^9$;

$B = 10^{-3}$;

$E_y = 10^{-4}$;

$E_z = 5 * 10^{-7}$;

$\alpha_0 = \dfrac{v_0 * B}{E_y} - 1$;

$\omega = 10^8$;

$e = -1.6 * 10^{-19}$;

$m = 9.1 * 10^{-31}$;

$x[t_-] := \dfrac{E_y}{\omega * B} * (\omega * t + \alpha_0 * \mathrm{Sin}[\omega * t])$;

$$y[t_-] := \alpha_0 * \frac{E_y}{\omega * B} * (1 - \text{Cos}[\omega * t]);$$

$$z[t_-] := -\frac{e * E_z}{2 * m} * t^2;$$

ParametricPlot3D[$\{x[t], y[t], z[t]\}, \{t, 0, 0.02\}$].

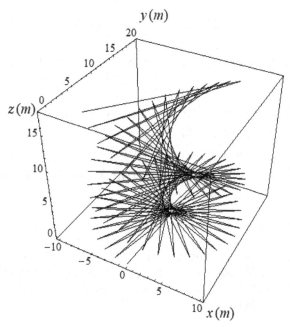

FIGURE 6.1
Trajectory of an electron, determined by the parametric equations (6.2.20).

6.3   Problem No. 49

A charged particle moves in a variable electromagnetic field, described by the electrodynamic potentials $V(x, y, z, t)$ and $\vec{A}(x, y, z, t)$. Show that the equations of motion of the particle can be put in a Lagrangian form, and find the Lagrange's function, L, neglecting the relativistic effects.

Solution

Let m be the mass and q the charge of the particle moving with velocity $\vec{v}$ in the external electromagnetic field $(\vec{E}, \vec{B})$. Since

$$\vec{E} = -\nabla V - \frac{\partial \vec{A}}{\partial t}, \qquad (6.3.21)$$

and

$$\vec{B} = \nabla \times \vec{A}, \tag{6.3.22}$$

the equation of motion of the particle,

$$m\ddot{\vec{r}} = q(\vec{E} + \vec{v} \times \vec{B}) \tag{6.3.23}$$

writes

$$m\ddot{\vec{r}} = q\left[-\nabla V - \frac{\partial \vec{A}}{\partial t} + \vec{v} \times \left(\nabla \times \vec{A}\right)\right]. \tag{6.3.24}$$

The last term of Eq. (6.3.24) can be transformed as follows

$$
\begin{aligned}
\vec{v} \times \left(\nabla \times \vec{A}\right) &= \varepsilon_{ijk} v_j \left(\nabla \times \vec{A}\right)_k \vec{u}_i = \varepsilon_{ijk} v_j \varepsilon_{klm} \left(\partial_l A_m\right) \vec{u}_i \\
&= \varepsilon_{ijk} \varepsilon_{klm} v_j \left(\partial_l A_m\right) \vec{u}_i = \varepsilon_{kij} \varepsilon_{klm} v_j \left(\partial_l A_m\right) \vec{u}_i \\
&= \left(\delta_{il}\delta_{jm} - \delta_{im}\delta_{jl}\right) v_j \left(\partial_l A_m\right) \vec{u}_i \\
&= \left(\partial_i A_j\right) v_j \vec{u}_i - v_j \left(\partial_j A_i\right) \vec{u}_i \\
&= \partial_i \left(v_j A_j\right) \vec{u}_i - \left(v_j \partial_j\right)\left(A_i \vec{u}_i\right) = \partial_i \left(\vec{v} \cdot \vec{A}\right) \vec{u}_i - \left(\vec{v} \cdot \nabla\right)\vec{A} \\
&= \nabla\left(\vec{v} \cdot \vec{A}\right) - \left(\vec{v} \cdot \nabla\right)\vec{A},
\end{aligned}
$$

where we used the fact that within the Lagrangian formalism the generalized coordinates — in our case the Cartesian coordinates x_i, $i = \overline{1,3}$ — and the generalized velocities — in this case, the ordinary velocities v_i, $i = \overline{1,3}$ — are independent variables, so that we are allowed to "move" the velocity v_i and the derivative symbol $\frac{\partial}{\partial x_i}$. In addition, we used the fact that $\vec{u}_i = const.$ Equation (6.3.24) then becomes

$$
\begin{aligned}
m\ddot{\vec{r}} &= q\left[-\nabla V - \frac{\partial \vec{A}}{\partial t} + \nabla\left(\vec{v} \cdot \vec{A}\right) - \left(\vec{v} \cdot \nabla\right)\vec{A}\right] \\
&= q\left\{-\nabla\left(V - \vec{v} \cdot \vec{A}\right) - \left[\frac{\partial \vec{A}}{\partial t} + \left(\vec{v} \cdot \nabla\right)\vec{A}\right]\right\} \\
&= -q\left[\nabla\left(V - \vec{v} \cdot \vec{A}\right) + \frac{d\vec{A}}{dt}\right], \tag{6.3.25}
\end{aligned}
$$

where we used the formula

$$\frac{\partial \vec{A}}{\partial t} + (\vec{v} \cdot \nabla)\vec{A} = \frac{d\vec{A}}{dt}.$$

Since

$$\ddot{\vec{r}} = \frac{d\dot{\vec{r}}}{dt} = \frac{d\vec{v}}{dt},$$

Eq. (6.3.25) can also be written as

$$\frac{d}{dt}\left(m\vec{v} + q\vec{A}\right) - \nabla\left\{q\left[-V + \left(\vec{v} \cdot \vec{A}\right)\right]\right\} = 0. \tag{6.3.26}$$

If this equation is compared to Lagrange's equations of the second kind

$$\frac{d}{dt}\left(\frac{\partial L}{\partial \vec{v}}\right) - \frac{\partial L}{\partial \vec{r}} = 0, \qquad (6.3.27)$$

then we obtain by identification

$$\frac{\partial L}{\partial \vec{v}} = m\vec{v} + q\vec{A}, \qquad (6.3.28)$$

and

$$\frac{\partial L}{\partial \vec{r}} = \nabla \left\{ q\left[-V + (\vec{v} \cdot \vec{A})\right] \right\} = \frac{\partial}{\partial \vec{r}} \left\{ q\left[-V + (\vec{v} \cdot \vec{A})\right] \right\}, \qquad (6.3.29)$$

where L is the Lagrange's function. In the above equations $\frac{\partial L}{\partial \vec{v}}$ is the formal writing of the vector of components $\left(\frac{\partial L}{\partial v_x}, \frac{\partial L}{\partial v_y}, \frac{\partial L}{\partial v_z}\right)$, while $\left(\frac{\partial L}{\partial x}, \frac{\partial L}{\partial y}, \frac{\partial L}{\partial z}\right)$ signify the components of $\frac{\partial L}{\partial \vec{r}}$ (being also a formal writing). According to Eq. (6.3.29), we then have

$$L(\vec{r}, \vec{v}, t) = -qV + q(\vec{v} \cdot \vec{A}) + f(\vec{v}),$$

where $f(\vec{v})$ is an arbitrary function, at least of the class $C^0(\mathbb{R})$, which can be determined by means of Eq. (6.3.28):

$$\frac{df}{d\vec{v}} = m\vec{v} \quad \Rightarrow \quad f(\vec{v}) = \frac{1}{2}m\vec{v}^2,$$

so that the Lagrange's function becomes

$$L(\vec{r}, \vec{v}, t) = -qV + q(\vec{v} \cdot \vec{A}) + \frac{m\vec{v}^2}{2}. \qquad (6.3.30)$$

Therefore, the equation of motion (6.3.23) can be put in a Lagrangian form

$$\frac{d}{dt}\left(\frac{\partial L}{\partial \vec{v}}\right) - \frac{\partial L}{\partial \vec{r}} = 0,$$

where L is given by Eq. (6.3.30). Since in general the force depends on velocity, one defines the so-called *generalized potential* $U(\vec{r}, \vec{v}, t)$, given by

$$U(\vec{r}, \vec{v}, t) = T(\vec{r}, \vec{v}, t) - L(\vec{r}, \vec{v}, t) = qV - q(\vec{v} \cdot \vec{A}),$$

where $T(\vec{r}, \vec{v}, t)$ is the most general form of the kinetic energy (which in our case is the very well known expression $\frac{1}{2}m\vec{v}^2$).

As is well-known, the electrodynamic potentials are not univocally determined, but only up to a gauge transformation, and the behaviour of Lagrange's function under such a transformation becomes interesting. Let us, therefore, determine the Lagrange's function $L'(V', \vec{A}')$, where

$$V' = V - \dot{\phi}, \quad \vec{A}' = \vec{A} + \nabla\phi,$$

where $\phi = \phi(\vec{r}, t)$ is a function of class C^1 on its definition domain, depending on space-time variables. Then, we have

$$L' = -qV' + q(\vec{v} \cdot \vec{A}') + \frac{mv^2}{2}$$

$$= -q(V - \dot{\phi}) + q\vec{v} \cdot (\vec{A} + \nabla\phi) + \frac{mv^2}{2} = L + q\dot{\phi} + q\vec{v} \cdot \nabla\phi$$

$$= L + q\left[\frac{\partial\phi}{\partial t} + (\vec{v} \cdot \nabla)\phi\right] = L + q\frac{d\phi}{dt} = L + \frac{d}{dt}(q\phi),$$

meaning that the two Lagrangians are equivalent (lead to the same differential equation of motion), because they differ only by a term which is the total derivative with respect to time of an arbitrary function $\phi = \phi(\vec{r}, t)$ that depends only on $\vec{r}$ and t.

6.4 Problem No. 50

Using the Lagrangian formalism, determine the equation of motion of a particle having the electric charge e, that moves in the electromagnetic field $(\vec{E}, \vec{B})$.

Solution

Within the framework of the Lagrangian formalism, the differential equations of motion of a particle with electric charge e moving in the electromagnetic field $(\vec{E}, \vec{B})$ are written as

$$\frac{d}{dt}\left(\frac{\partial T}{\partial \dot{q}^k}\right) - \frac{\partial T}{\partial q^k} = Q_k, \qquad (6.4.31)$$

where the number of values of the index k equals the number of degrees of freedom of the system (in our case, a charged particle), usually denoted by n.

In what follows we will consider that the active (applied) force derives from a generalized potential of the form

$$U(q^j, \dot{q}^j, t) = C_j(q^j, t)\dot{q}^j + U_0 = U_1 + U_0, \quad j = \overline{1, n},$$

i.e., $U(q^j, \dot{q}^j, t)$ depends not only on the generalized coordinates q^j and the time t, but also on the generalized velocities $\dot{q}^j$, while the functions $C_j(q^j, t)$ depend only on generalized coordinates q^j and time t. If to the both sides of Eq. (6.4.31) is added the expression

$$\frac{\partial U}{\partial q^k} - \frac{d}{dt}\left(\frac{\partial U}{\partial \dot{q}^k}\right),$$

then we obtain

$$\frac{d}{dt}\left[\frac{\partial(T-U)}{\partial \dot{q}^k}\right] - \frac{\partial(T-U)}{\partial q^k} = Q_k - \frac{d}{dt}\left(\frac{\partial U}{\partial \dot{q}^k}\right) + \frac{\partial U}{\partial q^k}. \tag{6.4.32}$$

If, in addition, we suppose that the generalized forces Q_k are given by

$$Q_k = \frac{d}{dt}\left(\frac{\partial U}{\partial \dot{q}^k}\right) - \frac{\partial U}{\partial q^k}, \tag{6.4.33}$$

then Eqs. (6.4.32) take the well-known form of Lagrange's equations of the second kind, for a natural system,

$$\frac{d}{dt}\left(\frac{\partial L}{\partial \dot{q}^k}\right) - \frac{\partial L}{\partial q^k} = 0, \quad k = \overline{1, n},$$

with the Lagrange function given by

$$L(q^k, \dot{q}^k, t) = T(q^k, \dot{q}^k, t) - U(q^k, \dot{q}^k, t). \tag{6.4.34}$$

The function $U(q^k, \dot{q}^k, t)$ is called *generalized potential* or *velocity-dependent potential*.

To establish the differential equation of motion for the particle with charge e, moving with velocity $\vec{v}$ in the electromagnetic field $(\vec{E}, \vec{B})$, one must first construct the Lagrangian of this system. The particle is free, therefore it has three degrees of freedom (corresponding to the motion of translation), and we can consider as generalized coordinates just the Cartesian coordinates, *i.e.*, $q^i = x_i$. Then, the "ordinary" velocities are considered as generalized velocities, $\dot{q}^i \equiv \dot{x}_i = v_i$, $i = \overline{1, 3}$, and, according to the Lagrangian formalism, the variables x_i and $\dot{x}_i$ are independent.

If the magnetic field is missing, then the Lagrangian is $L = T - U = T - eV$, where $V = V(\vec{r}, t)$ is the potential of the electric field. In order to take into account the magnetic field in the Lagrangian expression, we use the property of the Lagrangian of being an invariant, so that the vector potential $\vec{A}(\vec{r}, t)$ can appear in the Lagrangian only as a scalar product with another vector. There are three possibilities: $\vec{A} \cdot \vec{A}$, $\vec{A} \cdot \dot{\vec{r}}$ and $\vec{A} \cdot \ddot{\vec{r}}$. Since in the equation of motion does not appear $\vec{B}^2$, the first possibility is excluded. Then, as physics is currently understood, the equations of motion of particles/physical systems are differential equations of maximum second order (*e.g.*, Newton's second law, Lagrange's equations of the first kind, Lagrange's equations of the second kind, Hamilton's (canonical) equations, wave propagation equation, Schrödinger equation, Klein-Gordon equation, Dirac equation, Rarita-Schwinger equation, etc. − all of them are first order or at most second-order differential equations); this is why we will ask to have a differential equation of motion of at most the second order (the equation should not contain the derivative with respect to time of acceleration or other superior terms), so that the third possibility is also excluded. Therefore, using the remaining (second) possibility, the Lagrangian writes

$$L = \frac{1}{2}mv^2 - eV + e\vec{A} \cdot \dot{\vec{r}} = \frac{1}{2}mv^2 - eV + e\vec{A} \cdot \vec{v}. \qquad (6.4.35)$$

Since

$$\frac{\partial L}{\partial \dot{q}^i} \equiv \frac{\partial L}{\partial v_i} = mv_i + eA_i,$$

we have

$$\frac{d}{dt}\left(\frac{\partial L}{\partial \dot{q}^i}\right) \equiv \frac{d}{dt}\left(\frac{\partial L}{\partial v_i}\right) = \frac{d}{dt}\left(\frac{\partial L}{\partial \dot{x}_i}\right)$$

$$= m\ddot{x}_i + e\left(\frac{\partial A_i}{\partial t} + \dot{x}_k\frac{\partial A_i}{\partial x_k}\right),$$

where we took into account the fact that $\vec{A}(\vec{r}(t), t)$ depends on time both explicitly and implicitly, through spatial coordinates. Also,

$$\frac{\partial L}{\partial q^i} \equiv \frac{\partial L}{\partial x_i} = -\frac{\partial V}{\partial x_i} + ev_k\frac{\partial A_k}{\partial x_i}, \quad i, k = 1, 2, 3,$$

so that Lagrange's equations of the second kind

$$\frac{d}{dt}\left(\frac{\partial L}{\partial \dot{q}^i}\right) - \frac{\partial L}{\partial q^i} = 0 \quad \Leftrightarrow \quad \frac{d}{dt}\left(\frac{\partial L}{\partial \dot{x}_i}\right) - \frac{\partial L}{\partial x_i} = 0,$$

lead to

$$m\ddot{x}_i = e\left(-\frac{\partial A_i}{\partial t} - \frac{\partial V}{\partial x_i}\right) + ev_k\left(\frac{\partial A_k}{\partial x_i} - \frac{\partial A_i}{\partial x_k}\right), \qquad (6.4.36)$$

or

$$m\ddot{x}_i = eE_i + ev_k\underbrace{\left(\frac{\partial A_k}{\partial x_i} - \frac{\partial A_i}{\partial x_k}\right)}_{=T_{ik}=-T_{ki}} = eE_i + ev_k\varepsilon_{ikj}B_j = eE_i$$

$$- \varepsilon_{ijk}B_j v_k = eE_i - e(\vec{B}\times\vec{v})_i = eE_i + e(\vec{v}\times\vec{B})_i, \qquad (6.4.37)$$

where B_j are the components of the pseudovector associated with the second-order antisymmetric tensor $T_{ik} \equiv \frac{\partial A_k}{\partial x_i} - \frac{\partial A_i}{\partial x_k}$. According to the isomorphism between the set of second-order antisymmetric tensors, defined on a three-dimensional space, and the set of associated pseudovectors, we have

$$B_j = \frac{1}{2}\varepsilon_{jik}T_{ik} = \frac{1}{2}\varepsilon_{jik}\left(\frac{\partial A_k}{\partial x_i} - \frac{\partial A_i}{\partial x_k}\right)$$

$$= \frac{1}{2}\varepsilon_{jik}\frac{\partial A_k}{\partial x_i} - \frac{1}{2}\varepsilon_{jik}\frac{\partial A_i}{\partial x_k} = \frac{1}{2}\varepsilon_{jik}\frac{\partial A_k}{\partial x_i}$$

$$- \frac{1}{2}\varepsilon_{jki}\frac{\partial A_k}{\partial x_i} = \frac{1}{2}\varepsilon_{jik}\frac{\partial A_k}{\partial x_i} + \frac{1}{2}\varepsilon_{jik}\frac{\partial A_k}{\partial x_i}$$

$$= \varepsilon_{jik}\frac{\partial A_k}{\partial x_i} = \varepsilon_{jik}\frac{\partial A_k}{\partial x_i} = (\nabla\times\vec{A})_j,$$

meaning that the pseudovector (axial vector) whose components are B_j is no other vector than the magnetic induction (which justifies its notation, from the beginning, with the letter B).

Eq. (6.4.37) can be written in vector form as

$$m\ddot{\vec{r}} = e(\vec{E} + \vec{v} \times \vec{B}),\qquad(6.4.38)$$

which is the well-known equation of motion of a particle possessing the mass m and charge e which is moving into an electromagnetic field $(\vec{E}, \vec{B})$. According to the second principle of dynamics (the second Newton's law), the product between the mass and acceleration of the particle equals the force acting on the particle (in our case, the force due to electric component of the electromagnetic field, $\vec{F}_e = e\vec{E}$, in addition to the force corresponding to the magnetic component of the field, which is the Lorentz force $\vec{F}_L = e(\vec{v} \times \vec{B})$).

6.5 Problem No. 51

Using the Hamiltonian formalism, determine the equation of motion of a particle with electric charge e, moving in the (external) electromagnetic field $(\vec{E}, \vec{B})$.

Solution

The Lagrangian of a particle with mass m and electric charge e, moving in an external electromagnetic field $(\vec{E}, \vec{B})$ is written as follows

$$L = \frac{1}{2}v_j v_j - eV + ev_j A_j, \quad j = \overline{1,3},\qquad(6.5.39)$$

where $V = V(\vec{r}, t)$ and $\vec{A} = \vec{A}(\vec{r}, t)$ are the field potentials, while x_j and p_j are the generalized coordinates and, respectively, the components of the generalized momentum associated to the particle. According to the Hamiltonian method of solving a problem, the first step consists in determination of the Hamilton's function (the Hamiltonian) of the system. Since by definition

$$p_j = \frac{\partial L}{\partial \dot{q}^j} = \frac{\partial L}{\partial v_j} = mv_j + eA_j, \ j = \overline{1,3},$$

in our case, the Hamiltonian of the particle is

$$H = p_j v_j - L = \frac{1}{2}mv_j v_j + eV.\qquad(6.5.40)$$

As can be easily seen, as generalized coordinates and generalized velocities have been chosen the Cartesian coordinates and, respectively, the "usual" velocity components, associated to the particle, this choice being always possible when the particle/system is free (there are no constraints).

In order to use the Hamiltonian formalism, the Hamilton's function must be expressed in terms of canonical coordinates $(q_j, p_j) \equiv (x_j, p_j)$. Since

$$p_j = m v_j + e A_j,$$

one follows that

$$v_j = \frac{1}{m} (p_j - e A_j),$$

and the Hamiltonian given by Eq. (6.5.40) becomes

$$H = \frac{1}{2m} (p_j - e A_j)(p_j - e A_j) + eV. \tag{6.5.41}$$

In the frame of the Hamiltonian formalism, the motion of the particle is governed by the Hamilton's canonical equations,

$$\begin{cases} \dot{q}_k = \dfrac{\partial H}{\partial p_k}, \\ \dot{p}_k = -\dfrac{\partial H}{\partial q_k}, \quad k = \overline{1,3}, \end{cases} \Leftrightarrow \begin{cases} \dot{x}_k = \dfrac{\partial H}{\partial p_k}, \\ \dot{p}_k = -\dfrac{\partial H}{\partial x_k}, \quad k = \overline{1,3}. \end{cases}$$

By means of Eq. (6.5.41), this system receives the following form:

$$\begin{cases} \dot{x}_k = \dfrac{1}{m} (p_k - e A_k), \\ \dot{p}_k = \dfrac{e}{m} (p_j - e A_j) \dfrac{\partial A_j}{\partial x_k} - e \dfrac{\partial V}{\partial x_k}. \end{cases}$$

Taking the total derivative with respect to time of the first equation and introducing $\dot{p}_k$ obtained from the second equation of the system, we obtain

$$\begin{aligned} \ddot{x}_k &= \frac{1}{m} \left[\frac{e}{m} (p_j - e A_j) \frac{\partial A_j}{\partial x_k} - e \frac{\partial V}{\partial x_k} - e \dot{A}_k \right] \\ &= \frac{1}{m} \left(\frac{e}{m} m \dot{x}_j \frac{\partial A_j}{\partial x_k} - e \frac{\partial V}{\partial x_k} - e \dot{A}_k \right) \\ &= \frac{1}{m} \left[e \dot{x}_j \frac{\partial A_j}{\partial x_k} - e \frac{\partial V}{\partial x_k} - e \left(\frac{\partial A_k}{\partial t} + \dot{x}_j \frac{\partial A_k}{\partial x_j} \right) \right], \end{aligned} \tag{6.5.42}$$

where we used once more the first equation of the system and took into account that, since $A_k = A_k(\vec{r}, t)$, we have

$$\dot{A}_k = \frac{\partial A_k}{\partial t} + \dot{x}_j \frac{\partial A_k}{\partial x_j}.$$

After rearrangement of some terms, Eq. (6.5.42) becomes

$$m \ddot{x}_k = -e \frac{\partial V}{\partial x_k} - e \frac{\partial A_k}{\partial t} + e v_j \left(\frac{\partial A_j}{\partial x_k} - \frac{\partial A_k}{\partial x_j} \right). \tag{6.5.43}$$

But, algebraically, the quantity

$$\frac{\partial A_j}{\partial x_k} - \frac{\partial A_k}{\partial x_j}$$

is a second-order antisymmetric tensor, namely the tensor T_{kj} met in the **Problem No. 50**. Being defined on a three-dimensional space, its corresponding pseudovector writes

$$B_i = \frac{1}{2}\varepsilon_{ikj}T_{kj} = \frac{1}{2}\varepsilon_{ikj}\left(\frac{\partial A_j}{\partial x_k} - \frac{\partial A_k}{\partial x_j}\right)$$

$$= \frac{1}{2}\varepsilon_{ikj}\frac{\partial A_j}{\partial x_k} - \frac{1}{2}\varepsilon_{ikj}\frac{\partial A_k}{\partial x_j} = \frac{1}{2}\varepsilon_{ikj}\frac{\partial A_j}{\partial x_k}$$

$$- \frac{1}{2}\varepsilon_{ijk}\frac{\partial A_j}{\partial x_k} = \frac{1}{2}\varepsilon_{ikj}\frac{\partial A_j}{\partial x_k} + \frac{1}{2}\varepsilon_{ikj}\frac{\partial A_j}{\partial x_k}$$

$$= \varepsilon_{ikj}\frac{\partial A_j}{\partial x_k} = \varepsilon_{ikj}\partial_k A_j = \left(\nabla \times A\right)_i,$$

which is precisely the axial vector of magnetic induction of the field. Under these circumstances, according to the isomorphism between the set of antisymmetric tensors of the second order, defined on a three-dimensional space, and the set of associated pseudo-vectors (defined on the same space), we have

$$\frac{\partial A_j}{\partial x_k} - \frac{\partial A_k}{\partial x_j} \equiv T_{kj} = \varepsilon_{kji}B_i,$$

and Eq. (6.5.43) becomes

$$m\ddot{x}_k = -e\frac{\partial V}{\partial x_k} - e\frac{\partial A_k}{\partial t} + ev_j\varepsilon_{kji}B_i = e\left(-\frac{\partial V}{\partial x_k} - \frac{\partial A_k}{\partial t}\right)$$

$$+ \varepsilon_{kji}v_jB_i = eE_k + e\left(\vec{v} \times \vec{B}\right)_k = e\left[E_k + \left(\vec{v} \times \vec{B}\right)_k\right],$$

which is the projection on x_k-axis of the vector equation we were looking for,

$$m\ddot{\vec{r}} = e\left(\vec{E} + \vec{v} \times \vec{B}\right).$$

6.6   Problem No. 52

The Lagrangian of a charged particle, subject to an electromagnetic force is

$$L_{\text{em}} = \frac{1}{2}mv^2 - eV + e\vec{v} \cdot \vec{A}, \qquad (6.6.44)$$

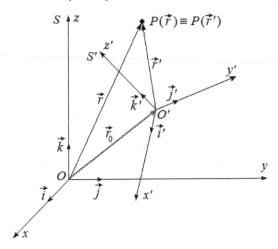

FIGURE 6.2
The position P of a particle relative to an inertial frame S, and a non-inertial frame S'.

while the Lagrangian of a particle of mass m, in motion with respect to a non-inertial frame S' (see Fig. 6.2) is

$$L_{\text{mech}} = \frac{1}{2}m\,|\vec{v}_r|^2 + \frac{1}{2}m\,|\vec{\omega} \times \vec{r}'|^2 + m\,\vec{v}_r \cdot (\vec{\omega} \times \vec{r}')$$
$$- m\,\vec{a}_0 \cdot \vec{r}' - W_{\text{pot}}, \qquad (6.6.45)$$

where $\vec{\omega}$ is the instantaneous vector of rotation, $\vec{v}_r$ – the relative velocity, and W_{pot} – the potential energy of the particle.

a) Using the correspondences $m \leftrightarrow e$ and $\vec{\omega} \leftrightarrow \nabla \times \vec{A}'$, show that the equation of motion of a massive and electrically neutral particle in a non-inertial system can be obtained from the Lagrangian of a charged particle moving under the action of a force which derives from a generalized potential (depending on velocities), where the generalized potential energy is of the form

$$W' = m\left(V' - 2\,\vec{v} \cdot \vec{A}'\right),$$

with

$$V' = \vec{r} \cdot \vec{a}_0 - \frac{1}{2}\left|\vec{\omega} \times \vec{r}\right|^2,$$

and

$$\vec{A}' = \frac{1}{2}\,\vec{\omega} \times \vec{r},$$

and the magnetic field is considered constant and homogeneous.

b) Show that the "new fields" $\vec{E}'$ and $\vec{B}'$ (corresponding to the potentials V' and $\vec{A}'$ defined above) satisfy the sourceless Maxwell equations.

Solution

a) Starting from the Lagrangian in Eq. (6.6.45) one obtains the equation of motion of the particle with respect to the non-inertial frame as the Lagrange equation:

$$m\,\vec{a}_r = \vec{F} - m\,\vec{a}_0 - m\,\dot{\vec{\omega}} \times \vec{r}' - m\,\vec{\omega} \times \left(\vec{\omega} \times \vec{r}'\right) - 2m\,\vec{\omega} \times \vec{v}_r. \qquad (6.6.46)$$

The analogy between the two systems whose Lagrangians are given by Eqs. (6.6.44) and (6.6.45) appears at various levels, as follows. First, one observes that both the *Coriolis* force $\vec{F}_c = 2m\,\vec{v}_r \times \vec{\omega}$ and the *Lorentz* force are *gyroscopic* (their power is zero). Since we deal with a motion in a non-inertial frame, we are allowed to use $\vec{r}$, $\vec{v}$, instead of $\vec{r}'$, $\vec{v}_r$, respectively. Secondly, each Lagrangian contains two velocity-dependent terms. Thirdly, if the field $\vec{B}$ is constant and homogeneous, then it is easy to show (see **Problem No. 10**) that $\vec{A}$ can be written as

$$\vec{A} = \frac{1}{2}\vec{B} \times \vec{r}.$$

This means that the terms

$$m\,\vec{v} \cdot \left(\vec{\omega} \times \vec{r}\right)$$

and

$$e\,\vec{v} \cdot \vec{A}$$

are equivalent, if we set the correspondence $m \leftrightarrow e$, and choose

$$\vec{A}' = \frac{1}{2}\,\vec{\omega} \times \vec{r}, \qquad (6.6.47)$$

i.e.,

$$\vec{\omega} = \nabla \times \vec{A}'. \qquad (6.6.48)$$

This analogy leads to the following Lagrangian of the particle, relative to the non-inertial frame:

$$L = \frac{1}{2}m\,|\vec{v}|^2 - m\,V' + 2\,m\,\vec{v} \cdot \vec{A}' - V(\vec{r}), \qquad (6.6.49)$$

where $\vec{A}'$ is defined by Eq. (6.6.47) and V' by

$$V' = \vec{r} \cdot \vec{a}_0 - \frac{1}{2}\left|\vec{\omega} \times \vec{r}\right|^2. \qquad (6.6.50)$$

We emphasize that the potentials $\vec{A}'$ and V' are functions of coordinates and time, while W_{pot} yields the potential force. By the above settled convention, $\dot{\vec{r}} = \vec{v}$ and $\ddot{\vec{r}} = \vec{a}$ are determined relative to the non-inertial frame. Using the Lagrange equations and performing calculations, we obtain the following equation of motion:

$$m\,\ddot{\vec{r}} + 2\,m\,\dot{\vec{A}}' + m\,\nabla V' - 2\,m\,\nabla\left(\vec{r} \cdot \vec{A}'\right) + \nabla W = 0,$$

or, if we use Eq. (B.3.55) and make some re-arrangements of the terms,

$$m\,\ddot{\vec{r}} = m\left(\vec{E}' + \vec{v} \times \vec{B}'\right) + \vec{F}, \qquad (6.6.51)$$

where

$$\begin{cases} \vec{E}' = -\nabla V' - \dfrac{\partial}{\partial t}\left(2\vec{A}'\right), \\ \vec{B}' = \nabla \times \left(2\vec{A}'\right). \end{cases} \qquad (6.6.52)$$

Equation (6.6.51) shows that the terms $\vec{E}'$ and $\vec{v} \times \vec{B}'$ have the units of an acceleration. It also shows that the particle moves in an inertial force field, defined by the potentials $\vec{A}'$, V', and in an applied force field $\vec{F} = -\nabla V$ as well. If the frame S' becomes inertial, and $\vec{F} = 0$, then $\vec{a}_0 = 0$, $\vec{\omega} = 0$ and Eq. (6.6.51) yields $\ddot{\vec{r}} = 0$, as expected. We reach the same result if the charged particle is neither accelerated by the electric field $\vec{E}$ nor rotated by the magnetic field $\vec{B}$.

b) We can still go further with this analogy, and observe that the fields $\vec{E}'$ and $\vec{B}'$ satisfy the source-free Maxwell equations,

$$\begin{cases} \nabla \times \vec{E}' = -\dfrac{\partial \vec{B}'}{\partial t}, \\ \nabla \cdot \vec{B}' = 0. \end{cases} \qquad (6.6.53)$$

Indeed, since $\nabla \times \left(\nabla \varphi\right) = 0$ for any scalar field $\varphi(\vec{r}, t)$ and $\nabla \cdot \left(\nabla \times \vec{a}\right) = 0$ for any vector field $\vec{a}(\vec{r}, t)$, both fields being of class C^1 on their domains of definition, from Eq. (6.6.52) the validity of Eq. (6.6.53) follows immediately.

We can then conclude that the study of a massive $(m \neq 0)$, non-charged particle, moving in a non-inertial frame, can be accomplished by using the same Lagrangian as for a charged particle, moving in a velocity-dependent force field, the generalized potential energy being given by

$$W = m\left(V' - 2\vec{v} \cdot \vec{A}'\right). \qquad (6.6.54)$$

6.7   Problem No. 53

Using the Hamiltonian formalism, study the motion of a charged particle in a stationary magnetic field with axial symmetry.

Solution

Since the field is stationary, the vector potential $\vec{A}$ does not explicitly depend on time and the magnetic field is given by

$$\vec{B} = \nabla \times \vec{A}. \qquad (6.7.55)$$

The geometry of the problem indicates to choose a cylindrical system of coordinates ρ, φ, z, with the z-axis directed along the symmetry axis, and assume that

$$A_\rho = 0, \quad A_\varphi \neq 0, \quad A_z = 0.$$

Since by definition of the field with axial symmetry $B_\varphi = 0$, the non-zero components of $\vec{B}$ are

$$\begin{cases} B_\rho = -\dfrac{\partial A_\varphi}{\partial z}, \\[2mm] B_z = \dfrac{1}{\rho}\dfrac{\partial}{\partial \rho}(\rho A_\varphi). \end{cases}$$

Observing that $\vec{A}$ is independent of φ, we may write the Hamiltonian as

$$H = \frac{1}{2m}\left\{ p_\rho^2 + \frac{1}{\rho^2}\left[p_\varphi - e\rho A_\varphi(\rho, z) \right]^2 + p_z^2 \right\}. \tag{6.7.56}$$

Hamilton's equation for the conjugate variables φ and p_φ are

$$\begin{cases} \dot{\varphi} = \dfrac{\partial H}{\partial p_\varphi} = \dfrac{1}{m\rho^2}(p_\varphi - e\rho A_\varphi), \\[3mm] \dot{p}_\varphi = -\dfrac{\partial H}{\partial \varphi} = 0. \end{cases}$$

The second equation leads to the first integral

$$p_\varphi = m\rho^2\dot{\varphi} + e\rho A_\varphi = \text{const.} \tag{6.7.57}$$

Since H does not explicitly depend on time, there exists also the energy first integral

$$E = \frac{1}{2}m(\dot{\rho}^2 + \rho^2\dot{\varphi}^2 + \dot{z}^2) = \text{const.} \tag{6.7.58}$$

The differential equations of motion for ρ and z are then

$$\begin{cases} m(\ddot{\rho} - \rho\dot{\varphi}^2) = e\dot{\rho}B_z, \\[2mm] m\ddot{z} = -e\rho\dot{\varphi}B_\rho. \end{cases} \tag{6.7.59}$$

Assume now that $\vec{B}$ is constant and homogeneous. Then, according to the result of **Problem No. 10**,

$$A_\varphi = \frac{1}{2}(\vec{B} \times \vec{r})_\varphi = \frac{1}{2}\rho B_z,$$

and the first integral in Eq. (6.7.57) reads

$$m\rho^2\dot{\varphi} + \frac{1}{2}e\rho^2 B_z = C \,(= \text{const.}), \tag{6.7.60}$$

known as *Busch's relation*, after the German physicist Hans Busch (1884–1973). If at the initial moment $t = 0$ the particle is at the origin

O of the reference frame, where $\rho = 0$ and $v_\varphi = \rho\dot{\varphi} = 0$, then $C = 0$ and Eq. (6.7.60) yields

$$\dot{\varphi} = -\frac{eB_z}{2m}, \qquad (6.7.61)$$

meaning that the particle performs a motion of precession around the z-axis. This effect is applied in the construction of magnetic focusing devices, called *magnetic lenses*. Substituting the quantity in Eq. (6.7.61) into the energy first integral given by Eq. (6.7.58) (and noting that for $\vec{B} = (0, 0, B_z)$, the second equation in the system (6.7.59) leads to $\dot{z} = $ const.), one obtains the following second-order differential equation:

$$\ddot{\rho} + \left(\frac{eB_z}{2m}\right)^2 \rho = 0. \qquad (6.7.62)$$

Assume that the particle moves close to the z-axis, while the magnetic field acts only over a small portion of the beam (paraxial beam). The components of the velocity along ρ and φ are then negligible as compared to the component along z. As $\dot{z} = v_z = $ const., we have

$$\ddot{\rho} = \dot{z}^2 \frac{d^2\rho}{dz^2} = v_z^2 \frac{d^2\rho}{dz^2},$$

and Eq. (6.7.63) yields

$$\frac{d^2\rho}{dz^2} + \left(\frac{eB_z}{2mv_z}\right)^2 \rho = 0. \qquad (6.7.63)$$

Since $d^2\rho/dz^2 < 0$, the magnetic lens is *convergent*, independently of the sign of charged particles.

The *electric lens* is based on a similar focusing principle. Both electric and magnetic lenses are used in electronic microscopy, old television devices, etc.

6.8   Problem No. 54

Let N_0 be the electron density of an anisotropic dielectric material. Determine the tensor of relative permittivity of the dielectric, $(\varepsilon_r)_{ik}$, $i, k = \overline{1,3}$, if it is placed in a static and homogeneous magnetic field of magnetic induction $\vec{B}_0$, admitting the model of elastic/harmonic bonding of the electrons and ions in the anisotropic dielectric material. For the sake of simplicity, it will be considered that all electrons have the same own frequency, while the relativistic effects and bremsstrahlung (braking radiation) will be neglected.

Solution

Without restricting the generality of the problem, let us choose the z-axis along the magnetic field $\vec{B}_0$. The equation of motion of an electron in the field with monochromatic electric component of frequency $\nu(=\omega/2\pi)$ is

$$m\frac{d^2\vec{r}}{dt^2} = -m\omega_0^2\vec{r} + e\vec{E} + e\vec{v} \times \vec{B}_0$$

$$= -m\omega_0^2\vec{r} + e\vec{E}_0 e^{-i\omega t} + e\frac{d\vec{r}}{dt} \times \vec{B}_0, \tag{6.8.64}$$

where $\vec{r}$ represents the displacement of the electron with respect to its equilibrium position, m is the electron mass, and $\nu_0 = \omega_0/2\pi$ is the frequency of the oscillatory motion of the electron in the material (the elastic restoring force is $\vec{F} = -m\omega_0^2\vec{r}$). By expanding the vector product in Eq. (6.8.64) one obtains

$$m\frac{d^2\vec{r}}{dt^2} = -m\omega_0^2\vec{r} + e\vec{E}_0 e^{-i\omega t} + e\frac{dy}{dt}B_0\vec{u}_x - e\frac{dx}{dt}B_0\vec{u}_y, \tag{6.8.65}$$

where $\vec{u}_x$ and $\vec{u}_y$ are the versors (unit vectors) of the coordinate axes Ox and Oy, respectively. Projecting the vector equation (6.8.65) on the three coordinate axes we get

$$m\frac{d^2x}{dt^2} = -m\omega_0^2 x + eE_{0x}e^{-i\omega t} + e\frac{dy}{dt}B_0$$

$$= -m\omega_0^2 x + eE_x + e\frac{dy}{dt}B_0, \tag{6.8.66.a}$$

$$m\frac{d^2y}{dt^2} = -m\omega_0^2 y + eE_{0y}e^{-i\omega t} - e\frac{dx}{dt}B_0$$

$$= -m\omega_0^2 y + eE_y - e\frac{dx}{dt}B_0, \tag{6.8.66.b}$$

$$m\frac{d^2z}{dt^2} = -m\omega_0^2 z + eE_{0z}e^{-i\omega t} = -m\omega_0^2 z + eE_z. \tag{6.8.66.c}$$

To facilitate the calculation, let us introduce the pair of "cyclic" coordinates

$$\xi = -\frac{1}{\sqrt{2}}(x + iy), \ \eta = \frac{1}{\sqrt{2}}(x - iy),$$

in the plane perpendicular to direction of the magnetic field $\vec{B}_0 = B_0\vec{u}_z$, where $\vec{u}_z$ is the versor of z-axis.

Let us now rewrite Eqs. (6.8.65) in the new coordinates ξ, η and z. To determine the equation in variable ξ, we multiply Eq. (6.8.66.a) by $\left(-1/\sqrt{2}\right)$ and Eq. (6.8.66.b) by $\left(-i/\sqrt{2}\right)$, then add them member by member. We obtain

$$m\frac{d^2}{dt^2}\left[-\frac{1}{\sqrt{2}}(x+iy)\right] = -m\omega_0^2\left[-\frac{1}{\sqrt{2}}(x+iy)\right]$$
$$+ e\left[-\frac{1}{\sqrt{2}}(E_x+iE_y)\right] - ieB_0\frac{d}{dt}\left[-\frac{1}{\sqrt{2}}(x+iy)\right].$$

Since $\xi = -\frac{1}{\sqrt{2}}(x+iy)$, we still have

$$m\frac{d^2\xi}{dt^2} = -m\omega_0^2\xi + eE_+ - ieB_0\frac{d\xi}{dt},$$

so that

$$\frac{d^2\xi}{dt^2} + 2i\Omega\frac{d\xi}{dt} + \omega_0^2\xi = \frac{e}{m}E_+, \qquad (6.8.67.a)$$

where the following notations have been used:

$$\Omega \equiv \frac{eB_0}{2m}, \quad E_+ \equiv -\frac{1}{\sqrt{2}}(E_x+iE_y).$$

In a similar way the equation in the η variable can be obtained: Eq. (6.8.66.a) is multiplied by $(1/\sqrt{2})$ and Eq. (6.8.66.b) by $(-i/\sqrt{2})$, then the obtained equations are added member by member. The result is

$$m\frac{d^2}{dt^2}\left[\frac{1}{\sqrt{2}}(x-iy)\right] = -m\omega_0^2\left[\frac{1}{\sqrt{2}}(x-iy)\right]$$
$$+ e\left[\frac{1}{\sqrt{2}}(E_x-iE_y)\right] + ieB_0\frac{d}{dt}\left[\frac{1}{\sqrt{2}}(x-iy)\right].$$

Since $\eta = \frac{1}{\sqrt{2}}(x-iy)$, we still have

$$m\frac{d^2\eta}{dt^2} = -m\omega_0^2\eta + eE_- + ieB_0\frac{d\eta}{dt},$$

or

$$\frac{d^2\eta}{dt^2} - 2i\Omega\frac{d\eta}{dt} + \omega_0^2\eta = \frac{e}{m}E_-, \qquad (6.8.67.b)$$

where

$$E_- \equiv \frac{1}{\sqrt{2}}(E_x-iE_y)$$

is a new notation.

Since the variable z remains unchanged, and Eq. (6.8.66.c) contains only the variable z, this equation keeps its form,

$$\frac{d^2z}{dt^2} + \omega_0^2 z = \frac{e}{m}E_z. \qquad (6.8.67.c)$$

Because the time variable interferes only in the exponential $e^{-i\omega t}$, the total derivative with respect to time is equivalent to multiplication by $(-i\omega)$, that is

$$\frac{d*}{dt} = -i\omega * .$$

In this case, Eqs. (6.8.67) become

$$-\omega^2\xi + 2\Omega\omega\xi + \omega_0^2\xi = \frac{e}{m}E_+ ,$$

$$-\omega^2\eta - 2\Omega\omega\eta + \omega_0^2\eta = \frac{e}{m}E_- ,$$

$$-\omega^2 z + \omega_0^2 z = \frac{e}{m}E_z ,$$

or

$$\xi = \frac{e}{m}\frac{E_+}{\omega_0^2 - \omega^2 + 2\omega\Omega} ,\qquad (6.8.68.a)$$

$$\eta = \frac{e}{m}\frac{E_-}{\omega_0^2 - \omega^2 - 2\omega\Omega} ,\qquad (6.8.68.b)$$

$$z = \frac{e}{m}\frac{E_z}{\omega_0^2 - \omega^2} .\qquad (6.8.68.c)$$

In order to go back to the old variables x, y and z, we recall the connections between the variables:

$$\begin{cases} x = -\dfrac{1}{\sqrt{2}}(\xi - \eta), \\[2mm] y = \dfrac{i}{\sqrt{2}}(\xi + \eta), \\[2mm] z = z . \end{cases}$$

Therefore, the equation in variable x is obtained by multiplying Eq. (6.8.68.a) with $\left(-1/\sqrt{2}\right)$ and Eq. (6.8.68.b) with $\left(1/\sqrt{2}\right)$, and then adding them member by member. It results that

$$\begin{aligned} -\frac{1}{\sqrt{2}}(\xi - \eta) &= \frac{e}{m}\left[-\frac{1}{\sqrt{2}}\frac{E_+}{\omega_0^2 - \omega^2 + 2\omega\Omega} + \frac{1}{\sqrt{2}}\frac{E_-}{\omega_0^2 - \omega^2 - 2\omega\Omega}\right] \\ &= \frac{e}{m}\frac{(a+b)E_- - (a-b)E_+}{\sqrt{2}\left(a^2 - b^2\right)} \\ &= \frac{e}{m}\frac{(a+b)\frac{1}{\sqrt{2}}\left(E_x - iE_y\right) + (a-b)\frac{1}{\sqrt{2}}\left(E_x + iE_y\right)}{\sqrt{2}\left(a^2 - b^2\right)} \\ &= \frac{e}{m}\frac{aE_x - ibE_y}{a^2 - b^2} , \end{aligned}$$

that is

$$x = \frac{e}{m}\frac{aE_x - ibE_y}{a^2 - b^2} ,\qquad (6.8.69.a)$$

where the following notations have been used:

$$a \equiv \omega_0^2 - \omega^2, \quad b \equiv 2\omega\Omega.$$

In its turn, the equation in variable y can be obtained by multiplying Eqs. (6.8.68.a) and (6.8.68.b) with the same complex number $(i/\sqrt{2})$, then adding the results. One finds

$$\frac{i}{\sqrt{2}}(\xi + \eta) = \frac{e}{m}\left[\frac{i}{\sqrt{2}}\frac{E_+}{\omega_0^2 - \omega^2 + 2\omega\Omega} + \frac{i}{\sqrt{2}}\frac{E_-}{\omega_0^2 - \omega^2 - 2\omega\Omega}\right]$$

$$= \frac{e}{m} \cdot \frac{(a-b)iE_+ + (a+b)iE_-}{\sqrt{2}\left(a^2 - b^2\right)}$$

$$= \frac{e}{m}\frac{-(a-b)\frac{i}{\sqrt{2}}\left(E_x + iE_y\right) + (a+b)\frac{i}{\sqrt{2}}\left(E_x - iE_y\right)}{\sqrt{2}\left(a^2 - b^2\right)}$$

$$= \frac{e}{m}\frac{aE_y + ibE_x}{a^2 - b^2},$$

or

$$y = \frac{e}{m}\frac{aE_y + ibE_x}{a^2 - b^2}. \tag{6.8.69.b}$$

Finally, the equation for variable z results directly from Eq. (6.8.68.c) by means of notation $a \equiv \omega_0^2 - \omega^2$:

$$z = \frac{e}{m}\frac{E_z}{a}. \tag{6.8.69.c}$$

Within this simple model of polarization of a dielectric, the dipolar moment induced in a single atomic/molecular system is $\vec{p} = e\vec{r}$, which means that the electric polarization of the medium (the dipole moment per unit volume) is

$$\vec{P} = N_0\vec{p} = eN_0\vec{r},$$

where N_0 is the number of dipoles (electron-ion pairs) per unit volume (equal to the number of electrons per unit volume). Therefore, in this simple model one supposes that the dielectric material is composed by atomic/molecular systems as pairs of electron-ion ("small elementary dipoles"). In this case, according to the definition of electric displacement field (electric induction), we have

$$\vec{D} = \varepsilon_0\vec{E} + \vec{P} = \varepsilon_0\vec{E} + eN_0\vec{r},$$

or, written by components (and using the relations (6.8.69)):

$$D_x = \varepsilon_0 E_x + eN_0 x = \varepsilon_0 E_x + eN_0\frac{e}{m}\frac{aE_x - ibE_y}{a^2 - b^2}$$

$$= \left[\varepsilon_0 + \frac{ae^2 N_0}{m\left(a^2 - b^2\right)}\right]E_x - \frac{ibe^2 N_0}{m\left(a^2 - b^2\right)}E_y, \tag{6.8.70.a}$$

$$D_y = \varepsilon_0 E_y + eN_0 y = \varepsilon_0 E_y + eN_0 \frac{e}{m} \frac{aE_y + ibE_x}{a^2 - b^2}$$

$$= \left[\varepsilon_0 + \frac{ae^2 N_0}{m(a^2 - b^2)} \right] E_y + \frac{ibe^2 N_0}{m(a^2 - b^2)} E_x, \qquad (6.8.70.\mathrm{b})$$

$$D_z = \varepsilon_0 E_z + eN_0 z = \varepsilon_0 E_z + eN_0 \frac{e}{m} \frac{E_z}{a} = \left(\varepsilon_0 + \frac{e^2 N_0}{ma} \right) E_z. \qquad (6.8.70.\mathrm{c})$$

Comparing relations (6.8.70) with the general tensor relation

$$D_i = \varepsilon_{ik} E_k = \varepsilon_0 (\varepsilon_r)_{ik} E_k, \quad i, k = \overline{1, 3},$$

it follows that the tensor of relative permittivity of the dielectric is

$$(\varepsilon_r)_{ik} = \begin{pmatrix} 1 + \dfrac{ae^2 N_0}{\varepsilon_0 m(a^2 - b^2)} & -\dfrac{ibe^2 N_0}{\varepsilon_0 m(a^2 - b^2)} & 0 \\[3mm] \dfrac{ibe^2 N_0}{\varepsilon_0 m(a^2 - b^2)} & 1 + \dfrac{ae^2 N_0}{\varepsilon_0 m(a^2 - b^2)} & 0 \\[3mm] 0 & 0 & 1 + \dfrac{e^2 N_0}{\varepsilon_0 ma} \end{pmatrix},$$

where $\varepsilon_0 = 8,854 \times 10^{-12}\,\mathrm{F \cdot m^{-1}}$ is the electric permittivity of free space.

7

Magnetohydrodynamics. Plasma

7.1 Problem No. 55

Using the expression of the electromagnetic field in terms of *generalized anti-potentials* and appealing to the formalism offered by the variational calculus, establish the equation of motion of a compressible, non-viscous, infinitely conducting fluid, performing an isentropic motion in the electromagnetic field $(\vec{E}, \vec{B})$.

Complements

1) As is well known (see the reference [15]), the electromagnetic field can be expressed in terms of the anti-potentials (improper denomination, but tolerated) using source Maxwell's equations, when the sources are absent. As Calkin observed (1963), the equation of continuity for a moving electrized fluid,

$$\frac{\partial \rho_e}{\partial t} + \nabla \cdot \left(\vec{j} + \rho_e \vec{v} \right) = 0, \tag{7.1.1}$$

where ρ_e is the electric charge density, $\vec{v}$ is the velocity of the fluid, and $\rho_e \vec{v}$ is the convection current density, is identically satisfied if we choose

$$\begin{cases} \vec{j} = \dfrac{\partial \vec{P}}{\partial t} + \nabla \times \left(\vec{P} \times \vec{v} \right) + \vec{v} \, \nabla \cdot \vec{P}, \\ \rho_e = -\nabla \cdot \vec{P}, \end{cases} \tag{7.1.2}$$

where the vector field $\vec{P}$ is called *pseudo-polarization*, and Eq. (7.1.2)$_1$ defines its *Lorentzian derivative* (see the above cited reference). Using Eq. (7.1.2), the Maxwell's source equations write

$$\begin{cases} \nabla \times \left(\dfrac{1}{\mu_0} \vec{B} - \vec{P} \times \vec{v} \right) = \dfrac{\partial}{\partial t} \left(\varepsilon_0 \vec{E} + \vec{P} \right), \\ \nabla \cdot \left(\varepsilon_0 \vec{E} + \vec{P} \right) = 0. \end{cases} \tag{7.1.3}$$

DOI: 10.1201/9781003402602-7

We therefore deduce that there are two fields, a vector field $\vec{M}$ and a scalar field ψ, so that we have

$$\begin{cases} \varepsilon_0 \vec{E} + \vec{P} = \nabla \times \vec{M}, \\ \dfrac{1}{\mu_0} \vec{B} - \vec{P} \times \vec{v} - \dfrac{\partial \vec{M}}{\partial t} = \nabla \psi, \end{cases} \qquad (7.1.4)$$

which yields

$$\begin{cases} \vec{E} = \dfrac{1}{\varepsilon_0} \left(\nabla \times \vec{M} - \vec{P} \right), \\ \vec{B} = \mu_0 \left(\nabla \psi + \vec{P} \times \vec{v} + \dfrac{\partial \vec{M}}{\partial t} \right). \end{cases} \qquad (7.1.5)$$

Consequently, by replacing the four scalar quantities $\vec{j}$ and ρ_e (the sources) with three scalar quantities $\vec{P}$, the electromagnetic field can be expressed in terms of the potentials $\vec{M}$ and ψ, called *generalized anti-potentials*.

2) Thermodynamically, the quantities describing the magneto-fluid are connected by the equation of continuity,

$$\frac{d\rho}{dt} + \rho \nabla \cdot \vec{v} = 0 \qquad (7.1.6)$$

and the fundamental equation of thermodynamics for the quasistatic and reversible processes (the entropy equation):

$$T ds = d\varepsilon + p \, d(\rho^{-1}), \qquad (7.1.7)$$

where s is the entropy per unit mass, and $\varepsilon = \varepsilon(\rho, s)$ is the internal energy per unit mass. Since the motion of the fluid is isentropic, the last equation reduces to

$$\frac{ds}{dt} = 0. \qquad (7.1.8)$$

Solution

To deduce the equation of motion, we will use the analytical formalism. As is well-known, the choice of the Lagrangian density is not unique, so that we must choose those relations which contain all characteristic quantities (electrical and thermodynamical) in order to form an appropriate Lagrangian. In this respect, we will use the method of Lagrange multipliers, which consists in finding certain convenient equations ("zeros"), multiply them by some dependent on point/coordinates and time parameters, and introduce the result in the "composition" of the Lagrangian density. The subsequent application of an appropriate variational principle allows obtaining a set of relations which, by eliminating the multipliers, leads to the desired equations.

In view of the above described procedure, we will write the Lagrangian density as

$$L = \frac{1}{2\mu_0}B^2 - \frac{1}{2}\varepsilon_0 E^2 + \frac{1}{2}\rho v^2 - \rho\varepsilon - \alpha\left(\frac{d\rho}{dt} + \rho\nabla\cdot\vec{v}\right) - \beta\rho\frac{ds}{dt}. \qquad (7.1.9)$$

Here all quantities are functions of coordinates and time, denoted by x_k, ($x_1 = x$, $x_2 = y$, $x_3 = z$, $x_4 = t$). Expressing the field $(\vec{E}, \vec{B})$ in terms of the anti-potentials $\vec{M}$ and ψ, and choosing as variational parameters $\vec{M}$, ψ, $\tilde{P}$, $\vec{v}$, ρ and s (all of them generically denoted by σ_ν, $\nu = \overline{1,6}$), the Euler-Lagrange equations

$$(L)_\nu \equiv \frac{\partial L}{\partial \sigma_\nu} - \frac{\partial}{\partial x_k}\left(\frac{\partial L}{\partial \sigma_{\nu,k}}\right) = 0, \qquad (7.1.10)$$

lead to

$$\begin{cases} (L)_{\vec{M}} = -\dfrac{\partial \vec{B}}{\partial t} - \nabla \times \vec{E} = 0, \\[2mm] (L)_\psi = \nabla\cdot\vec{B} = 0, \\[2mm] (L)_{\vec{P}} = \vec{E} + \vec{v}\times\vec{B} = 0, \\[2mm] (L)_{\vec{v}} = \rho\vec{v} + \vec{B}\times\vec{\tilde{P}} + \rho\nabla\alpha - \beta\rho\nabla s = 0, \\[2mm] (L)_\rho = \dfrac{1}{2}v^2 - \varepsilon - \dfrac{p}{\rho} + \vec{v}\cdot\nabla\alpha + \dfrac{\partial\alpha}{\partial t} = 0, \\[2mm] (L)_s = \vec{v}\cdot\nabla\beta + \dfrac{\partial\beta}{\partial t} - T = 0. \end{cases} \qquad (7.1.11)$$

We found, as expected, the Maxwell's source-free equations $(7.1.11)_{1,2}$ and the Ohm's law for infinite conductivity expressed by Eq. $(7.1.11)_3$. Eq. $(7.1.11)_4$ is a generalization of the representation of velocity field $\vec{v}$ in terms of the *Clebsh potentials* which, in our case are α, β and s.

In order to find the equation of motion of the magneto-fluid, we will eliminate the Lagrangian multipliers α and β from Eqs. $(7.1.11)_{4,5,6}$. To start, we will apply the "gradient" to Eq. $(7.1.11)_5$. By using of Eq. $(7.1.11)_4$ we have

$$\nabla(\beta\,\vec{v}\cdot\nabla s) + \nabla\left(\frac{\partial\alpha}{\partial t}\right) - \nabla\left(\frac{v^2}{2}\right)$$
$$= \nabla\left[\varepsilon + \frac{p}{\rho} + \frac{1}{\rho}\vec{v}\cdot\left(\vec{B}\times\vec{\tilde{P}}\right)\right]. \qquad (7.1.12)$$

Next, we will use Eqs. (7.1.8) and $(7.1.11)_6$, as well as the well-known vector relation (B.3.55):

$$\nabla(\vec{a}\cdot\vec{b}) = \vec{a}\times(\nabla\times\vec{b}) + \vec{b}\times(\nabla\times\vec{a}) + (\vec{a}\cdot\nabla)\vec{b} + (\vec{b}\cdot\nabla)\vec{a}.$$

We then find

$$\frac{\partial\vec{v}}{\partial t} + (\vec{v}\cdot\nabla)\vec{v} = -\rho^{-1}\nabla p + \frac{\partial}{\partial t}\left(\rho^{-1}\vec{B}\times\vec{\tilde{P}}\right)$$
$$+ \vec{v}\times\left[\nabla\times\left(\rho^{-1}\vec{B}\times\vec{\tilde{P}}\right)\right] + \nabla\left[\rho^{-1}\vec{v}\cdot\left(\vec{\tilde{P}}\times\vec{B}\right)\right]. \qquad (7.1.13)$$

If we use now the vector identity

$$\nabla\left[\vec{A}\cdot\left(\vec{B}\times\vec{C}\right)\right] = \vec{A}\times\left[\nabla\times\left(\vec{B}\times\vec{C}\right)\right]$$
$$+ \vec{B}\times\left[\nabla\times\left(\vec{C}\times\vec{A}\right)\right] + \vec{C}\times\left[\nabla\times\left(\vec{A}\times\vec{B}\right)\right]$$
$$-\left(\vec{A}\times\vec{B}\right)\nabla\cdot\vec{C} - \left(\vec{B}\times\vec{C}\right)\nabla\cdot\vec{A}$$
$$-\left(\vec{C}\times\vec{A}\right)\nabla\cdot\vec{B},$$

one observes that some terms disappear and Eq. (7.1.13) leads to the equation we are looking for,

$$\rho\left[\frac{\partial\vec{v}}{\partial t} + (\vec{v}\cdot\nabla)\vec{v}\right] = -\nabla p + \vec{j}\times\vec{B}. \tag{7.1.14}$$

7.2   Problem No. 56

The equation of motion of an electronic plasma, including a phenomenological collision term, but neglecting the hydrostatic pressure (*i.e.*, using the zeroth-order approximation) is

$$\frac{\partial\vec{v}}{\partial t} + (\vec{v}\cdot\nabla)\vec{v} = \frac{e}{m}\left(\vec{E} + \vec{v}\times\vec{B}\right) - \nu\vec{v},$$

where ν is the frequency of collisions.

a) Show that in the presence of the external electric and magnetic fields, static and uniform, the linearized stationary expression of the Ohm's law becomes

$$J_i = \lambda_{ij}E_j, \quad i,j = \overline{1,3},$$

where the conductivity tensor λ_{ij} is

$$\lambda_{ij} = \frac{\omega_p^2}{\nu\left(1 + \dfrac{\omega_B^2}{\nu^2}\right)}\begin{pmatrix} 1 & \dfrac{\omega_B}{\nu} & 0 \\ -\dfrac{\omega_B}{\nu} & 1 & 0 \\ 0 & 0 & 1 + \dfrac{\omega_B^2}{\nu^2} \end{pmatrix},$$

where ω_p is the pulsation of the electron plasma, ω_B is the pulsation of the Larmor precession, and $\vec{B}$ is oriented along the z-axis.

b) Supposing that at the moment $t = 0$ an electric field $\vec{E}$ is suddenly applied in the direction of x-axis, in addition to the magnetic field $\vec{B}$, while at $t = 0$ the current is zero, determine the current components as functions of time, including the transient behaviour.

Solution

a) The *fundamental system of equations of magneto-fluido-dynamics* is composed by:
- the equations describing the electromagnetic field;
- the equations of continuity (for the electric and mechanical components);
- the force equation (equation of motion);
- Ohm's law;
- the equation of state.

These equations are written as follows:

$$
\begin{cases}
\nabla \times \vec{E} = -\dfrac{\partial \vec{B}}{\partial t}, \\[2mm]
\nabla \times \vec{B} = \mu \vec{J}, \\[2mm]
\nabla \cdot \vec{E} = \dfrac{1}{\varepsilon}\rho_e, \\[2mm]
\nabla \cdot \vec{B} = 0, \\[2mm]
\dfrac{\partial \rho_e}{\partial t} + \nabla \cdot \left(\rho_e \vec{v}\right) = 0, \\[2mm]
\dfrac{\partial \rho}{\partial t} + \nabla \cdot (\rho \vec{v}) = 0, \\[2mm]
\rho \dfrac{d\vec{v}}{dt} = \rho\vec{F} - \nabla p + \left(\xi + \eta\right)\nabla(\nabla \cdot \vec{v}) + \eta \Delta \vec{v} + \vec{J} \times \vec{B}, \\[2mm]
\rho T \dfrac{ds}{dt} = T'_{ik}\dfrac{\partial v_i}{\partial x_k} + \dfrac{\partial}{\partial x_k}\left(\kappa \dfrac{\partial T}{\partial x_k}\right) + \vec{J}\cdot\vec{E}, \\[2mm]
\vec{J} = \vec{J}(\vec{E}, \vec{B}), \\[2mm]
f(p, \rho, t) = 0,
\end{cases}
\qquad (7.2.15)
$$

where $\vec{F}$ is the specific mass force of non-electromagnetic nature, ξ and η are the coefficients of dynamic viscosity of the Newtonian fluid, T is the absolute temperature, T'_{ik} is the viscous stress tensor, κ is the thermal conductivity, s is the entropy per unit mass, and p is the pressure. For a simple, unicomponent and non-relativistic conducting fluid, the Ohm's law writes

$$
\vec{J} = \lambda(\vec{E} + \vec{v} \times \vec{B}),
$$

where λ is the electrical conductivity of the fluid. If the motion is isentropic, the state equation takes the simple form $ds/dt = 0$.

The system (7.2.15) can take various forms, depending on the particular cases approached in the study. The simplest case corresponds to a perfect fluid ($\xi = 0$, $\eta = 0$), infinitely conducting $\lambda \to \infty$, which performs an isentropic motion $(ds/dt = 0)$.

In order to solve the requirement **a)** of the present problem (determination of the linearized, steady form of Ohm's law) the above system of equations can be reduced to three, by omitting equations four, five and six. So, we are left with

$$
\begin{cases}
\nabla \times \vec{E} = -\dfrac{\partial \vec{B}}{\partial t}, \\[2mm]
\nabla \times \vec{B} = \mu \vec{J}, \\[2mm]
\nabla \cdot \vec{E} = \dfrac{1}{\varepsilon}\rho_e, \\[2mm]
\nabla \cdot \vec{B} = 0, \\[2mm]
\dfrac{\partial \rho_e}{\partial t} + \nabla \cdot \left(\rho_e \vec{v}\right) = 0, \\[2mm]
\dfrac{\partial \rho}{\partial t} + \nabla \cdot \left(\rho \vec{v}\right) = 0, \\[2mm]
\dfrac{\partial \vec{v}}{\partial t} + \left(\vec{v} \cdot \nabla\right)\vec{v} = \dfrac{e}{m}\left(\vec{E} + \vec{v} \times \vec{B}\right) - \nu \vec{v}.
\end{cases}
\tag{7.2.16}
$$

Multiplying the last equation of the system (7.2.16) by the charge density ρ_e and taking into account that $\vec{J} = \rho_e \vec{v}$, one obtains

$$
\rho_e \left[\frac{\partial \vec{v}}{\partial t} + \left(\vec{v} \cdot \nabla\right)\vec{v}\right] = \frac{\rho_e e}{m}\vec{E} + \frac{e}{m}\vec{J} \times \vec{B} - \nu \vec{J},
$$

or

$$
\rho_e \frac{d\vec{v}}{dt} = \frac{\rho_e e}{m}\vec{E} + \frac{e}{m}\vec{J} \times \vec{B} - \nu \vec{J}.
\tag{7.2.17}
$$

In order to calculate the single term in the left-hand side of the above equation, we will use the equation

$$
\nabla \times \vec{B} = \mu \vec{J},
$$

written in the form

$$
\nabla \times \vec{B} = \mu \rho_e \vec{v}.
$$

We then have

$$
\vec{v} = \frac{1}{\mu \rho_e}\nabla \times \vec{B} \quad \Rightarrow \quad \frac{d\vec{v}}{dt} = \frac{d}{dt}\left(\frac{1}{\mu \rho_e}\nabla \times \vec{B}\right)
$$

$$
= \frac{1}{\mu}\frac{\rho_e \nabla \times \frac{d\vec{B}}{dt} - \frac{d\rho_e}{dt}\nabla \times \vec{B}}{\rho_e^2}
$$

$$
\Rightarrow \quad \rho_e \frac{d\vec{v}}{dt} = \frac{1}{\mu \rho_e}\left(\rho_e \nabla \times \frac{d\vec{B}}{dt} - \frac{d\rho_e}{dt}\nabla \times \vec{B}\right)
$$

$$
= \frac{1}{\mu}\nabla \times \frac{d\vec{B}}{dt} - \frac{\nabla \times \vec{B}}{\mu}\frac{1}{\rho_e}\frac{d\rho_e}{dt}
$$

$$
= -\frac{\nabla \times \vec{B}}{\mu}\frac{1}{\rho_e}\frac{d\rho_e}{dt} = -\vec{J}\frac{1}{\rho_e}\frac{d\rho_e}{dt} = \vec{J}\nabla \cdot \vec{v},
$$

where, the property of the magnetic field of being *static*, $d\vec{B}/dt = 0$, has been used, and the charge conservation law, expressed by the equation of continuity, has been also taken into account:

$$\frac{\partial \rho_e}{\partial t} + \nabla \cdot \left(\rho_e \vec{v} \right) = \frac{d\rho_e}{dt} + \rho_e \nabla \cdot \vec{v} = 0 \ \Rightarrow \ \frac{d\rho_e}{dt} = -\rho_e \nabla \cdot \vec{v}.$$

Because we are working within the zeroth-order approximation (which implies linearization of all equations), we can write

$$\nabla \cdot \vec{v} = \nabla \cdot \left(\frac{1}{\mu \rho_e} \nabla \times \vec{B} \right) \simeq \nabla \cdot \left(\frac{1}{\mu n_0 e} \nabla \times \vec{B} \right)$$

$$= \frac{1}{\mu n_0 e} \left[\nabla \cdot \left(\nabla \times \vec{B} \right) \right] = 0,$$

where we have used the approximation $\rho_e \simeq \rho_{e0}$. Here $\rho_{e0} = e n_0$ is the equilibrium electric charge density of the electrons.

Under these conditions, Eq. (7.2.17) becomes

$$\omega_p^2 \vec{E} + \frac{e}{m} \vec{J} \times \vec{B} - \nu \vec{J} = 0,$$

where

$$\omega_p^2 \equiv \frac{n_0 e^2}{m}$$

is called the *plasma pulsation*. By components, the above vector equation writes

$$\begin{cases} \omega_p^2 E_x + \omega_B J_y - \nu J_x = 0, \\ \omega_p^2 E_y - \omega_B J_x - \nu J_y = 0, \\ \omega_p^2 E_z - \nu J_z = 0, \end{cases} \qquad (7.2.18)$$

where a new notation has been introduced, namely,

$$\omega_B \equiv \frac{e B_z}{m} = \frac{e B_0}{m},$$

which is called the *pulsation of the Larmor precession*, and we have used the statement of the problem that the magnetic field is static and uniform along the z-axis: $\vec{B} = (0, 0, B_0)$.

The last relation of the system (7.2.18) easily yields

$$J_z = \frac{\omega_p^2}{\nu} E_z,$$

or

$$J_z = 0 \cdot E_x + 0 \cdot E_y + \frac{\omega_p^2}{\nu} E_z$$

$$= \frac{\omega_p^2}{\nu \left(1 + \frac{\omega_B^2}{\nu^2} \right)} \left[0 \cdot E_x + 0 \cdot E_y + \left(1 + \frac{\omega_B^2}{\nu^2} \right) E_z \right]. \qquad (7.2.19.a)$$

The first relation of the same system gives J_y which, introduced in the second relation, gives the x-component of the current

$$\omega_p^2 E_x + \omega_B J_y - \nu J_x = 0 \quad \Rightarrow \quad J_y = \frac{\nu J_x - \omega_0^2 E_x}{\omega_B}, \qquad (7.2.20)$$

$$\omega_p^2 E_y - \omega_B J_x - \nu J_y = 0 \quad \Rightarrow \quad \omega_p^2 E_y - \omega_B J_x - \nu \frac{\nu J_x - \omega_p^2 E_x}{\omega_B} = 0$$

$$\Rightarrow J_x(\omega_B^2 + \nu^2) = \nu \omega_p^2 E_x + \omega_B \omega_p^2 E_y \quad \Rightarrow \quad J_x = \frac{\nu \omega_p^2 E_x + \omega_B \omega_p^2 E_y}{\nu^2 + \omega_B^2},$$

or

$$J_x = \frac{\omega_p^2}{\nu \left(1 + \frac{\omega_B^2}{\nu^2}\right)} E_x + \frac{\omega_p^2}{\nu \left(1 + \frac{\omega_B^2}{\nu^2}\right)} \frac{\omega_B}{\nu} E_y + 0 \cdot E_z$$

$$= \frac{\omega_p^2}{\nu \left(1 + \frac{\omega_B^2}{\nu^2}\right)} \left(1 \cdot E_x + \frac{\omega_B}{\nu} E_y + 0 \cdot E_z\right). \qquad (7.2.19.b)$$

Finally, using Eq. (7.2.20) we obtain

$$J_y = \frac{\nu J_x - \omega_p^2 E_x}{\omega_B} = \frac{1}{\omega_B} \nu \frac{\nu \omega_p^2 E_x + \omega_B \omega_p^2 E_y}{\nu^2 + \omega_B^2} - \omega_p^2 E_x$$

$$= \frac{-\omega_B^2 \omega_p^2 E_x + \nu \omega_B \omega_p^2 E_y}{\omega_B(\nu^2 + \omega_B^2)},$$

or

$$J_y = -\frac{\omega_p^2}{\nu \left(1 + \frac{\omega_B^2}{\nu^2}\right)} \frac{\omega_B}{\nu} E_x + \frac{\omega_p^2}{\nu \left(1 + \frac{\omega_B^2}{\nu^2}\right)} E_y + 0 \cdot E_z$$

$$= \frac{\omega_p^2}{\nu \left(1 + \frac{\omega_B^2}{\nu^2}\right)} \left(-\frac{\omega_B}{\nu} E_x + 1 \cdot E_y + 0 \cdot E_z\right). \qquad (7.2.19.c)$$

Relations (7.2.19) can be written in a single formula

$$J_i = \lambda_{ij} E_j, \quad i, j = \overline{1, 3}, \qquad (7.2.21)$$

where

$$\lambda_{ij} = \frac{\omega_p^2}{\nu \left(1 + \frac{\omega_B^2}{\nu^2}\right)} \begin{pmatrix} 1 & \frac{\omega_B}{\nu} & 0 \\ -\frac{\omega_B}{\nu} & 1 & 0 \\ 0 & 0 & 1 + \frac{\omega_B^2}{\nu^2} \end{pmatrix} \qquad (7.2.22)$$

is the *electric conductivity tensor*.

b) In this particular case, namely $\vec{E} = (E, 0, 0)$ and $\vec{B} = (0, 0, B)$, the linearized equation of motion writes, by components, as follows:

$$\begin{cases} \dfrac{\partial v_x}{\partial t} = \dfrac{eE}{m} + \dfrac{eB}{m} v_y - \nu v_x, \\ \dfrac{\partial v_y}{\partial t} = -\dfrac{eB}{m} v_x - \nu v_y, \\ \dfrac{\partial v_z}{\partial t} = -\nu v_z. \end{cases} \tag{7.2.23}$$

The last equation of the system (7.2.23) can be easily integrated and gives

$$v_z = C_1(x, y, z) e^{-\nu t}. \tag{7.2.24}$$

By taking the derivative with respect to time of the first equation of the system (7.2.23) and using the second equation of the same system in order to eliminate $\partial v_y / \partial t$, then again the first equation for a final elimination of v_y, one obtains

$$\frac{\partial^2 v_x}{\partial t^2} = \omega_B \frac{\partial v_y}{\partial t} - \nu \frac{\partial v_x}{\partial t} = \omega_B \left(-\omega_B v_x - \nu v_y \right) - \nu \frac{\partial v_x}{\partial t}$$

$$= -\omega_B^2 v_x - \nu \omega_B v_y - \nu \frac{\partial v_x}{\partial t} = -\omega_B^2 v_x - \nu \left(\frac{\partial v_x}{\partial t} + \nu v_x - \frac{eE}{m} \right)$$

$$- \nu \frac{\partial v_x}{\partial t} = -\omega_B^2 v_x - 2\nu \frac{\partial v_x}{\partial t} - \nu^2 v_x + \frac{\nu e E}{m},$$

or

$$\frac{\partial^2 v_x}{\partial t^2} + 2\nu \frac{\partial v_x}{\partial t} + \left(\nu^2 + \omega_B^2 \right) v_x = \frac{\nu e E}{m}. \tag{7.2.25}$$

Viewed only in terms of the time variable, this is a non-homogeneous, second order differential equation with constant coefficients. Thus, if we are interested only in the time variation of v_x, then we can replace the partial derivative symbols in Eq. (7.2.25) by the total derivative symbols, that is, in this case we can treat this equation as an ordinary differential equation and we can apply the corresponding method of solving of such differential equations. Under these circumstances, the general solution of this differential equation is obtained by adding up the general solution of the homogeneous differential equation with a particular solution of the non-homogeneous differential equation, which, as usual, can be determined by the general method, *i.e.*, the Lagrange's method of variation of parameters, or much easier — as in our case, when this way of finding a particular solution of the non-homogeneous differential equation can indeed be applied — we can search for a particular solution of the non-homogeneous differential equation of the form of the term that gives the non-homogeneity. The homogeneous differential equation attached to the non-homogeneous differential equation (7.2.25) writes

$$\frac{\partial^2 v_x}{\partial t^2} + 2\nu \frac{\partial v_x}{\partial t} + \left(\nu^2 + \omega_B^2 \right) v_x = 0,$$

with the attached characteristic equation

$$r^2 + 2\nu r + \left(\nu^2 + \omega_B^2\right) = 0,$$

which has the solutions

$$r_{1,2} = -\nu \pm i\omega_B.$$

The general solution of the homogeneous differential equation therefore is

$$v_x^{(ho)} = C_2(x,y,z)e^{-\nu t}e^{i\omega_B t} + C_3(x,y,z)e^{-\nu t}e^{-i\omega_B t}. \qquad (7.2.26)$$

Let us choose a particular solution of the non-homogeneous differential equation (7.2.25) of the form

$$v_x^{(p)} = K_p = const.$$

The requirement that $v_x^{(p)}$ must verify Eq. (7.2.25) leads to

$$K_p\left(\nu^2 + \omega_B^2\right) = \frac{\nu e E}{m} \quad \Rightarrow \quad K_p = v_x^{(p)} = \frac{\nu e E}{m\left(\nu^2 + \omega_B^2\right)},$$

so that the general solution of the non-homogeneous differential equation is

$$v_x = v_x^{(ho)} + v_x^{(p)} = C_2(x,y,z)e^{-\nu t}e^{i\omega_B t}$$
$$+ C_3(x,y,z)e^{-\nu t}e^{-i\omega_B t} + \frac{\nu e E}{m\left(\nu^2 + \omega_B^2\right)}, \qquad (7.2.27)$$

which can also be written as

$$v_x = v_x^{(ho)} + v_x^{(p)} = e^{-\nu t}\Big[C_2(x,y,z)\sin\omega_B t$$
$$+ C_3(x,y,z)\cos\omega_B t\Big] + \frac{\nu e E}{m\left(\nu^2 + \omega_B^2\right)}. \qquad (7.2.28)$$

The first equation of the system (7.2.23) also gives the component v_y of the velocity, which is

$$v_y = \frac{1}{\omega_B}\frac{\partial v_x}{\partial t} + \frac{\nu v_x}{\omega_B} - \frac{eE}{m\omega_B} = -\frac{\nu e^{-\nu t}}{\omega_B}$$
$$\times \Big[C_2(x,y,z)\sin\omega_B t + C_3(x,y,z)\cos\omega_B t\Big]$$
$$+ e^{-\nu t}\Big[C_2(x,y,z)\cos\omega_B t - C_3(x,y,z)\sin\omega_B t\Big]$$
$$+ \frac{\nu e^{-\nu t}}{\omega_B}\Big[C_2(x,y,z)\sin\omega_B t + C_3(x,y,z)\cos\omega_B t\Big]$$
$$+ \frac{\nu^2 e E}{m\omega_B\left(\nu^2 + \omega_B^2\right)} - \frac{eE}{m\omega_B} = e^{-\nu t}\Big[C_2(x,y,z)\cos\omega_B t$$
$$- C_3(x,y,z)\sin\omega_B t\Big] - \frac{eE\omega_B}{m\left(\nu^2 + \omega_B^2\right)}.$$

Therefore, the velocity vector $\vec{v} = (v_x, v_y, v_z)$ of the fluid is

$$
\begin{cases}
v_x(x,y,z,t) = e^{-\nu t}\Big[C_2(x,y,z)\sin\omega_B t \\
\qquad\qquad +C_3(x,y,z)\cos\omega_B t\Big] + \dfrac{eE\nu}{m\big(\nu^2 + \omega_B^2\big)}, \\[2mm]
v_y(x,y,z,t) = e^{-\nu t}\Big[C_2(x,y,z)\cos\omega_B t \\
\qquad\qquad -C_3(x,y,z)\sin\omega_B t\Big] - \dfrac{eE\omega_B}{m\big(\nu^2 + \omega_B^2\big)}, \\[2mm]
v_z(x,y,z,t) = C_1(x,y,z)e^{-\nu t}.
\end{cases}
\tag{7.2.29}
$$

Consider now the simplest case, when the three integration "constants" intervening in the system (7.2.29) are true constants, and use again the approximation $\rho \simeq \rho_{e0} = en_0$. In this case, the time dependence of the current is the same as that of velocity, and we have

$$
\begin{cases}
J_x = en_0\Big[C_2\sin\omega_B t + C_3\cos\omega_B t\Big]e^{-\nu t} + \dfrac{n_0 e^2 E\nu}{m\big(\nu^2 + \omega_B^2\big)}, \\[2mm]
J_y = en_0\Big[C_2\cos\omega_B t - C_3\sin\omega_B t\Big]e^{-\nu t} - \dfrac{n_0 e^2 E\omega_B}{m\big(\nu^2 + \omega_B^2\big)}, \\[2mm]
J_z = en_0 C_1 e^{-\nu t},
\end{cases}
\tag{7.2.30}
$$

where the three arbitrary constants of integration C_i, $i = \overline{1,3}$, are determined from the initial conditions: $J_i(t=0) = 0$, $i = \overline{1,3}$. If these initial conditions are imposed, one obtains the following system of three algebraic equations for the three unknowns C_i, $i = \overline{1,3}$:

$$
\begin{cases}
0 = en_0 C_3 + \dfrac{n_0 e^2 E\nu}{m\big(\nu^2 + \omega_B^2\big)}, \\[2mm]
0 = en_0 C_2 - \dfrac{n_0 e^2 E\omega_B}{m\big(\nu^2 + \omega_B^2\big)}, \\[2mm]
0 = en_0 C_1,
\end{cases}
$$

with the solution

$$
\begin{cases}
C_3 = -\dfrac{eE\nu}{m\big(\nu^2 + \omega_B^2\big)}, \\[2mm]
C_2 = \dfrac{eE\omega_B}{m\big(\nu^2 + \omega_B^2\big)}, \\[2mm]
C_1 = 0.
\end{cases}
$$

By introducing these constants in Eq. (7.2.30), we obtain the final form of the time-dependence of the current $\vec{J} = (J_x,\, J_y,\, J_z)$:

$$
\begin{cases}
J_x = en_0 \left[\dfrac{eE\omega_B}{m(\nu^2 + \omega_B^2)} \sin\omega_B t - \dfrac{eE\nu}{m(\nu^2 + \omega_B^2)} \cos\omega_B t \right] e^{-\nu t} + \dfrac{n_0\nu e^2 E}{m(\nu^2 + \omega_B^2)}, \\[4mm]
J_y = en_0 \left[\dfrac{eE\omega_B}{m(\nu^2 + \omega_B^2)} \cos\omega_B t + \dfrac{eE\nu}{m(\nu^2 + \omega_B^2)} \sin\omega_B t \right] e^{-\nu t} - \dfrac{n_0 e^2 E\omega_B}{m(\nu^2 + \omega_B^2)}, \\[4mm]
J_z = 0,
\end{cases}
$$

or

$$
\begin{cases}
J_x = \dfrac{e^2 E n_0}{m(\nu^2 + \omega_B^2)} \left[\omega_B \sin\omega_B t + \nu\left(e^{\nu t} - \cos\omega_b t\right) \right] e^{-\nu t}, \\[4mm]
J_y = \dfrac{e^2 E n_0}{m(\nu^2 + \omega_B^2)} \left[\nu \sin\omega_B t - \omega_B\left(e^{\nu t} - \cos\omega_B t\right) \right] e^{-\nu t}, \\[4mm]
J_z = 0.
\end{cases}
\qquad (7.2.31)
$$

Below we show the graphical representations of the relations (7.2.31). The drawings were depicted by the help of the following command lines written using the Mathematica 5.0 software package:

$q = -1.6 * 10^{-19};$

$n_0 = 10^{20};$

$m = 9.1 * 10^{-31};$

$EE = 10^{-4};$

$B = 10^{-7};$

$\omega_B = \dfrac{q * B}{m};$

$\nu = 10^2;$

$K = \dfrac{n_0 * q^2 * EE}{m * (\nu^2 + \omega_B^2)};$

$J_x[t_-] := K * e^{-\nu * t}\left(\omega_B * \mathrm{Sin}[\omega_B * t] + \nu * \left(e^{\nu * t} - \mathrm{Cos}[\omega_B * t]\right)\right);$

$J_y[t_-] := K * e^{-\nu * t}\left(\nu * \mathrm{Sin}[\omega_B * t] - \omega_B * \left(e^{\nu * t} - \mathrm{Cos}[\omega_B * t]\right)\right);$

$t_{i1} = 0.0;$

$t_{f1} = 0.9 * 10^{-1};$

The figures are given below. In addition, we graphically give the time dependence of the "total" current

$$
J(t) = \sqrt{J_x^2 + J_y^2 + J_z^2} = \frac{n_0 e^2 E}{m\sqrt{\nu^2 + \omega_B^2}} \sqrt{2e^{-\nu t}\left[\cosh(\nu t) - \cos(\omega_B t)\right]}.
$$

$$
(7.2.32)
$$

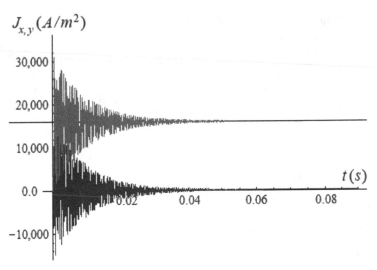

FIGURE 7.1
Graphical representation of time variation of currents $J_x = J_x(t)$ (the curve below) and $J_y = J_y(t)$ given by relations $(7.2.31)_{1,2}$, for $0\,s \le t \le 0.09\,s$.

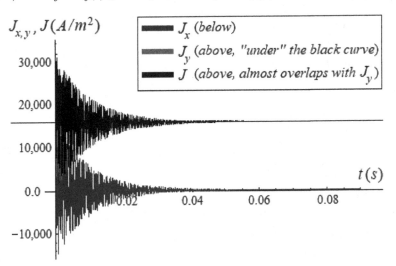

FIGURE 7.2
Graphical representation of time variation of currents $J_x = J_x(t)$ (the curve below), $J_y = J_y(t)$, and $J(t) = \sqrt{J_x(t)^2 + J_y(t)^2 + J_z(t)^2}$ (almost overlaps with J_y) for $0\,s \le t \le 0.09\,s$.

7.3   Problem No. 57

A plasma is placed in a static and homogeneous magnetic field with magnetic induction $\vec{B}_0 = \text{const.}$

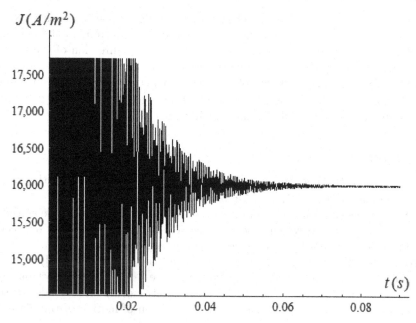

FIGURE 7.3

Graphical representation of time variation of "total" current $J(t) = \sqrt{J_x(t)^2 + J_y(t)^2 + J_z(t)^2}$ for $0\,s \leq t \leq 0.09\,s$.

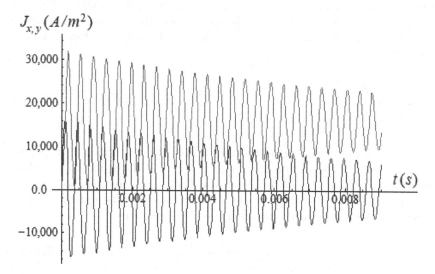

FIGURE 7.4

Graphical representation of time variation of currents $J_x = J_x(t)$ (the curve below), and $J_y = J_y(t)$ given by relations $(7.2.31)_{1,2}$, for $0\,s \leq t \leq 0.009\,s$.

a) Considering the plasma model in which the plasma can be considered a rarefied electronic gas, determine the motion of the electrons in the field of a plane monochromatic wave which propagates in the direction of the static field $\vec{B}_0$ and find the corresponding dispersion relation;

b) Calculate the group velocity of an electromagnetic wave which propagates in plasma, in the direction of the static and homogeneous magnetic field $\vec{B}_0$, and particularize the result for the case in which $|n - 1| \ll 1$.

Solution

a) Without restricting the generality of the problem, let us consider the magnetic field of induction $\vec{B}_0$ as being oriented along the z-axis.

Since the electronic gas is rarefied, the electrons of plasma can be considered as being free, in which case the quasi-elastic force of bonding between electrons and the positive nuclei (ions) can be neglected. In addition, we will also neglect the braking force due to the own electromagnetic field, as well as the magnetic component of the Lorentz force − due to the electromagnetic field of the wave − as compared to the electric component. Under these assumptions, we will write the differential equation of motion of the plasma electrons as follows:

$$m\frac{d^2\vec{r}}{dt^2} = e\vec{E} + e\frac{d\vec{r}}{dt} \times \vec{B}_0, \qquad (7.3.33)$$

where $\vec{E}$ is the electric component of the electromagnetic field of the wave propagating in plasma, and $\vec{B}_0$ is the induction of the external magnetic field in which plasma is placed. Expanding the vector product in Eq. (7.3.33), it becomes

$$m\frac{d^2\vec{r}}{dt^2} = e\vec{E} + e\begin{vmatrix} \vec{i} & \vec{j} & \vec{k} \\ \dot{x} & \dot{y} & \dot{z} \\ 0 & 0 & B_0 \end{vmatrix} = e\vec{E} + e\dot{y}B_0\vec{i} - e\dot{x}B_0\vec{j}, \qquad (7.3.34)$$

where $\vec{i}, \vec{j}, \vec{k}$ are the versors of the coordinate axes. Since the vector product does not have component on z-axis and $E_z = 0$ (see the explanations below), we will leave aside the motion along the z-axis, and continue to deal with the projections on x- and y-axes, of the equations of motion of the electron, written in the form in Eq. (7.3.34). These projections are

$$\ddot{x} = \frac{e}{m}E_x + \frac{eB_0}{m}\dot{y}, \qquad (7.3.37.\text{a})$$

and

$$\ddot{y} = \frac{e}{m}E_y - \frac{eB_0}{m}\dot{x}. \qquad (7.3.37.\text{b})$$

Let us now make the change of variables: $u = x + iy$, $\mathcal{E} = E_x + iE_y$ and

$\mathcal{B} = B_x + iB_y$. Multiplying Eq. (7.3.37.b) by $i = \sqrt{-1}$ and adding member by member the obtained equation with Eq. (7.3.37.a), we obtain

$$\ddot{x} + i\ddot{y} = \frac{e}{m}\left(E_x + iE_y\right) - \frac{eB_0}{m}\left(i\dot{x} - \dot{y}\right) \quad \Leftrightarrow$$

$$\ddot{x} + i\ddot{y} = \frac{e}{m}\left(E_x + iE_y\right) - i\frac{eB_0}{m}\left(\dot{x} + i\dot{y}\right),$$

that is

$$\ddot{u} + i\omega_0\dot{u} = \frac{e}{m}\mathcal{E}, \tag{7.3.38}$$

where we have used the notation

$$\frac{eB_0}{m} \equiv \omega_0.$$

On the other side, the projections of Maxwell's equations

$$\nabla \times \vec{E} = -\frac{\partial \vec{B}}{\partial t}, \tag{7.3.39}$$

and

$$\nabla \times \vec{B} = \mu_0\vec{j} + \frac{1}{c^2}\frac{\partial \vec{E}}{\partial t}, \tag{7.3.40}$$

on the coordinate axes Ox and Oy, are

$$Ox: \left(\nabla \times \vec{E}\right)_x \left(= \frac{\partial E_z}{\partial y} - \frac{\partial E_y}{\partial z} = -\frac{\partial E_y}{\partial z}\right) = -\frac{\partial B_x}{\partial t}, \tag{7.3.39'.a}$$

and

$$Oy: \left(\nabla \times \vec{E}\right)_y \left(= \frac{\partial E_x}{\partial z} - \frac{\partial E_z}{\partial x} = \frac{\partial E_x}{\partial z}\right) = -\frac{\partial B_y}{\partial t}, \tag{7.3.39'.b}$$

(because $E_z = 0$) and, respectively,

$$Ox: \left(\nabla \times \vec{B}\right)_x \left(= \frac{\partial B_z}{\partial y} - \frac{\partial B_y}{\partial z} = -\frac{\partial B_y}{\partial z}\right) = \mu_0 j_x + \frac{1}{c^2}\frac{\partial E_x}{\partial t}, \tag{7.3.40'.a}$$

and

$$Oy: \left(\nabla \times \vec{B}\right)_y \left(= \frac{\partial B_x}{\partial z} - \frac{\partial B_z}{\partial x} = \frac{\partial B_x}{\partial z}\right) = \mu_0 j_y + \frac{1}{c^2}\frac{\partial E_y}{\partial t}, \tag{7.3.40'.b}$$

(because $B_z = B_0 = \text{const.}$).

As specified in the statement of the problem, the plane and monochromatic electromagnetic wave which propagates in the plasma is oriented along the static field $\vec{B}_0$ (as chosen from the beginning). Since the plane and monochromatic electromagnetic wave is a transverse wave, it follows that the electric field intensity of the wave, $\vec{E}$, is oriented in a plane orthogonal to the direction of propagation, that is in the xy−plane; therefore, $E_z = 0$, as we already applied in obtaining the equations (7.3.39′).

Let us now use again the change of variable procedure. Multiplying Eqs. (7.3.39′.b) and (7.3.40′.b) by $i = \sqrt{-1}$ and adding them conveniently to Eqs. (7.3.39′.a) and (7.3.40′.a), we obtain the following equations for the new variables:

$$-\frac{\partial E_y}{\partial z} + \frac{\partial(iE_x)}{\partial z} = -\frac{\partial B_x}{\partial t} - \frac{\partial(iB_y)}{\partial t} \quad \Leftrightarrow$$

$$i\frac{\partial(E_x + iE_y)}{\partial z} = -\frac{\partial(B_x + iB_y)}{\partial t} \quad \Leftrightarrow$$

$$i\frac{\partial \mathcal{E}}{\partial z} = -\frac{\partial \mathcal{B}}{\partial t} \quad \Leftrightarrow \quad \frac{\partial \mathcal{E}}{\partial z} = -i\frac{\partial \mathcal{B}}{\partial t} = 0, \tag{7.3.41}$$

and

$$-\frac{\partial B_y}{\partial z} + \frac{\partial(iB_x)}{\partial z} = \mu_0 j_x + \mu_0 i j_y + \frac{1}{c^2}\frac{\partial E_x}{\partial t} + \frac{1}{c^2}\frac{\partial(iE_y)}{\partial t} \quad \Leftrightarrow$$

$$i\frac{\partial(B_x + iB_y)}{\partial z} = \mu_0(j_x + ij_y) + \frac{1}{c^2}\frac{\partial(E_x + iE_y)}{\partial t} \quad \Leftrightarrow$$

$$i\frac{\partial \mathcal{B}}{\partial z} = \mu_0 \mathcal{J} + \frac{1}{c^2}\frac{\partial \mathcal{E}}{\partial t} \quad \Leftrightarrow \quad \frac{\partial \mathcal{B}}{\partial z} + \frac{i}{c^2}\frac{\partial \mathcal{E}}{\partial t} = -i\mu_0 \mathcal{J}, \tag{7.3.42}$$

where, by repeating the procedure, we have introduced the new variable

$$\mathcal{J} \equiv j_x + ij_y = Ne\dot{x} + iNe\dot{y} = Ne(\dot{x} + i\dot{y}) = eN\dot{u}, \tag{7.3.43}$$

where N is the electron density of the plasma; in the first approximation, if plasma is rarefied, these electrons can be considered as being free.

This way, we have obtained the following system of three coupled differential equations for u, $\mathcal{E}$ and $\mathcal{B}$:

$$\begin{cases} \ddot{u} + i\omega_0 \dot{u} = \dfrac{e}{m}\mathcal{E}, \\[2mm] \dfrac{\partial \mathcal{E}}{\partial z} - i\dfrac{\partial \mathcal{B}}{\partial t} = 0, \\[2mm] \dfrac{\partial \mathcal{B}}{\partial z} + \dfrac{i}{c^2}\dfrac{\partial \mathcal{E}}{\partial t} = -i\mu_0 Ne\dot{u}, \end{cases} \tag{7.3.44}$$

where the relation (7.3.43) have been used.

Taking into account the problem requirements, we will look for the unknowns u, $\mathcal{E}$ and $\mathcal{B}$ of the system (7.3.44) in the form of plane waves of the same pulsation ω and the same wave number k (monochromatic plane waves):

$$\begin{cases} u = Ue^{\pm i(kz - \omega t)}, \\[1mm] \mathcal{E} = Ve^{\pm i(kz - \omega t)}, \\[1mm] \mathcal{B} = We^{\pm i(kz - \omega t)}. \end{cases} \tag{7.3.45}$$

As required by the problem statement, in the relations (7.3.45) we have considered the particular case of propagation of the monochromatic plane wave in the direction of the static field $\vec{B}_0$ (that is oriented along the z-axis). Requiring to the relations (7.3.45) to verify the system of differential equations (7.3.44), one obtains the following linear and homogeneous system of algebraic equations for the unknowns U, V and W:

$$\begin{cases} -\omega^2 U + i\omega_0(\mp i\omega U) = \dfrac{e}{m}V, \\ \pm ikV \mp \omega W = 0, \\ \pm ikW + \dfrac{i}{c^2}(\mp i\omega V) = -i\mu_0 Ne(\mp i\omega U), \end{cases}$$

or

$$\begin{cases} -\omega^2 U \pm \omega\omega_0 U = \dfrac{e}{m}V, \\ ikV = \omega W, \\ ikW + \dfrac{\omega}{c^2}V = -\mu_0 Ne\omega U, \end{cases}$$

which gives

$$\begin{cases} \left(-\omega^2 \pm \omega\omega_0\right)U - \dfrac{e}{m}V + 0 \cdot W = 0, \\ 0 \cdot U + ikV - \omega W = 0, \qquad\qquad\qquad (7.3.46) \\ \mu_0 Ne\omega U + \dfrac{\omega}{c^2}V + ikW = 0. \end{cases}$$

In order to have solutions different from the trivial solution, the determinant of the matrix of the system (7.3.46) must be zero, that is,

$$\begin{vmatrix} -\omega^2 \pm \omega\omega_0 & \dfrac{-e}{m} & 0 \\ 0 & ik & -\omega \\ \mu_0 Ne\omega & \dfrac{\omega}{c^2} & ik \end{vmatrix} = 0,$$

or

$$\left(\dfrac{\omega^2}{c^2} - k^2\right)\left(-\omega^2 \pm \omega\omega_0\right) + \dfrac{\mu_0 Ne^2\omega^2}{m} = 0,$$

and, still

$$\left(1 - \dfrac{k^2c^2}{\omega^2}\right)\left(\omega^2 \mp \omega\omega_0\right) = \dfrac{Ne^2}{\varepsilon_0 m},$$

that is

$$\dfrac{k^2c^2}{\omega^2}\left(= \dfrac{c^2}{v_f^2} = n^2\right) = 1 - \dfrac{\dfrac{Ne^2}{\varepsilon_0 m}}{\omega^2 \mp \omega\omega_0} \quad\Rightarrow$$

$$n^2 = 1 - \dfrac{Ne^2}{\varepsilon_0 m\omega(\omega \mp \omega_0)}, \qquad\qquad (7.3.47)$$

or, by introducing the notation

$$\frac{Ne^2}{\varepsilon_0 m} \equiv \omega_p^2$$

(the square of the plasma pulsation),

$$n^2 = 1 - \frac{\omega_p^2}{\omega^2 \mp \omega\omega_0},$$

which is the required *dispersion relation*. In formula (7.3.47), $v_f = \omega/k$ is the *phase velocity* of the wave.

To find the equation of motion of the electrons of plasma, we have to determine u. Taking the ratio of the first two relations of the system (7.3.45), we obtain

$$u = \frac{U}{V}\mathcal{E},$$

or, if the ratio U/V is expressed by means of the first relation of the system (7.3.46), then we have

$$u = \frac{U}{V}\mathcal{E} = \frac{e/m}{-\omega^2 \pm \omega\omega_0}\mathcal{E} = -\frac{e\mathcal{E}}{m\omega(\omega \mp \omega_0)}. \qquad (7.3.48)$$

Since $u = x + iy$ and $\mathcal{E} = E_x + iE_y$, by separating the real and imaginary parts of Eq. (7.3.48) one finds

$$x = -\frac{eE_x}{m\omega(\omega \mp \omega_0)},$$

and

$$y = -\frac{eE_y}{m\omega(\omega \mp \omega_0)}.$$

Including now the motion along the static magnetic field, it follows that the vector form of the equation of motion of the plasma electrons is

$$\vec{r} = -\frac{e\vec{E}}{m\omega(\omega \mp \omega_0)} = -\frac{e}{m}\frac{\vec{E}}{\omega^2 \mp \omega\frac{eB_0}{m}}. \qquad (7.3.49)$$

Observation. If the induction of the static magnetic field in which the plasma is placed equals zero, $\vec{B}_0 = 0$, the relations (7.3.47) and (7.3.49) reduce to their form corresponding to a plasma which is placed in a domain where there is no external field,

$$\varepsilon_r(\simeq n^2) = 1 - \frac{Ne^2}{\varepsilon_0 m\omega^2},$$

(because for rarefied gases one can approximate $\mu_r \simeq 1$), in which case

$$\vec{r} = -\frac{e\vec{E}}{m\omega^2}.$$

b) According to the definition of the *group velocity*,

$$v_g = \frac{\partial \omega}{\partial k},$$

we have

$$\frac{1}{v_g} = \frac{\partial k}{\partial \omega} \frac{\partial}{\partial \omega} \left(\frac{n\omega}{c}\right) = \frac{n}{c} + \frac{\omega}{c}\frac{\partial n}{\partial \omega}. \tag{7.3.50}$$

From relation (7.3.47) it follows that

$$n = \sqrt{1 - \frac{Ne^2}{\varepsilon_0 m\omega(\omega \mp \omega_0)}} \simeq 1 - \frac{Ne^2}{2\varepsilon_0 m\omega(\omega \mp \omega_0)},$$

where the approximation

$$\sqrt{1 - \xi} = (1 - \xi)^{\frac{1}{2}} \simeq 1 - \frac{\xi}{2}, \quad |\xi| < 1$$

was used. The validity of this approximation is ensured in this case by the fact that the density of the free electrons (in fact, the density of the polarized atoms) of a gas, N, is relatively small. This approximation is given in the problem statement, under the following equivalent mathematical formulation: $|n - 1| \ll 1$. We have

$$\frac{\partial n}{\partial \omega} = \frac{\partial}{\partial \omega}\left[1 - \frac{Ne^2}{2\varepsilon_0 m\omega(\omega \mp \omega_0)}\right]$$

$$= -\frac{Ne^2}{2\varepsilon_0 m}\frac{\partial}{\partial \omega}\left[\frac{1}{\omega(\omega \mp \omega_0)}\right] = -\frac{\omega_p^2}{2}\frac{\partial}{\partial \omega}\left[\frac{1}{\omega(\omega \mp \omega_0)}\right]$$

$$= -\frac{\omega_p^2}{2}\left[-\frac{(\omega \mp \omega_0) + \omega}{\omega^2(\omega \mp \omega_0)^2}\right] = \frac{\omega_p^2}{2}\frac{2\omega \mp \omega_0}{\omega^2(\omega \mp \omega_0)^2},$$

and, if this result is introduced into relation (7.3.50), then one obtains

$$\frac{1}{v_g} = \frac{n}{c} + \frac{\omega}{c}\frac{\partial n}{\partial \omega} = \frac{1}{c}\left[1 - \frac{Ne^2}{2\varepsilon_0 m\omega(\omega \mp \omega_0)}\right] + \frac{\omega\omega_p^2}{2c}\frac{2\omega \mp \omega_0}{\omega^2(\omega \mp \omega_0)^2}$$

$$= \frac{1}{c}\left[1 - \frac{\omega_p^2}{2\omega(\omega \mp \omega_0)}\right] + \frac{\omega\omega_p^2}{2c}\frac{2\omega \mp \omega_0}{\omega^2(\omega \mp \omega_0)^2}$$

$$= \frac{1}{c}\left[1 - \frac{\omega_p^2}{2\omega(\omega \mp \omega_0)} + \frac{\omega_p^2(2\omega \mp \omega_0)}{2\omega(\omega \mp \omega_0)^2}\right]$$

$$= \frac{1}{c}\left[1 - \frac{\omega_p^2(\omega \mp \omega_0)}{2\omega(\omega \mp \omega_0)^2} + \frac{\omega_p^2(2\omega \mp \omega_0)}{2\omega(\omega \mp \omega_0)^2}\right] = \frac{1}{c}\left[1 + \frac{\omega_p^2}{2(\omega \mp \omega_0)^2}\right],$$

which finally leads to

$$v_g = c \frac{1}{1 + \dfrac{\omega_p^2}{2(\omega \mp \omega_0)^2}} = c \frac{1}{1 + \frac{1}{2}\left(\dfrac{\omega_p}{\omega \mp \omega_0}\right)^2} < c. \qquad (7.3.51)$$

7.4 Problem No. 58

A conducting fluid, homogeneous, viscous and incompressible is placed between the conducting $z = 0$ and $z = d$ planes. The $z = d$ plane moves along
x-axis with the constant velocity v_0. Determine the velocity distribution in the
stationary moving fluid, if a uniform magnetic field $\vec{H}_0$ acts along the z-axis,
and a uniform electric field $\vec{E}_0$ acts along the y-axis.

Solution

According to the elementary theory of magneto-fluido-dynamics, the equation
of motion of a conductor, perfectly viscous (Newtonian) fluid is

$$\rho_m \vec{a} = \rho_m \vec{F} - \operatorname{grad} p + (\xi + \eta)\operatorname{grad}\theta' + \eta\Delta\vec{v} + \vec{j} \times \vec{B}, \qquad (7.4.52)$$

where ρ_m is the mass density of the fluid,

$$\vec{a} = \frac{\partial \vec{v}}{\partial t} + (\vec{v} \cdot \nabla)\vec{v}$$

is the fluid (particle) acceleration, $\vec{F}$ is the specific mass force of non-
electromagnetic nature (if, e.g., the fluid is in the gravitational field, this is
just the gravitational acceleration, $\vec{g}$), ξ and η are the coefficients of dynamic
viscosity, and

$$\theta' \equiv \operatorname{div}\vec{v} \equiv \nabla \cdot \vec{v}.$$

Using the equation of continuity for the fluid mass, it can be easily shown
that the equation

$$\nabla \cdot \vec{v} = \theta' = 0$$

expresses the fact that the fluid is homogeneous and incompressible. Indeed,
if $\rho_m(\vec{r}, t) = const.$, the equation of continuity written in the form

$$\frac{d\rho_m}{dt} + \rho_m \operatorname{div}\vec{v} = 0,$$

yields $\nabla \cdot \vec{v} \equiv \operatorname{div}\vec{v} = 0$.

According to the problem statement, it follows that the acceleration of the
fluid particle is also considered zero and, in addition, if we neglect the effect

of gravitation as compared to that of the electric and magnetic fields (a very plausible assumption, if we think of the huge difference between the coupling constants of the gravitational and electromagnetic fields), then the equation of motion (7.4.52) receives a simpler form

$$\operatorname{grad} p = \eta \Delta \vec{v} + \vec{j} \times \vec{B}. \tag{7.4.53}$$

Here the current density of the conductive fluid, $\vec{j}$, is connected to its electrical conductivity λ by the relation

$$\vec{j} = \lambda(\vec{E} + \vec{v} \times \vec{B}),$$

(the Ohm's law). Written by components, this relation leads to

$$\begin{cases} j_x = \lambda E_x + \lambda(\vec{v} \times \vec{B})_x = \lambda E_x + \lambda(v_y B_z - v_z B_y) = 0, \\ j_y = \lambda E_y + \lambda(\vec{v} \times \vec{B})_y = \lambda E_y + \lambda(v_z B_x - v_x B_z) \\ \quad = \lambda E_0 - \mu_0 \lambda v_x(z) H_0, \\ j_z = \lambda E_z + \lambda(\vec{v} \times \vec{B})_z = \lambda E_z + \lambda(v_x B_y - v_y B_x) = 0, \end{cases}$$

where we took into account that the fluid moves along the x-axis, while its velocity at any point depends only on the position of that point between the two planes, that is – considering the symmetry of the problem – only on coordinate z:

$$\vec{v} = v_x \vec{u}_x = v_x(z) \vec{u}_x.$$

Since

$$\begin{cases} (\vec{j} \times \vec{B})_x = j_y B_z - j_z B_y = j_y B_0 = -\mu_0 j_y H_0 \\ \quad = \mu_0 \lambda H_0 \Big[E_0 - \mu_0 v_x(z) H_0 \Big], \\ (\vec{j} \times \vec{B})_y = j_z B_x - j_x B_z = 0, \\ (\vec{j} \times \vec{B})_z = j_x B_y - j_y B_x = -j_y \mu_0 H_x \\ \quad = -\mu_0 \lambda H_x \Big[E_0 - \mu_0 v_x(z) H_0 \Big], \end{cases}$$

the projections on the coordinate axes of the equation of motion (7.4.53) are

$$\begin{cases} \dfrac{\partial p}{\partial x} = \eta \underbrace{\dfrac{\partial^2 v_x}{\partial z^2}}_{=\frac{d^2 v_x}{dz^2}} + \mu_0 \lambda H_0 (E_0 - \mu_0 v_x H_0), \\ \dfrac{\partial p}{\partial y} = 0, \\ \dfrac{\partial p}{\partial z} = -\mu_0 \lambda H_0 (E_0 - \mu_0 v_x H_0). \end{cases} \tag{7.4.54}$$

Since, due to the symmetry related to the geometry of the problem, there is no pressure gradient along the x-axis, $\frac{\partial p}{\partial x} = 0$, the formula $(7.4.54)_1$ gives the equation for the velocity distribution in the fluid, $v \equiv v_x = v_x(z)$:

$$\frac{d^2 v_x}{dz^2} - \frac{\lambda \mu_o^2 H_0^2}{\eta} v_x = -\frac{\lambda \mu_0 H_0 E_0}{\eta},$$

or, if the notation

$$\frac{\lambda \mu_0^2 H_0^2}{\eta} \equiv \frac{1}{d_0^2},$$

is introduced, then we have

$$\frac{d^2 v_x}{dz^2} - \frac{1}{d_0^2} v_x = -\frac{1}{d_0^2} \frac{E_0}{\mu_0 H_0}. \tag{7.4.55}$$

We therefore have to solve the Cauchy's problem for Eq. (7.4.55) with the "initial" conditions $v_x(z = 0) = 0$ and $v_x(z = d) = v_0$. Equation (7.4.55) is an ordinary, non-homogeneous, second-order differential equation with constant coefficients. The general solution of this equation is given by the sum between the general solution of the attached homogeneous differential equation and a particular solution of the non-homogeneous differential equation which, in the general case, can be determined by the Lagrange's method of variation of parameters. But in our case this particular solution can be found much easier, by choosing it in the form of the term that gives the non-homogeneity (the right member of the equation), that is as a constant. The homogeneous differential equation attached to the non-homogeneous differential equation writes

$$\frac{d^2 v_x}{dz^2} - \frac{1}{d_0^2} v_x = 0, \tag{7.4.56}$$

and it has the following attached characteristic equation:

$$r^2 - \frac{1}{d_0^2} = 0.$$

The above algebraic equation has two real solutions, namely,

$$r_{1,2} = \pm \frac{1}{d_0}.$$

Therefore, the general solution of the homogeneous differential equation (7.4.56) can be written as

$$v_x^{(ho)}(z) = V_1 e^{\frac{z}{d_0}} + V_2 e^{-\frac{z}{d_0}}, \tag{7.4.57}$$

where V_1 and V_2 are two arbitrary integration constants, or, under an equivalent form,

$$v_x^{(ho)}(z) = W_1 \sinh \frac{z}{d_0} + W_2 \cosh \frac{z}{d_0}, \tag{7.4.58}$$

where W_1 and W_2 are two other arbitrary integration constants, related to V_1 and V_2 by the definition relationships,

$$\sinh x = \frac{e^x - e^{-x}}{2},$$

and

$$\cosh x = \frac{e^x + e^{-x}}{2}.$$

Choosing the form in Eq. (7.4.58) for the solution of the homogeneous differential equation (7.4.56), we will look for a particular solution of the non-homogeneous equation of the form $v_x^{(p)} = W_3$, where W_3 is a nonzero, real constant. Asking this particular solution to verify the non-homogeneous differential equation (7.4.55), the result is

$$-\frac{1}{d_0^2} v_x^{(p)} \left(= -\frac{1}{d_0^2} W_3 \right) = -\frac{1}{d_0^2} \frac{E_0}{\mu_0 H_0} \quad \Rightarrow \quad W_3 = \frac{E_0}{\mu_0 H_0}.$$

The general solution of the non-homogeneous differential equation (7.4.55) then is

$$v_x(z) = v_x^{(ho)}(z) + v_x^{(p)} = W_1 \sinh \frac{z}{d_0} + W_2 \cosh \frac{z}{d_0} + \frac{E_0}{\mu_0 H_0}, \qquad (7.4.59)$$

where the arbitrary constants of integration are deduced from the boundary conditions $v_x(z = 0) = 0$ and $v_x(z = d) = v_0$:

$$\begin{cases} v_x(0) = W_2 + \dfrac{E_0}{\mu_0 H_0} = 0, \\ v_x(d) = W_1 \sinh \frac{d}{d_0} + W_2 \cosh \frac{d}{d_0} + \dfrac{E_0}{\mu_0 H_0} = v_0. \end{cases}$$

The first equation of the above system gives

$$W_2 = -\frac{E_0}{\mu_0 H_0},$$

in which case the second equation leads to

$$W_1 \sinh \frac{d}{d_0} - \frac{E_0}{\mu_0 H_0} \cosh \frac{d}{d_0} + \frac{E_0}{\mu_0 H_0} = v_0,$$

from where we get

$$W_1 = \frac{v_0 - \dfrac{E_0}{\mu_0 H_0} \left(1 - \cosh \dfrac{d}{d_0} \right)}{\sinh \dfrac{d}{d_0}}.$$

Introducing the just determined constants W_1 and W_2 into relation (7.4.59), we have

$$
v_x(z) = \frac{v_0 - \dfrac{E_0}{\mu_0 H_0}\left(1 - \cosh\dfrac{d}{d_0}\right)}{\sinh\dfrac{d}{d_0}}\sinh\frac{z}{d_0}
$$

$$
- \frac{E_0}{\mu_0 H_0}\cosh\frac{z}{d_0} + \frac{E_0}{\mu_0 H_0} = \frac{v_0}{\sinh\frac{d}{d_0}}\sinh\frac{z}{d_0}
$$

$$
+ \frac{E_0}{\mu_0 H_0}\left(1 - \cosh\frac{z}{d_0} - \frac{1 - \cosh\frac{d}{d_0}}{\sinh\frac{d}{d_0}}\sinh\frac{z}{d_0}\right)
$$

$$
= \frac{v_0}{\sinh\frac{d}{d_0}}\sinh\frac{z}{d_0} + \frac{E_0}{\mu_0 H_0}
$$

$$
\times\left(1 - \frac{\sinh\frac{d}{d_0}\cosh\frac{z}{d_0} - \sinh\frac{z}{d_0}\cosh\frac{d}{d_0} + \sinh\frac{z}{d_0}}{\sinh\frac{d}{d_0}}\right)
$$

$$
= \frac{v_0}{\sinh\frac{d}{d_0}}\sinh\frac{z}{d_0} + \frac{E_0}{\mu_0 H_0}\left(1 - \frac{\sinh\frac{d-z}{d_0} + \sinh\frac{z}{d_0}}{\sinh\frac{d}{d_0}}\right),
$$

where we have used the formula of trigonometry of hyperbolic functions

$$
\sinh(x - y) = \sinh x \cosh y - \sinh y \cosh x.
$$

Therefore, the expression of the fluid velocity we are looking for is

$$
v_x(z) = v_0\frac{\sinh\frac{z}{d_0}}{\sinh\frac{d}{d_0}} + \frac{E_0}{\mu_0 H_0}\left(1 - \frac{\sinh\frac{d-z}{d_0} + \sinh\frac{z}{d_0}}{\sinh\frac{d}{d_0}}\right). \qquad (7.4.60)
$$

For

$$
d \ll d_0 = \sqrt{\frac{\eta}{\lambda}\frac{1}{\mu_0 H_0}},
$$

(which means weak magnetic fields), by using the Mac-Laurin series expansion formula of the hyperbolic sinus,

$$
\sinh x = \frac{x}{1!} + \frac{x^3}{3!} + \cdots + \frac{x^{2n+1}}{(2n+1)!} + \cdots, \quad \forall x \in \mathbb{R},
$$

(which is convergent for any $x \in \mathbb{R}$), we can approximate $\sinh x \simeq x$, in which case, the relationship (7.4.60) yields $v_x(z) = v_0 z/d$, which means that in the case of weak magnetic fields, the velocity increases linearly along the distance between the two planes, from $v_x = 0$ for $z = 0$, to $v_x = v_0$ for $z = d$.

If, on the contrary, the magnetic field is strong, meaning that

$$
d \gg d_0 = \sqrt{\frac{\eta}{\lambda}\frac{1}{\mu_0 H_0}},
$$

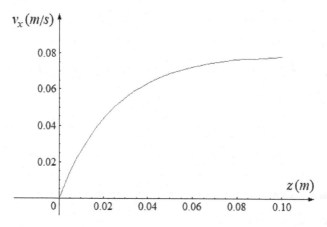

FIGURE 7.5

Graphical representation of the fluid velocity $v_x = v_x(z)$ as a function of the distance between the planes $z = 0$ and $z = d$.

then

$$\frac{\sinh \frac{z}{d_0}}{\sinh \frac{d}{d_0}} \simeq 0,$$

$$\frac{\sinh \frac{d-z}{d_0}}{\sinh \frac{d}{d_0}} = \frac{e^{\frac{d-z}{d_0}} - e^{\frac{z-d}{d_0}}}{e^{\frac{d}{d_0}} - e^{-\frac{d}{d_0}}} = \frac{e^{-\frac{z}{d_0}} - e^{-\frac{2d-z}{d_0}}}{1 - e^{-\frac{2d}{d_0}}} \simeq e^{-\frac{z}{d_0}},$$

in which case the relationship (7.4.60) gives

$$v_x(z) = \frac{E_0}{\mu_0 H_0} \left(1 - e^{-\frac{z}{d_0}}\right) = \frac{E_0}{\mu_0 H_0} \left(1 - e^{-\frac{\mu_0 H_0 \sqrt{\lambda}}{\sqrt{\eta}} z}\right),$$

which means that in the case of strong magnetic fields the velocity varies exponentially with the distance z, measured from the $z = 0$, where $v_x = 0$ (see Fig. 7.5, which shows a graphic representation of the velocity as a function of the distance between the two planes, $v_x = v_x(z)$, for the particular case when $E_0 = 10^{-3} \mathrm{V} \cdot \mathrm{m}^{-1}$, $H_0 = 10^4 \mathrm{A} \cdot \mathrm{m}^{-1}$, $\lambda = 10^4 \Omega^{-1} \cdot \mathrm{m}^{-1}$, and $\eta = 10^{-3} \mathrm{N} \cdot \mathrm{s} \cdot \mathrm{m}^{-2}$).

8

Special Theory of Relativity: Relativistic Kinematics

8.1 Problem No. 59

Show that the wave equation

$$\frac{\partial^2 \psi}{\partial x^2} - \frac{1}{c^2}\frac{\partial^2 \psi}{\partial x^2} = 0 \qquad (8.1.1)$$

is covariant with respect to the Lorentz transformations

$$\begin{cases} x = \Gamma\left(x' + V\,t'\right), \\ y = y', \\ z = z', \\ t = \Gamma\left(t' + \dfrac{V}{c^2}\,x'\right). \end{cases} \qquad (8.1.2)$$

Solution

We have

$$\frac{\partial \psi}{\partial x'} = \frac{\partial \psi}{\partial x}\frac{\partial x}{\partial x'} + \frac{\partial \psi}{\partial t}\frac{\partial t}{\partial x'} = \Gamma\frac{\partial \psi}{\partial x} + \Gamma\frac{V}{c^2}\frac{\partial \psi}{\partial t}.$$

Thus, the operator "correspondence" is

$$\frac{\partial}{\partial x'} = \Gamma\frac{\partial}{\partial x} + \Gamma\frac{V}{c^2}\frac{\partial}{\partial t}, \qquad (8.1.3)$$

which leads to

$$\frac{\partial^2 \psi}{\partial x'^2} = \left(\Gamma\frac{\partial}{\partial x} + \Gamma\frac{V}{c^2}\frac{\partial}{\partial t}\right)\left(\Gamma\frac{\partial \psi}{\partial x} + \Gamma\frac{V}{c^2}\frac{\partial \psi}{\partial t}\right)$$

$$= \Gamma^2\frac{\partial^2 \psi}{\partial x^2} + 2\Gamma^2\frac{V}{c^2}\frac{\partial^2 \psi}{\partial x\partial t} + \Gamma^2\frac{V^2}{c^4}\frac{\partial^2 \psi}{\partial t^2}. \qquad (8.1.4)$$

In the same way

$$\frac{\partial \psi}{\partial t'} = \frac{\partial \psi}{\partial x}\frac{\partial x}{\partial t'} + \frac{\partial \psi}{\partial t}\frac{\partial t}{\partial t'} = \Gamma V\frac{\partial \psi}{\partial x} + \Gamma\frac{\partial \psi}{\partial t},$$

DOI: 10.1201/9781003402602-8

with the operator "correspondence"

$$\frac{\partial}{\partial t'} = \Gamma V \frac{\partial}{\partial x} + \Gamma \frac{\partial}{\partial t}. \tag{8.1.5}$$

Next,

$$\frac{\partial^2 \psi}{\partial t'^2} = \left(\Gamma V \frac{\partial}{\partial x} + \Gamma \frac{\partial}{\partial t} \right) \left(\Gamma V \frac{\partial \psi}{\partial x} + \Gamma \frac{\partial \psi}{\partial t} \right)$$

$$= \Gamma^2 V^2 \frac{\partial^2 \psi}{\partial x^2} + 2 \Gamma^2 V \frac{\partial^2 \psi}{\partial x \partial t} + \Gamma^2 \frac{\partial^2 \psi}{\partial t^2}. \tag{8.1.6}$$

It is easily seen that

$$\frac{\partial^2 \psi}{\partial x'^2} - \frac{1}{c^2} \frac{\partial^2 \psi}{\partial t'^2} = \frac{\partial^2 \psi}{\partial x^2} - \frac{1}{c^2} \frac{\partial^2 \psi}{\partial t^2},$$

which concludes the proof.

8.2   Problem No. 60

Denoting $\frac{V}{c} = \tanh \theta$ and $u = ct$, show that the Lorentz transformation

$$\begin{cases} x' = \Gamma \left(x - Vt \right), \\ y' = y, \\ z' = z, \\ t' = \Gamma \left(t - \dfrac{V}{c^2} x \right), \end{cases}$$

can be written as

$$\begin{cases} x' = x \cosh \theta - u \sinh \theta, \\ u' = - x \sinh \theta + u \cosh \theta. \end{cases} \tag{8.2.7}$$

Using this result, show that a succession of Lorentz transformations, performed in the same direction (see Fig. 8.1), is also a Lorentz transformation.

Solution

Using the formulas

$$\cosh^2 \theta - \sinh^2 \theta = 1,$$

$$\cosh \theta = \frac{1}{\sqrt{1 - \tanh^2 \theta}} = \frac{1}{\sqrt{1 - \frac{V^2}{c^2}}} = \Gamma,$$

$$\sinh \theta = \sqrt{\cosh^2 \theta - 1} = \frac{V}{c} \Gamma, \tag{8.2.8}$$

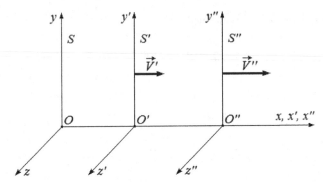

FIGURE 8.1
Schematic representation of three inertial reference frames referred to in
Problem No. 60.

we can write

$$x' = \Gamma\left(x - \frac{V}{c}ct\right) = x\cosh\theta - u\sinh\theta,$$

$$ct' = u' = \Gamma\left(u - \frac{V}{c}x\right) = -x\sinh\theta + u\cosh\theta,$$

and relations (8.2.7) are proved.

Next, let us consider three inertial frames S, S', and S'', and suppose that
the velocity of S' with respect to S is V, while S'' moves with velocity V'
with respect to S', and V'' with respect to S.

One observes that the Lorentz transformation given by Eq. (8.2.7) can be
written in a matrix form as

$$X' = BX,$$

where

$$X = \begin{pmatrix} x \\ u \end{pmatrix}, \quad X' = \begin{pmatrix} x' \\ u' \end{pmatrix}, \quad B = \begin{pmatrix} \cosh\theta & -\sinh\theta \\ -\sinh\theta & \cosh\theta \end{pmatrix}, \qquad (8.2.9)$$

Using this representation, we also have $X'' = AX'$, so that the total transformation is

$$X'' = AX' = ABX = CX,$$

where C is given by

$$C = AB = \begin{pmatrix} \cosh\theta' & -\sinh\theta' \\ -\sinh\theta' & \cosh\theta' \end{pmatrix} \begin{pmatrix} \cosh\theta & -\sinh\theta \\ -\sinh\theta & \cosh\theta \end{pmatrix}.$$

Since the element c_{ij} of the matrix product AB is

$$c_{ij} = \sum_{k=1}^{2} a_{ik}b_{kj}, \quad i, j = 1, 2,$$

and in view of the trigonometric formulas

$$\sinh\left(\theta + \theta'\right) = \sinh\theta\cosh\theta' + \cosh\theta\sinh\theta',$$
$$\cosh\left(\theta + \theta'\right) = \cosh\theta\cosh\theta' + \sinh\theta\sinh\theta',$$

we obtain

$$C = \begin{pmatrix} \cosh\left(\theta + \theta'\right) & -\sinh\left(\theta + \theta'\right) \\ -\sinh\left(\theta + \theta'\right) & \cosh\left(\theta + \theta'\right) \end{pmatrix}$$

$$= \begin{pmatrix} \cosh\theta'' & -\sinh\theta'' \\ -\sinh\theta'' & \cosh\theta'' \end{pmatrix}, \tag{8.2.10}$$

where $\theta'' = \theta + \theta'$. The parameter θ is called *rapidity*. This name was given in 1911 by the English physicist Alfred Robb (1873–1936), being an alternative to speed as a method of measuring motion. For low speeds, rapidity and speed are proportional, but for high speeds rapidity becomes very large and becomes infinite for $v = c$.

Returning to our problem, formula (8.2.10) shows that the successive application of two Lorentz transformations is also a Lorentz transformation. Note that the rapidity does not obey the rule of relativistic transformation of coordinates and time.

8.3   Problem No. 61

A particle moves with respect to the inertial frame S' with the constant velocity $\vec{v}\,'$ that makes an angle θ' with $Ox \equiv O'x'$. Determine the angle θ between the trajectory of the particle and $Ox \equiv O'x'$ in the inertial frame S.

Solution

Without any loss of generality, we can study the motion of the particle in the xy-plane (see Fig. 8.2).

The components of the velocity of the particle in the two frames are as follows:

$$\begin{cases} v_x = v\,\cos\theta, \\ v_y = v\,\sin\theta, \end{cases}$$

in S and

$$\begin{cases} v'_x = v'\,\cos\theta', \\ v'_y = v'\,\sin\theta', \end{cases}$$

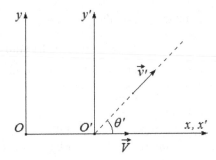

FIGURE 8.2
A particle moving with the constant velocity $\vec{v}\,'$ that makes the angle θ' with $Ox \equiv O'x'$.

in S'. But

$$\begin{cases} v_x = \dfrac{v_x' + V}{1 + \frac{V}{c^2}v_x'}, \\[4mm] v_y = \dfrac{v_y'}{\Gamma\left(1 + \frac{V}{c^2}v_x'\right)}, \end{cases}$$

which gives

$$\begin{cases} \tan\theta = \dfrac{v_y}{v_x} = \dfrac{v_y'}{\Gamma(v_x' + V)} = \dfrac{1}{\Gamma}\dfrac{\sin\theta'}{\cos\theta' + \frac{V}{v'}}, \\[4mm] \tan\theta' = \dfrac{1}{\Gamma}\dfrac{\sin\theta}{\cos\theta - \frac{V}{v}}. \end{cases} \qquad (8.3.11)$$

Then

$$\cos\theta = \dfrac{1}{\sqrt{1 + \tan^2\theta}} = \dfrac{\cos\theta' + \frac{V}{c}}{1 + \frac{V}{c}\cos\theta'}, \qquad (8.3.12)$$

which is the formula expressing the *relativistic Doppler effect*. We also have

$$\sin\theta' = \dfrac{\tan\theta'}{\sqrt{1 + \tan^2\theta'}} = \dfrac{\sqrt{1 - \frac{V^2}{c^2}}\sin\theta}{1 - \frac{V}{c}\cos\theta}. \qquad (8.3.13)$$

If $\dfrac{V}{c} \ll 1$, we may take $\Gamma \approx 1$ and write

$$\sin\theta' \simeq \sin\theta\left(1 - \frac{V}{c}\cos\theta\right)^{-1} \simeq \sin\theta + \frac{V}{c}\sin\theta\cos\theta,$$

or

$$2\sin\dfrac{\theta' - \theta}{2}\cos\dfrac{\theta' + \theta}{2} = \dfrac{V}{c}\sin\theta\cos\theta. \qquad (8.3.14)$$

Since $\theta' \simeq \theta$, one can approximate

$$\theta' + \theta = 2\theta.$$

Denoting

$$\theta' - \theta \equiv \alpha,$$

we also realize that α is very small, so that we may take

$$\sin \frac{\alpha}{2} \approx \frac{\alpha}{2},$$

and Eq. (8.3.14) becomes

$$\alpha \simeq \frac{V}{c} \sin \theta. \tag{8.3.15}$$

If the light is coming from an object situated at the zenith of the observer ($\theta = \pi/2$), we finally arrive at the formula for the *aberration of light*,

$$\alpha \simeq \frac{V}{c}.$$

8.4   Problem No. 62

Maxwell's equations are covariant under the Lorentz transformation of frames

$$\begin{cases} x' = \Gamma(x - Vt), \\ y' = y, \\ z' = z, \\ t' = \Gamma\left(t - \dfrac{V}{c^2}x\right). \end{cases}$$

Find the transformation properties of the electromagnetic field vectors $\vec{E}$ and $\vec{B}$, in vacuum.

Solution

Following Einstein, let us use Maxwell's source-free equations

$$\frac{\partial \vec{B}}{\partial t} = -\nabla \times \vec{E}, \quad \nabla \cdot \vec{B} = 0. \tag{8.4.16}$$

If the motion of the inertial frame takes place along the $Ox \equiv O'x'$ axis, then

$$\frac{\partial}{\partial z} = \frac{\partial}{\partial z'}, \quad \frac{\partial}{\partial y} = \frac{\partial}{\partial y'},$$

and so, in the frame S', we can write

$$\begin{cases} \dfrac{\partial B'_x}{\partial t'} = \dfrac{\partial E'_y}{\partial z} - \dfrac{\partial E'_z}{\partial y}, \\[2mm] \dfrac{\partial B'_y}{\partial t'} = \dfrac{\partial E'_z}{\partial x'} - \dfrac{\partial E'_x}{\partial z}, \\[2mm] \dfrac{\partial B'_z}{\partial t'} = \dfrac{\partial E'_x}{\partial y} - \dfrac{\partial E'_y}{\partial x'}. \end{cases} \tag{8.4.17}$$

Since $\vec{E}'$ and $\vec{B}'$ are functions of x' and t', we can use the operators in Eqs. (8.1.3) and (8.1.5) to re-write Eq. (8.4.17); we obtain

$$
\begin{cases}
\Gamma V \dfrac{\partial B'_x}{\partial x} + \Gamma \dfrac{\partial B'_x}{\partial t} = \dfrac{\partial E'_y}{\partial z} - \dfrac{\partial E'_z}{\partial y}, \\[2mm]
\Gamma V \dfrac{\partial B'_y}{\partial x} + \Gamma \dfrac{\partial B'_y}{\partial t} = \Gamma \dfrac{\partial E'_z}{\partial x} + \dfrac{V}{c^2} \Gamma \dfrac{\partial E'_z}{\partial t} - \dfrac{\partial E'_x}{\partial z}, \\[2mm]
\Gamma V \dfrac{\partial B'_z}{\partial x} + \Gamma \dfrac{\partial B'_z}{\partial t} = \dfrac{\partial E'_x}{\partial y} - \Gamma \dfrac{\partial E'_y}{\partial x} - \dfrac{V}{c^2} \Gamma \dfrac{\partial E'_y}{\partial t}.
\end{cases} \tag{8.4.18}
$$

From the last two Eqs. (8.4.18), we get

$$
\begin{cases}
\dfrac{\partial}{\partial t} \left[\Gamma \left(B'_y - \dfrac{V}{c^2} E'_z \right) \right] = \dfrac{\partial}{\partial x} \left[\Gamma (E'_z - V B'_y) \right] - \dfrac{\partial E'_x}{\partial z}, \\[3mm]
\dfrac{\partial}{\partial t} \left[\Gamma \left(B'_z + \dfrac{V}{c^2} E'_y \right) \right] = \dfrac{\partial E'_x}{\partial y} - \dfrac{\partial}{\partial x} \left[\Gamma (E'_y + V B'_z) \right].
\end{cases} \tag{8.4.19}
$$

Equations (8.4.19) are covariant with respect to Lorentz transformation, that is they preserve their form expressed by Eqs. (8.4.17)$_{2,3}$, if the field components E_x, E_y, E_z, B_y and B_z satisfy the following relations:

$$
\begin{cases}
E_x = E'_x, \\[1mm]
E_y = \Gamma (E'_y + V B'_z), \\[1mm]
E_z = \Gamma (E'_z - V B'_y), \\[1mm]
B_y = \Gamma \left(B'_y - \dfrac{V}{c^2} E'_z \right), \\[2mm]
B_z = \Gamma \left(B'_z + \dfrac{V}{c^2} E'_y \right).
\end{cases} \tag{8.4.20}
$$

Replacing V by $-V$, and interchanging the "primed" components with the "unprimed" ones, we also have the inverse transformations

$$
\begin{cases}
E'_x = E_x, \\[1mm]
E'_y = \Gamma (E_y - V B_z), \\[1mm]
E'_z = \Gamma (E_z + V B_y), \\[1mm]
B'_y = \Gamma \left(B_y + \dfrac{V}{c^2} E_z \right), \\[2mm]
B'_z = \Gamma \left(B_z - \dfrac{V}{c^2} E_y \right).
\end{cases} \tag{8.4.21}
$$

Introducing these results into Eq. (8.4.17)$_1$ and conveniently grouping the terms, we find

$$
\frac{\partial B'_x}{\partial t} = V \left(\frac{\partial B'_x}{\partial x} + \frac{\partial B_y}{\partial y} + \frac{\partial B_z}{\partial z} \right) = \frac{\partial E_y}{\partial z} - \frac{\partial E_z}{\partial y}.
$$

Maxwell's equation $\nabla \cdot \vec{B} = 0$ preserves its form if we take

$$B'_x = B_x. \tag{8.4.22}$$

This relation, together with Eq. (8.4.19), or Eq. (8.4.18), give the whole picture of the transformation relations for all the field components. One observes that the field components oriented along the direction of motion of frames *do not change*, while the components orthogonal to this direction *change* according to the above formulas. These formulas can synthetically be written as follows:

$$
\begin{cases}
\vec{E}'_\parallel = \vec{E}_\parallel, \\
\vec{E}'_\perp = \Gamma\left(\vec{E} + \vec{V} \times \vec{B}\right)_\perp, \\
\vec{B}'_\parallel = \vec{B}_\parallel, \\
\vec{B}'_\perp = \Gamma\left(\vec{B} - \dfrac{1}{c^2}\vec{V} \times \vec{E}\right)_\perp.
\end{cases} \tag{8.4.23}
$$

To summarize, Maxwell's source-free equations

$$B_{i,t} + \epsilon_{ijk}E_{k,j} = 0, \quad i,j,k = 1,2,3, \tag{8.4.24}$$

are covariant if the space coordinates and time obey the Lorentz transformations

$$
\begin{cases}
x' = \Gamma\left(x - Vt\right), \\
y' = y, \\
z' = z, \\
t' = \Gamma\left(t - \dfrac{V}{c^2}x\right),
\end{cases}
$$

while the field components $\vec{E}$, $\vec{B}$ transform according to Eq. (8.4.23). The reader is invited to resume the calculation for Maxwell's equations with sources (taking $\vec{j} = 0$, $\rho = 0$), and show that the result is the same.

In case of small velocities, $V \ll c$, one can approximate

$$
\begin{cases}
\vec{E}' \simeq \vec{E} + \vec{V} \times \vec{B}, \\
\vec{B}' \simeq \vec{B},
\end{cases} \tag{8.4.25}
$$

in accordance with the formulas first discovered by H.A. Lorentz and H. Hertz in their theory concerning the electrodynamics of moving media at low speeds compared to the speed of light in vacuum (see the two main notions introduced by Lorentz, namely, the *Lorentzian derivative*

$$\frac{D\vec{a}}{Dt} = \frac{\partial \vec{a}}{\partial t} + \nabla \times \left(\vec{a} \times \vec{v}\right) + \vec{v}\nabla \cdot \vec{a},$$

where $\vec{a}$ is an arbitrary vector field, and the *effective electric field*

$$\vec{E}'' \equiv \vec{v} \times \vec{B}.)$$

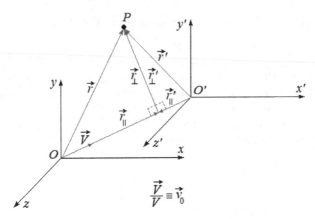

FIGURE 8.3
Two inertial frames, one moving with respect to the other in an arbitrary
direction, with the velocity $\vec{V}$.

8.5 Problem No. 63

Show that the generalized Lorentz transformations (see Fig. 8.3):

$$\begin{cases} \vec{r}' = (\Gamma - 1)(\vec{r} \times \vec{v}_0) \times \vec{v}_0 + \Gamma(\vec{r} - \vec{V}t), \\ t' = \Gamma\left[t - \dfrac{V}{c^2}(\vec{r} \cdot \vec{v}_0)\right], \end{cases} \qquad (8.5.26)$$

where the unit vector $\vec{v}_0$ is defined by

$$\vec{v}_0 = \frac{\vec{V}}{V}, \quad \left(V = |\vec{V}|\right),$$

can be derived by three successive proper space-time transformations.

Solution

Let us consider the inertial frames S and S', $\vec{V}$ being their relative velocity.
Without loss of generality, we may assume that $\vec{V}$ is situated in the xy-plane
(see Fig. 8.4). The transition from S to S' can be performed in three steps,
as follows:

(i) A counterclockwise rotation of angle θ, in the xy-plane, until the x-axis
becomes parallel to OO'. By this transition one passes from $S\left(Oxyzt\right)$ to

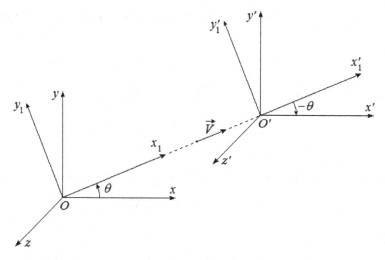

FIGURE 8.4
Transition from inertial frame S to inertial frame S' in three steps (**Problem No. 63**).

$S_1\left(Ox_1y_1z_1t_1\right) \equiv S_1\left(Ox_1y_1zt\right)$. The transformation relations are

$$\begin{cases} x_1 = \quad x\cos\theta + y\sin\theta, \\ y_1 = -x\sin\theta + y\cos\theta, \\ z_1 = z, \\ t_1 = t. \end{cases} \tag{8.5.27}$$

(ii) A Lorentz boost, along the x_1-axis, until O coincides with O', and the frame S_1 becomes $S'_1\left(O'x'_1y'_1z'_1t'_1\right) \equiv S'_1\left(O'x'_1y_1z_1t'_1\right)$. This transformation is

$$\begin{cases} x'_1 = \Gamma\left(x_1 - V t_1\right), \\ y'_1 = y_1, \\ z'_1 = z_1, \\ t'_1 = \Gamma\left(t_1 - \dfrac{V}{c^2}x_1\right). \end{cases} \tag{8.5.28}$$

(iii) The last transformation is a clockwise rotation of angle $-\theta$ about the x'-axis, until S'_1 coincides with S'. This transformation is given by

$$\begin{cases} x' = \quad x'_1\cos\theta - y'_1\sin\theta, \\ y' = -x'_1\sin\theta + y'_1\cos\theta, \\ z' = z'_1, \\ t' = t'_1. \end{cases} \tag{8.5.29}$$

Introducing Eq. (8.5.27) into Eq. (8.5.28), and the result into Eq. (8.5.29), we find

$$
\begin{cases}
x' = \left[1 + (\Gamma - 1)\cos^2\theta\right]x + (\Gamma - 1)y\sin\theta\cos\theta - V\Gamma t\cos\theta, \\
y' = (\Gamma - 1)x\sin\theta\cos\theta + \left[1 + (\Gamma - 1)\sin^2\theta\right]y - V\Gamma t\sin\theta, \\
z' = z, \\
t' = -\dfrac{V}{c^2}\Gamma x\cos\theta - \dfrac{V}{c^2}\Gamma y\sin\theta + \Gamma t.
\end{cases}
\tag{8.5.30}
$$

These relations can be also obtained by taking the matrix product

$$
X' = R_1\, B\, R_2\, X = A\, X\,,
\tag{8.5.31}
$$

where

$$
X = \begin{pmatrix} x \\ y \\ z \\ t \end{pmatrix}, \quad
X' = \begin{pmatrix} x' \\ y' \\ z' \\ t' \end{pmatrix}, \quad
R_1 = \begin{pmatrix}
\cos\theta & -\sin\theta & 0 & 0 \\
\sin\theta & +\cos\theta & 0 & 0 \\
0 & 0 & 1 & 0 \\
0 & 0 & 0 & 1
\end{pmatrix},
$$

$$
B = \begin{pmatrix}
\Gamma & 0 & 0 & -V\Gamma \\
0 & 1 & 0 & 0 \\
0 & 0 & 1 & 0 \\
-\dfrac{V}{c^2}\Gamma & 0 & 0 & \Gamma
\end{pmatrix}, \quad
R_2 = \begin{pmatrix}
+\cos\theta & \sin\theta & 0 & 0 \\
-\sin\theta & \cos\theta & 0 & 0 \\
0 & 0 & 1 & 0 \\
0 & 0 & 0 & 1
\end{pmatrix}.
$$

Let us now return to Eq. (8.5.30) and observe that

$$
\vec{r}\cdot\vec{V} = xV_x + yV_y = V\left(x\cos\theta + y\sin\theta\right),
$$

that is

$$
x\cos\theta + y\sin\theta = \frac{1}{V}\left(\vec{r}\cdot\vec{V}\right) = \vec{r}\cdot\vec{v}_0,
$$

where, as we already mentioned at the very beginning, $\vec{v}_0$ is the unit vector of $\vec{V}$. Then we can write Eq. (8.5.30) as follows:

$$
\begin{cases}
x' = x + (\Gamma - 1)(\vec{r}\cdot\vec{v}_0)\cos\theta - V\Gamma t\cos\theta, \\
y' = y + (\Gamma - 1)(\vec{r}\cdot\vec{v}_0)\sin\theta - V\Gamma t\sin\theta, \\
z' = z, \\
t' = \Gamma\left[t - \dfrac{V}{c^2}(\vec{r}\cdot\vec{v}_0)\right].
\end{cases}
\tag{8.5.32}
$$

If we now multiply relations (8.5.32) by the unit vectors of the axes x', y' and z', which are the same as those of the axes x, y and z (the frames are inertial), we arrive at the desired result:

$$\begin{cases} \vec{r}' = \vec{r} + (\Gamma - 1)(\vec{r} \cdot \vec{v}_0)\vec{v}_0 - \Gamma \vec{V} t, \\ t' = \Gamma \left[t - \dfrac{V}{c^2}(\vec{r} \cdot \vec{v}_0) \right]. \end{cases}$$

In a nutshell, we proved that a boost in an arbitrary direction can be obtained by rotating first the coordinates into a boost along a direction for which the Lorentz transformation is simple and known, perform that Lorentz transformation and then rotate the coordinates back.

9

Relativistic Dynamics

9.1 Problem No. 64

A particle of rest mass m_0 moves under the action of a force $\vec{F}$. Express its acceleration, $\vec{a}$, as a function of force and velocity.

Solution

Using the general relation

$$\vec{F} = \frac{d\vec{p}}{dt} \tag{9.1.1}$$

with

$$\vec{p} = \frac{m_0 \vec{v}}{\sqrt{1 - \frac{v^2}{c^2}}} = \gamma \, m_0 \vec{v}, \tag{9.1.2}$$

it follows that

$$\vec{F} = \frac{d}{dt}\left(\gamma \, m_0 \vec{v}\right) = \gamma \, m_0 \vec{a} + m_0 \vec{v} \, \frac{d\gamma}{dt}. \tag{9.1.3}$$

But

$$\frac{d\gamma}{dt} = \frac{d}{dt}\left(\frac{1}{\sqrt{1 - \frac{v^2}{c^2}}}\right) = -\frac{\frac{1}{2\sqrt{1 - \frac{v^2}{c^2}}}}{1 - \frac{v^2}{c^2}}$$

$$\times \left(-\frac{1}{c^2}\right) 2\,\vec{v} \cdot \frac{d\vec{v}}{dt} = \frac{1}{c^2}\gamma^3\left(\vec{v} \cdot \vec{a}\right), \tag{9.1.4}$$

so that Eq. (9.1.3) becomes

$$\vec{F} = \gamma \, m_0 \vec{a} + \gamma^3 m_0 \left(\frac{\vec{v} \cdot \vec{a}}{c}\right)\frac{\vec{v}}{c}. \tag{9.1.5}$$

Let us now perform the scalar product between Eq. (9.1.5) and the vector

DOI: 10.1201/9781003402602-9

$\vec{v}/c$. We obtain

$$
\begin{aligned}
\frac{\vec{v}}{c} \cdot \vec{F} &= \gamma \, m_0 \left(\frac{\vec{v} \cdot \vec{a}}{c} \right) + \gamma^3 m_0 \left(\frac{\vec{v} \cdot \vec{a}}{c} \right) \frac{\vec{v}}{c} \cdot \frac{\vec{v}}{c} \\
&= \gamma \, m_0 \left(\frac{\vec{v} \cdot \vec{a}}{c} \right) + \gamma^3 m_0 \left(\frac{\vec{v} \cdot \vec{a}}{c} \right) \frac{v^2}{c^2} \\
&= \gamma \, m_0 \left(\frac{\vec{v} \cdot \vec{a}}{c} \right) + \gamma^3 m_0 \left(\frac{\vec{v} \cdot \vec{a}}{c} \right) \left(1 - \frac{1}{\gamma^2} \right) \\
&= \gamma^3 m_0 \left(\frac{\vec{v} \cdot \vec{a}}{c} \right).
\end{aligned}
$$

Replacing now $\gamma^3 m_0 \left(\frac{\vec{v} \cdot \vec{a}}{c} \right)$ from the above relation in Eq. (9.1.5), we have

$$
\vec{F} = \gamma \, m_0 \vec{a} + \left(\frac{\vec{v}}{c} \cdot \vec{F} \right) \frac{\vec{v}}{c}, \tag{9.1.6}
$$

or

$$
\gamma \, m_0 \vec{a} = \vec{F} - \left(\frac{\vec{v} \cdot \vec{F}}{c} \right) \frac{\vec{v}}{c}, \tag{9.1.7}
$$

which yields

$$
\vec{a} = \frac{1}{m_0} \sqrt{1 - \frac{v^2}{c^2}} \left[\vec{F} - \left(\frac{\vec{v} \cdot \vec{F}}{c} \right) \frac{\vec{v}}{c} \right]. \tag{9.1.8}
$$

This relation shows the fact that generally, within the frame of relativistic mechanics, the acceleration produced by a force $\vec{F}$ is not collinear with the force, but is a vector contained in the plane determined by the vectors force and velocity. In addition, the acceleration depends on velocity both directly and indirectly, by means of mass m.

If the directions of the force and velocity are orthogonal, the scalar product $\vec{v} \cdot \vec{F}$ vanishes and we obtain the newtonian relation

$$
\vec{a}_\perp = \vec{F}_\perp / m,
$$

where

$$
m(\equiv m_\perp) = \frac{m_0}{\sqrt{1 - \frac{v^2}{c^2}}}
$$

can be defined as a "transverse mass" of the body during the motion.

If the force and the velocity of the body are parallel, $\vec{F} \| \vec{v}$, then, denoting by $\hat{v}$ the versor of the common direction of the two vectors, we obtain from Eq. (9.1.8) that

$$
\vec{a} = \frac{F}{m_0} \left(1 - \frac{v^2}{c^2} \right)^{3/2} \hat{v} = \frac{1}{m_0} \left(1 - \frac{v^2}{c^2} \right)^{3/2} \vec{F}, \tag{9.1.9}
$$

that is, $\vec{a} \parallel \vec{F}$. Therefore, in this particular case the three vectors are collinear and, furthermore, one can write that

$$\vec{F}_{\parallel} = \frac{m_0}{\left(1 - \frac{v^2}{c^2}\right)^{3/2}} \, \vec{a}_{\parallel}, \tag{9.1.10}$$

which allows defining a "longitudinal" mass through the relation

$$m_{\parallel} = \frac{m_0}{\left(1 - \frac{v^2}{c^2}\right)^{3/2}}.$$

9.2   Problem No. 65

Determine the one-dimensional motion of a particle of rest mass m_0, moving under the action of a constant force $\vec{F}$.

Solution

Since the motion is supposed to be one-dimensional, in the following we will give up the arrow used to designate the vector quantities. Therefore, we will simply write

$$F = \frac{dp}{dt} = \frac{d}{dt}\left[m_0 v \left(1 - \frac{v^2}{c^2}\right)^{-1/2}\right]. \tag{9.2.11}$$

Calculating the derivative, we have

$$F = \left[m_0 \frac{dv}{dt}\sqrt{1 - \frac{v^2}{c^2}} - \frac{d}{dt}\left(\sqrt{1 - \frac{v^2}{c^2}}\right)m_0 v\right]\frac{1}{1 - \frac{v^2}{c^2}}$$

$$= \left(\sqrt{1 - \frac{v^2}{c^2}} + \frac{v^2}{c^2\sqrt{1 - \frac{v^2}{c^2}}}\right)\frac{m_0}{1 - \frac{v^2}{c^2}}\frac{dv}{dt}$$

$$= \left(1 - \frac{v^2}{c^2} + \frac{v^2}{c^2}\right)\frac{m_0}{\left(1 - \frac{v^2}{c^2}\right)^{3/2}}\frac{dv}{dt} = m_0\left(1 - \frac{v^2}{c^2}\right)^{-3/2}\frac{dv}{dt},$$

which means that

$$\frac{F}{m_0}dt = \frac{dv}{\left(1 - \frac{v^2}{c^2}\right)^{3/2}}. \tag{9.2.12}$$

Taking as initial condition $v_0 = v(t_0 = 0) = 0$ and integrating Eq. (9.2.12), we can write

$$\frac{F}{m_0}\int_{t_0}^{t} dt = \int_{0}^{v}\left(1 - \frac{v^2}{c^2}\right)^{-3/2} dv. \tag{9.2.13}$$

In order to calculate the integral on the right side, we will make the change of variable

$$v = c \sin \varphi \;\Rightarrow\; dv = c \cos \varphi \, d\varphi,$$

and Eq. (9.2.13) becomes

$$\frac{F}{m_0}(t - t_0) = c \int_0^{\arcsin(v/c)} \frac{1}{\cos^2 \varphi} d\varphi = c \tan \varphi \Big|_0^{\arcsin(v/c)}$$

$$= c \tan \left[\arcsin \left(\frac{v}{c} \right) \right] = c \frac{\sin \left[\arcsin \left(\frac{v}{c} \right) \right]}{\cos \left[\arcsin \left(\frac{v}{c} \right) \right]} = \frac{v}{\sqrt{1 - \left(\frac{v}{c} \right)^2}},$$

or, squaring and performing some elementary calculations,

$$v^2 \left[1 + \frac{(F/m_0)^2}{c^2} (t - t_0)^2 \right] = (F/m_0)^2 (t - t_0)^2,$$

which leads to

$$v = \frac{(F/m_0)(t - t_0)}{\sqrt{1 + \frac{(F/m_0)^2}{c^2} (t - t_0)^2}} = \frac{dx}{dt}. \tag{9.2.14}$$

This relation can also be written as

$$dx = \frac{c^2}{(F/m_0)} d \left[\sqrt{1 + \frac{(F/m_0)^2}{c^2} (t - t_0)^2} \right],$$

and can be easily integrated with the initial condition $x(t_0 = 0) = x_0$:

$$x - x_0 = \frac{c^2}{(F/m_0)} \sqrt{1 + \frac{(F/m_0)^2}{c^2} (t - t_0)^2}. \tag{9.2.15}$$

It is more convenient for physical interpretation to write the relationship (9.2.15) in the form

$$\frac{(x - x_0)^2}{\left(\frac{c^2}{F/m_0} \right)^2} - \frac{(t - t_0)^2}{\left(\frac{c}{F/m_0} \right)^2} = 1, \tag{9.2.16}$$

which shows that the space-time diagram (the world line) of the motion is a hyperbola. For this reason, the one-dimensional motion of a particle under the action of a constant force is called *hyperbolic motion*. This motion tends asymptotically to a uniform motion with velocity c.

In the non-relativistic case, with the same initial conditions and $m = m_0 = $ const., we have

$$x - x_0 = \frac{F}{m} \frac{(t - t_0)^2}{2},$$

so that, obviously, the corresponding space-time diagram is an arc of parabola.

9.3   Problem No. 66

Considering the disintegration reaction $A \rightarrow B + C$,
a) Calculate the energy E_B of the particle B, if the particle that disintegrates is at rest in the laboratory reference frame. The rest masses of the particles, m_A, m_B and m_C, are supposed to be known, and one considers the natural unit system (with $c = 1$).
b) Show that the kinetic energy T_i of the particle i, where $i = B, C$, in the proper reference system of the particle that disintegrates is

$$T_i = \Delta m \left(1 - \frac{m_i}{m_A} - \frac{\Delta m}{2m_A} \right),$$

where

$$\Delta m = m_A - m_B - m_C$$

is the "mass excess" or the "Q value" of the process.
c) The π^- meson with negative charge ($m_{\pi^-} = 139.6\,\text{MeV}$) disintegrates into a negative muon ($m_{\mu^-} = 105.7\,\text{MeV}$) and a muon antineutrino, whose mass is supposed to be $m_{\bar{\nu}_\mu} \simeq 0$. Calculate the kinetic energies of the muon and antineutrino, in the reference frame of the π^- meson. The "unique" kinetic energy of the muon is the result of a two-particle disintegration. This fact played an important role in the experimental discovery of the π^- meson in the photographic emulsions, performed by Cecil Frank Powell (1903−1969) and Giuseppe P.S. Occhialini (1907−1993) in 1947.

Solution

To solve the problem, the conservation laws of the relativistic energy and relativistic momentum are used.

a) Since the particle A is at rest, its *initial* momentum (momentum before disintegration) is zero, $\left| \vec{p}_A^{\,i} \right| = 0$. Obviously, since after reaction (interaction) the particle A "disappears", we also have $\left| \vec{p}_A^{\,f} \right| = 0$. The relations $\left| \vec{p}_B^{\,i} \right| = 0$ and $\left| \vec{p}_C^{\,i} \right| = 0$ are also obvious. The conservation of the total momentum of the system writes

$$\vec{p}_A^{\,i} + \vec{p}_B^{\,i} + \vec{p}_C^{\,i} = \vec{p}_A^{\,f} + \vec{p}_B^{\,f} + \vec{p}_C^{\,f} \quad \Leftrightarrow$$
$$0 = \vec{p}_B^{\,f} + \vec{p}_C^{\,f} \Rightarrow \vec{p}_B^{\,f} = -\vec{p}_C^{\,f} \Rightarrow \left| \vec{p}_B^{\,f} \right| = \left| \vec{p}_C^{\,f} \right|. \tag{9.3.17}$$

The momentum conservation can be graphically represented as in Fig. 9.1. The common direction of the two momenta $\vec{p}_B^{\,f}$ and $\vec{p}_C^{\,f}$ can be any. The energy conservation law in reaction is expressed as

$$E_A^i + E_B^i + E_C^i = E_A^f + E_B^f + E_C^f,$$

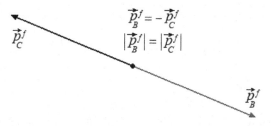

FIGURE 9.1
Graphical illustration of momentum conservation in disintegration reaction
$A \to B + C$, particle A being at rest.

where

$$E_A^i \left(= m_A c^2 \right) = m_A,$$

where m_A represents the rest mass of particle A (the particle A is initially at
rest), $E_B^i = 0$, $E_C^i = 0$,

$$\left(E_B^f \right)^2 \left[= m_B^2 c^4 + \left(p_B^f \right)^2 c^2 \right] = m_B^2 + \left(p_B^f \right)^2,$$

and

$$\left(E_C^f \right)^2 \left[= m_C^2 c^4 + \left(p_C^f \right)^2 c^2 \right] = m_C^2 + \left(p_C^f \right)^2,$$

where between parentheses has been written the corresponding relations in the
LMT unit system. The transition from the system of fundamental dimensions
LSV (length, action and velocity) to the system LMT (length, mass and time)
is performed by the relations

$$L = L; \ S = ML^2 T^{-1}; \ V = LT^{-1},$$

while the reverse transition is achieved by means of relations

$$L = L; \ M = SL^{-1} V^{-1}; \ T = LV^{-1}.$$

The energy conservation law then becomes

$$m_A = E_B^f + \sqrt{m_C^2 + \left(p_C^f \right)^2}. \tag{9.3.18}$$

By using the relations (9.3.17) and (9.3.18), we have

$$m_A = E_B^f + \sqrt{m_C^2 + \left(p_B^f \right)^2} = E_B^f + \sqrt{m_C^2 + \left(E_B^f \right)^2 - m_B^2},$$

or

$$m_A - E_B^f = \sqrt{m_C^2 + \left(E_B^f \right)^2 - m_B^2}.$$

Squaring the last relation, we have

$$m_A^2 + \left(E_B^f\right)^2 - 2m_A E_B^f = m_C^2 + \left(E_B^f\right)^2 - m_B^2,$$

so that

$$E_B^f = \frac{m_A^2 + m_B^2 - m_C^2}{2m_A}. \tag{9.3.19}$$

In a completely analogous way, the total energy of the particle C can be written; this can be easily done by means of the change $B \leftrightarrow C$ in Eq. (9.3.19):

$$E_C^f = \frac{m_A^2 + m_C^2 - m_B^2}{2m_A}. \tag{9.3.20}$$

b) In the frame of relativistic mechanics, the kinetic energy of a particle is defined as the difference between the total energy and the rest energy of the particle i:

$$T_i = E_i - m_i. \tag{9.3.21}$$

Taking into account the relations (9.3.19) and/or (9.3.20), the kinetic energy of the particle i given by Eq. (9.3.21) can be successively written as

$$T_i = E_i - m_i = \frac{m_A^2 + m_i^2 - m_j^2}{2m_A} - m_i = \frac{m_A^2 + m_i^2 - m_j^2 - 2m_A m_i}{2m_A}$$

$$= \frac{\left(m_A^2 - 2m_A m_i + m_i^2\right) - m_j^2}{2m_A} = \frac{\left(m_A - m_i\right)^2 - m_j^2}{2m_A}$$

$$= \frac{\left(m_A - m_i - m_j\right)\left(m_A - m_i + m_j\right)}{2m_A} = \frac{\Delta m\left(m_A - m_i + m_j\right)}{2m_A}$$

$$= \frac{\Delta m\left[m_A - m_i - \left(m_A - m_i - m_j\right) + m_A - m_i\right]}{2m_A}$$

$$= \frac{\Delta m\left(2m_A - 2m_i - \Delta m\right)}{2m_A} = \Delta m\left(1 - \frac{m_i}{m_A} - \frac{\Delta m}{2m_A}\right), \tag{9.3.22}$$

that is, what was to be demonstrated. While calculating T_i from Eq. (9.3.22), the definition of the "mass excess",

$$\Delta m = m_A - m_i - m_j,$$

has been used, where $j = C$ if $i = B$ and $j = B$ if $i = C$, these "values" of the indices i and j being valid everywhere in Eq. (9.3.22). In addition, one observes that the sum of kinetic energies of the two reaction "products" is just the "mass excess"

$$T_B + T_C = \Delta m.$$

This relation can also be obtained by using the definition of the kinetic energy,

$$T_B + T_C = \left(E_B - m_B\right) + \left(E_C - m_C\right) = \left(E_B + E_C\right) - m_B - m_C$$

$$= E_A - m_B - m_C = m_A - m_B - m_C \equiv \Delta m,$$

where, in writing the last equation, we used the fact that $\vec{p}_A = 0$.

c) The nuclear reaction of disintegration of the π^- meson writes as follows:

$$\pi^- \;\rightarrow\; \mu^- + \;\bar{\nu}_\mu$$

$$\downarrow$$

$$e^- \;+\; \nu_\mu \;+\; \bar{\nu}_e$$

According to relation (9.3.22), the kinetic energy of the negative muon is

$$T_{\mu^-} = \Delta m \left(1 - \frac{m_{\mu^-}}{m_{\pi^-}} - \frac{\Delta m}{2m_{\pi^-}} \right), \tag{9.3.23}$$

and the kinetic energy of the muon antineutrino is

$$T_{\bar{\nu}_\mu} = \Delta m \left(1 - \frac{m_{\bar{\nu}_\mu}}{m_{\pi^-}} - \frac{\Delta m}{2m_{\pi^-}} \right)$$

$$\simeq \Delta m \left(1 - \frac{\Delta m}{2m_{\pi^-}} \right), \tag{9.3.24}$$

where

$$\Delta m = m_{\pi^-} - m_{\mu^-} - m_{\bar{\nu}_\mu} \simeq m_{\pi^-} - m_{\mu^-}$$
$$= \left(139.6 - 105.7 \right) \text{MeV} = 33.9\,\text{MeV}.$$

The numerical calculations give the following values for the kinetic energy of the two particles:

$$T_{\mu^-} = \Delta m \left(1 - \frac{m_{\mu^-}}{m_{\pi^-}} - \frac{\Delta m}{2m_{\pi^-}} \right)$$

$$= 33.9 \left(1 - \frac{105.7}{139.6} - \frac{33.9}{279.2} \right) \text{MeV} \simeq 4.116\,\text{MeV},$$

and, respectively

$$T_{\bar{\nu}_\mu} \simeq \Delta m \left(1 - \frac{\Delta m}{2m_{\pi^-}} \right) = 33.9 \left(1 - \frac{33.9}{279.2} \right) \text{MeV} \simeq 29.784\,\text{MeV},$$

which verifies the relation

$$T_B + T_C = \Delta m,$$

written for this case as

$$T_{\mu^-} + T_{\bar{\nu}_\mu} = \left(4.116 + 29.784 \right) \text{MeV} = 33.9\,\text{MeV} = \Delta m.$$

9.4 Problem No. 67

A particle A disintegrates while being in motion, through the reaction

$$A \to B + C.$$

Knowing the rest masses of all particles as well as the energy of particle A just before disintegrating, and the energy of particle B as soon as it appeared, determine the angle θ between the direction of "emerging" particle B and the direction of "incident" particle A, by working in the natural unit system $(c = 1)$.

Solution

Since all particles implied in reaction are moving (the disintegration takes place with the initial particle A in motion), one can write

$$\left(E_A^i\right)^2 \left[= m_A^2 c^4 + \left(p_A^i\right)^2 c^2\right] = m_A^2 + \left(p_A^i\right)^2, \quad E_A^f = 0, \quad E_B^i = 0,$$

$$\left(E_B^f\right)^2 \left[= m_B^2 c^4 + \left(p_B^f\right)^2 c^2\right] = m_B^2 + \left(p_B^f\right)^2, \quad E_C^i = 0,$$

$$\left(E_C^f\right)^2 \left[= m_C^2 c^4 + \left(p_C^f\right)^2 c^2\right] = m_c^2 + \left(p_C^f\right)^2,$$

where the index "i" concerns the "initial" state (immediately before the reaction) and "f" the "final" state (immediately after disintegration). Regarding the particle momenta before and after disintegration of the particle A, using the problem statement, one can write

$$p_B^i = 0, \ p_C^i = 0, \text{ and } p_A^f = 0.$$

The relations between the other momenta, $p_A^i \neq 0$, $p_B^f \neq 0$ and $p_C^f \neq 0$ can be correctly obtained by means of the graphic representation of the momentum conservation law regarding the studied process, as shown in Fig. 9.2, where, without restricting the generality of the problem, the direction of the momentum of "initial" particle A has been considered as being horizontal.

The problem is easy to solve, with the help of the conservation laws of relativistic momentum and energy

$$\vec{p}_A^i + \vec{p}_B^i + \vec{p}_C^i = \vec{p}_A^f + \vec{p}_B^f + \vec{p}_C^f, \tag{9.4.25}$$

$$E_A^i + E_B^i + E_C^i = E_A^f + E_B^f + E_C^f, \tag{9.4.26}$$

In view of the previous discussion and the corresponding relations, the two laws expressed by Eqs. (9.4.25) and (9.4.26) become

$$\vec{p}_A^i = \vec{p}_B^f + \vec{p}_C^f, \tag{9.4.27}$$

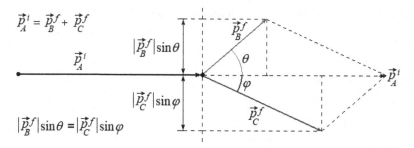

FIGURE 9.2
Decay of the particle A into two "fragments" B and C. The scattering angle of particle B is denoted by θ.

and

$$E_A^i = E_B^f + E_C^f, \qquad (9.4.28)$$

respectively.

Projecting the vector relation (9.4.27) on the direction of the momentum of the "initial" particle A, and on a direction orthogonal to the first, we have (see Fig. 9.2):

$$\begin{cases} \left|\vec{p}_A^i\right| = \left|\vec{p}_B^f\right| \cos\theta + \left|\vec{p}_C^f\right| \cos\varphi, \\ \left|\vec{p}_B^f\right| \sin\theta = \left|\vec{p}_C^f\right| \sin\varphi, \end{cases}$$

or, in a simpler writing

$$\begin{cases} p_A^i = p_B^f \cos\theta + p_C^f \cos\varphi, \\ p_B^f \sin\theta = p_C^f \sin\varphi, \end{cases} \qquad (9.4.29)$$

where the new notations are obvious. Taking the square of $\cos\varphi$ which appears in the first relation (9.4.29) and adding the result to the square of $\sin\varphi$ obtained from the second relation (9.4.29), one obtains

$$\left(p_C^f\right)^2 = \left(p_A^i\right)^2 + \left(p_B^f\right)^2 - 2p_A^i p_B^f \cos\theta,$$

leading to

$$\cos\theta = \frac{\left(p_A^i\right)^2 + \left(p_B^f\right)^2 - \left(p_C^f\right)^2}{2p_A^i p_B^f}. \qquad (9.4.30)$$

Formula (9.4.30) gives the answer to the problem. Nevertheless, this result can be put in a different form, in terms of masses and energies of involved particles. To do this, we start with the relativistic energy conservation law, written as

$$E_C^f = E_A^i - E_B^f,$$

or, by squaring the above relation,

$$\left(E_C^f\right)^2 \left[= m_C^2 + \left(p_C^f\right)^2\right] = \left(E_A^i\right)^2 + \left(E_B^f\right)^2 - 2E_A^i E_B^f,$$

which allows to get the momentum of the particle C:

$$\left(p_C^f\right)^2 = \left(E_A^i\right)^2 + \left(E_B^f\right)^2 - 2E_A^i E_B^f - m_C^2. \tag{9.4.31}$$

By introducing $\left(p_C^f\right)^2$ from the relation (9.4.31) into relation (9.4.30), we finally obtain

$$\begin{aligned}
\cos\theta &= \frac{\left(p_A^i\right)^2 + \left(p_B^f\right)^2 - \left(p_C^f\right)^2}{2p_A^i p_B^f} \\[2mm]
&= \frac{\left(p_A^i\right)^2 + \left(p_B^f\right)^2 - \left(E_A^i\right)^2 - \left(E_B^f\right)^2 + 2E_A^i E_B^f + m_C^2}{2p_A^i p_B^f} \\[2mm]
&= \frac{\left(E_A^i\right)^2 - m_A^2 + \left(E_B^f\right)^2 - m_B^2 - \left(E_A^i\right)^2 - \left(E_B^f\right)^2}{2\sqrt{\left[\left(E_A^i\right)^2 - m_A^2\right]\left[\left(E_B^f\right)^2 - m_B^2\right]}} \\[2mm]
&\quad + \frac{2E_A^i E_B^f + m_C^2}{2\sqrt{\left[\left(E_A^i\right)^2 - m_A^2\right]\left[\left(E_B^f\right)^2 - m_B^2\right]}} \\[2mm]
&= \frac{m_C^2 - m_A^2 - m_B^2 + 2E_A^i E_B^f}{2\sqrt{\left[\left(E_A^i\right)^2 - m_A^2\right]\left[\left(E_B^f\right)^2 - m_B^2\right]}}. \tag{9.4.32}
\end{aligned}$$

9.5   Problem No. 68

The Λ^0 particle is a neutral baryon (the quark content is uds) of mass $M = 1115.683\,\text{MeV}$ ($\simeq 1115\,\text{MeV}$) (the natural system of units, with $c = 1$, is considered) which disintegrates into a proton of mass $m_p \simeq 939\,\text{MeV}$ and a "pi-minus" meson (negative pion) of mass $m_{\pi^-} \simeq 140\,\text{MeV}$, having the mean lifetime of $\tau_{\Lambda^0} = 2.631 \times 10^{-10}\,\text{s}$. This baryon was first observed on its flight through the characteristic mode of disintegration in charged particles,

$$\Lambda^0 \to p + \pi^-,$$

inside the cloud chambers. The traces left by the charged particles resulting from disintegration start from a single point and look very much like a turned "V", or a Greek letter Λ. By the way, that is why this particle was called "lambda"-zero baryon. The identity and momenta of these particles can be deduced by observing the length of their traces and their curvatures in the magnetic field of the cloud chambers.

a) Using the momentum and energy conservation laws show that, if the opening angle θ between the two quasi-rectilinear traces is determined by measurement, the mass of the disintegrating particle (the lambda-zero baryon) can be obtained by means of the formula

$$M_{\Lambda^0}^2 = m_p^2 + m_{\pi^-}^2 + 2\left(E_p E_{\pi^-}\right) - 2\left(p_p p_{\pi^-}\right)\cos\theta,$$

where p_p and p_{π^-} are the moduli of the three-dimensional momenta of particles resulting from disintegration, while E_p and E_{π^-} are their total energies.

b) As a result of a collision at the roof of a cloud chamber, a lambda-zero particle with the total energy of 10 GeV is created. How far the particle will travel before disintegrating?

Solution

Being electrically neutral, the lambda-zero baryons leave no trace in a cloud chamber (Wilson chamber). The charged particles resulting from the disintegration leave such traces, so that they appear – as specified in the problem statement – in the form of the letter Λ from the Greek alphabet. In Fig. 9.3 this disintegration "diagram" is represented as part of a more complex diagram, which puts into evidence (for the first time) a relatively recently discovered particle named "the double strange baryon" *omega b minus* (the quark content is ssb). In this diagram, the path of the proton has the letter p associated with it, and the path of the negative pion resulting from disintegration of the neutral lambda baryon has the letter π^- associated with it. The path of the neutral lambda baryon is represented by a dotted thin line to indicate that this part of the disintegration "diagram" is not observed in the cloud chamber.

As seen in Fig. 9.3, the lambda zero particle results (together with the K^- meson – the negative kaon) as a result of disintegration of omega minus baryon, which particle is composed by three quarks s ("strange"), being considered the first indisputable proof of the quark existence. This was performed by means of the bubble chamber from Brookhaven National Laboratory, in 1964. But much "stranger" is the "omega b minus" particle (the Ω_b^- baryon, discovered in the DZero Experiment, by means of the Tevatron from the Fermilab, as related by the newspaper ScienceDaily of September 4, 2008 (for details click on the link below):

http://www.sciencedaily.com/releases/2008/09/080903172201.htm

Once produced, the *omega b minus* particle disintegrates as a jet after a path of about one millimetre in the cloud chamber. It first "decomposes" into two intermediate particles, namely J/Ψ and omega minus. Right after, the J/Ψ particle disintegrates into two muons with different electric charges. The omega minus baryon can go through a path of a few centimetres, then it also disintegrates into an unstable particle, the lambda zero baryon, and a negative kaon (the K^- meson) with a lifetime much longer than all particles "appeared" until now. In its turn, the lambda zero baryon, which is charge-free, can travel a distance of a few centimetres before disintegrating into a proton and a negative pion (the π^- meson).

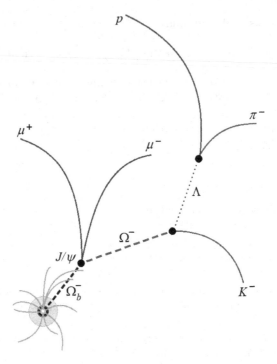

FIGURE 9.3
The disintegration "diagram" of Λ baryon, represented as part of a more complex diagram, which puts into evidence "the double strange baryon" Ω_b^-. The path of the neutral lambda baryon is represented by a dotted line to indicate that this part of the disintegration "diagram" is not observed in the cloud chamber.

a) To prove the formula by which the mass of lambda zero particle can be obtained, we will appeal to the four-momentum conservation law in the analysed process (as recommended in the statement of the problem), written separately for its spatial and temporal parts.

Using the notations in Fig. 9.4, the spatial part of the four-momentum conservation law writes

$$\vec{p}_{\Lambda^0}^{\,i} + \vec{p}_p^{\,i} + \vec{p}_{\pi^-}^{\,i} = \vec{p}_{\Lambda^0}^{\,f} + \vec{p}_p^{\,f} + \vec{p}_{\pi^-}^{\,f}\,, \qquad (9.5.33)$$

where the index "i" designates the initial state, and "f" the final state.

The temporal part of the same conservation law (in fact, the law of conservation of the relativistic total energy) is

$$E_{\Lambda^0}^{i} + E_p^{i} + E_{\pi^-}^{i} = E_{\Lambda^0}^{f} + E_p^{f} + E_{\pi^-}^{f}\,. \qquad (9.5.34)$$

In the above relations $\vec{p}_p^{\,i} = 0$, $\vec{p}_{\pi^-}^{\,i} = 0$, $\vec{p}_{\Lambda^0}^{\,f} = 0$, and for the energies

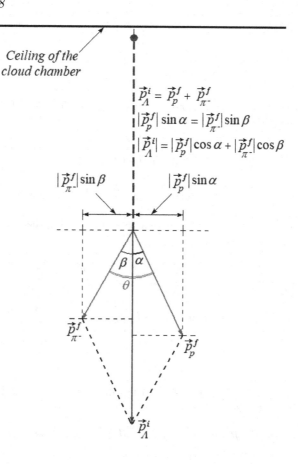

FIGURE 9.4
Decay of Λ baryon into a proton and a negative pion. The scattering angle of negative pion is denoted by β, while that corresponding to the proton by α.

as well, $E_p^i = 0$, $E_{\pi^-}^i = 0$, and $E_{\Lambda^0}^i = 0$, so that the relations (9.5.33) and (9.5.34) receive a simpler form,

$$\vec{p}_{\Lambda^0}^i = \vec{p}_p^f + \vec{p}_{\pi^-}^f , \qquad (9.5.35)$$

and

$$E_{\Lambda^0}^i = E_p^f + E_{\pi^-}^f , \qquad (9.5.36)$$

respectively.

Projecting the vector relation (9.5.35) on the direction of the initial momentum of the Λ^0 particle and on a direction orthogonal to the first, one obtains

$$\begin{cases} |\vec{p}_{\Lambda^0}^i| = |\vec{p}_p^f| \cos \alpha + |\vec{p}_{\pi^-}^f| \cos \beta, \\ |\vec{p}_p^f| \sin \alpha = |\vec{p}_{\pi^-}^f| \sin \beta, \end{cases}$$

or, in a simpler form,

$$\begin{cases} p_{\Lambda^0} = p_p \cos \alpha + (p_{\pi^-}) \cos \beta, \\ p_p \sin \alpha = (p_{\pi^-}) \sin \beta, \end{cases} \tag{9.5.37}$$

where the notations are obvious. In turn, the relation (9.5.36) can be written in a simpler form

$$E_{\Lambda^0} = E_p + E_{\pi^-} . \tag{9.5.38}$$

Squaring this relation and using the fact that

$$E_{\Lambda^0}^2 = M_{\Lambda^0}^2 c^4 + p_{\Lambda^0}^2 c^2 \overset{c=1}{=} M_{\Lambda^0}^2 + p_{\Lambda^0}^2,$$

together with the two analogous relations written for the proton and the pi-minus meson, we have

$$M_{\Lambda^0}^2 + p_{\Lambda^0}^2 = m_p^2 + p_p^2 + m_{\pi^-}^2 + p_{\pi^-}^2 + 2 E_p E_{\pi^-},$$

which yields

$$M_{\Lambda^0}^2 = m_p^2 + m_{\pi^-}^2 + 2(E_p E_{\pi^-}) - p_{\Lambda^0}^2 + p_p^2 + p_{\pi^-}^2 .$$

This way, the point **a)** of the problem comes to demonstrate that

$$-p_{\Lambda^0}^2 + p_p^2 + p_{\pi^-}^2 = -2(p_p p_{\pi^-}) \cos \theta,$$

where $\theta = \alpha + \beta$. This purpose turns out easily from relations (9.5.37) as follows: the first relation gives

$$p_{\Lambda^0}^2 = p_p^2 \cos^2 \alpha + p_{\pi^-}^2 \cos^2 \beta + 2(p_p p_{\pi^-}) \cos \alpha \cos \beta,$$

so that

$$-p_{\Lambda^0}^2 + p_p^2 + p_{\pi^-}^2 = - p_p^2 \cos^2 \alpha - p_{\pi^-}^2 \cos^2 \beta \\ - 2(p_p p_{\pi^-}) \cos \alpha \cos \beta + p_p^2 + p_{\pi^-}^2,$$

or

$$-p_{\Lambda^0}^2 + p_p^2 + p_{\pi^-}^2 = p_p^2 \sin^2 \alpha + (p_{\pi^-}^2) \sin^2 \beta - 2(p_p p_{\pi^-}) \cos \alpha \cos \beta,$$

which gives (by using the relation (9.5.37)$_2$),

$$- p_{\Lambda^0}^2 + p_p^2 + p_{\pi^-}^2 = (p_p \sin \alpha) \underbrace{(p_p \sin \alpha)}_{=(p_{\pi^-}) \sin \beta}$$

$$+ [(p_{\pi^-}) \sin \beta] \underbrace{[(p_{\pi^-}) \sin \beta]}_{=p_p \sin \alpha}$$

$$- 2(p_p p_{\pi^-}) \cos \alpha \cos \beta = 2(p_p p_{\pi^-}) (\sin \alpha \sin \beta - \cos \alpha \cos \beta)$$

$$= -2(p_p p_{\pi^-}) (\cos \alpha \cos \beta - \sin \alpha \sin \beta)$$

$$= -2(p_p p_{\pi^-}) \cos (\alpha + \beta) = -2(p_p p_{\pi^-}) \cos \theta,$$

that is the result we were looking for.

b) The momentum of the lambda-zero particle, whose energy is $10\,\text{GeV}$, is

$$p_{\Lambda^0} = \sqrt{E_{\Lambda^0}^2 - M_{\Lambda^0}^2} = \sqrt{10^{20} - (1.115)^2 \times 10^{18}}\ \text{eV}$$

$$= \sqrt{10^{18}\left(100 - 1.243225\right)}\ \text{eV} = 9.937\,\text{GeV}, \tag{9.5.39}$$

meaning that the velocity of this particle – based on the relativistic relation between momentum and energy,

$$p_{\Lambda^0} = \frac{M_{\Lambda^0} v_{\Lambda^0}}{\sqrt{1 - \dfrac{v_{\Lambda^0}^2}{c^2}}},$$

in the *LMT* (length, mass and time) system of units, or

$$p_{\Lambda^0} = \frac{M_{\Lambda^0} v_{\Lambda^0}}{\sqrt{1 - v_{\Lambda^0}^2}},$$

in the *LSV* (length, action and velocity) natural system of units – is

$$v_{\Lambda^0} = \frac{1}{\sqrt{1 + \left(\dfrac{M_{\Lambda^0}}{p_{\Lambda^0}}\right)^2}} = \sqrt{1 - \frac{M_{\Lambda^0}^2}{E_{\Lambda^0}^2}}$$

$$= \sqrt{1 - \left(\frac{M_{\Lambda^0}}{E_{\Lambda^0}}\right)^2} = \sqrt{1 - \left(\frac{1.225}{10}\right)^2} = 0.994. \tag{9.5.40}$$

This value of the velocity is very close to the velocity limit in the natural unit system, $v(\equiv c) = 1$, so that the relativistic treatment of the problem is required.

As an observation, we mention that the particle velocity can be also deduced from the relativistic relation between momentum and velocity,

$$p_{\Lambda^0} = \frac{E_{\Lambda^0}}{c^2} v_{\Lambda^0},$$

in the *LMT* system, or

$$p_{\Lambda^0} = E_{\Lambda^0} v_{\Lambda^0},$$

in the natural unit system. Indeed, in this case one (re)obtains

$$v_{\Lambda^0} = \frac{p_{\Lambda^0}}{E_{\Lambda^0}} = \frac{\sqrt{E_{\Lambda^0}^2 - M_{\Lambda^0}^2}}{E_{\Lambda^0}} = \sqrt{1 - \frac{M_{\Lambda^0}^2}{E_{\Lambda^0}^2}}$$

$$= \sqrt{1 - \left(\frac{1.115}{10}\right)^2} \simeq 0.994\ \left(= 0.994\,c = 2.982 \times 10^8\,\text{m}\cdot\text{s}^{-1}\right).$$

In view of the fact that the mean lifetime of a particle is always reported in its own reference system (in which the particle is at rest; usually, this reference system is called the "proper reference system"), it follows that the particle itself plays the role of "origin" of the reference system S', which moves relative to the inertial frame of reference of the laboratory, S_{lab}, with velocity v_{Λ^0}. Taking into account the relativistic effect of time dilation, when passing from the proper inertial referential S' to any other inertial referential S, it follows that in the laboratory referential, S_{lab}, from the moment of creation until the moment of disintegration, the lambda zero particle covers a distance of

$$d_{\Lambda^0}^{(\mathrm{RIS}_{lab})} = (v_{\Lambda^0} t_{\Lambda^0})^{(\mathrm{RIS}_{lab})} = \sqrt{1 - \left(\frac{M_{\Lambda^0}}{E_{\Lambda^0}}\right)^2} \; \frac{\tau_{\Lambda^0}}{\sqrt{1 - v_{\Lambda^0}^2}}$$

$$= \sqrt{1 - \left(\frac{M_{\Lambda^0}}{E_{\Lambda^0}}\right)^2} \; \frac{\tau_{\Lambda^0}}{\sqrt{\left(\frac{M_{\Lambda^0}}{E_{\Lambda^0}}\right)^2}} = \sqrt{1 - \left(\frac{M_{\Lambda^0}}{E_{\Lambda^0}}\right)^2} \; \frac{E_{\Lambda^0}}{M_{\Lambda^0}} \tau_{\Lambda^0}$$

$$= \sqrt{1 - \left(\frac{1.115}{10}\right)^2} \; \frac{10}{1.115} \, 2.6 \times 10^{-10} \, \mathrm{s}$$

$$= 23.173 \times 10^{-10} \, \mathrm{s} \left(= 69.519 \times 10^{-2} \, \mathrm{m} = 69.519 \, \mathrm{cm}\right).$$

10

Relations of Field Transformations

Show that, if the relative displacement of the inertial reference frames S and S' takes place in some direction of versor $\vec{v}_0$ (the velocity versor), than the transformation laws of the electromagnetic field components $\vec{E}$ and $\vec{B}$ are

$$\begin{cases} \vec{E}' = (1 - \gamma)(\vec{E} \cdot \vec{v}_0)\vec{v}_0 + \gamma(\vec{E} + \vec{v} \times \vec{B}), \\ \vec{B}' = (1 - \gamma)(\vec{B} \cdot \vec{v}_0)\vec{v}_0 + \gamma\left(\vec{B} - \frac{1}{c^2}\vec{v} \times \vec{E}\right), \end{cases} \tag{10.1.1}$$

where $\gamma = (1 - v^2/c^2)^{-1/2} = (1 - \beta^2)^{-1/2}$.

Solution

For the beginning, we will write the matrix of the Lorentz-Herglotz transformation

$$\vec{r}' = \vec{r} + (\gamma - 1)(\vec{r} \cdot \vec{v}_0)\vec{v}_0 - \gamma\vec{v}t, \tag{10.1.2}$$

which is

$$\overline{X} \equiv (\overline{x}_i^k)$$

$$= \begin{pmatrix} 1 + (\gamma - 1)\alpha_1^2 & (\gamma - 1)\alpha_1\alpha_2 & (\gamma - 1)\alpha_1\alpha_3 & -\frac{v}{c}\gamma\alpha_1 \\ (\gamma - 1)\alpha_2\alpha_1 & 1 + (\gamma - 1)\alpha_2^2 & (\gamma - 1)\alpha_2\alpha_3 & -\frac{v}{c}\gamma\alpha_2 \\ (\gamma - 1)\alpha_3\alpha_1 & (\gamma - 1)\alpha_3\alpha_2 & 1 + (\gamma - 1)\alpha_3^2 & -\frac{v}{c}\gamma\alpha_3 \\ -\frac{v}{c}\gamma\alpha_1 & -\frac{v}{c}\gamma\alpha_2 & -\frac{v}{c}\gamma\alpha_3 & \gamma \end{pmatrix}, \tag{10.1.3}$$

where α_1, α_2 and α_3 are direction cosines (or directional cosines) of the direction defined by the versor $\vec{v}_0 = \vec{v}/|\vec{v}|$.

Next, we will develop the transformation law of the electromagnetic field tensor F^{ik},

$$F'^{ik} = \overline{x}_l^i \overline{x}_m^k F^{lm}, \tag{10.1.4}$$

DOI: 10.1201/9781003402602-10

where (see §A.2 from **Appendix A**),

$$\overline{x}_i^k = \frac{\partial x'^k}{\partial x^i}.$$

Choosing, for example, $i = 1$ and $k = 2$, we have

$$F'^{12} = \overline{x}_l^1 \overline{x}_m^2 F^{lm}$$
$$= \left(\overline{x}_1^1 \overline{x}_2^2 - \overline{x}_2^1 \overline{x}_1^2\right) F^{12} + \left(\overline{x}_1^1 \overline{x}_3^2 - \overline{x}_3^1 \overline{x}_1^2\right) F^{13} + \left(\overline{x}_1^1 \overline{x}_4^2 - \overline{x}_4^1 \overline{x}_1^2\right) F^{14}$$
$$+ \left(\overline{x}_2^1 \overline{x}_3^2 - \overline{x}_3^1 \overline{x}_2^2\right) F^{23} + \left(\overline{x}_2^1 \overline{x}_4^2 - \overline{x}_4^1 \overline{x}_2^2\right) F^{24} + \left(\overline{x}_3^1 \overline{x}_4^2 - \overline{x}_4^1 \overline{x}_3^2\right) F^{34}.$$

Using the matrix in relation (10.1.3) and performing some term reductions, we obtain

$$F'^{12} = \left[1 + (\gamma - 1)(\alpha_1^2 + \alpha_2^2)\right] F^{12} + (\gamma - 1)\alpha_2\alpha_3 F^{13}$$
$$- (\gamma - 1)\alpha_1\alpha_3 F^{23} + \frac{v}{c}\gamma\left(\alpha_1 F^{24} - \alpha_2 F^{14}\right).$$

Thus, if we take into account the significance of the components of F^{ik} and group the terms, we have

$$B'_z = B'_3 = (1 - \gamma)(\vec{B} \cdot \vec{v}_0)\alpha_3 + \gamma\left(\vec{B} - \frac{1}{c^2}\vec{v} \times \vec{E}\right)_3,$$

which is the z-axis component of the relation

$$\vec{B}' = (1 - \gamma)(\vec{B} \cdot \vec{v}_0)\vec{v}_0 + \gamma\left(\vec{B} - \frac{1}{c^2}\vec{v} \times \vec{E}\right).$$

The other components of $\vec{B}$, together with the components of $\vec{E}$, are obtained in a similar way.

10.2   Problem No. 70

If in a fixed inertial reference system the electric field intensity $\vec{E}$ and the magnetic field induction $\vec{B}$ are orthogonal and not equal in modulus,
a) determine the velocity $\vec{v}_B$ of an inertial reference system in which there is only a magnetic field;
b) determine the velocity $\vec{v}_E$ of an inertial reference system in which there is only an electric field.

Solution

If the relative displacement of the inertial reference systems S and S' takes place with the relative velocity $\vec{v}$, in an arbitrary direction of versor $\hat{v} = \vec{v}/v$, where $v = |\vec{v}|$, then the (general) relations of the field transformations are (see the previous problem):

$$\begin{cases} \vec{E}' = (1-\gamma)(\vec{E} \cdot \vec{v})\dfrac{\vec{v}}{v^2} + \gamma(\vec{E} + \vec{v} \times \vec{B}), \\[2mm] \vec{B}' = (1-\gamma)(\vec{B} \cdot \vec{v})\dfrac{\vec{v}}{v^2} + \gamma\left(\vec{B} - \dfrac{\vec{v}}{c^2} \times \vec{E}\right), \end{cases} \qquad (10.2.5)$$

where

$$\gamma = \left(1 - \frac{v^2}{c^2}\right)^{-1/2} = (1 - \beta^2)^{-1/2}.$$

a) Since in the inertial reference system S' there is no electric field, that is $\vec{E}' = 0$, in agreement with Eq. $(10.2.5)_1$ we have

$$(1-\gamma)(\vec{E} \cdot \vec{v})\frac{\vec{v}}{v^2} + \gamma(\vec{E} + \vec{v} \times \vec{B}) = 0. \qquad (10.2.6)$$

Supposing that S is fixed, the point **a)** of the problem rests in determination of the relative velocity $\vec{v}$ (which is now the velocity of the system S') that satisfies the condition expressed by Eq. (10.2.6). Taking the scalar product of this relation with the vector $\vec{B}$ and using the problem assumption that $\vec{E} \cdot \vec{B} = 0$, one obtains

$$(1-\gamma)(\vec{E} \cdot \vec{v})\frac{\vec{v} \cdot \vec{B}}{v^2} + \gamma\vec{B} \cdot (\vec{E} + \vec{v} \times \vec{B}) = 0 \quad \Rightarrow$$

$$(1-\gamma)(\vec{E} \cdot \vec{v})\frac{\vec{v} \cdot \vec{B}}{v^2} + \gamma\vec{B} \cdot (\vec{v} \times \vec{B}) = 0, \qquad (10.2.7)$$

and, since $\vec{B} \cdot (\vec{v} \times \vec{B}) = 0$, we still have

$$\frac{(1-\gamma)}{v^2}(\vec{E} \cdot \vec{v})(\vec{v} \cdot \vec{B}) = 0.$$

Excluding the case $\gamma = 1$, which would mean $v = 0$, the last relation is mathematically possible if

$$i)\ \vec{v} \cdot \vec{E} = 0 \text{ and } \vec{v} \cdot \vec{B} \neq 0, \text{ or}$$

$$ii)\ \vec{v} \cdot \vec{B} = 0 \text{ and } \vec{v} \cdot \vec{E} \neq 0, \text{ or}$$

$$iii)\ \vec{v} \cdot \vec{E} = 0 \text{ and } \vec{v} \cdot \vec{B} = 0.$$

In the $i)$ sub-case, the velocity $\vec{v}$ must be as $\vec{v} = \lambda\vec{n}$, with $\vec{n} \cdot \vec{v} = 0$, which means that $\vec{n}$ must be a versor placed in the plane orthogonal on the direction of the electric field $\vec{E}$, i.e., $\vec{n} \perp \vec{E}$, and also in the same plane, making an angle different from $\pi/2$ with vector $\vec{B}$. In this situation, the problem is to determine λ, since the direction and sense of velocity have been already determined (the direction and sense of the versor $\vec{n}$).

By introducing $\vec{v} = \lambda \vec{n}$ into relation (10.2.6) and taking into account that $\vec{n} \cdot \vec{E} = 0$, it turns out that $\vec{E} + \lambda \vec{n} \times \vec{B} = 0$. Taking the scalar product of this relation with the vector $\vec{n} \times \vec{B}$, one finds that

$$\lambda = -\frac{\vec{E} \cdot (\vec{n} \times \vec{B})}{(\vec{n} \times \vec{B}) \cdot (\vec{n} \times \vec{B})} = \frac{\vec{E} \cdot (\vec{B} \times \vec{n})}{\left|\vec{n} \times \vec{B}\right|^2} = \vec{n} \cdot \frac{\vec{E} \times \vec{B}}{\left|\vec{n} \times \vec{B}\right|^2},$$

therefore the searched velocity is

$$\vec{v}_B^{(i)} = \lambda \vec{n} = -\frac{\vec{E} \cdot (\vec{n} \times \vec{B})}{\left|\vec{n} \times \vec{B}\right|^2} \, \vec{n}. \tag{10.2.8}$$

In the *ii)* sub-case the velocity $\vec{v}$ must be of the form $\vec{v} = \lambda \vec{n}$ with $\vec{n} \cdot \vec{B} = 0$, meaning that $\vec{n}$ must be a versor placed in the plane perpendicular to direction of the magnetic induction $\vec{B}$, *i.e.*, $\vec{n} \perp \vec{B}$, and also in the same plane with the vector of electric field intensity $\vec{E}$, the angle between the two vectors being different from $\pi/2$. Introducing $\vec{v} = \lambda \vec{n}$ into relation (10.2.6), we have

$$(1 - \gamma)(\vec{E} \cdot \vec{n})\vec{n} + \gamma \vec{E} + \lambda \gamma \vec{n} \times \vec{B} = 0,$$

or, as a result of scalar multiplication by vector $\vec{n}$,

$$(1 - \gamma)(\vec{E} \cdot \vec{n}) + \gamma(\vec{E} \cdot \vec{n}) = 0 \iff \vec{E} \cdot \vec{n} = 0.$$

This result contradicts the assumption according to which $\vec{v} \cdot \vec{E} = \lambda \vec{n} \cdot \vec{E} \neq 0$. It then follows that this sub-case is not physically achievable.

Finally, in the *iii)* sub-case the vector $\vec{v}$ must be perpendicular on both $\vec{E}$ and $\vec{B}$, in other words the vectors $\vec{E}$, $\vec{B}$ and $\vec{v}$ must form an orthogonal trihedron, which implies the fact that the velocity $\vec{v}$ must be of the form $\vec{v} = \lambda \vec{E} \times \vec{B}$, where λ is, for the moment, an undetermined scalar. By replacing $\vec{v} = \lambda \vec{E} \times \vec{B}$ in relation (10.2.6), one results

$$(1 - \gamma) \underbrace{\left[\vec{E} \cdot (\vec{E} \times \vec{B})\right]}_{=0} \frac{\vec{E} \times \vec{B}}{\left|\vec{E} \times \vec{B}\right|^2} + \gamma \left[\vec{E} + \lambda(\vec{E} \times \vec{B}) \times \vec{B}\right] = 0 \iff$$

$$\vec{E} - \lambda \vec{B} \times (\vec{E} \times \vec{B}) = 0 \iff \vec{E} - \lambda B^2 \vec{E} + \lambda \underbrace{\vec{B}(\vec{B} \cdot \vec{E})}_{=0} = 0 \iff \lambda = \frac{1}{B^2}.$$

Therefore, in this particular sub-case, the searched velocity is

$$\vec{v}_B^{(iii)} = \frac{\vec{E} \times \vec{B}}{B^2}. \tag{10.2.9}$$

b) In order to have only the electric field in S', the magnetic field $\vec{B}'$ must be
– obviously – equal to zero. Using Eq. (10.2.5)$_2$, this means that the following
condition must be fulfilled:

$$\left(1 - \gamma\right)\left(\vec{B} \cdot \vec{v}\right)\frac{\vec{v}}{v^2} + \gamma\left(\vec{B} - \frac{\vec{v}}{c^2} \times \vec{E}\right) = 0. \qquad (10.2.10)$$

Supposing again that the inertial reference system S is fixed, the requirement of the point **b)** of the problem comes to determine the relative velocity
$\vec{v}$ (which now is the velocity of the referential S') that satisfies the condition
expressed by Eq. (10.2.10). Performing the dot product of this relation by $\vec{E}$,
recalling that by the problem assumptions $\vec{E} \cdot \vec{B} = 0$, and using the fact that
a scalar triple product with at least two of them being collinear is zero, we
obtain the same condition as in the case **a)**, namely

$$\frac{\left(1 - \gamma\right)}{v^2}\left(\vec{E} \cdot \vec{v}\right)\left(\vec{v} \cdot \vec{B}\right) = 0,$$

with the same mathematical possibilities of realization, *i.e.*,

    i) $\vec{v} \cdot \vec{E} = 0$ and $\vec{v} \cdot \vec{B} \neq 0$, or

    ii) $\vec{v} \cdot \vec{B} = 0$ and $\vec{v} \cdot \vec{E} \neq 0$, or

    iii) $\vec{v} \cdot \vec{E} = 0$ and $\vec{v} \cdot \vec{B} = 0$.

Proceeding in the same way as in the case **a)**, for the present sub-case *i)*,
that is, for $\vec{v} \cdot \vec{E} = 0$ and $\vec{v} \cdot \vec{B} \neq 0$ to be valid, it is necessary that $\vec{v} = \lambda\vec{n}$,
with $\vec{n} \cdot \vec{E} = 0$, that is $\vec{n}$ must be a versor placed in the plane perpendicular
to $\vec{E}$, and also in the same plane with $\vec{B}$, the angle between $\vec{n}$ and $\vec{B}$ being
different from $\pi/2$. Introducing $\vec{v} = \lambda\vec{n}$ into relation (10.2.10), we obtain

$$\left(1 - \gamma\right)\left(\vec{B} \cdot \vec{n}\right)\vec{n} + \gamma\left(\vec{B} - \frac{\lambda}{c^2}\vec{n} \times \vec{E}\right) = 0,$$

and, by scalar multiplication with the versor $\vec{n}$, we have

$$\left(1 - \gamma\right)\left(\vec{B} \cdot \vec{n}\right) + \gamma\left(\vec{B} \cdot \vec{n}\right) = 0 \iff \vec{B} \cdot \vec{n} = 0 \iff \vec{B} \cdot \vec{v} = 0,$$

which contradicts the starting hypothesis according to which $\vec{v} \cdot \vec{B} \neq 0$. Therefore, this sub-case is physically unfeasible.

If $\vec{v} \cdot \vec{B} = 0$ and $\vec{v} \cdot \vec{E} \neq 0$, (the present sub-case *ii)*), then we must choose
$\vec{v} = \lambda\vec{n}$, with $\vec{n} \cdot \vec{B} = 0$, meaning that $\vec{n}$ must be a versor placed in the plane
perpendicular on $\vec{B}$ and, at the same time, it must be in the plane of $\vec{E}$, and
the angle between $\vec{E}$ and $\vec{n}$ must be different from $\pi/2$. Introducing $\vec{v} = \lambda\vec{n}$
defined this way into relation (10.2.10), one obtains

$$\vec{B} - \frac{\lambda}{c^2}\vec{n} \times \vec{E} = 0.$$

Performing a dot product between this relation and $\vec{n} \times \vec{E}$, one finds

$$\vec{B} \cdot \left(\vec{n} \times \vec{E}\right) - \frac{\lambda}{c^2}\left(\vec{n} \times \vec{E}\right) \cdot \left(\vec{n} \times \vec{E}\right) = 0 \ \Rightarrow \ \lambda = c^2\frac{\vec{B} \cdot \left(\vec{n} \times \vec{E}\right)}{\left|\vec{n} \times \vec{E}\right|^2},$$

meaning that in this sub-case the searched velocity is

$$\vec{v} \equiv \vec{v}_E^{(ii)} = c^2\frac{\vec{B} \cdot \left(\vec{n} \times \vec{E}\right)}{\left|\vec{n} \times \vec{E}\right|^2}\,\vec{n}. \tag{10.2.11}$$

Finally, the present sub-case *iii*), *i.e.*, $\vec{v}\cdot\vec{E} = 0$ and $\vec{v}\cdot\vec{B} = 0$ implies the fact that the velocity forms an orthogonal trihedron with $\vec{E}$ and $\vec{B}$, meaning that it must be of the form $\vec{v} = \lambda\vec{E} \times \vec{B}$. Using this result into relation (10.2.10), we have

$$(1-\gamma)\underbrace{\left[\vec{B} \cdot \left(\vec{E} \times \vec{B}\right)\right]}_{=0}\frac{\vec{E} \times \vec{B}}{\left|\vec{E} \times \vec{B}\right|^2} + \gamma\left[\vec{B} - \frac{\lambda}{c^2}(\vec{E} \times \vec{B}) \times \vec{E}\right] = 0 \ \Leftrightarrow$$

$$c^2\vec{B} + \lambda\vec{E} \times \left(\vec{E} \times \vec{B}\right) = 0 \ \Leftrightarrow \ c^2\vec{B} + \lambda\vec{E}\underbrace{\left(\vec{E} \cdot \vec{B}\right)}_{=0} - \lambda E^2\vec{B} = 0 \ \Rightarrow$$

$$\lambda = \frac{c^2}{E^2}.$$

Therefore, in this last sub-case, the searched velocity is

$$\vec{v} \equiv \vec{v}_E^{(iii)} = c^2\frac{\vec{E} \times \vec{B}}{E^2}. \tag{10.2.12}$$

10.3   Problem No. 71

The electric field intensity $\vec{E}$ and the magnetic induction $\vec{B}$ of a uniform electromagnetic field are given in a fixed reference system and, in addition, $\vec{E} \cdot \vec{B} > 0$. Determine the velocity $\vec{v}$ of the inertial reference system in which the two vectors are parallel.

Solution

The transformation relations of the electric and magnetic fields at the transition from the inertial reference system S to another one S' are

$$\begin{cases} \vec{E}' = (1-\gamma)\left(\vec{E} \cdot \vec{v}\right)\dfrac{\vec{v}}{v^2} + \gamma\left(\vec{E} + \vec{v} \times \vec{B}\right), \\[2mm] \vec{B}' = (1-\gamma)\left(\vec{B} \cdot \vec{v}\right)\dfrac{\vec{v}}{v^2} + \gamma\left(\vec{B} - \dfrac{1}{c^2}\vec{v} \times \vec{E}\right), \end{cases} \tag{10.3.13}$$

where $\vec{v}$ is the relative velocity of the two frames (here, the velocity of S' with respect to S). The reverse transition can be done using the transformations: $\vec{E} \leftrightarrow \vec{E}'$, $\vec{B} \leftrightarrow \vec{B}'$ and $\pm\vec{v} \leftrightarrow \mp\vec{v}$ in the relations (10.3.13).

According to the problem statement, in the inertial reference frame S the fields are uniform and, in addition, they satisfy the relation $\vec{E} \cdot \vec{B} > 0$. To be parallel in S', the two fields must satisfy the obvious condition $\vec{E}' \times \vec{B}' = 0$. In view of relations (10.3.13), this condition writes

$$\left[(1-\gamma)(\vec{E} \cdot \vec{v})\frac{\vec{v}}{v^2} + \gamma(\vec{E} + \vec{v} \times \vec{B}) \right]$$
$$\times \left[(1-\gamma)(\vec{B} \cdot \vec{v})\frac{\vec{v}}{v^2} + \gamma\left(\vec{B} - \frac{1}{c^2}\vec{v} \times \vec{E} \right) \right] = 0,$$

or, by applying the properties of the triple scalar product,

$$\gamma(1-\gamma)(\vec{E} \cdot \vec{v})\frac{\vec{v} \times \vec{B}}{v^2} - \gamma(1-\gamma)(\vec{E} \cdot \vec{v})\frac{\vec{v} \times (\vec{v} \times \vec{E})}{v^2 c^2}$$
$$- \gamma(1-\gamma)(\vec{B} \cdot \vec{v})\frac{\vec{v} \times \vec{E}}{v^2} + \gamma^2(\vec{E} \times \vec{B})$$
$$- \frac{\gamma^2}{c^2}\left[\vec{E} \times (\vec{v} \times \vec{E}) \right] - \gamma(1-\gamma)(\vec{B} \cdot \vec{v})\frac{\vec{v} \times (\vec{v} \times \vec{B})}{v^2}$$
$$- \gamma^2\left[\vec{B} \times (\vec{v} \times \vec{B}) \right] - \frac{\gamma^2}{c^2}(\vec{v} \times \vec{B}) \times (\vec{v} \times \vec{E}) = 0. \qquad (10.3.14)$$

Considering the fact that

$$\left[(\vec{v} \times \vec{B}) \times (\vec{v} \times \vec{E}) \right]_i = \varepsilon_{ijk}\varepsilon_{jlm}\varepsilon_{kpq}v_l v_p E_q B_m$$
$$= (\varepsilon_{jki}\varepsilon_{jlm})\varepsilon_{kpq}v_l v_p E_q B_m$$
$$= (\delta_{kl}\delta_{im} - \delta_{km}\delta_{il})\varepsilon_{kpq}v_l v_p E_q B_m$$
$$= \varepsilon_{lpq}v_l v_p E_q B_i - \varepsilon_{mpq}v_i v_p E_q B_m$$
$$= -(\varepsilon_{mpq}B_m v_p E_q)v_i = -\vec{B} \cdot (\vec{v} \times \vec{E})v_i,$$

or, in vector form,

$$(\vec{v} \times \vec{B}) \times (\vec{v} \times \vec{E}) = -\left[\vec{B} \cdot (\vec{v} \times \vec{E}) \right]\vec{v},$$

and using the formula for the vector triple product,

$$\vec{a} \times (\vec{b} \times \vec{c}) = \vec{b}(\vec{a} \cdot \vec{c}) - \vec{c}(\vec{a} \cdot \vec{b}),$$

by performing some simplifications, Eq. (10.3.14) rewrites as

$$\gamma(1-\gamma)(\vec{E}\cdot\vec{v})\frac{\vec{v}\times\vec{B}}{v^2} - \gamma(1-\gamma)(\vec{E}\cdot\vec{v})^2\frac{\vec{v}}{v^2c^2}$$

$$+\gamma(1-\gamma)(\vec{E}\cdot\vec{v})\frac{\vec{E}}{c^2} - \gamma(1-\gamma)(\vec{B}\cdot\vec{v})\frac{\vec{v}\times\vec{E}}{v^2}$$

$$+\gamma^2(\vec{E}\times\vec{B}) - \gamma^2E^2\frac{\vec{v}}{c^2} + \gamma^2(\vec{E}\cdot\vec{v})\frac{\vec{E}}{c^2}$$

$$-\gamma(1-\gamma)(\vec{B}\cdot\vec{v})^2\frac{\vec{v}}{v^2} + \gamma(\vec{B}\cdot\vec{v})\vec{B} - \gamma^2B^2\vec{v}$$

$$+\gamma^2\left[\vec{v}\cdot(\vec{E}\times\vec{B})\right]\frac{\vec{v}}{c^2} = 0. \tag{10.3.15}$$

Performing a vector multiplication by $\vec{v}$ on the left of the above relation, one obtains

$$\gamma(1-\gamma)(\vec{E}\cdot\vec{v})\frac{\vec{v}\times(\vec{v}\times\vec{B})}{v^2} + \gamma(1-\gamma)(\vec{E}\cdot\vec{v})\frac{\vec{v}\times\vec{E}}{c^2}$$

$$-\gamma(1-\gamma)(\vec{B}\cdot\vec{v})\frac{\vec{v}\times(\vec{v}\times\vec{E})}{v^2} + \gamma^2\left[\vec{v}\times(\vec{E}\times\vec{B})\right]$$

$$+\gamma^2(\vec{E}\cdot\vec{v})\frac{\vec{v}\times\vec{E}}{c^2} + \gamma(\vec{B}\cdot\vec{v})(\vec{v}\times\vec{B}) = 0.$$

Developing the vector triple products, we still have

$$\gamma(1-\gamma)(\vec{E}\cdot\vec{v})(\vec{B}\cdot\vec{v})\frac{\vec{v}}{v^2} - \gamma(1-\gamma)(\vec{E}\cdot\vec{v})\vec{B}$$

$$+\gamma(1-\gamma)(\vec{E}\cdot\vec{v})\frac{\vec{v}\times\vec{E}}{c^2} - \gamma(1-\gamma)(\vec{B}\cdot\vec{v})(\vec{E}\cdot\vec{v})\frac{\vec{v}}{v^2}$$

$$+\gamma(1-\gamma)(\vec{B}\cdot\vec{v})\vec{E} + \gamma^2(\vec{B}\cdot\vec{v})\vec{E} - \gamma^2(\vec{E}\cdot\vec{v})\vec{B}$$

$$+\gamma^2(\vec{E}\cdot\vec{v})\frac{\vec{v}\times\vec{E}}{c^2} + \gamma(\vec{B}\cdot\vec{v})(\vec{v}\times\vec{B}) = 0,$$

or

$$-\gamma(\vec{E}\cdot\vec{v})\vec{B} + \gamma(\vec{E}\cdot\vec{v})\frac{\vec{v}\times\vec{E}}{c^2} + \gamma(\vec{B}\cdot\vec{v})\vec{E} + \gamma(\vec{B}\cdot\vec{v})(\vec{v}\times\vec{B}) = 0,$$

which yields

$$(\vec{E}\cdot\vec{v})\left(\vec{B} - \frac{\vec{v}\times\vec{E}}{c^2}\right) = (\vec{B}\cdot\vec{v})(\vec{E} + \vec{v}\times\vec{B}).$$

This relation can be true for any $\vec{v} \neq 0$, with $\vec{E}\cdot\vec{B} > 0$, only if $\vec{E}\cdot\vec{v} = 0$ and $\vec{B}\cdot\vec{v} = 0$, that is, if $\vec{v} \perp \vec{E}$ and $\vec{v} \perp \vec{B}$, and this situation can take place only if the velocity vector is placed in the plane determined by the direction of

$\vec{E} \times \vec{B}$ (since, as well known, the direction of the vector product of two vectors is perpendicular to the plane determined by the two vectors). Therefore one necessarily results that $\vec{v} = \lambda \vec{E} \times \vec{B}$, and the problem rests in determination of the non-zero scalar λ. To do this, let us take the scalar product of Eq. (10.3.15) with the vector $\vec{v}$ and remember that $\vec{v} \cdot (\vec{v} \times \vec{E}) = 0$ and $\vec{v} \cdot (\vec{v} \times \vec{B}) = 0$. The result is

$$- \gamma(1-\gamma)\frac{(\vec{E}\cdot\vec{v})^2}{c^2} + \gamma(1-\gamma)\frac{(\vec{E}\cdot\vec{v})^2}{c^2} + \gamma^2\vec{v}\cdot(\vec{E}\times\vec{B})$$

$$- \gamma^2 E^2\frac{v^2}{c^2} + \gamma^2\frac{(\vec{E}\cdot\vec{v})^2}{c^2} - \gamma(1-\gamma)(\vec{B}\cdot\vec{v})^2 + \gamma(\vec{B}\cdot\vec{v})^2$$

$$- \gamma^2 B^2 v^2 + \gamma^2\left[\vec{v}\cdot(\vec{E}\times\vec{B})\right]\frac{v^2}{c^2} = 0,$$

or, performing the necessary simplifications,

$$\vec{v}\cdot(\vec{E}\times\vec{B})\left(1+\frac{v^2}{c^2}\right) - v^2\left(B^2+\frac{E^2}{c^2}\right) + \frac{(\vec{E}\cdot\vec{v})^2}{c^2} + (\vec{B}\cdot\vec{v})^2 = 0.$$

Since $\vec{E}\cdot\vec{v} = 0$, $\vec{B}\cdot\vec{v} = 0$ and $\vec{v} = \lambda\vec{E}\times\vec{B}$, the last relation becomes

$$\lambda|\vec{E}\times\vec{B}|^2\left(1+\frac{\lambda^2|\vec{E}\times\vec{B}|^2}{c^2}\right) - \lambda^2|\vec{E}\times\vec{B}|^2\left(B^2+\frac{E^2}{c^2}\right) = 0.$$

Simplifying by $\lambda|\vec{E}\times\vec{B}|^2 \neq 0$ and grouping the remaining terms, we obtain the following second degree algebraic equation in λ:

$$\lambda^2|\vec{E}\times\vec{B}|^2 - \lambda(E^2+c^2B^2) + c^2 = 0,$$

with the solutions

$$\lambda_{1,2} = \frac{E^2+c^2B^2 \pm \sqrt{(E^2+c^2B^2)^2 - 4c^2|\vec{E}\times\vec{B}|^2}}{2|\vec{E}\times\vec{B}|^2}$$

$$= \frac{E^2+c^2B^2 \pm \sqrt{E^4+c^4B^4+2c^2E^2B^2 - 4c^2E^2B^2\sin^2\left(\widehat{\vec{E},\vec{B}}\right)}}{2|\vec{E}\times\vec{B}|^2}$$

$$= \frac{E^2+c^2B^2 \pm \sqrt{E^4+c^4B^4-2c^2E^2B^2 + 4c^2E^2B^2\cos^2\left(\widehat{\vec{E},\vec{B}}\right)}}{2|\vec{E}\times\vec{B}|^2}$$

$$= \frac{E^2+c^2B^2 \pm \sqrt{(E^2-c^2B^2)^2 + 4c^2(\vec{E}\cdot\vec{B})^2}}{2|\vec{E}\times\vec{B}|^2}.$$

Therefore, from the mathematical point of view, the problem has two solutions, namely

$$\vec{v}_{1,2} = \frac{E^2 + c^2B^2 \pm \sqrt{\left(E^2 - c^2B^2\right)^2 + 4c^2\left(\vec{E} \cdot \vec{B}\right)^2}}{2\left|\vec{E} \times \vec{B}\right|^2} \vec{E} \times \vec{B}. \qquad (10.3.16)$$

To realize which of the solutions is physically acceptable, let us write relation (10.3.16) in the form

$$\frac{\left|\vec{v}_{1,2}\right|}{c} = \frac{E^2 + c^2B^2 \pm \sqrt{\left(E^2 - c^2B^2\right)^2 + 4c^2\left(\vec{E} \cdot \vec{B}\right)^2}}{2c\left|\vec{E} \times \vec{B}\right|}.$$

Since always we must have $\left|\vec{v}\right| \leq c$, the ratio $\frac{\left|\vec{v}_{1,2}\right|}{c}$ in the above relation must be less than or equal to 1. Consequently, it must be checked whether

$$E^2 + c^2B^2 + \sqrt{\left(E^2 - c^2B^2\right)^2 + 4c^2\left(\vec{E} \cdot \vec{B}\right)^2} \leq 2c\left|\vec{E} \times \vec{B}\right|, \qquad (10.3.17)$$

or/and

$$E^2 + c^2B^2 - \sqrt{\left(E^2 - c^2B^2\right)^2 + 4c^2\left(\vec{E} \cdot \vec{B}\right)^2} \leq 2c\left|\vec{E} \times \vec{B}\right|. \qquad (10.3.18)$$

For example, in the case of a plane electromagnetic wave for which $\left|\vec{E}\right| = E = c\left|\vec{B}\right| = cB$, relations (10.3.17) and (10.3.18) become

$$\sin\left(\widehat{\vec{E}, \vec{B}}\right) - \cos\left(\widehat{\vec{E}, \vec{B}}\right) \geq 1, \qquad (10.3.19)$$

and

$$\sin\left(\widehat{\vec{E}, \vec{B}}\right) + \cos\left(\widehat{\vec{E}, \vec{B}}\right) \geq 1, \qquad (10.3.20)$$

respectively. As can be seen, in the variation range of the angle between $\vec{E}$ and $\vec{B}$, $\left(\widehat{\vec{E}, \vec{B}}\right) \in [0, 2\pi]$, the inequality (10.3.19) is satisfied only in the subinterval $\left(\widehat{\vec{E}, \vec{B}}\right) \in \left[\frac{\pi}{2}, \pi\right]$, while the inequality (10.3.20) only in the subinterval $\left(\widehat{\vec{E}, \vec{B}}\right) \in \left[0, \frac{\pi}{2}\right]$. Since the problem requires $\vec{E} \cdot \vec{B} > 0$, from the physical point of view, only that subinterval where the cosine of the angle between the two vectors is positive must be chosen, which is the one corresponding to inequality (10.3.20), namely $\left(\widehat{\vec{E}, \vec{B}}\right) \in \left[0, \frac{\pi}{2}\right]$. Consequently, only solution with minus sign in front of the radical is physically acceptable, namely,

$$\vec{v} = \frac{E^2 + c^2B^2 - \sqrt{\left(E^2 - c^2B^2\right)^2 + 4c^2\left(\vec{E} \cdot \vec{B}\right)^2}}{2\left|\vec{E} \times \vec{B}\right|^2} \vec{E} \times \vec{B},$$

which also gives the final answer to the problem.

10.4   Problem No. 72

In the inertial reference system S an electromagnetic field is described by the vectors $\vec{E}$ and $\vec{B}$ which make the angle θ between them, with $|\vec{B}| = k|\vec{E}|/c$, where c is the speed of light in vacuum. Let S' be the inertial referential in which $\vec{E}'$ and $\vec{B}'$ are parallel and moves with velocity $\vec{v}$ with respect to S, so that $\vec{v} \perp \vec{E}$ and $\vec{v} \perp \vec{B}$.

a) Establish the equation $\beta = \beta(k, \theta)$, where $\beta = v/c$;

b) Show that the problem always admits a solution and calculate β in terms of θ, if $k = 1$. What happens when $\theta = \pi/2$?

Solution

a) The velocity $\vec{v}$ being orthogonal on both $\vec{E}$ and $\vec{B}$, it follows that the vectors $\vec{E}'$ and $\vec{B}'$ remain in the plane determined by $\vec{E}$ and $\vec{B}$. Let choose in this plane the x-axis parallel to $\vec{E}_\perp = \vec{E}$.

Then the intervening vectors have the components

$$\vec{E} = (E, 0),$$

$$\vec{B} = (B\cos\theta,\ B\sin\theta),$$

$$\vec{E}'_\perp = \gamma(E - vB\sin\theta,\ vB\cos\theta),$$

$$\vec{B}'_\perp = \gamma\left(B\cos\theta,\ B\sin\theta - \frac{v}{c^2}E\right),$$

where

$$\gamma = \frac{1}{\sqrt{1 - \frac{v^2}{c^2}}} = \frac{1}{\sqrt{1 - \beta^2}}. \tag{10.4.21}$$

Expressing the parallelism between $\vec{E}'$ and $\vec{B}'$ through the obvious relationship

$$\vec{u}_z \cdot (\vec{E}' \times \vec{B}') = 0,$$

where $\vec{u}_z$ is the versor of z-axis, that is through

$$E'_x B'_y - E'_y B'_x = 0, \tag{10.4.22}$$

and keeping into account that $|\vec{B}| = k|\vec{E}|/c$, one obtains the searched equation,

$$\beta^2 - \beta\frac{1 + k^2}{k\sin\theta} + 1 = 0. \tag{10.4.23}$$

This equation can also be found by means of more general considerations, appealing to the transformation relations of the electromagnetic field,

$$\begin{cases} \vec{E}' = (1 - \gamma)(\vec{E} \cdot \vec{v})\dfrac{\vec{v}}{v^2} + \gamma(\vec{E} + \vec{v} \times \vec{B}), \\ \vec{B}' = (1 - \gamma)(\vec{B} \cdot \vec{v})\dfrac{\vec{v}}{v^2} + \gamma\left(\vec{B} - \dfrac{1}{c^2}\vec{v} \times \vec{E}\right). \end{cases} \tag{10.4.24}$$

In vector form, the parallelism between $\vec{E}'$ and $\vec{B}'$ is expressed by the well-known relation

$$\vec{E}' \times \vec{B}' = 0. \tag{10.4.25}$$

Since $\vec{v}$ is perpendicular to both $\vec{E}$ and $\vec{B}$, i.e., $(\vec{v} \cdot \vec{E}) = 0$ and $(\vec{v} \cdot \vec{B}) = 0$, relations (10.4.24) become

$$\vec{E}' = \gamma(\vec{E} + \vec{v} \times \vec{B}),$$

$$\vec{B}' = \gamma\left(\vec{B} - \frac{1}{c^2}\vec{v} \times \vec{E}\right),$$

and, if these relations are introduced into the parallelism condition expressed by Eq. (10.4.25), we still have

$$\vec{E} \times \vec{B} - \frac{1}{c^2}\vec{E} \times (\vec{v} \times \vec{E}) - \vec{B} \times (\vec{v} \times \vec{B}) - \frac{1}{c^2}(\vec{v} \times \vec{B}) \times (\vec{v} \times \vec{E}) = 0,$$

or

$$\vec{E} \times \vec{B} - \frac{1}{c^2}\vec{E} \times (\vec{v} \times \vec{E})$$
$$- \vec{B} \times (\vec{v} \times \vec{B}) + \frac{\vec{v}}{c^2}\left[\vec{v} \cdot (\vec{E} \times \vec{B})\right] = 0, \tag{10.4.26}$$

where we took into account the relation

$$(\vec{v} \times \vec{B}) \times (\vec{v} \times \vec{E}) = -\vec{v}\left[\vec{B} \cdot (\vec{v} \times \vec{E})\right],$$

and the circular-shift property of the scalar triple product was used.

If we take the scalar product of vector relation (10.4.26) with the velocity vector $\vec{v}$, one obtains

$$\vec{v} \cdot (\vec{E} \times \vec{B}) - \frac{\vec{v}}{c^2} \cdot \left[\vec{E} \times (\vec{v} \times \vec{E})\right]$$
$$- \vec{v} \cdot \left[\vec{B} \times (\vec{v} \times \vec{B})\right] + \frac{v^2}{c^2}\left[\vec{v} \cdot (\vec{E} \times \vec{B})\right] = 0.$$

Using the relation

$$\vec{a} \times (\vec{b} \times \vec{c}) = \vec{b}(\vec{a} \cdot \vec{c}) - \vec{c}(\vec{a} \cdot \vec{b}),$$

we still have

$$\left(1+\frac{v^2}{c^2}\right)\left[\vec{v}\cdot(\vec{E}\times\vec{B})\right]-\frac{v^2}{c^2}E^2+\frac{\vec{v}\cdot\vec{E}}{c^2}\underbrace{(\vec{E}\cdot\vec{v})}_{=0}-v^2B^2+\underbrace{(\vec{v}\cdot\vec{B})}_{=0}$$

$$\times(\vec{B}\cdot\vec{v})=0 \Leftrightarrow \left(1+\frac{v^2}{c^2}\right)\left[\vec{v}\cdot(\vec{E}\times\vec{B})\right]-\frac{v^2}{c^2}E^2-v^2B^2=0,$$

because, as it is specified in the problem statement, $\vec{v}\perp\vec{E}$ and $\vec{v}\perp\vec{B}$. Since the angle between $\vec{E}$ and $\vec{B}$ is θ, and $|\vec{B}|=k|\vec{E}|/c$, it results that

$$\vec{E}\times\vec{B}=\vec{\nu}EB\sin\theta=\vec{\nu}\frac{k}{c}E^2\sin\theta,$$

where $\vec{\nu}$ is the versor of the vector $\vec{E}\times\vec{B}$, meaning that the unit vector $\vec{\nu}$ is perpendicular to both $\vec{E}$ and $\vec{B}$, that is, it is parallel to the velocity vector $\vec{v}$, $(\vec{v}\parallel\vec{\nu}\Rightarrow\vec{v}\cdot\vec{\nu}=v)$, as specified in the problem statement.

Under these circumstances, the previously found relation, *i.e.*,

$$\left(1+\frac{v^2}{c^2}\right)\left[\vec{v}\cdot(\vec{E}\times\vec{B})\right]-\frac{v^2}{c^2}E^2-v^2B^2=0,$$

becomes

$$\left(1+\frac{v^2}{c^2}\right)\frac{vk}{c}E^2\sin\theta-\frac{v^2}{c^2}E^2-v^2\frac{k^2}{c^2}E^2=0,$$

or, since $E\neq0$,

$$\left(1+\frac{v^2}{c^2}\right)\frac{vk}{c}\sin\theta-\left(1+k^2\right)\frac{v^2}{c^2}=0.$$

Using the notation $\beta\equiv v/c$, we still have

$$\left(1+\beta^2\right)k\sin\theta-\left(1+k^2\right)\beta=0,$$

or

$$\beta^2-\beta\frac{1+k^2}{k\sin\theta}+1=0,$$

that is, the relation (10.4.23).

b) As can be seen, the relation between β, k and θ is a second degree algebraic equation in $\beta=\frac{v}{c}$. The discriminant of this equation is

$$\Delta=\left(\frac{1+k^2}{k\sin\theta}\right)^2-4=\frac{1+2k^2+k^4-4k^2\sin^2\theta}{k^2\sin^2\theta}.$$

Since

$$1+2k^2+k^4-4k^2\sin^2\theta\geq1+2k^2+k^4-4k^2=\left(1-k^2\right)^2\geq0,$$

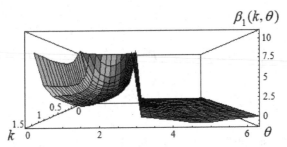

FIGURE 10.1
3D graphical representation of the solution $\beta_1 = \beta_1(k, \theta)$ for $k \in [0, 1.5]$ and $\theta \in [0, 2\pi]$ (a certain perspective).

it follows that the equation has always real solutions. These solutions are given by

$$\beta_{1,2} = \frac{\frac{1+k^2}{k\sin\theta} \pm \sqrt{\Delta}}{2} = \frac{\frac{1+k^2}{k\sin\theta} \pm \sqrt{\left(\frac{1+k^2}{k\sin\theta}\right)^2 - 4}}{2}$$

$$= \frac{1 + k^2}{2k\sin\theta} \pm \sqrt{\left(\frac{1 + k^2}{2k\sin\theta}\right)^2 - 1}. \qquad (10.4.27)$$

But, since $\beta = \frac{v}{c}$ must satisfy the requirement of the special relativity, $0 \leq \beta \leq 1$, it follows that not for all values of k and θ the two solutions are physically acceptable.

For $k \in [0, 1.5]$ and $\theta \in [0, 2\pi]$, the 3D graphic representation of the solution

$$\beta_1 = \frac{1 + k^2}{2k\sin\theta} + \sqrt{\left(\frac{1 + k^2}{2k\sin\theta}\right)^2 - 1}$$

shows as in Fig. 10.1 (a certain perspective), and Fig. 10.2, and Fig. 10.3 (two other perspectives); for comparison of 3D viewing parameters see the command lines written using Mathematica 5.0 software, corresponding to the three cases:

$$\text{Plot3D}[\frac{1 + k^2}{2 * k * \text{Sin}[\theta]} + \sqrt{\left(\frac{1 + k^2}{2 * k * \text{Sin}[\theta]}\right)^2 - 1}, \{k, 0., 1.5\}, \{\theta, 0., 2\pi\},$$

$$\text{ViewPoint} \rightarrow \{1.5, 0., 0.\}]$$

$$\text{Plot3D}[\frac{1 + k^2}{2 * k * \text{Sin}[\theta]} + \sqrt{\left(\frac{1 + k^2}{2 * k * \text{Sin}[\theta]}\right)^2 - 1}, \{k, 0., 1.5\}, \{\theta, 0., 2\pi\},$$

$$\text{ViewPoint} \rightarrow \{1.5, -1., 1.5\}]$$

$$\text{Plot3D}[\frac{1+k^2}{2*k*\text{Sin}[\theta]} + \sqrt{\left(\frac{1+k^2}{2*k*\text{Sin}[\theta]}\right)^2 - 1}, \{k, 0., 1.5\}, \{\theta, 0., 2\pi\},$$

$$\text{ViewPoint} \rightarrow \{1., 1., 1.\}]$$

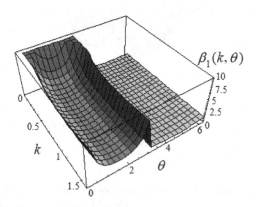

FIGURE 10.2
3D graphical representation of the solution $\beta_1 = \beta_1(k, \theta)$ for $k \in [0, 1.5]$ and $\theta \in [0, 2\pi]$ (another/a 2nd perspective).

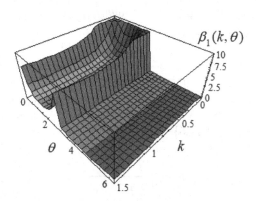

FIGURE 10.3
3D graphical representation of the solution $\beta_1 = \beta_1(k, \theta)$ for $k \in [0, 1.5]$ and $\theta \in [0, 2\pi]$ (another/a 3rd perspective).

For the same values of the variables, $k \in [0, 1.5]$ and $\theta \in [0, 2\pi]$, the 3D graphic representation of the solution

$$\beta_2 = \frac{1+k^2}{2k\sin\theta} - \sqrt{\left(\frac{1+k^2}{2k\sin\theta}\right)^2 - 1}$$

looks like in the Fig. 10.4 (a certain perspective), and Fig. 10.5, and Fig. 10.6

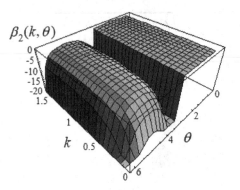

FIGURE 10.4
3D graphical representation of the solution $\beta_2 = \beta_2(k, \theta)$ for $k \in \left[0, 1.5\right]$ and $\theta \in \left[0, 2\pi\right]$ (a certain perspective).

(two other perspectives); for comparison of 3D viewing parameters see the command lines written using Mathematica 5.0 software, corresponding to the three cases:

$$\text{Plot3D}[\frac{1 + k^2}{2 * k * \text{Sin}[\theta]} - \sqrt{\left(\frac{1 + k^2}{2 * k * \text{Sin}[\theta]}\right)^2 - 1}, \{k, 0., 1.5\}, \{\theta, 0., 2\pi\},$$
$$\text{ViewPoint} \to \{-1., 1., 1.\}]$$

$$\text{Plot3D}[\frac{1 + k^2}{2 * k * \text{Sin}[\theta]} - \sqrt{\left(\frac{1 + k^2}{2 * k * \text{Sin}[\theta]}\right)^2 - 1}, \{k, 0., 1.5\}, \{\theta, 0., 2\pi\},$$
$$\text{ViewPoint} \to \{1.5, -1., 1.5\}]$$

$$\text{Plot3D}[\frac{1 + k^2}{2 * k * \text{Sin}[\theta]} - \sqrt{\left(\frac{1 + k^2}{2 * k * \text{Sin}[\theta]}\right)^2 - 1}, \{k, 0., 1.5\}, \{\theta, 0., 2\pi\},$$
$$\text{ViewPoint} \to \{1.5, 0., 0.\}]$$

If $k = 1$, Eq. (10.4.23) becomes

$$\beta^2 - \frac{2\beta}{\sin \theta} + 1 = 0, \tag{10.4.28}$$

with the solutions

$$\beta_{1,2} = \frac{1}{\sin \theta} \pm \sqrt{\left(\frac{1}{\sin \theta}\right)^2 - 1} = \frac{1}{\sin \theta} \pm \frac{\cos \theta}{\sin \theta},$$

that is, $\beta_1 = \cot \frac{\theta}{2}$ and $\beta_2 = \tan \frac{\theta}{2}$, whose graphic representations are given in the Fig. 10.7 (for $\beta_1 = \frac{1}{\sin \theta} + \cot \theta = \cot \frac{\theta}{2}$; the curve with vertical asymptotes) and Fig. 10.8 (for $\beta_2 = \frac{1}{\sin \theta} - \cot \theta = \tan \frac{\theta}{2}$; the curve with vertical asymptotes).

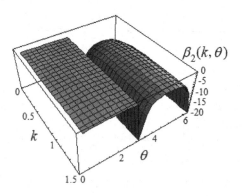

FIGURE 10.5
3D graphical representation of the solution $\beta_2 = \beta_2(k, \theta)$ for $k \in [0, 1.5]$ and $\theta \in [0, 2\pi]$ (another/a 2nd perspective).

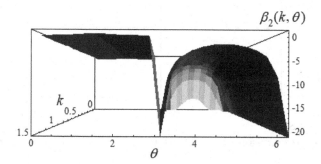

FIGURE 10.6
3D graphical representation of the solution $\beta_2 = \beta_2(k, \theta)$ for $k \in [0, 1.5]$ and $\theta \in [0, 2\pi]$ (another/a 3rd perspective).

b.1) The solution $\beta = \beta_1 = \cot \frac{\theta}{2}$. To visualize the range of values of the angle θ corresponding to the physically acceptable solution, on the graphic was also depicted the horizontal straight line $\beta_1^* = 1$. As can be seen, for $\theta > \pi$, $\beta_1 < 0$, which is not physically acceptable.

In the interval $\theta \in [0, 2\pi]$, the solution of the equation $\beta_1 = 1$ is $\theta = \frac{\pi}{2}$, that is, the range of values that are physically acceptable for the angle θ, corresponding to solution β_1, is $\theta \in \left(\frac{\pi}{2}, \pi\right)$. This can be graphically observed in Fig. 10.9, where the portion of Fig. 10.7 corresponding to the interval $\theta \in \left[\frac{\pi}{2} - 0.1, \pi + 0.1\right]$ is represented on the enlarged scale.

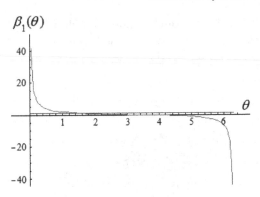

FIGURE 10.7
Graphical representation of the solution $\beta_1 = \beta_1(k = 1, \theta) \equiv \beta_1(\theta)$ for $\theta \in [0, 2\pi]$.

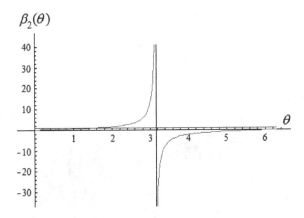

FIGURE 10.8
Graphical representation of the solution $\beta_2 = \beta_2(k = 1, \theta) \equiv \beta_2(\theta)$ for $\theta \in [0, 2\pi]$.

b.2) The solution $\beta = \beta_2 = \tan \frac{\theta}{2}$. To visualize (in this case as well) the range of values of the angle θ corresponding to the physically acceptable solution, on the graphic was also depicted the horizontal straight line $\beta_2^* = 1$. As can be observed in this case too, for $\theta > \pi$, $\beta_2 < 0$, which is not acceptable from the physical point of view.

Within the interval $\theta \in [0, 2\pi]$, equation $\beta_2 = 1$ has the solution $\theta = \frac{\pi}{2}$, therefore the range of values physically acceptable for the angle θ, corresponding to solution β_2, as also observed from the Fig. 10.10 (where the portion of Fig. 10.8 corresponding to the interval $\theta \in [-0.1, \frac{\pi}{2} + 0.1]$ is represented on a larger scale) is $\theta \in [0, \frac{\pi}{2}]$.

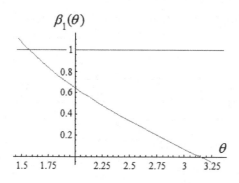

FIGURE 10.9
Detailed image of the portion of Fig. 10.7 corresponding to the interval $\theta \in \left[\frac{\pi}{2} - 0.1, \pi + 0.1\right]$.

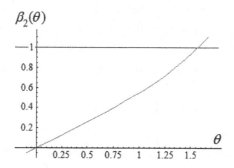

FIGURE 10.10
Detailed image of the portion of Fig. 10.8 corresponding to the interval $\theta \in \left[-0.1, \frac{\pi}{2} + 0.1\right]$.

To conclude, if $k = 1$, then

$$
\begin{cases}
\beta = \cot \dfrac{\theta}{2}, & \text{if } \theta \in \left[\dfrac{\pi}{2}, \pi\right], \\
\beta = \tan \dfrac{\theta}{2}, & \text{if } \theta \in \left[0, \dfrac{\pi}{2}\right].
\end{cases}
\tag{10.4.29}
$$

If $\theta = \frac{\pi}{2}$, then

$$
\begin{aligned}
(\beta_{1,2})_{\theta=\frac{\pi}{2}} &= \left[\frac{1+k^2}{2k\sin\theta} \pm \sqrt{\left(\frac{1+k^2}{2k\sin\theta}\right)^2 - 1}\right]_{\theta=\frac{\pi}{2}} \\
&= \frac{1+k^2}{2k} \pm \sqrt{\left(\frac{1+k^2}{2k}\right)^2 - 1} = \frac{1+k^2}{2k} \pm \sqrt{\left(\frac{1-k^2}{2k}\right)^2} \\
&= \frac{1+k^2}{2k} \pm \frac{1-k^2}{2k},
\end{aligned}
$$

therefore $\beta_1 = \frac{1}{k}$ if $k > 1$, and $\beta_2 = k$ if $k < 1$.

If $\beta = (\beta_1)_{\theta=\frac{\pi}{2}, k>1} = \frac{1}{k}$, then

$$\gamma\Big|_{\theta=\frac{\pi}{2}, k>1} = \frac{1}{\sqrt{1-\beta^2}} = \frac{1}{\sqrt{1-\frac{1}{k^2}}} = \frac{k}{\sqrt{k^2-1}},$$

and

$$(\vec{E}')_{\theta=\frac{\pi}{2}, k>1} = \left[\gamma(\vec{E} + \vec{v} \times \vec{B})\right]_{\gamma=\frac{k}{\sqrt{k^2-1}}}$$

$$= \frac{k}{\sqrt{k^2-1}}(\vec{E} + \vec{v} \times \vec{B})_{\beta=\frac{1}{k}}$$

$$= \frac{k}{\sqrt{k^2-1}}(\vec{E} + v\vec{\nu} \times \vec{B})_{\beta=\frac{1}{k}}$$

$$= \frac{k}{\sqrt{k^2-1}}(\vec{E} - \beta k \vec{E})_{\beta=\frac{1}{k}} = -\left(\frac{\beta k^2 - k}{\sqrt{k^2-1}}\right)_{\beta=\frac{1}{k}} \vec{E} = 0, (10.4.30)$$

where $\vec{\nu}$ is the versor of $\vec{E} \times \vec{B}$, and

$$(\vec{B}')_{\theta=\frac{\pi}{2}, k>1} = \left[\gamma\left(\vec{B} - \frac{\vec{v}}{c^2} \times \vec{E}\right)\right]_{\gamma=\frac{k}{\sqrt{k^2-1}}}$$

$$= \left[\gamma\left(\vec{B} - \frac{v}{c^2}\vec{\nu} \times \vec{E}\right)\right]_{\gamma=\frac{k}{\sqrt{k^2-1}}}$$

$$= \left[\gamma\left(\vec{B} - \frac{\beta}{k}\vec{B}\right)\right]_{\gamma=\frac{k}{\sqrt{k^2-1}}, \beta=\frac{1}{k}}$$

$$= \frac{k}{\sqrt{k^2-1}}\left(1 - \frac{1}{k^2}\right)\vec{B} = \frac{\sqrt{k^2-1}}{k}\vec{B}. \quad (10.4.31)$$

If $\beta = (\beta_2)_{\theta=\frac{\pi}{2}, k<1} = k$, then

$$\gamma\Big|_{\theta=\frac{\pi}{2}, k<1} = \frac{1}{\sqrt{1-\beta^2}} = \frac{1}{\sqrt{1-k^2}}$$

and

$$(\vec{E}')_{\theta=\frac{\pi}{2}, k<1} = \left[\gamma(\vec{E} + \vec{v} \times \vec{B})\right]_{\gamma=\frac{1}{\sqrt{1-k^2}}}$$

$$= \frac{1}{\sqrt{1-k^2}}(\vec{E} + \vec{v} \times \vec{B})_{\beta=k}$$

$$= \frac{1}{\sqrt{1-k^2}}(\vec{E} + v\vec{\nu} \times \vec{B})_{\beta=k}$$

$$= \frac{1}{\sqrt{1-k^2}}(\vec{E} - \beta k \vec{E})_{\beta=k} = \sqrt{1-k^2}\,\vec{E}, \quad (10.4.32)$$

and

$$
\begin{aligned}
(\vec{B}')_{\theta=\frac{\pi}{2},\,k<1} &= \left[\gamma\left(\vec{B} - \frac{\vec{v}}{c^2}\times\vec{E}\right)\right]_{\gamma=\frac{1}{\sqrt{1-k^2}}} \\
&= \left[\gamma\left(\vec{B} - \frac{v}{c^2}\vec{\nu}\times\vec{E}\right)\right]_{\gamma=\frac{1}{\sqrt{1-k^2}}} \\
&= \left[\gamma\left(\vec{B} - \frac{\beta}{k}\vec{B}\right)\right]_{\gamma=\frac{1}{\sqrt{1-k^2}},\,\beta=k} \\
&= \left[\frac{1-\frac{\beta}{k}}{\sqrt{1-k^2}}\right]_{\beta=k}\vec{B} = 0.
\end{aligned}
\tag{10.4.33}
$$

Finally, if $k=1$ and $\theta=\frac{\pi}{2}$, it follows that $\beta=\beta_1=\beta_2=1$, and according to relations (10.4.31) and (10.4.32) we have $\vec{B}'=0$ and $\vec{E}'=0$. It remains to show that relations (10.4.30) and (10.4.33) lead to the same result, that is

$$
\lim_{k\to 1}\left[\frac{\beta k^2 - k}{\sqrt{k^2-1}}\right]_{\beta=\frac{1}{k}} = 0,
$$

and

$$
\lim_{k\to 1}\left[\frac{1-\frac{\beta}{k}}{\sqrt{1-k^2}}\right]_{\beta=k} = 0.
$$

Indeed, using the L'Hôpital's rule, we have

$$
\begin{aligned}
\lim_{k\to 1}\left[\frac{\beta k^2 - k}{\sqrt{k^2-1}}\right]_{\beta=\frac{1}{k}} &= \lim_{k\to 1}\left[\frac{k-k}{\sqrt{k^2-1}}\right] \\
&= \lim_{k\to 1}\left[\frac{1-1}{\frac{k}{\sqrt{k^2-1}}}\right] = \lim_{k\to 1}\frac{(1-1)\sqrt{k^2-1}}{k} = 0
\end{aligned}
$$

and

$$
\begin{aligned}
\lim_{k\to 1}\left[\frac{1-\frac{\beta}{k}}{\sqrt{1-k^2}}\right]_{\beta=k} &= \lim_{k\to 1}\left[\frac{k-\beta}{k\sqrt{1-k^2}}\right]_{\beta=k} \\
&= \lim_{k\to 1}\frac{k-k}{k\sqrt{1-k^2}} = \lim_{k\to 1}\frac{1-1}{\sqrt{1-k^2}-\frac{k^2}{\sqrt{1-k^2}}} \\
&= \lim_{k\to 1}\frac{(1-1)\sqrt{1-k^2}}{1-2k^2} = 0.
\end{aligned}
$$

10.5   Problem No. 73

Determine the electric and magnetic components of the electromagnetic field created by an electric point-charge, q, which is in uniform and rectilinear motion with velocity $\vec{v}$.

Solution

We assume that the electric charge q is located at the origin O' of the inertial reference system S' (proper frame), being in rectilinear and uniform motion in the direction of the common $Ox \equiv O'x'$ axis. In the frame S' we obviously have

$$\vec{E}' = \frac{q}{4\pi\varepsilon_0}\frac{\vec{r}'}{r'^3}; \quad \vec{B}' = 0. \tag{10.5.34}$$

Since the components of $\vec{E}$ and $\vec{B}$ along the xx'-axis are not affected by the relative motion of the two inertial reference systems, considering the general relations of transformation (10.1.1), and relations (10.5.34) as well, in the frame S we have

$$\begin{cases} E_x = E_x' = \dfrac{1}{4\pi\varepsilon_0}\dfrac{q}{r'^3}x' = \gamma\dfrac{1}{4\pi\varepsilon_0}\dfrac{q}{s^3}(x - vt), \\[2mm] E_y = \gamma\big(E_y' + vB_z'\big) = \gamma\dfrac{1}{4\pi\varepsilon_0}\dfrac{q}{s^3}\,y, \\[2mm] E_z = \gamma\big(E_z' - vB_y'\big) = \gamma\dfrac{1}{4\pi\varepsilon_0}\dfrac{q}{s^3}\,z, \end{cases} \tag{10.5.35}$$

where $s^2 = \big[\gamma(x - vt)\big]^2 + y^2 + z^2$. Also,

$$\begin{cases} B_x = B_x' = 0, \\[2mm] B_y = \gamma\left(B_y' - \dfrac{v}{c^2}E_z'\right) = -\dfrac{v}{c^2}E_z = -\dfrac{\mu_0}{4\pi}\gamma\dfrac{qv}{s^3}\,z, \\[2mm] B_z = \gamma\left(B_z' + \dfrac{v}{c^2}E_y'\right) = \dfrac{v}{c^2}E_y = \dfrac{\mu_0}{4\pi}\gamma\dfrac{qv}{s^3}\,y. \end{cases} \tag{10.5.36}$$

Consequently,

$$\begin{cases} \vec{E} = \dfrac{1}{4\pi\varepsilon_0}q\gamma\,\dfrac{\vec{r} - \vec{v}t}{s^3}, \\[2mm] \vec{B} = \dfrac{\mu_0}{4\pi}q\gamma\dfrac{\vec{r}}{s^3} = \dfrac{1}{c^2}\,\vec{v}\times\vec{E}. \end{cases} \tag{10.5.37}$$

11

Relativistic-Covariant Dynamics

11.1 Problem No. 74

Study the relativistic motion of an electron placed in a uniform electric field $\vec{E}$, using the four-dimensional formalism.

Solution

We will use the hyperbolic representation of Minkowski space and we choose

$$\begin{cases} x^0 = ct, \\ x^1 = x, \\ x^2 = y, \\ x^3 = z, \end{cases}$$

as the components of the position four-vector x^μ, $\mu = 0, 1, 2, 3$. This choice leads to the metric

$$ds^2 = \left(dx^0\right)^2 - \left(dx^1\right)^2 - \left(dx^2\right)^2 - \left(dx^3\right)^2.$$

The components of the metric tensor therefore are

$$g_{\mu\nu} = \operatorname{diag}\left(+1, -1, -1, -1, \right),$$

i.e., we will work with the so-called -2 signature metric. The contravariant and covariant components of the momentum four-vector of a particle of total energy W are

$$\begin{cases} p^\mu = \left(\dfrac{W}{c}, \ \vec{p} \right), \\ p_\mu = \left(\dfrac{W}{c}, \ -\vec{p} \right), \end{cases}$$

while the relationships between the momentum four-vector and the velocity four-vector write as follows:

$$\begin{cases} p^\mu = m_0 c u^\mu, \\ p_\mu = m_0 c u_\mu, \quad \mu = \overline{0,3}, \end{cases}$$

DOI: 10.1201/9781003402602-11

with

$$u^{\mu} = \frac{1}{c}\frac{dx^{\mu}}{d\tau}.$$

Of course, $u_{\mu} = g_{\mu\nu}u^{\nu}$.

Suppose that at the initial moment the particle has the momentum $\vec{p}_0$ and is situated at the common origin of the proper reference frame and laboratory frame, whose $Ox \equiv Ox'$-axes are oriented in the direction of the uniform electric field $\vec{E}$. To start, let us consider the tensor form of the equations of motion as

$$\frac{d}{d\tau}(m_0 u^{\mu}) = -eF^{\mu\nu}u_{\nu}, \quad \mu, \nu = \overline{0,3}, \qquad (11.1.1)$$

where the proper time τ is chosen as parameter; besides, we took into account that the electron has negative charge (we consider $e > 0$). The electromagnetic field tensor $F_{\mu\nu}$, in the particular case of our problem, writes

$$(F_{\mu\nu}) = \begin{pmatrix} 0 & \dfrac{E}{c} & 0 & 0 \\ -\dfrac{E}{c} & 0 & 0 & 0 \\ 0 & 0 & 0 & 0 \\ 0 & 0 & 0 & 0 \end{pmatrix}, \qquad (11.1.2)$$

while $F^{\mu\nu}$ is

$$(F^{\mu\nu}) = \begin{pmatrix} 0 & -\dfrac{E}{c} & 0 & 0 \\ \dfrac{E}{c} & 0 & 0 & 0 \\ 0 & 0 & 0 & 0 \\ 0 & 0 & 0 & 0 \end{pmatrix}. \qquad (11.1.3)$$

Thus, the components of the tensor equation (11.1.1) are

$$\begin{cases} m_0 \dfrac{du^0}{d\tau} = m_0 \dfrac{du_0}{d\tau} = (-e)F^{01}u_1 = e\dfrac{E}{c}u_1, \\ m_0 \dfrac{du^1}{d\tau} = -m_0 \dfrac{du_1}{d\tau} = (-e)F^{10}u_0 = -e\dfrac{E}{c}u_0, \\ m_0 \dfrac{du^2}{d\tau} = -m_0 \dfrac{du_2}{d\tau} = 0, \\ m_0 \dfrac{du^3}{d\tau} = -m_0 \dfrac{du_3}{d\tau} = 0. \end{cases} \qquad (11.1.4)$$

These four equations form a system for the four components of the velocity four-vector u_{μ}, $\mu = \overline{0,3}$, expressed as functions of proper time $\tau : u_{\mu} = u_{\mu}(\tau)$. The first equation of the system successively gives

$$m_0 \frac{du_0}{d\tau} = e\frac{E}{c}u_1 \Rightarrow u_1 = \frac{c}{eE}m_0\frac{du_0}{d\tau} \Rightarrow \frac{du_1}{d\tau} = \frac{c}{eE}m_0\frac{d^2u_0}{d\tau^2}, \qquad (11.1.5)$$

and, if this result is introduced into the second equation of the system, we still have

$$\frac{cm_0^2}{eE}\frac{d^2u_0}{d\tau^2} = \frac{eE}{c}u_0 \;\Rightarrow\; \frac{d^2u_0}{d\tau^2} = \left(\frac{eE}{m_0c}\right)^2 u_0 = \omega_0^2 u_0, \qquad (11.1.6)$$

where we used the notation

$$\omega_0 \equiv \frac{eE}{m_0c}. \qquad (11.1.7)$$

Therefore we have

$$u_0 = C_0\cosh\omega_0\tau + C_1\sinh\omega_0\tau, \qquad (11.1.8)$$

where C_0 and C_1 are two arbitrary integration constants. Introducing Eq. (11.1.8) into the second equation of the system, it follows that

$$m_0\frac{du_1}{d\tau} = e\frac{E}{c}u_0 = e\frac{E}{c}(C_0\cosh\omega_0\tau + C_1\sinh\omega_0\tau) \;\Leftrightarrow$$

$$\frac{du_1}{d\tau} = \omega_0(C_0\cosh\omega_0\tau + C_1\sinh\omega_0\tau), \qquad (11.1.9)$$

so that

$$u_1 = C_0\sinh\omega_0\tau + C_1\cosh\omega_0\tau. \qquad (11.1.10)$$

The third and the fourth equations of the system easily give

$$\begin{cases} u_2 = C_2, \\ u_3 = C_3, \end{cases} \qquad (11.1.11)$$

where C_2 and C_3 are, also, arbitrary integration constants.

Supposing that the origin of the proper time τ coincides with the initial moment of the motion, the initial conditions are

$$u_0(0) \equiv u_{00} = \frac{W_0}{m_0c^2}; \quad \tau_0 = 0;\; t_0 = 0, \qquad (11.1.12)$$

$$u_1(0) \equiv u_{10} = -\frac{p_{0x}}{m_0c}; \quad x(0) \equiv x_0 = 0, \qquad (11.1.13)$$

$$u_2(0) \equiv u_{20} = -\frac{p_{0y}}{m_0c}; \quad y(0) \equiv y_0 = 0, \qquad (11.1.14)$$

$$u_3(0) \equiv u_{30} = -\frac{p_{0z}}{m_0c}; \quad z(0) \equiv z_0 = 0, \qquad (11.1.15)$$

because in the four-dimensional formalism, as we already mentioned, the momentum four-vector is

$$p_\mu = \left(\frac{W}{c}, -\vec{p}\right), \qquad (11.1.16)$$

and, since $p_\mu = m_0 c u_\mu$, we have

$$
\begin{cases}
u_0 = \dfrac{p_0}{m_0 c} = \dfrac{p^0}{m_0 c} = \dfrac{W}{m_0 c^2}, \\[2ex]
u_1 = \dfrac{p_1}{m_0 c} = -\dfrac{p^1}{m_0 c} = -\dfrac{p_x}{m_0 c}, \\[2ex]
u_2 = \dfrac{p_2}{m_0 c} = -\dfrac{p^2}{m_0 c} = -\dfrac{p_y}{m_0 c}, \\[2ex]
u_3 = \dfrac{p_3}{m_0 c} = -\dfrac{p^3}{m_0 c} = -\dfrac{p_z}{m_0 c}.
\end{cases}
\tag{11.1.17}
$$

In the above formulae, W_0 is the total energy of the particle at the initial moment, p_{0x}, p_{0y} and p_{0z} are the initial values of the momentum projections on the coordinate axes, and $u_{\mu 0} = u_\mu(\tau_0) = u_\mu(\tau = 0)$.

The integration constants C_μ, $\mu = \overline{0,3}$ are unequivocally determined by the considered initial conditions. It simply follows that

$$
\begin{cases}
C_0 = \dfrac{W_0}{m_0 c^2}, \\[2ex]
C_1 = -\dfrac{p_{0x}}{m_0 c}, \\[2ex]
C_2 = -\dfrac{p_{0y}}{m_0 c}, \\[2ex]
C_3 = -\dfrac{p_{0z}}{m_0 c},
\end{cases}
$$

in which case

$$
\begin{cases}
u_0(\tau) = \dfrac{W_0}{m_0 c^2} \cosh \omega_0 \tau - \dfrac{p_{0x}}{m_0 c} \sinh \omega_0 \tau, \\[2ex]
u_1(\tau) = \dfrac{W_0}{m_0 c^2} \sinh \omega_0 \tau - \dfrac{p_{0x}}{m_0 c} \cosh \omega_0 \tau, \\[2ex]
u_2(\tau) = -\dfrac{p_{0y}}{m_0 c}, \\[2ex]
u_3(\tau) = -\dfrac{p_{0z}}{m_0 c}.
\end{cases}
\tag{11.1.18}
$$

Given that

$$
\begin{cases}
u^0 = u_0 = \dfrac{1}{c}\dfrac{dx^0}{d\tau} = \dfrac{dt}{d\tau}, \\[2ex]
u^1 = -u_1 = \dfrac{1}{c}\dfrac{dx}{d\tau}, \\[2ex]
u^2 = -u_2 = \dfrac{1}{c}\dfrac{dy}{d\tau}, \\[2ex]
u^3 = -u_3 = \dfrac{1}{c}\dfrac{dz}{d\tau},
\end{cases}
\tag{11.1.19}
$$

and integrating, one obtains

$$
t = \int u_0 d\tau = \int \left[\frac{W_0}{m_0 c^2} \cosh\omega_0\tau - \frac{p_{0x}}{m_0 c} \sinh\omega_0\tau \right] d\tau
$$

$$
= \frac{W_0}{m_0\omega_0 c^2} \sinh\omega_0\tau - \frac{p_{0x}}{m_0\omega_0 c} \cosh\omega_0\tau + K_0. \qquad (11.1.20)
$$

Analogously, are obtained

$$
x = -c \int u_1 d\tau = -c \int \left[\frac{W_0}{m_0 c^2} \sinh\omega_0\tau - \frac{p_{0x}}{m_0 c} \cosh\omega_0\tau \right] d\tau
$$

$$
= -\frac{W_0}{m_0\omega_0 c} \cosh\omega_0\tau + \frac{p_{0x}}{m_0\omega_0} \sinh\omega_0\tau + K_1, \qquad (11.1.21)
$$

$$
y = -c \int u_2 d\tau = -c \int -\frac{p_{0y}}{m_0 c} d\tau = \frac{p_{0y}}{m_0}\tau + K_2, \qquad (11.1.22)
$$

$$
z = -c \int u_3 d\tau = -c \int -\frac{p_{0z}}{m_0 c} d\tau = \frac{p_{0z}}{m_0}\tau + K_3. \qquad (11.1.23)
$$

The arbitrary integration constants K_μ, $\mu = \overline{0,3}$ can be determined by means of the initial conditions for coordinates (written above). It follows that

$$
\begin{cases}
K_0 = \dfrac{p_{0x}}{m_0\omega_0 c}, \\[2mm]
K_1 = \dfrac{W_0}{m_0\omega_0 c}, \\[2mm]
K_2 = 0, \\[2mm]
K_3 = 0.
\end{cases} \qquad (11.1.24)
$$

Introducing these constants in the expressions of the spatial and temporal coordinates, one obtains

$$
\begin{cases}
t(\tau) = \dfrac{p_{0x}}{m_0\omega_0 c}\left(1 - \cosh\omega_0\tau\right) + \dfrac{W_0}{m_0\omega_0 c^2} \sinh\omega_0\tau, \\[3mm]
x(\tau) = \dfrac{p_{0x}}{m_0\omega_0} \sinh\omega_0\tau + \dfrac{W_0}{m_0\omega_0 c}\left(1 - \cosh\omega_0\tau\right), \\[3mm]
y(\tau) = \dfrac{p_{0y}}{m_0}\tau, \\[3mm]
z(\tau) = \dfrac{p_{0z}}{m_0}\tau.
\end{cases} \qquad (11.1.25)
$$

In order to determine the motion in the laboratory inertial frame, S, that is, $x = x(t)$, $y = y(t)$ and $z = z(t)$, the proper time τ must be expressed as a function of the time elapsed in the frame S. This can be done easily using the first relation from the above system. Considering that

$$
\sinh\xi = \frac{e^\xi - e^{-\xi}}{2},
$$

and

$$\cosh \xi = \frac{e^\xi + e^{-\xi}}{2},$$

the result is a second degree algebraic equation in $e^{\omega_0 t}$. Indeed,

$$t = \frac{p_{0x}}{m_0 \omega_0 c} \left(1 - \cosh \omega_0 \tau \right) + \frac{W_0}{m_0 \omega_0 c^2} \sinh \omega_0 \tau$$

$$= \frac{p_{0x}}{m_0 \omega_0 c} \left(1 - \frac{e^{\omega_0 \tau} + e^{-\omega_0 \tau}}{2}\right) + \frac{W_0}{m_0 \omega_0 c^2} \frac{e^{\omega_0 \tau} - e^{-\omega_0 \tau}}{2}, \quad (11.1.26)$$

or, using the notation $e^{\omega_0 \tau} \equiv \lambda$,

$$t = \frac{p_{0x}}{m_0 c \omega_0} \left(1 - \frac{\lambda + \lambda^{-1}}{2}\right) + \frac{W_0}{m_0 c^2 \omega_0} \frac{\lambda - \lambda^{-1}}{2}, \quad (11.1.27)$$

that is,

$$\lambda^2 (q - r) - 2(t - r)\lambda - (q + r) = 0, \quad (11.1.28)$$

where the obvious notations

$$r \equiv \frac{p_{0x}}{m_0 c \omega_0}$$

and

$$q \equiv \frac{W_0}{m_0 c^2 \omega_0}$$

have been used.

The solutions of the above second degree algebraic equation in λ are

$$\lambda_{1,2} = \frac{t - r \pm \sqrt{(t - r)^2 + (q + r)(q - r)}}{q - r}. \quad (11.1.29)$$

Returning to the initial quantities and retaining only the physically acceptable solution (λ must be positive), we have

$$\lambda = e^{\omega_0 \tau} = \frac{eEt - p_{0x} + \sqrt{\left(eEt - p_{0x}\right)^2 + m_0^2 c^2 + p_{0y}^2 + p_{0z}^2}}{\frac{W_0}{c} - p_{0x}}, \quad (11.1.30)$$

which yields

$$\tau = \frac{1}{\omega_0} \ln \frac{eEt - p_{0x} + \sqrt{\left(eEt - p_{0x}\right)^2 + m_0^2 c^2 + p_{0y}^2 + p_{0z}^2}}{\frac{W_0}{c} - p_{0x}}. \quad (11.1.31)$$

Introducing τ just determined into $x = x(\tau)$, $y = y(\tau)$ and $z = z(\tau)$, given by Eq. (11.1.25), we finally obtain

$$x(t) = \frac{c}{eE} \left[\frac{W_0}{c} - \sqrt{\left(eEt - p_{0x}\right)^2 + m_0^2 c^2 + p_{0y}^2 + p_{0z}^2}\right], \quad (11.1.32)$$

$$y(t) = \frac{cp_{0y}}{eE} \ln \frac{eEt - p_{0x} + \sqrt{\left(eEt - p_{0x}\right)^2 + m_0^2 c^2 + p_{0y}^2 + p_{0z}^2}}{\frac{W_0}{c} - p_{0x}}, \quad (11.1.33)$$

$$z(t) = \frac{cp_{0z}}{eE} \ln \frac{eEt - p_{0x} + \sqrt{\left(eEt - p_{0x}\right)^2 + m_0^2 c^2 + p_{0y}^2 + p_{0z}^2}}{\frac{W_0}{c} - p_{0x}}. \quad (11.1.34)$$

In the non-relativistic (N.R.) limit ($p_0 \ll m_0 c$ and $t \ll m_0 c/eE$), one obtains

$$x(t) = \frac{p_{0x}}{m_0} t - \frac{eE}{2m_0} t^2, \quad (11.1.35)$$

$$y(t) = \frac{p_{0y}}{m_0} t, \quad (11.1.36)$$

$$z(t) = \frac{p_{0z}}{m_0} t. \quad (11.1.37)$$

Indeed, in the considered approximation (*i.e.*, the N.R. approximation) we can write

$$\sqrt{\left(eEt - p_{0x}\right)^2 + m_0^2 c^2 + p_{0y}^2 + p_{0z}^2}$$
$$= \sqrt{e^2 E^2 t^2 - 2eEp_{0x}t + m_0^2 c^2 + p_0^2}$$
$$\simeq \sqrt{e^2 E^2 t^2 - 2eEp_{0x}t + m_0^2 c^2}$$
$$= m_0 c \sqrt{1 - \frac{2eEp_{0x}t}{m_0^2 c^2} + \frac{e^2 E^2 t^2}{m_0^2 c^2}} \equiv m_0 c \sqrt{1 - \xi},$$

where we have introduced the new notation

$$\xi \equiv \frac{2eEp_{0x}t}{m_0^2 c^2} - \frac{e^2 E^2 t^2}{m_0^2 c^2},$$

with $\xi \ll 1$.

In this case, considering the Mac-Laurin series expansion of the function

$$f(\xi) = \sqrt{1 - \xi} = 1 - \frac{\xi}{2} + \mathcal{O}(\xi^2),$$

and keeping only the zeroth and first-order terms, we have

$$m_0 c \sqrt{1 - \xi} \simeq m_0 c \left(1 - \frac{eEp_0 t}{m_0^2 c^2} + \frac{1}{2} \frac{e^2 E^2 t^2}{m_0^2 c^2}\right),$$

and then Eq. (11.1.32) gives

$$x(t) = \frac{c}{eE} \left[\frac{W_0}{c} - \sqrt{\left(eEt - p_{0x}\right)^2 + m_0^2 c^2 + p_{0y}^2 + p_{0z}^2}\right]$$
$$= \frac{W_0}{eE} - \frac{c}{eE} \sqrt{\left(eEt - p_{0x}\right)^2 + m_0^2 c^2 + p_{0y}^2 + p_{0z}^2}$$
$$\simeq \frac{W_0}{eE} - \frac{m_0 c^2}{eE} \left(1 - \frac{eEp_{0x}t}{m_0^2 c^2} + \frac{1}{2} \frac{e^2 E^2 t^2}{m_0^2 c^2}\right) = \frac{p_{0x}}{m_0} t - \frac{1}{2} \frac{eE}{m_0} t^2,$$

where we took into account that, within the considered approximation, $W_0 \simeq m_0 c^2$. This way, the relation (11.1.35) has been proved. In order to justify the other two relations, namely, (11.1.36) and (11.1.37), we observe that, if we consider only the first term (the zeroth-order term) of the series expansion of $f(\xi)$, we can write

$$\sqrt{\left(eEt - p_{0x}\right)^2 + m_0^2 c^2 + p_{0y}^2 + p_{0z}^2} \simeq m_0 c,$$

and then

$$\frac{eEt - p_{0x} + \sqrt{\left(eEt - p_{0x}\right)^2 + m_0^2 c^2 + p_{0y}^2 + p_{0z}^2}}{\frac{W_0}{c} - p_{0x}} \simeq \frac{eEt - p_{0x} + m_0 c}{\frac{W_0}{c} - p_{0x}}$$

$$\simeq \frac{eEt - p_{0x} + m_0 c}{m_0 c - p_{0x}} \simeq \frac{eEt + m_0 c}{m_0 c} = 1 + \frac{eEt}{m_0 c}.$$

Then, using the series expansion below,

$$\ln\left(1 + \xi\right) = \xi - \frac{\xi^2}{2} + \frac{\xi^3}{3} - \frac{\xi^4}{4} + \dots = -\sum_{n=1}^{\infty}(-1)^n \frac{\xi^n}{n},$$

where

$$\xi \equiv \frac{eEt}{m_0 c} \ll 1,$$

and considering only its first term, formula (11.1.33) gives

$$y(t) = \frac{c p_{0y}}{eE} \ln \frac{eEt - p_{0x} + \sqrt{\left(eEt - p_{0x}\right)^2 + m_0^2 c^2 + p_{0y}^2 + p_{0z}^2}}{\frac{W_0}{c} - p_{0x}}$$

$$\simeq \frac{c p_{0y}}{eE} \ln(1 + \xi) \simeq \frac{c p_{0y}}{eE} \xi = \frac{c p_{0y}}{eE} \frac{eEt}{m_0 c} = \frac{p_{0y}}{m_0} t,$$

which proves the relation (11.1.36). The relation (11.1.37) can be also verified in a very similar way.

It can be observed that in the non-relativistic case/limit the motion along the x-axis is uniformly accelerated, with acceleration $\frac{eE}{m_0}$. This motion is graphically represented in Fig. 11.1 (the first part of the full line).

After a long enough time $\left(t \gg m_0 c/eE\right)$, the velocity of the particle becomes relativistic $(v \simeq c)$, even if it was very small at the initial moment.

In the ultra-relativistic (U.R.) limit $(v \simeq c)$, we can write

$$x(t) = \frac{m_0 c^2}{eE} - ct, \tag{11.1.38}$$

$$y(t) = \frac{c p_{0y}}{eE} \ln \frac{2eEt}{m_0 c}, \tag{11.1.39}$$

$$z(t) = \frac{c p_{0z}}{eE} \ln \frac{2eEt}{m_0 c}. \tag{11.1.40}$$

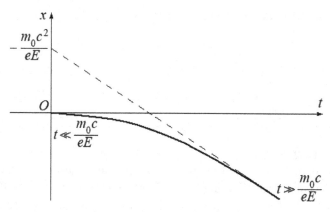

FIGURE 11.1

Graphical representation of the law of motion (the projection on the x-axis), $x = x(t)$, for an electron moving in a uniform electric field $\vec{E} = (E, 0, 0)$.

Indeed, in this case $(t \gg m_0 c/eE)$ we have

$$\sqrt{\left(eEt - p_{0x}\right)^2 + m_0^2 c^2 + p_{0y}^2 + p_{0z}^2} = \sqrt{e^2 E^2 t^2 - 2eEp_{0x}t + m_0^2 c^2 + p_0^2}$$
$$\simeq \sqrt{e^2 E^2 t^2} = eEt,$$

and, according to relation (11.1.32),

$$x(t) = \frac{c}{eE}\left[\frac{W_0}{c} - \sqrt{\left(eEt - p_{0x}\right)^2 + m_0^2 c^2 + p_{0y}^2 + p_{0z}^2}\right]$$
$$= \frac{c}{eE}\frac{W_0}{c} - \frac{c}{eE}\sqrt{\left(eEt - p_{0x}\right)^2 + m_0^2 c^2 + p_{0y}^2 + p_{0z}^2}$$
$$\simeq \frac{c}{eE}\frac{m_0 c^2}{c} - \frac{c}{eE}\sqrt{\left(eEt\right)^2} = \frac{m_0 c^2}{eE} - ct,$$

that is, the relation (11.1.38).

Next, using Eq. (11.1.33) and the inequality $t \gg m_0 c/eE$, then for $y(t)$ we can write

$$y(t) = \frac{cp_{0y}}{eE}\ln\frac{eEt - p_{0x} + \sqrt{\left(eEt - p_{0x}\right)^2 + m_0^2 c^2 + p_{0y}^2 + p_{0z}^2}}{\frac{W_0}{c} - p_{0x}}$$

$$\simeq \frac{cp_{0y}}{eE}\ln\frac{eEt - p_{0x} + \sqrt{\left(eEt\right)^2}}{\frac{m_0 c^2}{c} - p_{0x}} \simeq \frac{cp_{0y}}{eE}\ln\frac{2eEt}{m_0 c},$$

and, in a similar way, for $z(t)$ the following relation can be obtained:

$$z(t) \simeq \frac{cp_{0z}}{eE}\ln\frac{2eEt}{m_0 c}.$$

348 Relativistic-Covariant Dynamics

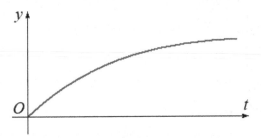

FIGURE 11.2
Graphical representation of the law of motion (the projection on the y-axis), $y = y(t)$, for an electron moving in a uniform electric field $\vec{E} = (E, 0, 0)$.

The graphical representation of this motion (*i.e.*, in the U.R. limit) is given in Fig. 11.1 (the dotted line, for $x(t)$), and Fig. 11.2 (for $y(t)$ and $z(t)$; since they are very similar, we have plotted only the first).

11.2   Problem No. 75

Study the relativistic motion of an electron placed in a uniform magnetic field $\vec{B}$, using the four-dimensional formalism.

Solution

As in the previous application, we suppose that at the initial moment the momentum $\vec{p}_0$ of the particle is given and the particle is situated at the common origin of the proper reference frame and laboratory frame, whose z-axis is oriented in the direction of the magnetic field.

We shall use again the covariant form of the equation of motion for the electron,

$$\frac{d}{d\tau}(m_0 u^\mu) = -eF^{\mu\nu}u_\nu, \quad \mu, \nu = \overline{0, 3}, \tag{11.2.41}$$

with the proper time τ as parameter. Besides, we took into account that the electron charge is negative and we have chosen $e > 0$. In this case, the electromagnetic field tensor $F_{\mu\nu}$ is

$$(F_{\mu\nu}) = \begin{pmatrix} 0 & 0 & 0 & 0 \\ 0 & 0 & -B_z & 0 \\ 0 & B_z & 0 & 0 \\ 0 & 0 & 0 & 0 \end{pmatrix} = \begin{pmatrix} 0 & 0 & 0 & 0 \\ 0 & 0 & -B & 0 \\ 0 & B & 0 & 0 \\ 0 & 0 & 0 & 0 \end{pmatrix}, \tag{11.2.42}$$

and $F^{\mu\nu}$ writes

$$(F^{\mu\nu}) = \begin{pmatrix} 0 & 0 & 0 & 0 \\ 0 & 0 & -B_z & 0 \\ 0 & B_z & 0 & 0 \\ 0 & 0 & 0 & 0 \end{pmatrix} = \begin{pmatrix} 0 & 0 & 0 & 0 \\ 0 & 0 & -B & 0 \\ 0 & B & 0 & 0 \\ 0 & 0 & 0 & 0 \end{pmatrix}. \tag{11.2.43}$$

Thus, the equation of motion are written like this

$$\begin{cases} m_0 \dfrac{du_0}{d\tau} = 0, \\ m_0 \dfrac{du_1}{d\tau} = -eBu_2, \\ m_0 \dfrac{du_2}{d\tau} = +eBu_1, \\ m_0 \dfrac{du_3}{d\tau} = 0. \end{cases} \tag{11.2.44}$$

The second equation from the above system gives

$$m_0 \frac{du_1}{d\tau} = -eBu_2 \ \Rightarrow \ u_2 = -\frac{m_0}{eB}\frac{du_1}{d\tau} \ \Rightarrow \ \frac{du_2}{d\tau} = -\frac{m_0}{eB}\frac{d^2u_1}{d\tau^2}, \tag{11.2.45}$$

and, if relation (11.2.45) is introduced into the third equation of the system, we obtain the following second-order differential equation for u_1:

$$-\frac{m_0^2}{eB}\frac{d^2u_1}{d\tau^2} = eBu_1 \ \Rightarrow \ \frac{d^2u_1}{d\tau^2} = -\left(\frac{eB}{m_0}\right)^2 u_1 = -\omega_0^2 u_1, \tag{11.2.46}$$

where we used the notation

$$\omega_0 \equiv \frac{eB}{m_0}.$$

In this case, we have

$$u_1 = A_1 \sin\left(\omega_0\tau + \alpha\right), \tag{11.2.47}$$

where A_1 and α are two arbitrary integration constants. If now we use this result (for u_1) in the second equation of the system, we obtain

$$m_0 \frac{du_1}{d\tau} = -eBu_2 \ \Rightarrow \ u_2 = -\frac{1}{\omega_0}\omega_0 A_1 \cos(\omega_0\tau + \alpha)$$
$$= -A_1 \cos(\omega_0\tau + \alpha). \tag{11.2.48}$$

The first and fourth equations of the system give in a simple manner

$$u_0 = A_0, \tag{11.2.49}$$

$$u_3 = A_3, \tag{11.2.50}$$

where A_0 and A_3 are, also, arbitrary integration constants.

Assuming that the origin of the proper time τ coincides with the initial moment of the motion, the initial conditions are

$$\begin{cases} u_0(0) \equiv u_{00} = \dfrac{W_0}{m_0 c^2}; & \tau_0 = 0, \ t_0 = 0, \\[2mm] u_1(0) \equiv u_{10} = -\dfrac{p_{0x}}{m_0 c}; & x(0) \equiv x_0 = 0, \\[2mm] u_2(0) \equiv u_{20} = -\dfrac{p_{0y}}{m_0 c}; & y(0) \equiv y_0 = 0, \\[2mm] u_3(0) \equiv u_{30} = -\dfrac{p_{0z}}{m_0 c}; & z(0) \equiv z_0 = 0, \end{cases} \tag{11.2.51}$$

because the momentum four vector is

$$p_\mu = \left(\frac{W}{c}, \, -\vec{p} \right), \tag{11.2.52}$$

and besides, since $p_\mu = m_0 c u_\mu$, we have

$$u_0 = \frac{W}{m_0 c^2}; \ u_1 = -\frac{p_1}{m_0 c}; \ u_2 = -\frac{p_2}{m_0 c}; \ u_3 = -\frac{p_3}{m_0 c}. \tag{11.2.53}$$

In the above relations W_0 is the total energy of the particle at the initial moment, p_{0x}, p_{0y} and p_{0z}, are the initial values of the momentum projections on the coordinate axes, and $u_{\mu 0} = u_\mu(\tau_0) = u_\mu(\tau = 0)$.

The integration constants A_μ, $\mu = 0, 1, 3$ and α are determined from the initial conditions, that is

$$\begin{cases} \dfrac{W_0}{m_0 c^2} = A_0, \\[2mm] -\dfrac{p_{0x}}{m_0 c} = A_1 \sin \alpha, \\[2mm] \dfrac{p_{0y}}{m_0 c} = A_1 \cos \alpha, \\[2mm] -\dfrac{p_{0z}}{m_0 c} = A_3. \end{cases} \tag{11.2.54}$$

The second and third relations in Eq. (11.2.54) yield

$$\tan \alpha = -\frac{p_{0x}}{p_{0y}} \quad \Rightarrow \quad \alpha = -\arctan \frac{p_{0x}}{p_{0y}}, \tag{11.2.55}$$

and

$$A_1 = \frac{1}{m_0 c} \sqrt{p_{0x}^2 + p_{0y}^2}, \tag{11.2.56}$$

respectively. The solution of the initial system of equations then is

$$
\begin{cases}
u_0(\tau) = \dfrac{W_0}{m_0 c^2}, \\[3mm]
u_1(\tau) = \dfrac{\sqrt{p_{0x}^2 + p_{0y}^2}}{m_0 c} \sin\left(\omega_0 \tau - \arctan \dfrac{p_{0x}}{p_{0y}}\right), \\[3mm]
u_2(\tau) = -\dfrac{\sqrt{p_{0x}^2 + p_{0y}^2}}{m_0 c} \cos\left(\omega_0 \tau - \arctan \dfrac{p_{0x}}{p_{0y}}\right), \\[3mm]
u_3(\tau) = -\dfrac{p_{0z}}{m_0 c}.
\end{cases}
\tag{11.2.57}
$$

Given that

$$
\begin{cases}
u^0 = u_0 = \dfrac{1}{c}\dfrac{dx^0}{d\tau} = \dfrac{dt}{d\tau}, \\[3mm]
u^1 = -u_1 = \dfrac{1}{c}\dfrac{dx}{d\tau}, \\[3mm]
u^2 = -u_2 = \dfrac{1}{c}\dfrac{dy}{d\tau}, \\[3mm]
u^3 = -u_3 = \dfrac{1}{c}\dfrac{dz}{d\tau},
\end{cases}
\tag{11.2.58}
$$

and integrating, we get

$$
t(\tau) = \int u_0(\tau) d\tau = \int \frac{W_0}{m_0 c^2} d\tau = \frac{W_0}{m_0 c^2}\tau + B_0.
\tag{11.2.59}
$$

Analogously, we obtain

$$
\begin{aligned}
x(\tau) &= -c \int u_1(\tau) d\tau = -c \int \frac{\sqrt{p_{0x}^2 + p_{0y}^2}}{m_0 c} \sin\left(\omega_0 \tau - \arctan \frac{p_{0x}}{p_{0y}}\right) d\tau \\[3mm]
&= \frac{\sqrt{p_{0x}^2 + p_{0y}^2}}{eB} \cos\left(\omega_0 \tau - \arctan \frac{p_{0x}}{p_{0y}}\right) + B_1,
\end{aligned}
\tag{11.2.60}
$$

$$
\begin{aligned}
y(\tau) &= -c \int u_2(\tau) d\tau = c \int \frac{\sqrt{p_{0x}^2 + p_{0y}^2}}{m_0 c} \cos\left(\omega_0 \tau - \arctan \frac{p_{0x}}{p_{0y}}\right) d\tau \\[3mm]
&= \frac{\sqrt{p_{0x}^2 + p_{0y}^2}}{eB} \sin\left(\omega_0 \tau - \arctan \frac{p_{0x}}{p_{0y}}\right) + B_2,
\end{aligned}
\tag{11.2.61}
$$

$$
z(\tau) = -c \int u_3(\tau) d\tau = c \int \frac{p_{0z}}{m_0 c} d\tau = \frac{p_{0z}}{m_0}\tau + B_3.
\tag{11.2.62}
$$

The arbitrary integration constants B_μ, $\mu = \overline{0,3}$, are determined from the initial conditions for coordinates, written above. So, we have

$$B_0 = 0, \tag{11.2.63}$$

$$0 = \frac{\sqrt{p_{0x}^2 + p_{0y}^2}}{eB} \cos\left(-\arctan\frac{p_{0x}}{p_{0y}}\right) + B_1, \tag{11.2.64}$$

$$0 = \frac{\sqrt{p_{0x}^2 + p_{0y}^2}}{eB} \sin\left(-\arctan\frac{p_{0x}}{p_{0y}}\right) + B_2, \tag{11.2.65}$$

$$B_3 = 0. \tag{11.2.66}$$

Taking into account the trigonometric formulas

$$\sin\xi = \frac{\tan\xi}{\sqrt{1 + \tan^2\xi}}; \quad \cos\xi = \frac{1}{\sqrt{1 + \tan^2\xi}},$$

we can write

$$
\begin{aligned}
B_1 &= -\frac{\sqrt{p_{0x}^2 + p_{0y}^2}}{eB} \cos\left(-\arctan\frac{p_{0x}}{p_{0y}}\right) \\
&= -\frac{\sqrt{p_{0x}^2 + p_{0y}^2}}{eB} \cos\left(\arctan\frac{p_{0x}}{p_{0y}}\right) \\
&= -\frac{\sqrt{p_{0x}^2 + p_{0y}^2}}{eB} \frac{1}{\sqrt{1 + \tan^2\left(\arctan\frac{p_{0x}}{p_{0y}}\right)}} \\
&= -\frac{\sqrt{p_{0x}^2 + p_{0y}^2}}{eB} \frac{1}{\sqrt{1 + \left(\frac{p_{0x}}{p_{0y}}\right)^2}} = -\frac{p_{0y}}{eB}
\end{aligned} \tag{11.2.67}
$$

and

$$
\begin{aligned}
B_2 &= -\frac{\sqrt{p_{0x}^2 + p_{0y}^2}}{eB} \sin\left(-\arctan\frac{p_{0x}}{p_{0y}}\right) \\
&= \frac{\sqrt{p_{0x}^2 + p_{0y}^2}}{eB} \sin\left(\arctan\frac{p_{0x}}{p_{0y}}\right) \\
&= \frac{\sqrt{p_{0x}^2 + p_{0y}^2}}{eB} \frac{\tan\left(\arctan\frac{p_{0x}}{p_{0y}}\right)}{\sqrt{1 + \tan^2\left(\arctan\frac{p_{0x}}{p_{0y}}\right)}}
\end{aligned}
$$

$$= \frac{\sqrt{p_{0x}^2 + p_{0y}^2}}{eB} \frac{\frac{p_{0x}}{p_{0y}}}{\sqrt{1 + \left(\frac{p_{0x}}{p_{0y}}\right)^2}} = \frac{p_{0x}}{eB}. \tag{11.2.68}$$

By replacing these constants in the expressions which give the spatial (x, y, z) ant temporal (t) coordinates, we finally obtain

$$t(\tau) = \frac{W_0}{m_0 c^2} \tau, \tag{11.2.69}$$

$$x(\tau) = \frac{\sqrt{p_{0x}^2 + p_{0y}^2}}{eB} \cos\left(\omega_0 \tau - \arctan \frac{p_{0x}}{p_{0y}}\right) - \frac{p_{0y}}{eB}, \tag{11.2.70}$$

$$y(\tau) = \frac{\sqrt{p_{0x}^2 + p_{0y}^2}}{eB} \sin\left(\omega_0 \tau - \arctan \frac{p_{0x}}{p_{0y}}\right) + \frac{p_{0x}}{eB}, \tag{11.2.71}$$

$$z(\tau) = \frac{p_{0z}}{m_0} \tau. \tag{11.2.72}$$

To determine the motion in the laboratory reference system, S, that is $x = x(t)$, $y = y(t)$ and $z = z(t)$, the proper time τ must be expressed in terms of the time elapsed in the frame S. This follows from the first relation written above,

$$\tau(t) = \frac{m_0 c^2}{W_0} t. \tag{11.2.73}$$

If $\tau(t)$ determined this way is introduced into $x = x(\tau)$, $y = y(\tau)$ and $z = z(\tau)$, one obtains

$$x(t) = \frac{\sqrt{p_{0x}^2 + p_{0y}^2}}{eB} \cos\left(\frac{m_0 \omega_0 c^2}{W_0} t - \arctan \frac{p_{0x}}{p_{0y}}\right) - \frac{p_{0y}}{eB}, \tag{11.2.74}$$

$$y(t) = \frac{\sqrt{p_{0x}^2 + p_{0y}^2}}{eB} \sin\left(\frac{m_0 \omega_0 c^2}{W_0} t - \arctan \frac{p_{0x}}{p_{0y}}\right) + \frac{p_{0x}}{eB}, \tag{11.2.75}$$

$$z(t) = \frac{p_{0z} c^2}{W_0} t. \tag{11.2.76}$$

By introducing the following notations:

$$\omega_1 \equiv \omega_0 \frac{m_0 c^2}{W_0}, \tag{11.2.77}$$

$$p_{01} \equiv \sqrt{p_{0x}^2 + p_{0y}^2}, \tag{11.2.78}$$

and taking into account that

$$\arctan \frac{p_{0x}}{p_{0y}} = -\alpha,$$

we finally obtain the electron law of motion, which is given by

$$\begin{cases} x(t) = \dfrac{p_{01}}{eB}\cos(\omega_1 t + \alpha) - \dfrac{p_{0y}}{eB}, \\[2mm] y(t) = \dfrac{p_{01}}{eB}\sin(\omega_1 t + \alpha) + \dfrac{p_{0x}}{eB}, \\[2mm] z(t) = \dfrac{p_{0z}c^2}{W_0}t. \end{cases} \qquad (11.2.79)$$

The trajectory of the electron is, therefore, a helix rolled up on a right cylinder with the axis oriented along z-axis (direction of the uniform magnetic field), of radius

$$R = \frac{p_{01}}{eB} = \frac{\sqrt{p_{0x}^2 + p_{0y}^2}}{eB}, \qquad (11.2.80)$$

and of constant pitch

$$\delta = v_z T = \frac{p_{0z}c^2}{W_0}\frac{2\pi}{\omega_1} = \frac{2\pi p_{0z}c^2 W_0}{W_0 \omega_0 m_0 c^2} = \frac{2\pi p_{0z}}{eB}. \qquad (11.2.81)$$

11.3   Problem No. 76

Study the relativistic motion of an electrized particle with rest mass m_0 and electric charge q placed in an external electromagnetic field, $(\vec{E}, \vec{B})$, within the four-dimensional formalism.

Analytical solution

We will use the same conventions as in the previous two problems, namely we will use the hyperbolic representation of Minkowski space, with -2 signature metric.

The equation of motion of an electrized particle placed in an external electromagnetic field, within the relativistically covariant formalism, writes

$$\frac{d}{d\tau}\big(m_0 \overline{u}_\mu\big) = q F_{\mu\nu}\overline{u}^\nu, \qquad (11.3.82)$$

where m_0 and q are the rest mass and the electric charge of the particle, respectively,

$$\overline{u}^\mu = \frac{dx^\mu}{d\tau}, \ \mu = \overline{0,3},$$

is the contravariant four-vector of the particle velocity (its components having the dimension of velocity), τ is the time elapsed in the proper reference frame of the particle (the frame S'), and $F_{\mu\nu}$ is the covariant tensor of the electromagnetic field,

$$
(F_{\mu\nu}) = \begin{pmatrix} 0 & \dfrac{E_x}{c} & \dfrac{E_y}{c} & \dfrac{E_z}{c} \\[2mm] -\dfrac{E_x}{c} & 0 & -B_z & B_y \\[2mm] -\dfrac{E_y}{c} & B_z & 0 & -B_x \\[2mm] -\dfrac{E_z}{c} & -B_y & B_x & 0 \end{pmatrix}. \tag{11.3.83}
$$

The equation of motion can also be written as

$$
\frac{d}{d\tau}(m_0 u_\mu) = q F_{\mu\nu} u^\nu, \tag{11.3.84}
$$

where

$$
u_\mu = \frac{dx_\mu}{ds}, \quad \mu = \overline{0,3},
$$

is the covariant four-vector of the particle velocity (its components being dimensionless). Since $m_0 = const.$, Eq. (11.3.84) can also be written as

$$
\frac{du_\mu}{d\tau} = \frac{q}{m_0 c} c F_{\mu\nu} u^\nu = \omega_0 c u^0 F_{\mu 0} + \omega_0 c u^\alpha F_{\mu\alpha}
$$
$$
= \omega_0 c u_0 F_{\mu 0} - \omega_0 c u_\alpha F_{\mu\alpha}, \quad \alpha = \overline{1,3}, \ \mu, \nu = \overline{0,3}, \tag{11.3.85}
$$

where the notation

$$
\omega_0 \equiv \frac{q}{m_0 c}
$$

has been used.

Therefore, for the moment, without making any suppositions or restrictions on the electromagnetic field, the equations for the velocity four-vector are

$$
\frac{du_\mu}{d\tau} = \omega_0 c u_0 F_{\mu 0} - \omega_0 c u_\alpha F_{\mu\alpha}, \quad \alpha = \overline{1,3}, \ \mu = \overline{0,3}. \tag{11.3.86}
$$

Giving values to the index μ, we obtain the following system of four differential equations of the first degree for the velocity four-vector

$$
\begin{cases} \dfrac{du_0}{d\tau} = \omega_0 \big(-E_x u_1 - E_y u_2 - E_z u_3 \big), \\[2mm] \dfrac{du_1}{d\tau} = \omega_0 \big(-E_x u_0 + B_z c u_2 - B_y c u_3 \big), \\[2mm] \dfrac{du_2}{d\tau} = \omega_0 \big(-E_y u_0 - B_z c u_1 + B_x c u_3 \big), \\[2mm] \dfrac{du_3}{d\tau} = \omega_0 \big(-E_z u_0 + B_y c u_1 - B_x c u_2 \big). \end{cases} \tag{11.3.87}
$$

Then, since

$$u^\mu = \frac{1}{c}\frac{dx^\mu}{d\tau}, \quad u_\mu = \frac{1}{c}\frac{dx_\mu}{d\tau}, \tag{11.3.88}$$

we obtain

$$\begin{cases} \dfrac{dt}{d\tau} = u_0, \\[2mm] \dfrac{dx}{d\tau} = -cu_1, \\[2mm] \dfrac{dy}{d\tau} = -cu_2, \\[2mm] \dfrac{dz}{d\tau} = -cu_3. \end{cases} \tag{11.3.89}$$

To determine the law of motion in the laboratory frame (the inertial reference frame S), that is, $x = x(t)$, $y = y(t)$ and $z = z(t)$, a set of eight first-order differential equations composed by the two systems given above have to be solved. To do this, let us consider the following initial conditions:

$$u_{00} = \frac{W_0}{m_0 c^2}; \quad \tau_0 = t_0 = 0, \tag{11.3.90}$$

$$u_{10} = -\frac{p_{0x}}{m_0 c}; \quad x_0 = 0, \tag{11.3.91}$$

$$u_{20} = -\frac{p_{0y}}{m_0 c}; \quad y_0 = 0, \tag{11.3.92}$$

$$u_{30} = -\frac{p_{0z}}{m_0 c}; \quad z_0 = 0, \tag{11.3.93}$$

where W_0 is the total energy of the particle at the initial moment $\tau_0 = t_0 = 0$.

Let us now justify the choice of the initial conditions. Since the momentum contravariant quadrivector is

$$p^\mu = \left(\frac{W}{c}, \vec{p}\right), \tag{11.3.94}$$

and the covariant one is given by

$$p_\mu = \left(\frac{W}{c}, -\vec{p}\right), \tag{11.3.95}$$

the last three relations which express the initial conditions for the spatial components of the four-velocity are self-evident. Next, using the definition

$$p_\mu = m_0 c u_\mu,$$

we can write

$$(p_0)_0 = W_0/c = m_0 c (u_0)_0, \tag{11.3.96}$$

so that

$$(u_0)_0 \equiv u_{00} = \frac{W_0}{m_0 c^2}, \tag{11.3.97}$$

which justifies the relation chosen for the initial condition of the fourth component of the velocity four-vector. Unfortunately, the analytical solution of the problem cannot continue, because, on the one side, the configuration of the electromagnetic field is not known, and on the other side, it is very difficult. Even if we would want to solve the first part of the problem — namely that concerning the four-velocity components — we should solve a system of four, coupled, first-order differential equations. This operation is extremely difficult in the general case, when all the six components of the electromagnetic field are non-zero, not to mention the case where these components would be non-uniform and/or non-stationary.

Solving the problem in the general case can only be done numerically, with the help of the computer. In the following exposure we will turn our attention to this issue.

Numerical Solution

In order to solve the previously deduced system of eight first-order differential equations, in the general case, it can be used the below presented algorithm of numerical integration of Runge-Kutta type of the fourth order, implemented using the software Mathematica 5.0 (specialized in both analytical and numerical calculation).

If the studied particle is a proton with the initial momentum of components

$$\begin{cases} p_{0x} = 5 \times 10^{-19}\,\mathrm{kg \cdot m \cdot s^{-1}}, \\ p_{0y} = 7.5 \times 10^{-19}\,\mathrm{kg \cdot m \cdot s^{-1}}, \\ p_{0z} = 10^{-18}\,\mathrm{kg \cdot m \cdot s^{-1}} \end{cases} \qquad (11.3.98)$$

which leaves at the initial moment $t_0 = \tau_0 = 0$ from the coordinate system origin ($x_0 = y_0 = z_0 = 0$) and moves for $\Delta\tau = 1.2\,\mathrm{s}$ (duration measured in its own reference system) in the electromagnetic field given by

$$\begin{cases} E_x = 5\cos\omega t, \\ E_y = 2 \times 10^2 \sin\omega t, \\ E_z = 0, \\ B_x = 1.25 \times 10^{-7}, \\ B_y = 0, \\ B_z = 1.2 \times 10^{-7}, \end{cases} \qquad (11.3.99)$$

where, for calculus convenience, we have taken $\omega = 1\,\mathrm{s^{-1}}$, the command lines written using the above-mentioned software package are given below:

Needs["Graphics'Colors'"]

Runge3D[$a0_-, b0_-, \alpha_-, \beta_-, \gamma_-, \delta_-, \lambda_-, \mu_-, \nu_-, \sigma_-, m0_-$] :=

  Module[{$a = a0, b = b0, j, m = m0$},

$$h = \frac{b - a}{m};$$

$P = \text{Table}[a + (j - 1)h, \{j, 1, m + 1\}];$

$Z1 = \text{Table}[\{\alpha, \beta, \gamma, \delta, \lambda, \mu, \nu, \sigma\}, \{j, 1, m + 1\}];$

$\quad \text{For}[j = 1, j \le m, j + +,$

$\quad\quad k_1 = hF[P_{[j]}, Z1_{[j]}];$

$\quad\quad k_2 = hF[P_{[j]} + \frac{h}{2}, Z1_{[j]} + \frac{k_1}{2}]$

$\quad\quad k_3 = hF[P_{[j]} + \frac{h}{2}, Z1_{[j]} + \frac{k_2}{2}]$

$\quad\quad k_4 = hF[P_{[j]} + h, Z1_{[j]} + k_3];$

$\quad\quad Z1_{[j+1]} = Z1_{[j]} + \frac{1}{6}(k_1 + 2k_2 + 2k_3 + k_4);];$

$\quad \text{Return}[Z1];];$

$m_0 = 1.67 * 10^{-27};$

$q = 1.6 * 10^{-19};$

$c = 3 * 10^{8};$

$p_{0x} = 5 * 10^{-19};$

$p_{0y} = 7.5 * 10^{-19};$

$p_{0z} = 10 * 10^{-19};$

$W_0 = \text{Sqrt}[m_0^2 * c^4 + (p_{0x}^2 + p_{0y}^2 + p_{0z}^2) * c^2];$

$$\omega_0 = \frac{q}{m_0 * c};$$

$E_x = 5 * \text{Cos}[t];$

$E_y = 2 * 10^2 * \text{Sin}[t];$

$E_z = 0;$

$B_x = 0.125 * 10^{-6};$

$B_y = 0;$

$B_z = 0.12 * 10^{-6};$

$f_1[\tau_-, u0_-, u1_-, u2_-, u3_-, t_-, x_-, y_-, z_-]$
$= \omega_0 * (-E_x * u1 - E_y * u2 - E_z * u3);$

$f_2[\tau_-, u0_-, u1_-, u2_-, u3_-, t_-, x_-, y_-, z_-]$
$= \omega_0 * (-E_x * u0 + B_z * c * u2 - B_y * c * u3);$

$f_3[\tau_-, u0_-, u1_-, u2_-, u3_-, t_-, x_-, y_-, z_-]$
$= \omega_0 * (-E_y * u0 - B_z * c * u1 + B_x * c * u3);$

$f_4[\tau_-, u0_-, u1_-, u2_-, u3_-, t_-, x_-, y_-, z_-]$

$= \omega_0 * (-E_z * u0 + B_y * c * u1 - B_x * c * u2);$

$f_5[\tau_-, u0_-, u1_-, u2_-, u3_-, t_-, x_-, y_-, z_-] = u0;$

$f_6[\tau_-, u0_-, u1_-, u2_-, u3_-, t_-, x_-, y_-, z_-] = -c * u1;$

$f_7[\tau_-, u0_-, u1_-, u2_-, u3_-, t_-, x_-, y_-, z_-] = -c * u2;$

$f_8[\tau_-, u0_-, u1_-, u2_-, u3_-, t_-, x_-, y_-, z_-] = -c * u3;$

$F[\tau_-, \{u0_-, u1_-, u2_-, u3_-, t_-, x_-, y_-, z_-\}] =$

$\{f_1[\tau, u0, u1, u2, u3, t, x, y, z], f_2[\tau, u0, u1, u2, u3, t, x, y, z],$

$f_3[\tau, u0, u1, u2, u3, t, x, y, z], f_4[\tau, u0, u1, u2, u3, t, x, y, z],$

$f_5[\tau, u0, u1, u2, u3, t, x, y, z], f_6[\tau, u0, u1, u2, u3, t, x, y, z],$

$f_7[\tau, u0, u1, u2, u3, t, x, y, z], f_8[\tau, u0, u1, u2, u3, t, x, y, z]\};$

Print$["u0' = ", f1[\tau, u0, u1, u2, u3, t, x, y, z]];$

Print$["u1' = ", f2[\tau, u0, u1, u2, u3, t, x, y, z]];$

Print$["u2' = ", f3[\tau, u0, u1, u2, u3, t, x, y, z]];$

Print$["u3' = ", f4[\tau, u0, u1, u2, u3, t, x, y, z]];$

Print$["t' = ", f5[\tau, u0, u1, u2, u3, t, x, y, z]];$

Print$["x' = ", f6[\tau, u0, u1, u2, u3, t, x, y, z]];$

Print$["y' = ", f7[\tau, u0, u1, u2, u3, t, x, y, z]];$

Print$["z' = ", f8[\tau, u0, u1, u2, u3, t, x, y, z]];$

Print$["F[\tau, \{u0, u1, u2, u3, t, x, y, z\}]$

$= ", F[\tau, \{u0, u1, u2, u3, t, x, y, z\}]];$

$u0' = -0.319361\,(5\,u1\,\text{Cos}[t] + 200\,u2\,\text{Sin}[t])$

$u1' = 0.319361\,(36.\,u2 - 5\,u0\,\text{Cos}[t])$

$u2' = 0.319361\,(-36.\,u1 + 37.5\,u3 - 200\,u0\,\text{Sin}[t])$

$u3' = -11.976\,u2$

$t' = u0$

$x' = -300000000\,u1$

$y' = -300000000\,u2$

$z' = -300000000\,u3$

$F[\tau\{u0, u1, u2, u3, x, y, z, t\}] = \{-0.319361(5\,u1\,\text{Cos}[t] + 200\,u2\,\text{Sin}[t]),$

$0.319361(36.\,u2 - 5\,u0\,\text{Cos}[t]),$

$0.319361(-36.\,u1 + 37.5\,u3 - 200\,u0\,\mathrm{Sin}[t]), -11.976\,u2,$

$u0, -300000000\,u1, -300000000\,u2, -300000000\,u3\}$

$a = 0;$

$b = 1.2;$

$u00 = \dfrac{W_0}{m_0 * c^2};$

$u10 = -\dfrac{p_{0x}}{m_0 * c};$

$u20 = -\dfrac{p_{0y}}{m_0 * c};$

$u30 = -\dfrac{p_{0z}}{m_0 * c};$

$t0 = 0;$

$x0 = 0;$

$y0 = 0;$

$z0 = 0;$

$n = 500;$

$\mathrm{Runge3D}[a, b, u00, u10, u20, u30, t0, x0, y0, z0, n];$

$u0s = \mathrm{Transpose}[\{\mathrm{P}, \mathrm{Transpose}[\mathrm{Z1}]_{[1]}\}];$

$gru0s1 = \mathrm{ListPlot}[u0s, \mathrm{PlotJoined} \to \mathrm{True}, \mathrm{PlotStyle} \to \mathrm{Black},$

$\mathrm{AxesLabel} \to \{''\tau(s)'', ''u0\ (\mathrm{adim.})''\}];$

$\mathrm{Print}[''u0 = u0[\tau]\ \mathrm{for}\ '', a, '' \leq \tau \leq '', b];$

The graphical dependences of the 4-velocity, time and three spatial coordinates as functions of proper time, as well as the projection of the proton trajectory (moving for 1.2 s in the electromagnetic field given by Eq. (11.3.99)) in the xy, yz and zx-planes, and also the proton 3D trajectory are shown in Figs. 11.3–11.14.

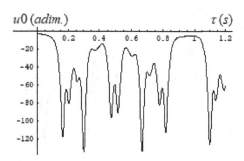

FIGURE 11.3
Graphical representation of the variation of $u0$-component of the covariant four-vector of dimensionless velocity of a proton moving for $\Delta\tau = 1.2$ s in the electromagnetic field given by Eq. (11.3.99), as a function of proper time, $u0 = u0(\tau)$.

$u1s = \mathrm{Transpose}[\{P, \mathrm{Transpose}[Z1]_{[2]}\}];$

$gru1s1 = \mathrm{ListPlot}[u1s, \mathrm{PlotJoined} \rightarrow \mathrm{True}, \mathrm{PlotStyle} \rightarrow \mathrm{Black}$

$\mathrm{AxesLabel} \rightarrow \{"\tau(s)", "u1(\mathrm{adim.})"\}];$

$\mathrm{Print}["u1 = u1[\tau]\ \mathrm{for}\ ", a, " \leq \tau \leq ", b];$

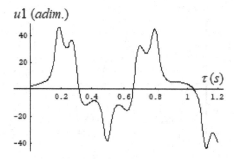

FIGURE 11.4
Graphical representation of the variation of $u1$-component of the covariant four-vector of dimensionless velocity of a proton moving for $\Delta\tau = 1.2$ s in the electromagnetic field given by Eq. (11.3.99), as a function of proper time, $u1 = u1(\tau)$.

$u2s = \mathrm{Transpose}[\{P, \mathrm{Transpose}\ [Z1]_{[3]}\}];$

$gru2s1 = \mathrm{ListPlot}[u2s, \mathrm{PlotJoined} \rightarrow \mathrm{True}, \mathrm{PlotStyle} \rightarrow \mathrm{Black},$

$\mathrm{AxesLabel} \rightarrow \{"\tau(s)", "u2(\mathrm{adim.})"\}];$

$\mathrm{Print}["u2 = u2[\tau]\ \mathrm{for}\ ", a, " \leq \tau \leq ", b];$

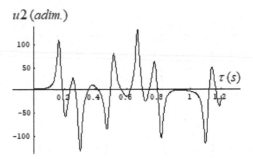

FIGURE 11.5
Graphical representation of the variation of $u2$-component of the covariant four-vector of dimensionless velocity of a proton moving for $\Delta\tau = 1.2$ s in the electromagnetic field given by Eq. (11.3.99), as a function of proper time, $u2 = u2(\tau)$.

$u3s=$ Transpose$[\{P,$ Transpose $[Z1]_{[\![4]\!]}\}];$

$gru3s1=$ListPlot$[u3s,$ PlotJoined $\rightarrow$ True, PlotStyle $\rightarrow$ Black,

AxesLabel $\rightarrow \{''\tau(s)'', ''u3(\text{adim.})''\}];$

Print$[''u3 = u3[\tau]$ for $'', a, '' \leq \tau \leq '', b];$

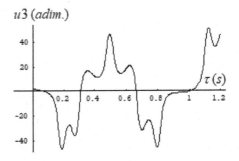

FIGURE 11.6
Graphical representation of the variation of $u3$-component of the covariant four-vector of dimensionless velocity of a proton moving for $\Delta\tau = 1.2$ s in the electromagnetic field given by Eq. (11.3.99), as a function of proper time, $u3 = u3(\tau)$.

$ts=$Transpose$[\{P,$ Transpose $[Z1]_{[\![5]\!]}\}];$

$grts1=$ListPlot$[ts,$ PlotJoined$\rightarrow$ True, PlotStyle$\rightarrow$ Black,

AxesLabel $\rightarrow \{''\tau(s)'', ''t(s)''\}];$

Print$[''t = t[\tau]$ for $'', a, '' \leq \tau \leq '', b];$

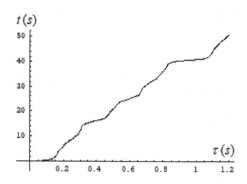

FIGURE 11.7

Graphical representation of dependence $t = t(\tau)$ for a proton moving for $\Delta\tau = 1.2\,$s in the electromagnetic field given by Eq. (11.3.99).

xs=Transpose[{P, Transpose [$Z1$]$_{[\![6]\!]}$}];

$grxs1$=ListPlot[xs, PlotJoined$\rightarrow$ True, PlotStyle $\rightarrow$ Black,

AxesLabel $\rightarrow$ {$"\tau(s)"$, $"x(m)"$}];

Print[$"x = x[\tau]$ for $"$, a, $" \leq \tau \leq "$, b];

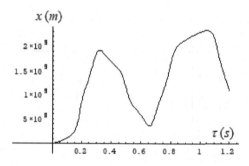

FIGURE 11.8

Graphical representation of the x-coordinate variation of a proton moving for $\Delta\tau = 1.2\,$s in the electromagnetic field given by Eq. (11.3.99), as a function of proper time, $x = x(\tau)$.

ys=Transpose[{P, Transpose [$Z1$]$_{[\![7]\!]}$}];

$grys1$=ListPlot[ys, PlotJoined$\rightarrow$ True, PlotStyle$\rightarrow$ Black,

AxesLabel $\rightarrow$ {$"\tau(s)"$, $"y(m)"$}];

Print[$"y = y[\tau]$ for $"$, a, $" \leq \tau \leq "$, b];

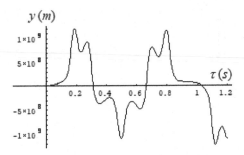

FIGURE 11.9

Graphical representation of the y-coordinate variation of a proton moving for $\Delta \tau = 1.2\,\text{s}$ in the electromagnetic field given by Eq. (11.3.99), as a function of proper time, $y = y(\tau)$.

zs=Transpose[$\{P$, Transpose $[Z1]_{[\![8]\!]}\}$];

$grzs1$=ListPlot[zs, PlotJoined$\to$ True, PlotStyle$\to$ Black,

AxesLabel $\to \{''\tau(s)'', ''z(m)''\}$];

Print[$''z = z[\tau]$ for $'', a, '' \leq \tau \leq '', b$];

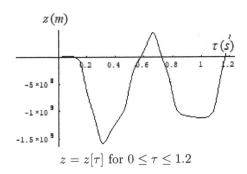

$z = z[\tau]$ for $0 \leq \tau \leq 1.2$

FIGURE 11.10

Graphical representation of the z-coordinate variation of a proton moving for $\Delta \tau = 1.2\,\text{s}$ in the electromagnetic field given by Eq. (11.3.99), as a function of proper time, $z = z(\tau)$.

scat=Table[Transpose[$\{$Transpose[$Z1]_{[\![6]\!]}$,Transpose[$Z1]_{[\![7]\!]}\}$]];

ListPlot[scat, PlotJoined $\to$ True, AxesLabel $\to \{''x(m)'', ''y(m)''\}$]

scat=Table[Transpose[$\{$Transpose[$Z1]_{[\![7]\!]}$,Transpose[$Z1]_{[\![8]\!]}\}$]];

ListPlot[scat, PlotJoined $\to$ True, AxesLabel $\to \{''y(m)'', ''z(m)''\}$]

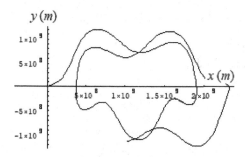

FIGURE 11.11
Projection in the yx-plane of trajectory of a proton moving for $\Delta\tau = 1.2\,\mathrm{s}$ in the electromagnetic field given by Eq. (11.3.99).

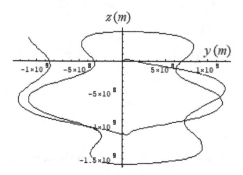

FIGURE 11.12
Projection in the zy-plane of trajectory of a proton moving for $\Delta\tau = 1.2\,\mathrm{s}$ in the electromagnetic field given by Eq. (11.3.99).

scat=Table[Transpose[{Transpose[$Z1$]$_{[\![6]\!]}$,Transpose[$Z1$]$_{[\![8]\!]}$}]];

ListPlot[scat, PlotJoined $\rightarrow$ True, AxesLabel $\rightarrow$ {$"x(m)"$, $"z(m)"$}]

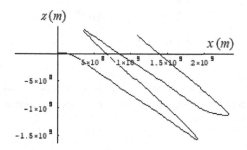

FIGURE 11.13
Projection in the zx-plane of trajectory of a proton moving for $\Delta\tau = 1.2\,\mathrm{s}$ in the electromagnetic field given by Eq. (11.3.99).

scat=Table[Transpose[{Transpose[$Z1$]$_{[6]}$, Transpose[$Z1$]$_{[7]}$,

Transpose[$Z1$]$_{[8]}$}]];

ScatterPlot3D[scat, PlotJoined→ True]

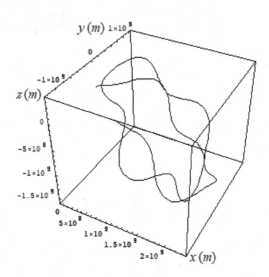

FIGURE 11.14
Trajectory of a proton moving for $\Delta\tau = 1.2$ s in the electromagnetic field given
by Eq. (11.3.99).

If the time interval measured in the proper reference frame grows up to
$\Delta\tau = 10$ s, and the electromagnetic field configuration is given by

$$\begin{cases} E_x = 50\cos\omega t, \\ E_y = 70\sin\omega t, \\ E_z = 0, \\ B_x = 1.25 \times 10^{-7}, \\ B_y = 0, \\ B_z = 1.25 \times 10^{-7} \end{cases} \qquad (11.3.100)$$

where, for calculus convenience, we have taken $\omega = 1\,\mathrm{s}^{-1}$, instead of the above
graphics, the following graphical representations are obtained, respectively
(see Figs. 11.15–11.26):

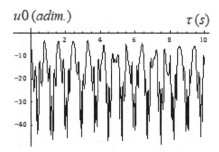

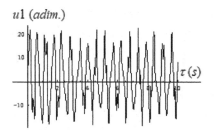

FIGURE 11.15

Graphical representation of the variation of $u0$-component of the covariant four-vector of dimensionless velocity of a proton moving for $\Delta\tau = 10\,\mathrm{s}$ in the electromagnetic field given by Eq. (11.3.100), as a function of proper time, $u0 = u0(\tau)$.

FIGURE 11.16

Graphical representation of the variation of $u1$-component of the covariant four-vector of dimensionless velocity of a proton moving for $\Delta\tau = 10\,\mathrm{s}$ in the electromagnetic field given by Eq. (11.3.100), as a function of proper time, $u1 = u1(\tau)$.

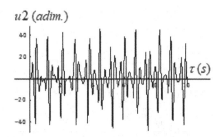

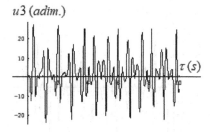

FIGURE 11.17

Graphical representation of the variation of $u2$-component of the covariant four-vector of dimensionless velocity of a proton moving for $\Delta\tau = 10\,\mathrm{s}$ in the electromagnetic field given by Eq. (11.3.100), as a function of proper time, $u2 = u2(\tau)$.

FIGURE 11.18

Graphical representation of the variation of $u3$-component of the covariant four-vector of dimensionless velocity of a proton moving for $\Delta\tau = 10\,\mathrm{s}$ in the electromagnetic field given by Eq. (11.3.100), as a function of proper time, $u3 = u3(\tau)$.

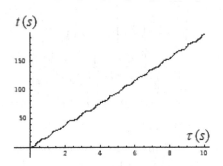

FIGURE 11.19
Graphical representation of dependence $t = t(\tau)$ for a proton moving for $\Delta\tau = 10\,\text{s}$ in the electromagnetic field given by Eq. (11.3.100).

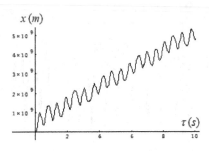

FIGURE 11.20
Graphical representation of the x-coordinate variation of a proton moving for $\Delta\tau = 10\,\text{s}$ in the electromagnetic field given by Eq. (11.3.100), as a function of proper time, $x = x(\tau)$.

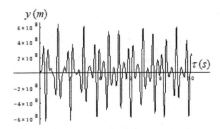

FIGURE 11.21
Graphical representation of the y-coordinate variation of a proton moving for $\Delta\tau = 10\,\text{s}$ in the electromagnetic field given by Eq. (11.3.100), as a function of proper time, $y = y(\tau)$.

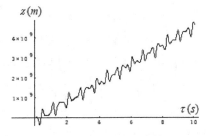

FIGURE 11.22
Graphical representation of the z-coordinate variation of a proton moving for $\Delta\tau = 10\,\text{s}$ in the electromagnetic field given by Eq. (11.3.100), as a function of proper time, $z = z(\tau)$.

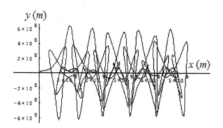

FIGURE 11.23
Projection in the yx-plane of trajectory of a proton moving for $\Delta\tau =$ 10 s in the electromagnetic field given by Eq. (11.3.100).

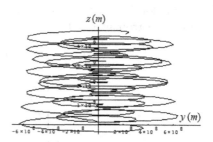

FIGURE 11.24
Projection in the zy-plane of trajectory of a proton moving for $\Delta\tau = 10$ s in the electromagnetic field given by Eq. (11.3.100).

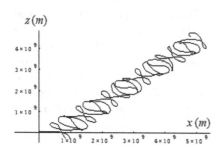

FIGURE 11.25
Projection in the zx-plane of trajectory of a proton moving for $\Delta\tau =$ 10 s in the electromagnetic field given by Eq. (11.3.100).

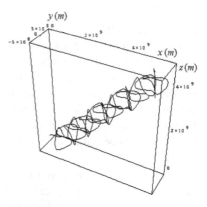

FIGURE 11.26
Trajectory of a proton moving for $\Delta\tau = 10$ s in the electromagnetic field given by Eq. (11.3.100).

For an electron with initial momentum $p_{0x} = 5 \times 10^{-25} \,\text{kg} \cdot \text{m} \cdot \text{s}^{-1}$, $p_{0y} = 7.5 \times 10^{-25} \,\text{kg} \cdot \text{m} \cdot \text{s}^{-1}$, $p_{0z} = 10^{-24} \,\text{kg} \cdot \text{m} \cdot \text{s}^{-1}$, leaving the origin of the coordinate system ($x_0 = y_0 = z_0 = 0$) at the initial moment $\tau_0 = t_0 = 0$, and moving for $\Delta\tau = 1.2\,\text{s}$ (time duration measured in the proper reference frame) in the electromagnetic field (with non-stationary electric component),

$$\begin{cases} E_x = 10^{-2}\cos\omega t, \\ E_y = 8 \times 10^{-1}\sin\omega t, \\ E_z = 0, \\ B_x = 8 \times 10^{-11}, \\ B_y = 0, \\ B_z = 9 \times 10^{-11}, \end{cases} \tag{11.3.101}$$

where, for calculus convenience, we have taken $\omega = 1\,\text{s}^{-1}$, the corresponding graphical representations are given in Figs. 11.27–11.38.

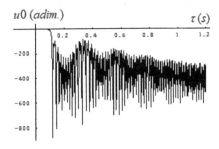

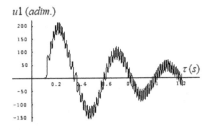

FIGURE 11.28
Graphical representation of the variation of $u1$-component of the covariant four-vector of dimensionless velocity of an electron moving for $\Delta\tau = 1.2\,\text{s}$ in the electromagnetic field given by Eq. (11.3.101), as a function of proper time, $u1 = u1(\tau)$.

FIGURE 11.27
Graphical representation of the variation of $u0$-component of the covariant four-vector of dimensionless velocity of an electron moving for $\Delta\tau = 1.2\,\text{s}$ in the electromagnetic field given by Eq. (11.3.101), as a function of proper time, $u0 = u0(\tau)$.

If the duration measured in the electron proper reference frame grows to $\Delta\tau = 10\,\text{s}$ and the electromagnetic field configuration remains unchanged, instead of the above graphics, the dependencies shown in Figs. 11.39–11.50 are obtained, respectively.

To complete the picture on this issue, we will next present the graphical representations of the same functional dependences for two more situations, namely:

i) The motion of a proton with the same momentum as the one considered at the beginning, during the time interval $\Delta\tau = 10\,\text{s}$ (determined in the proper

u2 (adim.)

FIGURE 11.29
Graphical representation of the variation of u2-component of the covariant four-vector of dimensionless velocity of an electron moving for $\Delta\tau = 1.2\,$s in the electromagnetic field given by Eq. (11.3.101), as a function of proper time, $u2 = u2(\tau)$.

u3 (adim.)

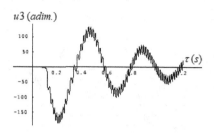

FIGURE 11.30
Graphical representation of the variation of u3-component of the covariant four-vector of dimensionless velocity of an electron moving for $\Delta\tau = 1.2\,$s in the electromagnetic field given by Eq. (11.3.101), as a function of proper time, $u3 = u3(\tau)$.

t (s)

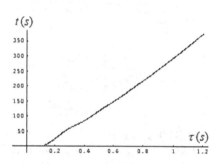

FIGURE 11.31
Graphical representation of dependence $t = t(\tau)$ for an electron moving for $\Delta\tau = 1.2\,$s in the electromagnetic field given by Eq. (11.3.101).

x (m)

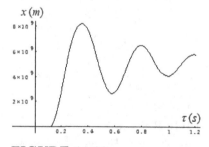

FIGURE 11.32
Graphical representation of the x-coordinate variation of an electron moving for $\Delta\tau = 1.2\,$s in the electromagnetic field given by Eq. (11.3.101), as a function of proper time, $x = x(\tau)$.

reference frame) in the non-uniform electromagnetic field with configuration

$$
\begin{cases}
E_x = 50 \cos \frac{\lambda x}{2} \sin \lambda y, \\
E_y = 70 \sin \frac{\lambda z}{3}, \\
E_z = 0, \\
B_x = 1.25 \times 10^{-7} \sin \lambda y \cos \frac{\lambda z}{3}, \\
B_y = 0, \\
B_z = 1.25 \times 10^{-7} \cos^2 \frac{\lambda y}{3},
\end{cases}
\tag{11.3.102}
$$

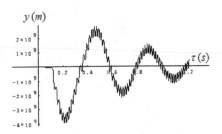

FIGURE 11.33
Graphical representation of the y-coordinate variation of an electron moving for $\Delta\tau = 1.2\,\text{s}$ in the electromagnetic field given by Eq. (11.3.101), as a function of proper time, $y = y(\tau)$.

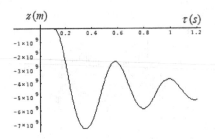

FIGURE 11.34
Graphical representation of the z-coordinate variation of an electron moving for $\Delta\tau = 1.2\,\text{s}$ in the electromagnetic field given by Eq. (11.3.101), as a function of proper time, $z = z(\tau)$.

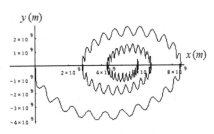

FIGURE 11.35
Projection in the yx-plane of trajectory of an electron moving for $\Delta\tau = 1.2\,\text{s}$ in the electromagnetic field given by Eq. (11.3.101).

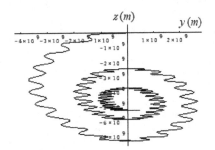

FIGURE 11.36
Projection in the zy-plane of trajectory of an electron moving for $\Delta\tau = 1.2\,\text{s}$ in the electromagnetic field given by Eq. (11.3.101).

where, for calculus convenience, we have taken $\lambda = 1\,\text{m}^{-1}$, in which case we obtained the (corresponding) dependencies/drawings shown in Figs. 11.51–11.62.

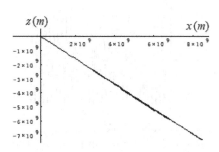

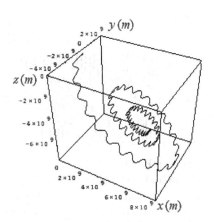

FIGURE 11.37
Projection in the zx-plane of trajectory of an electron moving for $\Delta\tau = 1.2\,\text{s}$ in the electromagnetic field given by Eq. (11.3.101).

FIGURE 11.38
Trajectory of an electron moving for $\Delta\tau = 1.2\,\text{s}$ in the electromagnetic field given by Eq. (11.3.101).

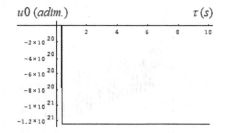

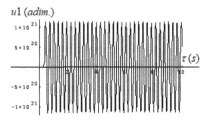

FIGURE 11.39
Graphical representation of the variation of $u0$-component of the covariant four-vector of dimensionless velocity of an electron moving for $\Delta\tau = 10\,\text{s}$ in the electromagnetic field given by Eq. (11.3.101), as a function of proper time, $u0 = u0(\tau)$.

FIGURE 11.40
Graphical representation of the variation of $u1$-component of the covariant four-vector of dimensionless velocity of an electron moving for $\Delta\tau = 10\,\text{s}$ in the electromagnetic field given by Eq. (11.3.101), as a function of proper time, $u1 = u1(\tau)$.

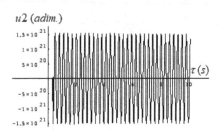

FIGURE 11.41
Graphical representation of the variation of $u2$-component of the covariant four-vector of dimensionless velocity of an electron moving for $\Delta\tau = 10$ s in the electromagnetic field given by Eq. (11.3.101), as a function of proper time, $u2 = u2(\tau)$.

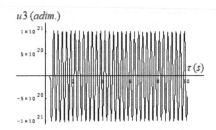

FIGURE 11.42
Graphical representation of the variation of $u3$-component of the covariant four-vector of dimensionless velocity of an electron moving for $\Delta\tau = 10$ s in the electromagnetic field given by Eq. (11.3.101), as a function of proper time, $u3 = u3(\tau)$.

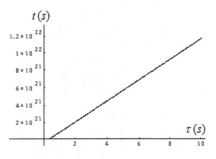

FIGURE 11.43
Graphical representation of dependence $t = t(\tau)$ for an electron moving for $\Delta\tau = 10$ s in the electromagnetic field given by Eq. (11.3.101).

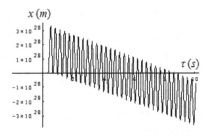

FIGURE 11.44
Graphical representation of the x-coordinate variation of an electron moving for $\Delta\tau = 10$ s in the electromagnetic field given by Eq. (11.3.101), as a function of proper time, $x = x(\tau)$.

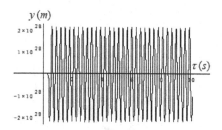

FIGURE 11.45
Graphical representation of the y-coordinate variation of an electron moving for $\Delta\tau = 10\,\text{s}$ in the electromagnetic field given by Eq. (11.3.101), as a function of proper time, $y = y(\tau)$.

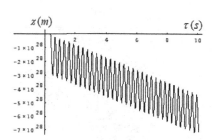

FIGURE 11.46
Graphical representation of the z-coordinate variation of an electron moving for $\Delta\tau = 10\,\text{s}$ in the electromagnetic field given by Eq. (11.3.101), as a function of proper time, $z = z(\tau)$.

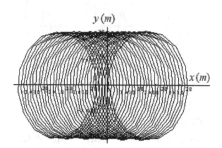

FIGURE 11.47
Projection in the yx-plane of trajectory of an electron moving for $\Delta\tau = 10\,\text{s}$ in the electromagnetic field given by Eq. (11.3.101).

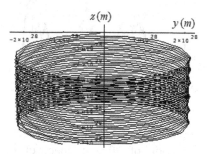

FIGURE 11.48
Projection in the zy-plane of trajectory of an electron moving for $\Delta\tau = 10\,\text{s}$ in the electromagnetic field given by Eq. (11.3.101).

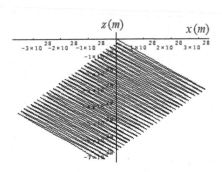

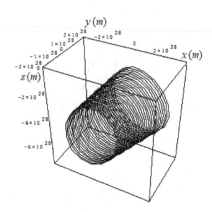

FIGURE 11.49
Projection in the zx-plane of trajectory of an electron moving for $\Delta\tau = 10\,\text{s}$ in the electromagnetic field given by Eq. (11.3.101).

FIGURE 11.50
Trajectory of an electron moving for $\Delta\tau = 10\,\text{s}$ in the electromagnetic field given by Eq. (11.3.101).

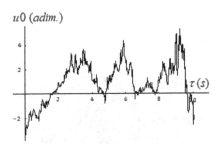

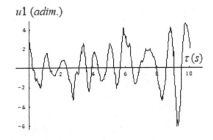

FIGURE 11.51
Graphical representation of the variation of $u0$-component of the covariant four-vector of dimensionless velocity of a proton moving for $\Delta\tau = 10\,\text{s}$ in the non-uniform and non-stationary electromagnetic field given by Eq. (11.3.102), as a function of proper time, $u0 = u0(\tau)$.

FIGURE 11.52
Graphical representation of the variation of $u1$-component of the covariant four-vector of dimensionless velocity of a proton moving for $\Delta\tau = 10\,\text{s}$ in the non-uniform and non-stationary electromagnetic field given by Eq. (11.3.102), as a function of proper time, $u1 = u1(\tau)$.

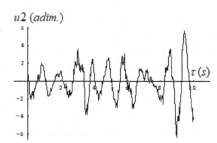

FIGURE 11.53
Graphical representation of the variation of $u2$-component of the covariant four-vector of dimensionless velocity of a proton moving for $\Delta\tau = 10\,\mathrm{s}$ in the non-uniform and non-stationary electromagnetic field given by Eq. (11.3.102), as a function of proper time, $u2 = u2(\tau)$.

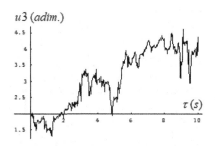

FIGURE 11.54
Graphical representation of the variation of $u3$-component of the covariant four-vector of dimensionless velocity of a proton moving for $\Delta\tau = 10\,\mathrm{s}$ in the non-uniform and non-stationary electromagnetic field given by Eq. (11.3.102), as a function of proper time, $u3 = u3(\tau)$.

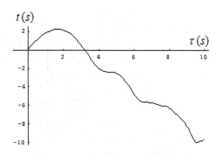

FIGURE 11.55
Graphical representation of dependence $t = t(\tau)$ for a proton moving for $\Delta\tau = 10\,\mathrm{s}$ in the non-uniform and non-stationary electromagnetic field given by Eq. (11.3.102).

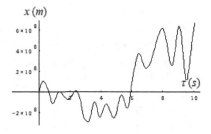

FIGURE 11.56
Graphical representation of the x-coordinate variation of a proton moving for $\Delta\tau = 10\,\mathrm{s}$ in the non-uniform and non-stationary electromagnetic field given by Eq. (11.3.102), as a function of proper time, $x = x(\tau)$.

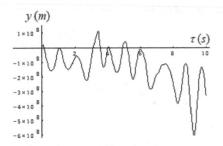

FIGURE 11.57
Graphical representation of the y-coordinate variation of a proton moving for $\Delta\tau = 10\,\text{s}$ in the non-uniform and non-stationary electromagnetic field given by Eq. (11.3.102), as a function of proper time, $y = y(\tau)$.

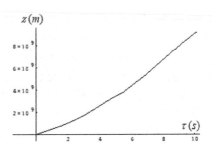

FIGURE 11.58
Graphical representation of the z-coordinate variation of a proton moving for $\Delta\tau = 10\,\text{s}$ in the non-uniform and non-stationary electromagnetic field given by Eq. (11.3.102), as a function of proper time, $z = z(\tau)$.

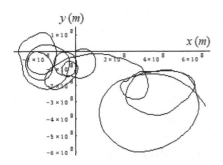

FIGURE 11.59
Projection in the yx-plane of trajectory of a proton moving for $\Delta\tau = 10\,\text{s}$ in the non-uniform and non-stationary electromagnetic field given by Eq. (11.3.102).

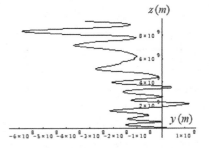

FIGURE 11.60
Projection in the zy-plane of trajectory of a proton moving for $\Delta\tau = 10\,\text{s}$ in the non-uniform and non-stationary electromagnetic field given by Eq. (11.3.102).

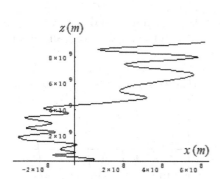

FIGURE 11.61
Projection in the zx-plane of trajectory of a proton moving for $\Delta\tau = 10\,\mathrm{s}$ in the non-uniform and non-stationary electromagnetic field given by Eq. (11.3.102).

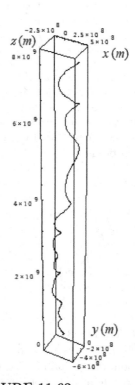

FIGURE 11.62
Trajectory of a proton moving for $\Delta\tau = 10\,\mathrm{s}$ in the non-uniform and non-stationary electromagnetic field given by Eq. (11.3.102).

and

ii) The motion of a proton with the same momentum as that considered at point *i*), during the time interval $\Delta\tau = 3\,\mathrm{s}$ (determined in the proper reference frame) in the non-uniform and non-stationary electromagnetic field with configuration

$$\begin{cases} E_x = 5\sin\lambda x\,\cos\lambda y, \\ E_y = 2\times 10^2\sin\lambda z\,\cos\omega t, \\ E_z = 0, \\ B_x = 1.25\times 10^{-7}\sin\frac{\lambda y}{2}\,\sin\omega t, \\ B_y = 0, \\ B_z = 1.25\times 10^{-7}\sin\omega t, \end{cases} \qquad (11.3.103)$$

where, for calculus convenience, we have taken $\lambda = 1\,\mathrm{m}^{-1}$ and $\omega = 1\,\mathrm{s}^{-1}$, in which case we obtained the (corresponding) dependencies/drawings shown in Figs. 11.63–11.74.

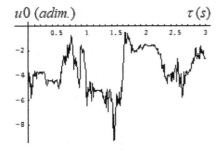

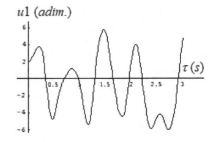

FIGURE 11.63
Graphical representation of the variation of $u0$-component of the covariant four-vector of dimensionless velocity of a proton moving for $\Delta\tau = 3\,\mathrm{s}$ in the non-uniform and non-stationary electromagnetic field given by Eq. (11.3.103), as a function of proper time, $u0 = u0(\tau)$.

FIGURE 11.64
Graphical representation of the variation of $u1$-component of the covariant four-vector of dimensionless velocity of a proton moving for $\Delta\tau = 3\,\mathrm{s}$ in the non-uniform and non-stationary electromagnetic field given by Eq. (11.3.103), as a function of proper time, $u1 = u1(\tau)$.

u2 (adim.)

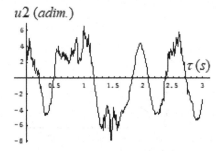

FIGURE 11.65
Graphical representation of the variation of $u2$-component of the covariant four-vector of dimensionless velocity of a proton moving for $\Delta\tau = 3\,\mathrm{s}$ in the non-uniform and non-stationary electromagnetic field given by Eq. (11.3.103), as a function of proper time, $u2 = u2(\tau)$.

u3 (adim.)

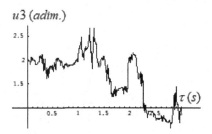

FIGURE 11.66
Graphical representation of the variation of $u3$-component of the covariant four-vector of dimensionless velocity of a proton moving for $\Delta\tau = 3\,\mathrm{s}$ in the non-uniform and non-stationary electromagnetic field given by Eq. (11.3.103), as a function of proper time, $u3 = u3(\tau)$.

t (s)

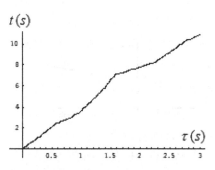

FIGURE 11.67
Graphical representation of dependence $t = t(\tau)$ for a proton moving for $\Delta\tau = 3\,\mathrm{s}$ in the non-uniform and non-stationary electromagnetic field given by Eq. (11.3.103).

x (m)

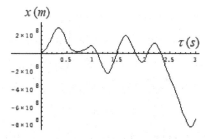

FIGURE 11.68
Graphical representation of the x-coordinate variation of a proton moving for $\Delta\tau = 3\,\mathrm{s}$ in the non-uniform and non-stationary electromagnetic field given by Eq. (11.3.103), as a function of proper time, $x = x(\tau)$.

11.4 Problem No. 77

Determine the relativistic, rectilinear motion of a uniformly accelerated particle, knowing that the magnitude w_0 of its acceleration remains constant in the proper reference frame.

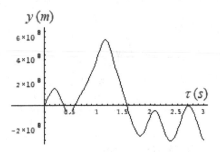

FIGURE 11.69

Graphical representation of the y-coordinate variation of a proton moving for $\Delta\tau = 3\,\text{s}$ in the non-uniform and non-stationary electromagnetic field given by Eq. (11.3.103), as a function of proper time, $y = y(\tau)$.

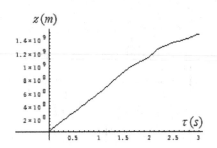

FIGURE 11.70

Graphical representation of the z-coordinate variation of a proton moving for $\Delta\tau = 3\,\text{s}$ in the non-uniform and non-stationary electromagnetic field given by Eq. (11.3.103), as a function of proper time, $z = z(\tau)$.

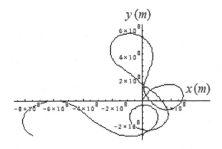

FIGURE 11.71

Projection in the yx-plane of trajectory of a proton moving for $\Delta\tau = 3\,\text{s}$ in the non-uniform and non-stationary electromagnetic field given by Eq. (11.3.103).

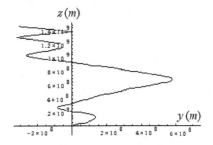

FIGURE 11.72

Projection in the zy-plane of trajectory of a proton moving for $\Delta\tau = 3\,\text{s}$ in the non-uniform and non-stationary electromagnetic field given by Eq. (11.3.103).

Solution

The acceleration four-vector is defined as

$$a^\mu = \frac{du^\mu}{ds} = \frac{\gamma}{c}\frac{du^\mu}{dt}, \tag{11.4.104}$$

where

$$u^\mu = \frac{dx^\mu}{ds} = \frac{\gamma}{c}\frac{dx^\mu}{dt}.$$

Note that with this definition, the dimension of the four-acceleration differs from the dimension of the ordinary acceleration w_0 by a factor c^{-2}. If the

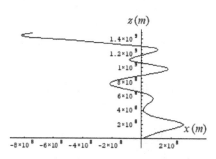

FIGURE 11.73
Projection in the zx-plane of trajectory of a proton moving for $\Delta\tau = 3\,\mathrm{s}$ in the non-uniform and non-stationary electromagnetic field given by Eq. (11.3.103).

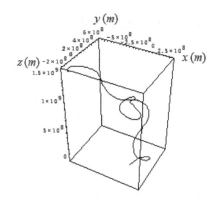

FIGURE 11.74
Trajectory of a proton moving for $\Delta\tau = 3\,\mathrm{s}$ in the non-uniform and non-stationary electromagnetic field given by Eq. (11.3.103).

particle moves along the $Ox \equiv O'x'$ axis, in the proper reference frame $(v = 0)$ we have

$$\begin{cases} a^0 = 0, \\ a^1 = \dfrac{w_0}{c^2}, \\ a^2 = 0, \\ a^3 = 0. \end{cases} \qquad (11.4.105)$$

Since

$$a^\mu a_\mu = \frac{w_0^2}{c^4} = \text{const.},$$

we can write

$$\frac{\gamma^2}{c^2}\left(\frac{d\gamma}{dt}\right)^2 - \frac{\gamma^2}{c^4}\left[\frac{d}{dt}(\gamma v)\right]^2 = \frac{w_0^2}{c^4},$$

where $v = dx^1/dt = dx/dt$. Some simple calculations give

$$\frac{d\gamma}{dt} = \frac{v}{c^2}\gamma^3\frac{dv}{dt},$$

$$\frac{d}{dt}(\gamma v) = \gamma\frac{dv}{dt}\left(1 + \frac{v^2}{c^2}\gamma^2\right) = \gamma^3\frac{dv}{dt},$$

$$\frac{\gamma^6}{c^4}\left(\frac{d\gamma}{dt}\right)^2 = \frac{\gamma^2}{c^4}\left[\frac{d}{dt}(\gamma v)\right]^2 - \frac{\gamma^2}{c^2}\left(\frac{d\gamma}{dt}\right)^2,$$

leading to

$$\frac{d}{dt}(\gamma v) = w_0. \qquad (11.4.106)$$

Integrating Eq. (11.4.106), we find

$$\frac{v}{\sqrt{1 - \frac{v^2}{c^2}}} = w_0 t + C. \tag{11.4.107}$$

Taking as initial conditions $v = 0$ at $t = 0$, one finds $C = 0$. Since $v = dx^1/dt = dx/dt$, we can separate the variables:

$$v = \frac{w_0 t}{\sqrt{1 + \frac{w_0^2 t^2}{c^2}}} = \frac{dx}{dt} \quad \Rightarrow \quad dx = \frac{w_0 t}{\sqrt{1 + \frac{w_0^2 t^2}{c^2}}} \, dt. \tag{11.4.108}$$

By integrating, with the initial condition $x = 0$ at $t = 0$, we obtain

$$x = \frac{c^2}{w_0} \left(\sqrt{1 + \frac{w_0^2 t^2}{c^2}} - 1 \right). \tag{11.4.109}$$

The universe line associated to the particle motion in the Minkowski xOt-plane is, therefore, a hyperbola. This thing can be more easily observed if we write Eq. (11.4.109) in the form

$$\frac{(x - x_0)^2}{p^2} - \frac{(t - t_0)^2}{q^2} = 1,$$

where

$$x_0 \equiv -\frac{c^2}{w_0}, \quad t_0 = 0, \quad p \equiv \frac{c^2}{w_0}, \quad q \equiv \frac{c}{w_0}.$$

Discussion

a) In the non-relativistic case ($w_0 t \ll c$), using the series expansion of the square root in Eq. (11.4.109), we have

$$x = \frac{c^2}{w_0} \left(1 + \frac{1}{2} \frac{w_0^2 t^2}{c^2} + \dots - 1 \right) \simeq \frac{1}{2} w_0 t^2; \quad v = w_0 t, \quad \text{(N.R.)}, \tag{11.4.110}$$

which is, as expected, a parabola.

b) If $w_0 t \to \infty$, then $v \to c$, *i.e.*, this result corresponds to the ultra-relativistic case. Indeed, by introducing the notation $\alpha \equiv w_0 t$, from relation (11.4.108) we have

$$\lim_{w_0 t \to \infty} v \equiv \lim_{\alpha \to \infty} v = \lim_{\alpha \to \infty} \frac{\alpha}{\sqrt{1 + \frac{\alpha^2}{c^2}}}$$

$$= c \lim_{\alpha \to \infty} \frac{\alpha}{\sqrt{c^2 + \alpha^2}} = c \lim_{\alpha \to \infty} \frac{1}{\sqrt{\frac{c^2}{\alpha^2} + 1}} = c, \quad \text{(U.R.)}.$$

c) The items deduced so far allow us to calculate the particle's proper time (the particle being in uniformly accelerated motion). By using the relationship between the proper time τ and the time t measured by an observer in the "laboratory reference frame", *i.e.*,

$$d\tau = dt\sqrt{1 - \frac{v^2}{c^2}},$$

and relation (11.4.108), we have

$$\tau = \int_0^t \sqrt{1 - \frac{v^2}{c^2}}\, dt = \frac{c}{w_0} \int_0^t \frac{dt}{\sqrt{t^2 + \frac{c^2}{w_0^2}}} = \frac{c}{w_0} \operatorname{arcsinh} \frac{w_0 t}{c}. \qquad (11.4.111)$$

Let us consider the U.R. case (which "corresponds" to $t \to \infty$). Denoting this time

$$\operatorname{arcsinh} \frac{w_0 t}{c} \equiv \rho,$$

we have

$$\frac{w_0 t}{c} = \sinh \rho = \frac{e^\rho - e^{-\rho}}{2} \simeq \frac{e^\rho}{2},$$

therefore, according to the relation (11.4.111), in the U.R. case the proper time is

$$\tau \simeq \frac{c}{w_0} \ln \frac{2 w_0 t}{c}. \qquad (11.4.112)$$

11.5 Problem No. 78

Using the formalism offered by the Special Theory of Relativity (more precisely, using the energy-momentum conservation law), show that a free electron (*i.e.*, an electron in vacuum) cannot emit (or absorb) a photon.

Solution

We shall prove that the relativistic energy-momentum conservation law, which is expressed by the relation

$$p_e^\mu = p_e'^\mu + p_p'^\mu \qquad (11.5.113)$$

cannot hold in vacuum. Here p_e^μ is the momentum four-vector of the incident particle (electron) before emission (absorption), and $p_e'^\mu$, $p_p'^\mu$ – the four-momenta of the emitted electron and photon, respectively.

To prove the requirement of the problem we will use the *reductio ad absurdum* method (the method of reduction to absurdity). For this we will assume that the process under study is possible, so we would have satisfied the relativistic energy-momentum conservation law, and this must lead us to an absurdity.

Squaring relation (11.5.113), we have

$$\left(p_e\right)^{\mu}\left(p_e\right)_{\mu} = \left(p'_e\right)^{\mu}\left(p'_e\right)_{\mu} + \left(p'_p\right)^{\mu}\left(p'_p\right)_{\mu} + 2\left(p'_e\right)^{\mu}\left(p'_p\right)_{\mu}. \tag{11.5.114}$$

Since

$$\left(p_e\right)^{\mu}\left(p_e\right)_{\mu} = \left(p'_e\right)^{\mu}\left(p'_e\right)_{\mu} = m_0^2 c^2,$$

and

$$\left(p'_p\right)^{\mu}\left(p'_p\right)_{\mu} = 0,$$

it follows that

$$\left(p'_e\right)^{\mu}\left(p'_p\right)_{\mu} = 0. \tag{11.5.115}$$

We use the following two relations that give the components of the contravariant and covariant momentum four-vectors, respectively:

$$p^{\mu} = \left(\frac{E}{c}, \vec{p}\right), \quad p_{\mu} = \left(\frac{E}{c}, -\vec{p}\right),$$

to write

$$\left(p'_e\right)^0 = \frac{E'_e}{c}, \quad \left|\vec{p}'_p\right| = \frac{E'_p}{c}, \quad \left(p'_p\right)^0 = \frac{E'_p}{c},$$

and, with the notation $p_e = \left|\vec{p}_e\right|$, relation (11.5.115) becomes

$$\frac{1}{c} E'_e E'_p - p'_e E'_p \cos\theta = 0.$$

Using now the energy-momentum "dispersion relation",

$$\frac{E^2}{c^2} = \vec{p}^2 + m_0^2 c^2$$

for the recoil electron,

$$\frac{E'_e}{c} = \sqrt{\left|\vec{p}'_e\right|^2 + m_0^2 c^2},$$

we obtain

$$E'_p \left[p'_e \cos\theta - \sqrt{\left|\vec{p}'_e\right|^2 + m_0^2 c^2} \right] = 0.$$

Since $E'_p \neq 0$, we must have

$$\cos\theta = \sqrt{1 + \frac{m_0^2 c^2}{\left|\vec{p}'_e\right|^2}} \geq 1, \tag{11.5.116}$$

which is absurd, and the demonstration is over.

11.6   Problem No. 79

Using the formalism offered by the Special Theory of Relativity study the one-dimensional relativistic motion of a particle of mass m_0, under the action of a quasi-elastic force.

Solution

Let $\vec{F} = -k_0\vec{x}$, with $k_0 > 0$, be the quasi-elastic force. According to Newtonian mechanics, the particle would perform a harmonic oscillatory motion of angular frequency

$$\omega_0 = \sqrt{\frac{k_0}{m_0}},$$

the solution of the equation of motion being

$$x = a \sin\left[\omega_0(t - t_0)\right]. \tag{11.6.117}$$

Since the particle is relativistic, the equation of motion − in projection on x-axis − is

$$\frac{d}{dt}\left(m_0\gamma v\right) = -k_0 x, \tag{11.6.118}$$

with

$$\gamma = \left(1 - \frac{v^2}{c^2}\right)^{-1/2}.$$

The system being conservative, Eq. (11.6.118) admits the *total energy first integral*

$$m_0 c^2\left(\gamma - 1\right) + \frac{1}{2}m_0\omega_0^2 x^2 = W_0. \tag{11.6.119}$$

To determine W_0, we make use of the initial conditions: at $t = 0$, $x = a$, and $v = 0$. Then

$$W_0 = \frac{1}{2}m_0\omega_0^2 a^2,$$

and the first integral expressed by Eq. (11.6.119) leads to

$$v = c\left\{1 - \left[1 + \frac{\omega_0^2}{2c^2}\left(a^2 - x^2\right)\right]^{-2}\right\}^{1/2} = \frac{dx}{dt}. \tag{11.6.120}$$

We note that the velocity of the particle obeys the condition $v \leq c$, while the constant a signifies the amplitude of the periodic motion $(-a \leq x \leq a)$.

To integrate the first-order differential equation (11.6.120), it is convenient to make the notations

$$\begin{cases} A^2 \equiv a^2 + \dfrac{2c^2}{\omega_0^2}, \\[2mm] A'^2 \equiv a^2 + \dfrac{4c^2}{\omega_0^2}, \end{cases} \tag{11.6.121}$$

and relation (11.6.120) becomes

$$\frac{\left(A^2 - x^2\right)dx}{\sqrt{\left(a^2 - x^2\right)\left(A'^2 - x^2\right)}} = c\,dt. \tag{11.6.122}$$

Setting

$$k^2 \equiv \frac{a^2}{A'^2} = \left(1 + \frac{4c^2}{\omega_0^2 a^2}\right)^{-1} < 1\,, \tag{11.6.123}$$

the integration of Eq. (11.6.122) gives

$$\left(\frac{A^2}{A'} - A'\right)F(\varphi, k) + A'\,E(\varphi, k) = c\big(t - t_0\big), \tag{11.6.124}$$

where

$$\varphi = \arcsin\left(\frac{x}{a}\right),$$

while

$$\begin{cases} F(\varphi, k) = \displaystyle\int_0^{\varphi} \frac{d\psi}{\sqrt{1 - k^2 \sin^2 \psi}}, \\[4mm] E(\varphi, k) = \displaystyle\int_0^{\varphi} \sqrt{1 - k^2 \sin^2 \psi}\; d\psi, \end{cases} \tag{11.6.125}$$

are the incomplete elliptic integral of the first and second kind, respectively.
If we denote

$$k'^2 \equiv 1 - k^2 = \frac{4c^2}{\omega_0^2 A'^2}, \tag{11.6.126}$$

and take into account Eqs. (11.6.122) and (11.6.123), then relation (11.6.124) can also be written as

$$\frac{1}{k'}\left[2E\left(\arcsin\frac{x}{a}, k\right) - k'^2 F\left(\arcsin\frac{x}{a}, k\right)\right] = \omega_0(t - t_0). \tag{11.6.127}$$

In the limit of small velocities ($\frac{v}{c} \to 0$), we have

$$k \to 0, \quad k' \to 1, \quad E(\varphi, 0) = F(\varphi, 0) = \varphi,$$

and Eq. (11.6.127) reduces to Eq. (11.6.117), as expected.
The period of the classical harmonic one-dimensional motion is

$$T_0 = \frac{2\pi}{\omega_0}.$$

The relativistic motion is also periodic, its period being given by the equation

$$\frac{1}{k'}\left[2E(\varphi + 2\pi, k) - k'^2 F(\varphi + 2\pi, k)\right] = \omega_0\big(t + T - t_0\big). \tag{11.6.128}$$

But

$$\begin{cases} F(\varphi + 2\pi, k) = F(\varphi, k) + F(2\pi, k), \\ E(\varphi + 2\pi, k) = E(\varphi, k) + E(2\pi, k), \end{cases}$$

and

$$\begin{cases} F(2\pi, k) = 4F(k), \\ E(2\pi, k) = 4E(k), \end{cases}$$

where

$$\begin{cases} F(k) = F\left(\frac{\pi}{2}, k\right), \\ E(k) = E\left(\frac{\pi}{2}, k\right), \end{cases}$$

are the complete elliptic integrals of the first and second kind, respectively. Then Eqs. (11.6.127) and (11.6.128) yield

$$\frac{\omega}{\omega_0} = \frac{\pi}{2} \frac{k'}{2E(k) - k'^2 F(k)}, \qquad (11.6.129)$$

where

$$\omega = \frac{2\pi}{T}$$

is the relativistic angular frequency.

Thus, the angular frequency ω is a function of m_0, k_0 and a. If ω_0 and a are chosen in such a way that

$$\omega_0 a \ll c,$$

then

$$k \simeq \frac{\omega_0 a}{2c} \ll 1,$$

and we can use the series expansions

$$E(k) \simeq \frac{\pi}{2}\left(1 - \frac{1}{4}k^2\right),$$

$$F(k) \simeq \frac{\pi}{2}\left(1 + \frac{1}{4}k^2\right),$$

$$k' \simeq 1 - \frac{1}{2}k^2.$$

In this case, formula (11.6.129) becomes

$$\frac{\omega}{\omega_0} \simeq 1 - \frac{3}{4}k^2 \simeq 1 - \frac{3}{16}\frac{\omega_0^2 a^2}{c^2}. \qquad (11.6.130)$$

Since the particle covers in a period the path $4a$, and its velocity cannot be greater than c, there exists a minimum time for the particle to move on that distance

$$T_{\min} = \frac{4a}{c},$$

or, equivalently, a *maximum angular frequency*,

$$\omega_{\max} = \frac{2\pi}{T_{\min}} = \frac{\pi}{2}\frac{c}{a}. \qquad (11.6.131)$$

11.7   Problem No. 80

If $\vec{S}$ is the intrinsic angular momentum (the spin) of the electron and $\vec{M}$ its magnetic moment, then in the proper reference frame of the electron one can write

$$\left(\frac{d\vec{S}}{dt}\right)_{\text{rest}} = \vec{M} \times \vec{B}'. \qquad (11.7.132)$$

Here

$$\vec{B}' \simeq \vec{B} - \frac{1}{c^2}\vec{v} \times \vec{E}$$

is the magnetic field in the rest frame of the electron, $\vec{v}$ is the velocity of the electron in an arbitrary inertial frame, and $\vec{M}$ is given by

$$\vec{M} = g_s \frac{e}{2m_0}\vec{S},$$

where g_s is the *spin gyromagnetic factor* and m_0 is the mass of the electron. Write the Eq. (11.7.132) in the relativistically covariant form.

Solution

In the following we shall denote by prime superscript the quantities in the rest frame. Let S^μ be an axial four-vector with three independent components, which in the rest frame of the electron reduces to the spin vector, *i.e.*, $S'^\mu = (0, \vec{S})$. As any four-vector, the time component of the spin four-vector transforms like

$$x'^0 = \Gamma\left(x^0 - \frac{1}{c}\vec{v}\cdot\vec{r}\right),$$

that is

$$S'^0 = \Gamma\left(S^0 - \frac{1}{c}\vec{v}\cdot\vec{S}\right). \qquad (11.7.133)$$

Take now the contracted product between S^μ and the velocity four-vector u^μ:

$$u_\mu S^\mu = u_0 S^0 + u_i S^i = \Gamma\left(S^0 - \frac{1}{c}\vec{v}\cdot\vec{S}\right), \qquad (11.7.134)$$

where

$$u^\mu = \left(\Gamma, \frac{\Gamma}{c}\vec{v}\right).$$

From the last two numbered relations, (11.7.133) and (11.7.134), we infer that

$$S'^0 = u_\mu S^\mu. \qquad (11.7.135)$$

The inverse transformation of Eq. (11.7.133) is

$$S^0 = \Gamma \left(S'^0 + \frac{1}{c} \vec{v} \cdot \vec{S} \right). \tag{11.7.136}$$

According to Eq. (11.7.135), the condition $S'^0 = 0$ can be imposed in a co-variant manner as

$$u_\mu S^\mu = 0, \tag{11.7.137}$$

while Eq. (11.7.136) yields

$$S^0 = \frac{\Gamma}{c} \vec{v} \cdot \vec{S}.$$

Using the relation

$$\vec{M} = g_s \frac{e}{2m_0} \vec{S},$$

we put Eq. (11.7.132) in the form

$$\left(\frac{d\vec{S}}{dt} \right)_{\text{rest}} = g_s \frac{e}{2m_0} \left(\vec{S} \times \vec{B}' \right). \tag{11.7.138}$$

To write this equation in a covariant form, we first observe that both sides of it have to be four-vectors. In the l.h.s., the derivative must be taken with respect to an invariant (either interval s, or the proper time τ), while the r.h.s. has to contain covariant combinations of only the four-velocity u^μ, four-acceleration a^μ, the spin four-vector S^μ, and the electromagnetic field tensor $F^{\mu\nu}$ and to be linear in S^μ and $F^{\mu\nu}$. Last but not least, in the rest reference frame the equation has to go to Eq. (11.7.138). The only non-vanishing combinations with four-vector structure, satisfying the above conditions, are the following:

$$F^{\mu\nu} S_\nu, \quad \left(S_\nu F^{\nu\lambda} u_\lambda \right) u^\mu, \quad \text{and} \quad \left(S_\nu \frac{du^\nu}{ds} \right) u^\mu.$$

Therefore, as a four-dimensional generalization of Eq. (11.7.138) we take the following equation:

$$\frac{dS^\mu}{ds} = C_1 F^{\mu\nu} S_\nu + C_2 \left(S_\nu F^{\nu\lambda} u_\lambda \right) u^\mu + C_3 \left(S_\nu \frac{du^\nu}{ds} \right) u^\mu, \tag{11.7.139}$$

where the constants C_1, C_2 and C_3 are to be determined. Taking the derivative with respect to s of the constraint expressed by Eq. (11.7.137), and using the relation

$$u^\mu u_\mu = 1,$$

we find

$$\left(C_3 + 1 \right) S_\mu \frac{du^\mu}{ds} + \left(C_2 - C_1 \right) F^{\nu\mu} S_\nu u_\mu = 0. \tag{11.7.140}$$

If the electromagnetic field is absent ($F^{\mu\nu} = 0$), while $du^\mu/ds \neq 0$, then $C_3 = -1$. Equation (11.7.140) therefore becomes

$$(C_2 - C_1) F^{\nu\mu} S_\nu u_\mu = 0,$$

which means that in the presence of the electromagnetic field, $C_1 = C_2$. Thus from Eq. (11.7.139) we obtain

$$\frac{dS^\mu}{ds} = C_1 F^{\mu\nu} S_\nu + C_1 S_\nu F^{\nu\lambda} u_\lambda u^\mu - S_\nu \frac{du^\nu}{ds} u^\mu. \qquad (11.7.141)$$

Recalling that in the rest reference frame

$$S'^\mu = (0, \vec{S}),$$

and $u'^\mu = (1, \vec{0})$, let us write the i-component of Eq. (11.7.141):

$$\frac{dS^i}{ds} = \frac{dS^i}{c\,dt'} = C_1 (\vec{S} \times \vec{B}')^i, \quad i = 1, 2, 3,$$

or

$$\left(\frac{d\vec{S}}{dt} \right)_{\text{rest}} = c\, C_1 (\vec{S} \times \vec{B}').$$

This equation is identical to Eq. (11.7.138) if one takes

$$C_1 = g_s \frac{1}{c} \frac{e}{2m_0},$$

and the equation of motion (11.7.139) becomes

$$\frac{dS^\mu}{ds} = g_s \frac{1}{c} \frac{e}{2m_0} \left(F^{\mu\nu} S_\nu + S_\nu F^{\nu\lambda} u_\lambda u^\mu \right) - S_\nu \frac{du^\nu}{ds} u^\mu.$$

Using the covariant form of the equation of motion of a charged particle in an external electromagnetic field,

$$\frac{dp^\nu}{ds} = m_0 c \frac{du^\nu}{ds} = e F^{\nu\lambda} u_\lambda,$$

we finally cast the equation of spin motion for the electron in the form

$$\frac{dS^\mu}{ds} = \frac{e}{m_0 c} \left[\frac{g_s}{2} F^{\mu\nu} S_\nu + \left(\frac{g_s}{2} - 1 \right) S_\nu F^{\nu\lambda} u_\lambda u^\mu \right]. \qquad (11.7.142)$$

This equation is known as *Bargmann-Michel-Telegdi* (BMT) *equation* and describes relativistically the spin precession of high-velocity particles.

The equation is named after Valentine Bargmann (1908–1989), Louis Michel (1923–1999) and Valentine Louis Telegdi (1922–2006). It is the key formula used in high-energy physics experiments to compute the $g_s - 2$ factor, or *anomalous magnetic moment* of the electron and muon. Its use is based on the fact that, according to the quantum mechanical treatment of the electron by Dirac equation, the gyromagnetic ratio g_s is exactly equal to 2. However, quantum field theory gives higher-order corrections, such that $g_s - 2 \neq 0$. Roughly, the idea of the experiment is to have electrons move in an orthogonal magnetic field. The electrons describe a circle with the cyclotron frequency

$$\omega_c = \frac{e}{m_0} \frac{1}{\Gamma} |\vec{B}|.$$

When the electrons complete one circle, their velocity returns to the initial direction, but the spin has precessed according to the first term in Eq. (11.7.142) by an amount proportional to $g_s - 2$. The experiment is done with electrons whose spins are initially polarized in the direction of motion. Due to the precession phenomenon, they develop (very slowly) a component of polarization transverse to the direction of motion. The g_s factor of the electron is measured in very high precision experiments to be 2.00231930436182, with an uncertainty of 5.2×10^{-13}. This is one of the most precisely measured values in physics, and a stringent test of quantum electrodynamics (QED).

11.8   Problem No. 81

A Lagrangian density leading directly to the equations of the electromagnetic field in covariant form,

$$\Box A^\mu = 0,$$

was proposed by Paul Dirac and Vladimir Fock and it is written as follows:

$$\mathcal{L} = -\frac{1}{2\mu_0} \partial_\mu A_\nu \partial^\mu A^\nu. \tag{11.8.143}$$

To obtain the source equations,

$$\Box A^\mu = \mu_0 j^\mu, \tag{11.8.144}$$

one has to consider the Lagrangian density

$$\mathcal{L} = -\frac{1}{2\mu_0} \partial_\mu A_\nu \partial^\mu A^\nu - j_\mu A^\mu. \tag{11.8.145}$$

a) Write the Euler-Lagrange equations for the Lagrangian density given by Eq. (11.8.145) and determine in which conditions they coincide with Maxwell's equations;

b) Show that the Lagrangian density given by Eq. (11.8.145) differs from the one given by

$$\mathcal{L} = -\frac{1}{4\mu_0} F^{\eta\lambda} F_{\eta\lambda} - j^\lambda A_\lambda, \tag{11.8.146}$$

by a four-divergence. Does this modify in any way the action or the equations of motion?

Solution

We start our solving by showing that $j^\mu A_\mu = j_\mu A^\mu$. Indeed, if $g_{\mu\nu}$ is the covariant fundamental metric tensor, and $g^{\mu\nu}$ is its contravariant counterpart, then we have

$$j_\mu A^\mu = g_{\mu\nu} g^{\mu\lambda} j^\nu A_\lambda = \delta^\lambda_\nu j^\nu A_\lambda = j^\nu A_\nu = j^\mu A_\mu.$$

a) The Euler-Lagrange equations write as follows:

$$\frac{\partial \mathcal{L}}{\partial A_\eta} - \frac{\partial}{\partial x^\lambda} \left(\frac{\partial \mathcal{L}}{\partial A_{\eta,\lambda}} \right) = 0. \tag{11.8.147}$$

If $\mathcal{L}$ is that given by Eq. (11.8.145) then we can write

$$\frac{\partial \mathcal{L}}{\partial A_\eta} = \frac{\partial}{\partial A_\eta} \left(- j_\mu A^\mu \right) = -\frac{\partial}{\partial A_\eta} \left(j^\mu A_\mu \right) = -j^\mu \frac{\partial A_\mu}{\partial A_\eta} = -j^\mu \delta^\eta_\mu = -j^\eta.$$

Then,

$$\frac{\partial \mathcal{L}}{\partial A_{\eta,\lambda}} = \frac{\partial}{\partial A_{\eta,\lambda}} \left(-\frac{1}{2\mu_0} \partial_\mu A_\nu \partial^\mu A^\nu \right) = -\frac{1}{2\mu_0} g^{\mu\rho} g^{\nu\sigma}$$

$$\times \frac{\partial}{\partial A_{\eta,\lambda}} \left(\partial_\mu A_\nu \partial_\rho A_\sigma \right) = -\frac{1}{2\mu_0} g^{\mu\rho} g^{\nu\sigma} \frac{\partial}{\partial A_{\eta,\lambda}} \left(A_{\nu,\mu} A_{\sigma,\rho} \right)$$

$$= -\frac{1}{2\mu_0} g^{\mu\rho} g^{\nu\sigma} \left(\delta^\eta_\nu \delta^\lambda_\mu A_{\sigma,\rho} + A_{\nu,\mu} \delta^\eta_\sigma \delta^\lambda_\rho \right)$$

$$= -\frac{1}{2\mu_0} g^{\lambda\rho} g^{\eta\sigma} A_{\sigma,\rho} - \frac{1}{2\mu_0} g^{\mu\lambda} g^{\nu\eta} A_{\nu,\mu}$$

$$= -\frac{1}{2\mu_0} A^{\eta,\lambda} - \frac{1}{2\mu_0} A^{\eta,\lambda} = -\frac{1}{\mu_0} A^{\eta,\lambda} = -\frac{1}{\mu_0} \partial^\lambda A^\eta.$$

Thus, Eq. (11.8.147) becomes

$$-j^\eta - \frac{\partial}{\partial x^\lambda} \left(-\frac{1}{\mu_0} \partial^\lambda A^\eta \right) = 0,$$

that is,

$$\partial_\lambda \partial^\lambda A^\eta = \mu_0 j^\eta \quad \Leftrightarrow \quad \Box A^\eta = \mu_0 j^\eta \quad \Leftrightarrow \quad \Box A^\mu = \mu_0 j^\mu. \tag{11.8.148}$$

This equation coincides with Maxwell's source equation

$$\partial_\nu F^{\nu\mu} = \mu_0 j^\mu, \quad \mu, \nu = \overline{0,3}, \tag{11.8.149}$$

if and only if the four-divergence of the four-vector A^μ vanishes:

$$\partial_\mu A^\mu = 0. \tag{11.8.150}$$

Indeed, if $\partial_\nu A^\nu = 0$, then $\Box A^\mu \equiv \partial_\nu F^{\nu\mu}$, because

$$\Box A^\mu = \partial_\nu \partial^\nu A^\mu = \partial_\nu \partial^\nu A^\mu - \partial^\mu(0) = \partial_\nu \partial^\nu A^\mu - \partial^\mu(\partial_\nu A^\nu)$$

$$= \partial_\nu \partial^\nu A^\mu - \partial^\mu \partial_\nu A^\nu = \partial_\nu \partial^\nu A^\mu - \partial_\nu \partial^\mu A^\nu$$

$$= \partial_\nu\left(\partial^\nu A^\mu - \partial^\mu A^\nu\right) = \partial_\nu F^{\nu\mu},$$

provided that the validity conditions of Schwarz's theorem are satisfied (and they are usually satisfied!), in order to have satisfied the equality of mixed partials (the symmetry of second derivatives) written in covariant formalism:

$$\partial^\mu \partial_\nu = \partial_\nu \partial^\mu.$$

b) Because the second part of the two Lagrangians is the same (*i.e.*, $-j_\mu A^\mu = -j^\lambda A_\lambda$), to prove the problem requirement related to this case we should prove that the first part of the two Lagrangians differs from each other by a four-divergence. Therefore, we should prove that

$$-\frac{1}{4\mu_0} F^{\eta\lambda} F_{\eta\lambda}$$

differs by a four-divergence from

$$-\frac{1}{2\mu_0} \partial_\mu A_\nu \partial^\mu A^\nu.$$

We successively have

$$-\frac{1}{4\mu_0} F^{\eta\lambda} F_{\eta\lambda} = -\frac{1}{4\mu_0} F^{\mu\nu} F_{\mu\nu} = -\frac{1}{4\mu_0}\left(\partial^\mu A^\nu - \partial^\nu A^\mu\right)\left(\partial_\mu A_\nu\right.$$

$$\left. - \partial_\nu A_\mu\right) = -\frac{1}{4\mu_0}\left(\partial^\mu A^\nu - \partial^\nu A^\mu\right)\left(\partial_\mu A_\nu - \partial_\nu A_\mu\right) - \frac{1}{2\mu_0}\partial_\nu A^\nu \partial_\mu A^\mu$$

$$= -\frac{1}{4\mu_0}\left(\partial^\mu A^\nu - \partial^\nu A^\mu\right)\left(\partial_\mu A_\nu - \partial_\nu A_\mu\right) - \frac{1}{2\mu_0}\partial_\nu A^\nu \partial^\mu A_\mu$$

$$= -\frac{1}{4\mu_0}\partial^\mu A^\nu \partial_\mu A_\nu - \frac{1}{4\mu_0}\partial^\nu A^\mu \partial_\nu A_\mu + \frac{1}{4\mu_0}\partial^\mu A^\nu \partial_\nu A_\mu$$

$$+ \frac{1}{4\mu_0}\partial^\nu A^\mu \partial_\mu A_\nu - \frac{1}{2\mu_0}\partial_\nu A^\nu \partial^\mu A_\mu = -\frac{1}{2\mu_0}\partial^\mu A^\nu \partial_\mu A_\nu$$

$$+ \frac{1}{2\mu_0}\partial^\mu A^\nu \partial_\nu A_\mu - \frac{1}{2\mu_0}\partial_\nu A^\nu \partial^\mu A_\mu = -\frac{1}{2\mu_0}\left(\partial^\mu A^\nu \partial_\mu A_\nu\right.$$

$$\left. - \partial^\mu A^\nu \partial_\nu A_\mu + \partial_\nu A^\nu \partial^\mu A_\mu\right) = -\frac{1}{2\mu_0}\partial^\mu A^\nu \partial_\mu A_\nu$$

$$+ \frac{1}{2\mu_0}\partial_\nu\left(A_\mu \partial^\mu A^\nu - A^\nu \partial^\mu A_\mu\right) - \frac{1}{2\mu_0}A_\mu \partial_\nu \partial^\mu A^\nu + \frac{1}{2\mu_0}A^\nu \partial_\nu \partial^\mu A_\mu$$

$$= -\frac{1}{2\mu_0}\partial_\mu A_\nu \partial^\mu A^\nu + \frac{1}{2\mu_0}\partial_\nu\left(A_\mu \partial^\mu A^\nu - A^\nu \partial^\mu A_\mu\right), \qquad (11.8.151)$$

where we used the equality

$$\partial_\mu A^\mu = \partial^\mu A_\mu,$$

and the obvious result

$$\partial^\mu A^\nu \partial_\mu A_\nu = \partial_\mu A_\nu \partial^\mu A^\nu.$$

Let now prove the last equality in relation (11.8.151). For this we must show that

$$-\frac{1}{2\mu_0} A_\mu \partial_\nu \partial^\mu A^\nu + \frac{1}{2\mu_0} A^\nu \partial_\nu \partial^\mu A_\mu = 0.$$

We have

$$-\frac{1}{2\mu_0} A_\mu \partial_\nu \partial^\mu A^\nu + \frac{1}{2\mu_0} A^\nu \partial_\nu \partial^\mu A_\mu = -\frac{1}{2\mu_0} A_\mu \partial_\nu \partial^\mu A^\nu$$

$$+\frac{1}{2\mu_0} A_\nu \partial^\nu \partial_\mu A^\mu = -\frac{1}{2\mu_0} A_\mu \partial_\nu \partial^\mu A^\nu + \frac{1}{2\mu_0} A_\mu \partial^\mu \partial_\nu A^\nu$$

$$= -\frac{1}{2\mu_0} A_\mu \partial_\nu \partial^\mu A^\nu + \frac{1}{2\mu_0} A_\mu \partial_\nu \partial^\mu A^\nu = 0,$$

where we made use of the followings:
 – the (proved from the very beginning) equality: $A^\nu \partial_\nu = A_\nu \partial^\nu$;
 – the interchange of summation/mute indices μ and ν, *i.e.*, $\mu \leftrightarrow \nu$;
 – the symmetry of second derivatives: $\partial^\mu \partial_\nu = \partial_\nu \partial^\mu$.
As can be easily seen from Eq. (11.8.151),

$$-\frac{1}{4\mu_0} F^{\mu\nu} F_{\mu\nu}$$

is equal to

$$-\frac{1}{2\mu_0} \partial_\mu A_\nu \partial^\mu A^\nu,$$

plus the four-divergence

$$\frac{1}{2\mu_0} \partial_\nu \left(A_\mu \partial^\mu A^\nu - A^\nu \partial^\mu A_\mu \right) \equiv \frac{1}{2\mu_0} \partial_\nu V^\nu,$$

where by V^ν we denoted the four-vector

$$V^\nu \equiv A_\mu \partial^\mu A^\nu - A^\nu \partial^\mu A_\mu.$$

Because the field equations remain unchanged if we drop (or add) a term which is a four-divergence in the Lagrangian of a system, we draw the conclusion that the two Lagrangians are equivalent. Of course, the two corresponding actions of the system are also equivalent.

11.9   Problem No. 82

The system composed of the zero rest-mass vector field (the electromagnetic field) and sources is described by the Lagrangian density

$$\mathcal{L} = -\frac{1}{4\mu_0}F^{\mu\nu}F_{\mu\nu} - j^\mu A_\mu, \qquad (11.9.152)$$

For the system composed of the massive vector field and sources, an appropriate Lagrangian density was first proposed in 1930 by Alexandru Proca (1897−1955) as

$$\mathcal{L} = -\frac{1}{4\mu_0}F^{\mu\nu}F_{\mu\nu} + \frac{m^2c^2}{2\mu_0\hbar^2}A^\mu A_\mu - j^\mu A_\mu. \qquad (11.9.153)$$

a) Using the Euler-Lagrange equations for the Lagrangian density given by Eq. (11.9.153), find the *Proca equations* (the equations of motion of the massive vector field in interaction with its sources):

$$\partial_\nu F^{\nu\mu} + \frac{m^2c^2}{\hbar^2}A^\mu = \mu_0 j^\mu, \quad \mu,\nu = 0,1,2,3, \qquad (11.9.154)$$

i.e., the analogues of Maxwell's source equations in covariant form,

$$\partial_\nu F^{\nu\mu} = \mu_0 j^\mu, \quad \mu,\nu = \overline{0,3}; \qquad (11.9.155)$$

b) Show that the conservation law for the four-current, expressed as the continuity equation in Lorentz covariant form, implies that the massive vector field satisfies *necessarily* the condition $\partial_\mu A^\mu = 0$. (Remark that for the massive field this condition is not an arbitrary gauge fixing, but a necessary constraint condition. Actually, the action of the massive vector field is not gauge invariant, due to the mass term.) Making use of this constraint, show that A^μ satisfies a non-homogeneous equation of the Klein-Gordon type:

$$\left(\Box + \frac{m^2c^2}{\hbar^2}\right)A^\mu = \mu_0 j^\mu, \quad \mu = 0,1,2,3; \qquad (11.9.156)$$

c) Starting from the static limit of the Proca equations with the constraint $\partial_\mu A^\mu = 0$, and considering that the sources are represented by a single point-like charge q at rest in the origin of the coordinate system (in which case only the time component of the four-potential, $A^0 = V/c$ is non-vanishing), show that one obtains the static potential with spherical symmetry, called *Yukawa potential*:

$$V(r) = \frac{q}{4\pi\varepsilon_0}\frac{e^{-\frac{mc}{\hbar}r}}{r}. \qquad (11.9.157)$$

Solution

We mention from the very beginning that, although the field function of the massive vector field was here denoted by the same letter as that customarily used for the zero rest-mass vector field (the electromagnetic field), namely the four-potential A^μ, $\mu = 0, 1, 2, 3$, we hope the reader will not make any confusion and will bear in mind that we are talking about two different field functions. For this reason some authors prefer to designate the field function of the massive vector field by B^μ instead of A^μ.

a) To get the Proca equations in μ and ν indices, we write the Euler-Lagrange equations in the same indices, *i.e.*,

$$\frac{\partial \mathcal{L}}{\partial A_\mu} - \frac{\partial}{\partial x^\nu}\left(\frac{\partial \mathcal{L}}{\partial A_{\mu,\nu}}\right) = 0, \tag{11.9.158}$$

where, according to Eq. (11.9.153), the Lagrangian of the system (the massive spin-1 field) is

$$\mathcal{L} = -\frac{1}{4\mu_0}F^{\mu\nu}F_{\mu\nu} + \frac{m^2 c^2}{2\mu_0 \hbar^2}A^\mu A_\mu - j^\mu A_\mu = -\frac{1}{4\mu_0}F^{\eta\lambda}F_{\eta\lambda}$$

$$+ \frac{m^2 c^2}{2\mu_0 \hbar^2}A^\lambda A_\lambda - j^\lambda A_\lambda = -\frac{1}{4\mu_0}\left(\partial^\eta A^\lambda - \partial^\lambda A^\eta\right)$$

$$\times \left(\partial_\eta A_\lambda - \partial_\lambda A_\eta\right) + \frac{m^2 c^2}{2\mu_0 \hbar^2}A^\lambda A_\lambda - j^\lambda A_\lambda$$

$$= -\frac{1}{4\mu_0}g^{\eta\rho}g^{\lambda\sigma}\left(\partial_\rho A_\sigma - \partial_\sigma A_\rho\right)\left(\partial_\eta A_\lambda - \partial_\lambda A_\eta\right)$$

$$+ \frac{m^2 c^2}{2\mu_0 \hbar^2}g^{\lambda\rho}A_\rho A_\lambda - j^\lambda A_\lambda, \tag{11.9.159}$$

which was written, as can be easily seen, only in terms of the covariant components of the four-vector A^μ.

Let us now calculate the two derivatives of the Lagrangian given by Eq. (11.9.159), that appear in Euler-Lagrange equations. We have

$$\frac{\partial \mathcal{L}}{\partial A_\mu} = \frac{m^2 c^2}{2\mu_0 \hbar^2}g^{\lambda\rho}\left(\frac{\partial A_\rho}{\partial A_\mu}A_\lambda + A_\rho\frac{\partial A_\lambda}{\partial A_\mu}\right) - j^\lambda\frac{\partial A_\lambda}{\partial A_\mu}$$

$$= \frac{m^2 c^2}{2\mu_0 \hbar^2}g^{\lambda\rho}\left(\delta_\rho^\mu A_\lambda + A_\rho\delta_\lambda^\mu\right) - j^\lambda\delta_\lambda^\mu$$

$$= \frac{m^2 c^2}{2\mu_0 \hbar^2}g^{\lambda\mu}A_\lambda + \frac{m^2 c^2}{2\mu_0 \hbar^2}g^{\mu\rho}A_\rho - j^\mu$$

$$= \frac{m^2 c^2}{\mu_0 \hbar^2}A^\mu - j^\mu, \tag{11.9.160}$$

and

$$\frac{\partial \mathcal{L}}{\partial A_{\mu,\nu}} = \frac{\partial}{\partial A_{\mu,\nu}} \left[-\frac{1}{4\mu_0} g^{\eta\rho} g^{\lambda\sigma} \left(\partial_\rho A_\sigma - \partial_\sigma A_\rho \right) \left(\partial_\eta A_\lambda - \partial_\lambda A_\eta \right) \right]$$

$$= \frac{\partial}{\partial A_{\mu,\nu}} \left[-\frac{1}{4\mu_0} g^{\eta\rho} g^{\lambda\sigma} \left(A_{\sigma,\rho} - A_{\rho,\sigma} \right) \left(A_{\lambda,\eta} - A_{\eta,\lambda} \right) \right]$$

$$= -\frac{1}{4\mu_0} g^{\eta\rho} g^{\lambda\sigma} \frac{\partial}{\partial A_{\mu,\nu}} \left[\left(A_{\sigma,\rho} - A_{\rho,\sigma} \right) \left(A_{\lambda,\eta} - A_{\eta,\lambda} \right) \right]$$

$$= -\frac{1}{4\mu_0} g^{\eta\rho} g^{\lambda\sigma} \left[\left(\frac{\partial A_{\sigma,\rho}}{\partial A_{\mu,\nu}} - \frac{\partial A_{\rho,\sigma}}{\partial A_{\mu,\nu}} \right) \left(A_{\lambda,\eta} - A_{\eta,\lambda} \right) \right.$$

$$\left. + \left(A_{\sigma,\rho} - A_{\rho,\sigma} \right) \left(\frac{\partial A_{\lambda,\eta}}{\partial A_{\mu,\nu}} - \frac{\partial A_{\eta,\lambda}}{\partial A_{\mu,\nu}} \right) \right] = -\frac{1}{4\mu_0} g^{\eta\rho} g^{\lambda\sigma} \left[\left(\delta^\mu_\sigma \delta^\nu_\rho \right. \right.$$

$$\left. \left. - \delta^\mu_\rho \delta^\nu_\sigma \right) \left(A_{\lambda,\eta} - A_{\eta,\lambda} \right) + \left(A_{\sigma,\rho} - A_{\rho,\sigma} \right) \left(\delta^\mu_\lambda \delta^\nu_\eta - \delta^\mu_\eta \delta^\nu_\lambda \right) \right]$$

$$= -\frac{1}{4\mu_0} \left(g^{\eta\nu} g^{\lambda\mu} - g^{\eta\mu} g^{\lambda\nu} \right) \left(A_{\lambda,\eta} - A_{\eta,\lambda} \right)$$

$$- \frac{1}{4\mu_0} \left(g^{\nu\rho} g^{\mu\sigma} - g^{\mu\rho} g^{\nu\sigma} \right) \left(A_{\sigma,\rho} - A_{\rho,\sigma} \right)$$

$$= -\frac{1}{4\mu_0} \left(A^{\mu,\nu} - A^{\nu,\mu} \right) + \frac{1}{4\mu_0} \left(A^{\nu,\mu} - A^{\mu,\nu} \right)$$

$$- \frac{1}{4\mu_0} \left(A^{\mu,\nu} - A^{\nu,\mu} \right) + \frac{1}{4\mu_0} \left(A^{\nu,\mu} - A^{\mu,\nu} \right)$$

$$= -\frac{1}{\mu_0} \left(A^{\mu,\nu} - A^{\nu,\mu} \right) = -\frac{1}{\mu_0} \left(\partial^\nu A^\mu - \partial^\mu A^\nu \right) = -\frac{1}{\mu_0} F^{\nu\mu}. \quad (11.9.161)$$

Replacing Eqs. (11.9.160) and (11.9.161) in Eq. (11.9.158) we find

$$\frac{m^2 c^2}{\mu_0 \hbar^2} A^\mu - j^\mu + \frac{\partial}{\partial x^\nu} \left(\frac{1}{\mu_0} F^{\nu\mu} \right) = 0,$$

or

$$\partial_\nu F^{\nu\mu} + \frac{m^2 c^2}{\hbar^2} A^\mu = \mu_0 j^\mu,$$

which are exactly the *Proca equations* (11.9.154).

b) The four-current j^μ can be expressed from Proca equations as

$$j^\mu = \frac{1}{\mu_0} \partial_\nu F^{\nu\mu} + \frac{m^2 c^2}{\mu_0 \hbar^2} A^\mu. \quad (11.9.162)$$

The conservation law of the four-current j^μ, expressed under the covariant form, is written as

$$\partial_\mu j^\mu = 0,$$

and, in view of Eq. (11.9.162) this law is equivalent to

$$\partial_\mu \partial_\nu F^{\nu\mu} + \frac{m^2 c^2}{\hbar^2} \partial_\mu A^\mu = 0,$$

or, having in view that

$$\partial_\mu \partial_\nu F^{\nu\mu} = \frac{1}{2}\left(\partial_\mu \partial_\nu F^{\nu\mu} + \partial_\mu \partial_\nu F^{\nu\mu}\right) = \frac{1}{2}\left(\partial_\mu \partial_\nu F^{\nu\mu} + \partial_\nu \partial_\mu F^{\mu\nu}\right)$$

$$= \frac{1}{2}\left(\partial_\mu \partial_\nu F^{\nu\mu} - \partial_\mu \partial_\nu F^{\nu\mu}\right) = 0,$$

(where we used the symmetry of second derivatives, $\partial_\nu \partial_\mu = \partial_\mu \partial_\nu$, and the antisymmetry property of the electromagnetic field tensor, $F^{\mu\nu} = -F^{\nu\mu}$ as well), it necessarily follows that

$$\partial_\mu A^\mu = 0. \tag{11.9.163}$$

Although the above relation is sometimes improperly called (only when referring to massive spin-1 field) the *generalized Lorenz gauge condition*, we stress the fact that for the massive vector field this condition is not a gauge fixing (as in the case of the electromagnetic field − which is a zero rest-mass field), but a necessary constraint condition.

As one knows, the gauge condition for the electromagnetic field is arbitrary (there are many gauge fixings, among them the Lorenz gauge condition and the Coulomb gauge condition are the most used and known; however, in general there are an infinity of such conditions), but in this case (for the massive vector field) it is not about a gauge condition, but a *constraint condition* which is *necessary to be imposed* and is unique in that no other "equivalent" constraint could be imposed.

Taking into account the expression of the electromagnetic field tensor, $F^{\nu\mu} = \partial^\nu A^\mu - \partial^\mu A^\nu$, and the symmetry of second derivatives, we have from the Proca equations:

$$\partial_\nu F^{\nu\mu} + \frac{m^2 c^2}{\hbar^2} A^\mu = \mu_0 j^\mu \;\Leftrightarrow\; \partial_\nu\left(\partial^\nu A^\mu - \partial^\mu A^\nu\right) + \frac{m^2 c^2}{\hbar^2} A^\mu = \mu_0 j^\mu$$

$$\Leftrightarrow\; \partial_\nu \partial^\nu A^\mu - \partial_\nu \partial^\mu A^\nu + \frac{m^2 c^2}{\hbar^2} A^\mu = \mu_0 j^\mu$$

$$\Leftrightarrow\; \Box A^\mu - \partial^\mu\left(\partial_\nu A^\nu\right) + \frac{m^2 c^2}{\hbar^2} A^\mu = \mu_0 j^\mu$$

$$\Leftrightarrow\; \Box A^\mu + \frac{m^2 c^2}{\hbar^2} A^\mu = \mu_0 j^\mu,$$

where we have also used the constraint condition given by Eq. (11.9.163). This way we found the required equation for the massive vector field function A^μ,

$$\Box A^\mu + \frac{m^2 c^2}{\hbar^2} A^\mu = \mu_0 j^\mu, \quad \mu = 0, 1, 2, 3,$$

which is a non-homogeneous *Klein-Gordon type equation*.

c) In the static limit the d'Alembert operator (d'Alembertian) becomes the

Laplace operator (Laplacian), so the relation (11.9.156) can be written as follows:

$$-\Delta A^\mu + \frac{m^2c^2}{\hbar^2} A^\mu = \mu_0 j^\mu. \tag{11.9.164}$$

Since the sources j^μ are represented by a single point-like charge q *at rest* in the origin of the coordinate system, which means that

$$j^\mu \to j^0 = c\rho(\vec{r}) = cq\delta(\vec{r}),$$

and

$$A^\mu \to A^0 = \frac{1}{c}V(\vec{r}),$$

(where $V(\vec{r})$ is the electrostatic potential), Eq. (11.9.164) becomes

$$\left(\Delta - \frac{m^2c^2}{\hbar^2}\right) V(\vec{r}) = -\frac{q}{\varepsilon_0}\delta(\vec{r}). \tag{11.9.165}$$

To solve this equation we will use the Fourier transform method. Given that

$$V(\vec{r}) = \frac{1}{(2\pi)^{3/2}} \int V(\vec{k})\,e^{i\vec{k}\cdot\vec{r}}\,d\vec{k}, \tag{11.9.166}$$

and

$$\delta(\vec{r}) = \frac{1}{(2\pi)^3} \int e^{i\vec{k}\cdot\vec{r}}\,d\vec{k},$$

we have

$$\Delta V(\vec{r}) = -\vec{k}^2\,V(\vec{r}) \equiv -k^2\,V(\vec{r}),$$

and Eq. (11.9.165) leads to

$$\left(k^2 + \frac{m^2c^2}{\hbar^2}\right) V(\vec{k}) = \frac{q}{(2\pi)^{3/2}\varepsilon_0}.$$

From the above relation we obtain

$$V(\vec{k}) = \frac{q}{(2\pi)^{3/2}\varepsilon_0} \frac{1}{k^2 + \frac{m^2c^2}{\hbar^2}},$$

therefore Eq. (11.9.166) will be written as

$$V(\vec{r}) = \frac{1}{(2\pi)^{3/2}} \int \frac{q}{(2\pi)^{3/2}\varepsilon_0} \frac{e^{i\vec{k}\cdot\vec{r}}}{k^2 + \frac{m^2c^2}{\hbar^2}}\,d\vec{k}, \tag{11.9.167}$$

which actually is the (searched) solution of Eq. (11.9.165).

Next we have

$$V(\vec{r}) = \frac{1}{(2\pi)^3} \frac{q}{\varepsilon_0} \int \frac{e^{i\vec{k}\cdot\vec{r}}}{k^2 + \frac{m^2c^2}{\hbar^2}}\,d\vec{k},$$

or, passing to spherical coordinates in $\vec{k}$-space,

$$V(\vec{r}) = \frac{1}{(2\pi)^3} \frac{q}{\varepsilon_0} \int_0^\infty \int_0^\pi \int_0^{2\pi} \frac{e^{ikr\cos\theta}}{k^2 + \frac{m^2c^2}{\hbar^2}} k^2 \sin\theta \, dk \, d\theta \, d\varphi$$

$$= \frac{1}{(2\pi)^2} \frac{q}{\varepsilon_0} \int_0^\infty \int_0^\pi \frac{e^{ikr\cos\theta}}{k^2 + \frac{m^2c^2}{\hbar^2}} k^2 \sin\theta \, dk \, d\theta, \qquad (11.9.168)$$

To perform the integration over θ we shall use the following change of variable:

$$kr\cos\theta = \xi.$$

Then,

$$\int_0^\pi e^{ikr\cos\theta} \sin\theta \, d\theta = \frac{1}{kr} \int_{-kr}^{kr} e^{i\xi} d\xi = \frac{2\sin(kr)}{kr},$$

and Eq. (11.9.168) takes the form

$$V(\vec{r}) = \frac{1}{(2\pi)^2} \frac{q}{\varepsilon_0} \frac{2}{r} \int_0^\infty \frac{k\sin(kr)}{k^2 + \frac{m^2c^2}{\hbar^2}} dk. \qquad (11.9.169)$$

To calculate the integral in Eq. (11.9.169) we first extend the integration interval over the entire real k-axis,

$$V(\vec{r}) = \frac{q}{2\pi^2\varepsilon_0 r} \int_0^\infty \frac{k\sin(kr)dk}{k^2 + \frac{m^2c^2}{\hbar^2}} \to V_1(\vec{r}) = \frac{q}{2\pi^2\varepsilon_0 r} \int_{-\infty}^\infty \frac{k\sin(kr)dk}{k^2 + \frac{m^2c^2}{\hbar^2}},$$

$$(11.9.170)$$

and we call upon the following helpful integral:

$$R(\vec{r}) = \frac{q}{2\pi^2\varepsilon_0 r} \int_{-\infty}^\infty \frac{k\cos(kr)}{k^2 + \frac{m^2c^2}{\hbar^2}} dk. \qquad (11.9.171)$$

Using $V_1(\vec{r})$ and $R(\vec{r})$ we now construct a new integral by the formula

$$C(\vec{r}) = R(\vec{r}) + iV_1(\vec{r}) = \frac{q}{2\pi^2\varepsilon_0 r} \int_{-\infty}^\infty \frac{ke^{ikr}}{k^2 + \frac{m^2c^2}{\hbar^2}} dk, \qquad (11.9.172)$$

and it is clear that

$$V(\vec{r}) = \frac{1}{2} \text{Im}\left[C(\vec{r})\right]. \qquad (11.9.173)$$

Passing to complex k-plane, and choosing the integration contour γ as presented in Fig. 11.75, the Cauchy's residue theorem gives the following result

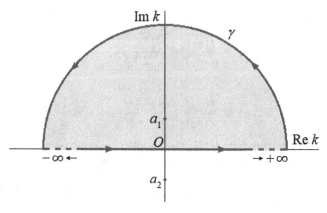

FIGURE 11.75

Integration contour γ in the complex k-plane, used to calculate the integral in formula (11.9.172).

for the integral in formula (11.9.172):

$$\int_{-\infty}^{\infty} \frac{k e^{ikr}}{k^2 + \frac{m^2 c^2}{\hbar^2}}\, dk \rightarrow \oint_{(\gamma)} w(k)dk = 2\pi i \sum_{j=1}^{n} \operatorname{Res}\{w, a_j\},$$

where k is now the new complex integration variable, n is the number of singular isolated points (denoted here by a_j) of the complex function $w(k)$, that are situated inside the simple closed curve γ, and $\operatorname{Res}\{w, a_j\}$ is the residue of the function $w(k)$ at the singular isolated point a_j. Since in our case the function $w(k)$ is a rational function of the form

$$w(k) = \frac{f(k)}{g(k)},$$

where

$$f(k) = k\,e^{ikr}, \quad \text{and} \quad g(k) = k^2 + \frac{m^2 c^2}{\hbar^2},$$

both functions $f(k)$ and $g(k)$ being holomorphic in a vicinity of the two simple poles

$$a_1 = i\frac{mc}{\hbar}, \quad \text{and} \quad a_2 = -i\frac{mc}{\hbar},$$

and, besides,

$$f(a_j) \neq 0, \ g(a_j) = 0, \ g'(a_j) \neq 0, \ j = 1, 2,$$

we have

$$\operatorname{Res}\{w, a_j\} = \frac{f(a_j)}{g'(a_j)}, \ j = 1, 2,$$

that is, explicitly,

$$\text{Res}\{w, a_1\} = \frac{f(a_1)}{g'(a_1)} = \frac{i\frac{mc}{\hbar} e^{i \cdot i \frac{mc}{\hbar} r}}{2i\frac{mc}{\hbar}} = \frac{1}{2} e^{-\frac{mc}{\hbar} r},$$

$$\text{Res}\{w, a_2\} = \frac{f(a_2)}{g'(a_2)} = \frac{-i\frac{mc}{\hbar} e^{-i \cdot i \frac{mc}{\hbar} r}}{-2i\frac{mc}{\hbar}} = \frac{1}{2} e^{\frac{mc}{\hbar} r}.$$

As can be seen from Fig. 11.75, the simple closed curve γ encloses only the simple pole a_1 of the complex function $w(k)$, so we have

$$C(\vec{r}) = \frac{q}{2\pi^2 \varepsilon_0 r} \left(2\pi i \, \text{Res}\{w, a_1\} \right) = \frac{q}{2\pi^2 \varepsilon_0 r} \left(2\pi i \frac{1}{2} e^{-\frac{mc}{\hbar} r} \right)$$

$$= i\pi \frac{q}{2\pi^2 \varepsilon_0 r} e^{-\frac{mc}{\hbar} r},$$

and Eq. (11.9.173) gives the final result of the point **c)** of the problem, namely

$$V(\vec{r}) = \frac{1}{2} \text{Im} \left[C(\vec{r}) \right] = \frac{\pi}{2} \frac{q}{2\pi^2 \varepsilon_0 r} e^{-\frac{mc}{\hbar} r} = \frac{q}{4\pi\varepsilon_0} \frac{e^{-\frac{mc}{\hbar} r}}{r} = \frac{q}{4\pi\varepsilon_0} \frac{e^{-\frac{r}{R}}}{r},$$

where

$$R = \frac{\hbar}{mc} = \frac{1}{2\pi} \frac{h}{mc} = \frac{\lambda_C}{2\pi},$$

is the *reduced Compton wavelength*.

This is a potential used in particle and atomic physics and it is also called a *screened Coulomb potential*. The name of this potential comes from the Japanese theoretical physicist (and the first Japanese Nobel laureate) Hideki Yukawa (1907−1981).

11.10   Problem No. 83

Find the solutions of the differential equations satisfied by electrodynamic potentials, in manifest covariant form, assuming that the electromagnetic perturbation propagates in vacuum.

Solution

Before explicitly solving the differential equations satisfied by electrodynamic potentials, let write down the covariant expressions of the main quantities that appear during the calculations.

First of all we specify that we will use the hyperbolic representation of the Minkowski space-time, with the signature −2 metric,

$$ds^2 = c^2 (dt)^2 - (dx)^2 - (dy)^2 - (dz)^2,$$

i.e., the covariant fundamental metric tensor has the form

$$g_{\mu\nu} = \mathrm{diag}\big(+1, -1, -1, -1\big).$$

Thus, we have:
 - the time t and the three spatial coordinate (x, y, z) form the *position four-vector*

$$x^{\mu}, \quad \mu = 0, 1, 2, 3, \text{with } x^0 = ct, \ x^1 = x, \ x^2 = y, \ x^3 = z,$$

which can be represented as

$$x^{\mu} = \big(ct, \vec{r}\big).$$

Of course,

$$x_{\mu} = \big(ct, -\vec{r}\big);$$

 - the Hamilton's operator (nabla operator), ∇, becomes the spatial part of the new derivative operator which now is a four-vector (instead of "normal" vector which is the nabla vector operator, ∇):

$$\partial_{\mu} = \frac{\partial}{\partial x^{\mu}}, \quad \partial^{\mu} = \frac{\partial}{\partial x_{\mu}},$$

with

$$\partial_{\mu} = \left(\frac{\partial}{\partial x^0}, \nabla\right), \quad \partial^{\mu} = \left(\frac{\partial}{\partial x_0}, -\nabla\right),$$

where

$$\nabla = \left(\frac{\partial}{\partial x^1}, \frac{\partial}{\partial x^2}, \frac{\partial}{\partial x^3}\right) = \left(\frac{\partial}{\partial x}, \frac{\partial}{\partial y}, \frac{\partial}{\partial z}\right);$$

 - the d'Alembertian operator now will be written as

$$\Box = \partial_{\mu}\partial^{\mu} = \partial^{\mu}\partial_{\mu} \equiv \frac{1}{c^2}\frac{\partial^2}{\partial t^2} - \Delta;$$

 - the two electrodynamic potentials, $V(\vec{r}, t)$ and $\vec{A}(\vec{r}, t)$ now form a four-vector, obviously named the *four-vector potential* A^{μ} of the electromagnetic field, defined as follows:

$$A^{\mu} = \left(\frac{V}{c}, \vec{A}\right), \quad A_{\mu} = \left(\frac{V}{c}, -\vec{A}\right);$$

 - the two sources of the electromagnetic field, namely, the spatial charge density $\rho(\vec{r}, t)$ and the current density $\vec{j}(\vec{r}, t)$ form now a four-vector which is denoted by j^{μ} and is named the *four-current*, being given by

$$j^{\mu} = \big(c\rho, \vec{j}\big), \quad j_{\mu} = \big(c\rho, -\vec{j}\big);$$

— the "usual" *Lorenz gauge condition*

$$\nabla \cdot \vec{A} + \varepsilon\mu\frac{\partial V}{\partial t} = 0$$

will be written as

$$\partial^\mu A_\mu = 0, \quad \text{or} \quad \partial_\mu A^\mu = 0,$$

because

$$\partial^\mu A_\mu = \left(g^{\mu\nu}\partial_\nu\right)\left(g_{\mu\sigma}A^\sigma\right) = g^{\mu\nu}g_{\mu\sigma}\partial_\nu A^\sigma = \delta_\sigma^\nu \partial_\nu A^\sigma = \partial_\nu A^\nu = \partial_\mu A^\mu;$$

— the "usual" *equation of continuity*

$$\frac{\partial\rho}{\partial t} + \nabla \cdot (\rho\vec{v}) = 0$$

becomes

$$\partial^\mu j_\mu = 0, \quad \text{or} \quad \partial_\mu j^\mu = 0;$$

— the two differential equations satisfied by the two electromagnetic potentials, $V(\vec{r}, t)$ and $\vec{A}(\vec{r}, t)$, namely

$$\Delta V(\vec{r}, t) - \varepsilon_0\mu_0\frac{\partial^2 V}{\partial t^2}(\vec{r}, t) = -\frac{\rho(\vec{r}, t)}{\varepsilon_0},$$

$$\Delta\vec{A}(\vec{r}, t) - \varepsilon_0\mu_0\frac{\partial^2\vec{A}}{\partial t^2}(\vec{r}, t) = -\mu_0\vec{j}(\vec{r}, t),$$

become

$$\varepsilon_0\mu_0\frac{\partial^2 V}{\partial t^2}(\vec{r}, t) - \Delta V(\vec{r}, t) = \frac{\rho(\vec{r}, t)}{\varepsilon_0},$$

$$\varepsilon_0\mu_0\frac{\partial^2\vec{A}}{\partial t^2}(\vec{r}, t) - \Delta\vec{A}(\vec{r}, t) = \mu_0\vec{j}(\vec{r}, t),$$

and can be written in covariant form as

$$\Box A^\mu = \mu_0 j^\mu, \quad \text{or} \quad \Box A_\mu = \mu_0 j_\mu.$$

Having said that, let's now move on to the problem itself. Unlike the case of non-covariant treatment, when the two electromagnetic potentials (scalar, $V(\vec{r}, t)$, and vector, $\vec{A}(\vec{r}, t)$) satisfy different equations (even if they are very similar), in the case of relativistic-covariant treatment, we have to solve only one equation, namely

$$\Box A^\mu = \mu_0 j^\mu \tag{11.10.174}$$

In order to solve the Eq. (11.10.174) we will use the *Green's function method* which is shortly described in **Appendix F**.

For the sake of the writing simplicity in the following we will use the notations x and x' for the position four-vectors of the two events in the Minkowski space-time (see Fig. 11.76).

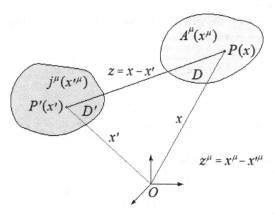

FIGURE 11.76
The domain D' of the sources (the cause) and the domain D of the fields (the effect).

According to Green's function method, the solution $A^\mu(x)$ of the Eq. (11.10.174) is given by

$$A^\mu(x) = A^\mu_{\text{hom}}(x) + \mu_0 \int G(x, x') j^\mu(x') \, d^4x',$$

where $A^\mu_{\text{hom}}(x)$ is the general solution of the homogeneous wave equation attached to the non-homogeneous wave equation

$$\Box A^\mu = \mu_0 \, j^\mu,$$

and $G(x, x')$ is the corresponding (*i.e.*, which is associated in a way that takes into account the concrete physical situation with an "appropriate" general solution $A^\mu_{\text{hom}}(x)$, as we shall see at the end) Green function of the problem. In our case (see **Appendix F**), $G(x, x')$ is the solution of the equation

$$\Box_x G(x, x') = \delta^{(4)}(x - x'), \qquad (11.10.175)$$

where $\delta^{(4)}(x - x')$ is the *four-dimensional Dirac delta function*,

$$\delta^{(4)}(x - x') = \delta\big(x^0 - x'^0\big)\delta\big(\vec{r} - \vec{r}'\big).$$

In the absence of discontinuity surfaces, the Green function can only depend on the difference

$$z^\mu = x^\mu - x'^\mu,$$

which actually is a four-vector, and thus Eq. (11.10.175) becomes

$$\Box_z G(z) = \delta^{(4)}(z). \qquad (11.10.176)$$

To solve this equation we will use the method of Fourier transform. Let $g(k)$ be the Fourier transform of $G(z)$, *i.e.*,

$$G(z) = \frac{1}{(2\pi)^2} \int g(k) e^{-ik \cdot z} d^4k, \qquad (11.10.177)$$

where

$$k \cdot z = k^\mu z_\mu = k^0 z_0 + k^l z_l = k^0 z_0 - \vec{k} \cdot (\vec{r} - \vec{r}') = k^0 z_0 - \vec{k} \cdot \vec{R}.$$

The Fourier transform of Dirac delta function $\delta^{(4)}(z)$ is

$$\delta^{(4)}(z) = \frac{1}{(2\pi)^4} \int e^{-ik \cdot z} d^4k, \qquad (11.10.178)$$

so that Eq. (11.10.176) becomes

$$\Box_z \left(\frac{1}{(2\pi)^2} \int g(k) e^{-ik \cdot z} d^4k \right) = \frac{1}{(2\pi)^4} \int e^{-ik \cdot z} d^4k,$$

or

$$\int \left[g(k) \Box_z e^{-ik \cdot z} - \frac{1}{(2\pi)^2} e^{-ik \cdot z} \right] d^4k = 0. \qquad (11.10.179)$$

Because

$$\begin{aligned}
\Box_z e^{-ik \cdot z} &= \partial^\mu \partial_\mu e^{-ik \cdot z} = \left(\partial^0 \partial_0 - \Delta \right)_z e^{-i\left(k^0 z_0 - \vec{k} \cdot \vec{R}\right)} \\
&= \left(\frac{\partial^2}{\partial z_0^2} - \Delta_{\vec{R}} \right) e^{-i\left(k^0 z_0 - \vec{k} \cdot \vec{R}\right)} = \frac{\partial^2}{\partial z_0^2} e^{-i\left(k^0 z_0 - \vec{k} \cdot \vec{R}\right)} \\
&\quad - \Delta_{\vec{R}} e^{-i\left(k^0 z_0 - \vec{k} \cdot \vec{R}\right)} = -\left(k^0\right)^2 e^{-i\left(k^0 z_0 - \vec{k} \cdot \vec{R}\right)} \\
&\quad + \left(\vec{k} \cdot \vec{k}\right) e^{-i\left(k^0 z_0 - \vec{k} \cdot \vec{R}\right)} = -\left[\left(k^0\right)^2 - \left(\vec{k} \cdot \vec{k}\right) \right] e^{-i\left(k^0 z_0 - \vec{k} \cdot \vec{R}\right)} \\
&= -\left(k \cdot k\right) e^{-i\left(k^0 z_0 - \vec{k} \cdot \vec{R}\right)} \equiv -k^2 e^{-ik \cdot z}, \qquad (11.10.180)
\end{aligned}$$

where

$$k \cdot k \equiv k^2 = k^\mu k_\mu = \left(k^0\right)^2 - \vec{k} \cdot \vec{k} = \left(k^0\right)^2 - \left|\vec{k}\right|^2. \qquad (11.10.181)$$

With Eq. (11.10.180) in Eq. (11.10.179) one obtains

$$\int \left[g(k) k^2 + \frac{1}{(2\pi)^2} \right] e^{-ik \cdot z} d^4k = 0, \qquad (11.10.182)$$

which implies

$$g(k) = -\frac{1}{(2\pi)^2} \frac{1}{k \cdot k} \equiv -\frac{1}{(2\pi)^2} \frac{1}{k^2},$$

therefore

$$G(z) = -\frac{1}{(2\pi)^4} \int \frac{e^{-ik \cdot z}}{k \cdot k} d^4k \equiv -\frac{1}{(2\pi)^4} \int \frac{e^{-ik \cdot z}}{k^2} d^4k. \qquad (11.10.183)$$

To calculate this integral we first proceed to separating the integration over the time component of k from the integration over its spatial components. Thus we have

$$G(z) = -\frac{1}{(2\pi)^4} \int e^{i\vec{k}\cdot\vec{z}} d\vec{k} \int_{-\infty}^{+\infty} \frac{e^{-ik^0 z_0}}{k^2} dk^0, \qquad (11.10.184)$$

where

$$d\vec{k} \equiv d^3k = dk_x dk_y dk_z,$$

if the Cartesian coordinates are used, or

$$d\vec{k} \equiv d^3k = \left|\vec{k}\right|^2 \sin\theta \, d\left|\vec{k}\right| d\theta \, d\varphi,$$

if the spherical coordinates are considered, both types of coordinates belonging to the $\vec{k}$-space, and $\vec{z} \equiv \vec{R}$.

Having in view Eq. (11.10.181), relation (11.10.184) can be written as

$$G(z) = -\frac{1}{(2\pi)^4} \int e^{i\vec{k}\cdot\vec{z}} d\vec{k} \int_{-\infty}^{+\infty} \frac{e^{-ik^0 z_0}}{\left(k^0\right)^2 - \left|\vec{k}\right|^2} dk^0, \qquad (11.10.185)$$

which shows us that the second integrand diverges at points $k^0 = -\left|\vec{k}\right|$ and $k^0 = +\left|\vec{k}\right|$, these isolated singular points being simple poles of this integrand (more precisely, of the rational complex function representing the second integrand).

Since the existence of the two poles makes the integral over k^0 divergent, to perform the integration we shall consider the integral in the complex k^0-plane (*i.e.*, k^0 becomes a complex variable: $k^0 \to k_r^0 + i\,k_i^0$, where $k_r^0 = \text{Re}(k^0)$ and $k_i^0 = \text{Im}(k^0)$) and treat the integral as a contour integral of a function of complex variable k^0, namely

$$w(k^0) = \frac{e^{-ik^0 z_0}}{\left(k^0\right)^2 - \left|\vec{k}\right|^2}, \qquad (11.10.186)$$

and use the *Cauchy's residue theorem*. Besides, to keep things simple, we assume that the "medium" extends to infinity and because it's about the vacuum, clearly it is also non-dispersive ($\left|\vec{k}\right| = \omega/c$).

The two simple poles of the integrand of

$$\int_{-\infty}^{+\infty} \frac{e^{-ik^0 z_0}}{\left(k^0\right)^2 - \left|\vec{k}\right|^2} dk^0$$

are on the real axis (k_r^0-axis), symmetrically with respect to the origin (see Fig. 11.77). By choosing various integration contours in the complex k^0-plane we shall obtain Green functions with different properties.

Specific to the analysis of complex functions is the fact that, if $w(\xi)$ is a

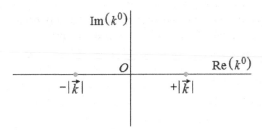

FIGURE 11.77
The two simple poles of the second integrand (that of the integral over k^0) in formula (11.10.185).

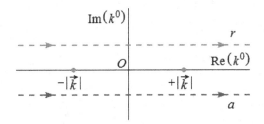

FIGURE 11.78
Two (unclosed yet) integration contours (r and a) for the integral over k^0 in Eq. (11.10.185), physically significant in the covariant theory of the electromagnetic radiation.

complex function of the complex variable ξ, the value of a contour integral of $w(\xi)$ from a fixed point ξ_1 to a fixed point ξ_2 depends, in general, on the path that is taken. In other words, the same integral (having the same integrand and the same integration "limits") can have several values, depending on how the poles are circumvented. Of course, the choice of integration contour is dictated by concrete physical reasons (unlike mathematics, in physics, particularly in the physics of electromagnetic radiation, these contours must be associated with situations of concrete physical significance).

In our case, various physically significant Green's functions are obtained by choosing suitably the integration contours and including one or the other pole, or even both. In Fig. 11.77 are given the two most important "contours", denoted by r and a. These "contours" can be closed at infinity by semicircles in the lower or upper half-planes, depending on the sign of z_0 on the boundary.

For example, in order to obtain the retarded (causal) Green's function, we choose the contour as in Fig. 11.78 (which comes from the "contour" r), while the contour in Fig. 11.79 (which comes from the "contour" a) will lead us to the so-called *advanced Green function*. Both of them are very important in the theory of electromagnetic radiation as we shall point out in the final discussion.

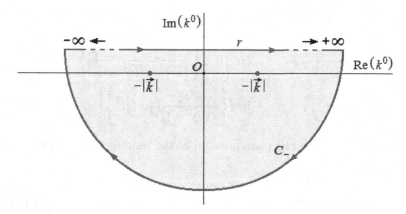

FIGURE 11.79
Integration contour for the retarded (causal) Green's function $G_{ret}(z)$.

When $z_0 > 0$, the exponential $\exp(-ik^0 z_0)$ grows unlimitedly in the upper half-plane, and for $\mathrm{Im}\, k^0 \to \infty$ it becomes infinite. If we wish to have no contribution to the integral from the semicircle that closes the contour at infinity, then we cannot choose the semicircle in the upper half-plane, but in the lower one, where it is easy to see that the integrand becomes zero for $z_0 > 0$. This way, the semicircle part of the integral will tend towards zero as the radius of the semicircle grows, leaving only the real-axis part of the integral, the one we were originally interested in.

The above requirements are fulfilled by the contour C_- in Fig. 11.78. A second but not less important reason for we cannot choose to close the r "contour" in the upper half-plane is that in this case inside the contour will be no singular points and consequently, according to Cauchy's residue theorem the integral equals zero.

On the contrary, choosing the contour C_- that closes in the lower half-plane, the contour integral with respect to k^0 on the closed at infinity curve C_- is given by the sum of the residues at the two simple poles which are enclosed by the integration contour:

$$
\int_{-\infty}^{+\infty} \frac{e^{-ik^0 z_0}}{\left(k^0\right)^2 - \left|\vec{k}\right|^2} dk^0 \to \oint_{(C_-)} \frac{e^{-ik^0 z_0}}{\left(k^0\right)^2 - \left|\vec{k}\right|^2} dk^0 \equiv \oint_{(C_-)} w(k^0) dk^0
$$

$$
= -2\pi i \sum_{j=1}^{2} \mathrm{Res}\{ w\left(k^0\right), k_j^0 \}, \qquad (11.10.187)
$$

the minus sign in front of the sum accounting for the fact that the contour C_- is reversed oriented (clockwise).

Because the two poles of $w(k^0)$ are $k_1^0 = -\left|\vec{k}\right|$ and $k_2^0 = +\left|\vec{k}\right|$, relation

(11.10.187) can be written more explicitly as

$$\oint_{C_-} \frac{e^{-ik^0 z_0}}{\left(k^0\right)^2 - \left|\vec{k}\right|^2} dk^0 = -2\pi i \left[\text{Res}\left\{ \frac{e^{-ik^0 z_0}}{\left(k^0\right)^2 - \left|\vec{k}\right|^2}, -\left|\vec{k}\right| \right\} \right.$$

$$\left. + \text{Res}\left\{ \frac{e^{-ik^0 z_0}}{\left(k^0\right)^2 - \left|\vec{k}\right|^2}, +\left|\vec{k}\right| \right\} \right]. \qquad (11.10.188)$$

In our case, since the two poles are simple, the residues can be calculated at least in two ways:
1) using the general formula

$$\text{Res}\left\{ w(k^0), k_j^0 \right\} = \lim_{k^0 \to k_j^0} (k^0 - k_j^0) w(k^0), \quad j = 1, 2; \qquad (11.10.189)$$

2) because the function $w(k^0)$ is a rational function of the form

$$w(k^0) = \frac{f(k^0)}{g(k^0)}, \text{ with } f(k^0) = e^{-ik^0 z_0} \text{ and } g(k^0) = \left(k^0\right)^2 - \left|\vec{k}\right|^2,$$

both functions $f(k^0)$ and $g(k^0)$ being holomorphic in a vicinity of the two simple poles k_j^0, $j = 1, 2$ and, besides,

$$f(k_j^0) \neq 0, \quad g(k_j^0) = 0, \quad g'(k_j^0) \neq 0, \quad j = 1, 2,$$

we have

$$\text{Res}\left\{ w(k^0), k_j^0 \right\} = \frac{f(k_j^0)}{g'(k_j^0)}, \quad j = 1, 2.$$

Following the second alternative, we have

$$\begin{cases} \text{Res}\left\{ w(k^0), k_1^0 \right\} = \dfrac{f(k_1^0)}{g'(k_1^0)} = \dfrac{e^{-ik_1^0 z_0}}{2k_1^0} = -\dfrac{e^{i\left|\vec{k}\right| z_0}}{2\left|\vec{k}\right|}, \\[4mm] \text{Res}\left\{ w(k^0), k_2^0 \right\} = \dfrac{f(k_2^0)}{g'(k_2^0)} = \dfrac{e^{-ik_2^0 z_0}}{2k_2^0} = \dfrac{e^{-i\left|\vec{k}\right| z_0}}{2\left|\vec{k}\right|}. \end{cases} \qquad (11.10.190)$$

Next, with Eq. (11.10.190) in Eq. (11.10.188) one obtains

$$\oint_{C_-} \frac{e^{-ik^0 z_0}}{\left(k^0\right)^2 - \left|\vec{k}\right|^2} dk^0$$

$$= -2\pi i \left[\text{Res}\left\{ \frac{e^{-ik^0 z_0}}{\left(k^0\right)^2 - \left|\vec{k}\right|^2}, -\left|\vec{k}\right| \right\} + \text{Res}\left\{ \frac{e^{-ik^0 z_0}}{\left(k^0\right)^2 - \left|\vec{k}\right|^2}, +\left|\vec{k}\right| \right\} \right]$$

$$= -2\pi i \left(-\frac{e^{i\left|\vec{k}\right| z_0}}{2\left|\vec{k}\right|} + \frac{e^{-i\left|\vec{k}\right| z_0}}{2\left|\vec{k}\right|} \right) = -\frac{2\pi}{\left|\vec{k}\right|} \sin\left(\left|\vec{k}\right| z_0\right). \qquad (11.10.191)$$

When $z_0 < 0$, the exponential $\exp\left(-ik^0 z_0\right)$ grows unlimitedly in the lower half-plane, and for $\text{Im}\, k^0 \to -\infty$ it becomes infinite. If we wish to have no contribution to the integral from the semicircle that closes the contour at infinity, then we cannot choose the semicircle in the lower half-plane, but in the upper one, where it is easy to see that the contour integral equals zero because inside the integration contour there are no singular points (more precisely, no poles).

Thereby, we can write

$$\oint_{C_-} \frac{e^{-ik^0 z_0}}{\left(k^0\right)^2 - \left|\vec{k}\right|^2} dk^0 = \begin{cases} 0, & z_0 < 0, \\ -\frac{2\pi}{\left|\vec{k}\right|} \sin\left(\left|\vec{k}\right| z_0\right), & z_0 > 0, \end{cases}$$

$$= -\theta\left(z^0\right) \frac{2\pi}{\left|\vec{k}\right|} \sin\left(\left|\vec{k}\right| z^0\right), \qquad (11.10.192)$$

where $\theta\left(z^0\right) = \theta\left(x^0 - x'^0\right)$ is the *Heaviside step function*, also called the *unit step function*:

$$\theta\left(z_0\right) = \theta\left(z^0\right) = \begin{cases} 0, & z^0 < 0, \\ 1, & z^0 > 0. \end{cases} \qquad (11.10.193)$$

Then the *retarded Green's function*, also called the *causal Green's function* has the following expression (which can be obtained by replacing Eq. (11.10.192) in Eq. (11.10.184)):

$$G(z) = \frac{\theta\left(z^0\right)}{(2\pi)^3} \int e^{i\vec{k}\cdot\vec{z}} \frac{\sin\left(\left|\vec{k}\right| z^0\right)}{\left|\vec{k}\right|} d^3 k, \qquad (11.10.194)$$

where we took into account that $z_0 = z^0$. The integral in Eq. (11.10.194) can be easier performed if we use the spherical coordinates in $\vec{k}$-plane, *i.e.*,

$$\begin{cases} \vec{k} \cdot \vec{z} = \left|\vec{k}\right| R \cos\theta, \\ d^3 k = \left|\vec{k}\right|^2 \sin\theta \, dk \, d\theta \, d\varphi. \end{cases}$$

Thus we successively have

$$G(z) = \frac{\theta\left(z^0\right)}{(2\pi)^3} \int\limits_0^\infty \left|\vec{k}\right| \sin\left(\left|\vec{k}\right| z^0\right) \left(\int\limits_0^\pi e^{i\left|\vec{k}\right| R \cos\theta} \sin\theta \, d\theta\right) d\left|\vec{k}\right| \int\limits_0^{2\pi} d\varphi$$

$$= \frac{\theta\left(z^0\right)}{(2\pi)^2} \int\limits_0^\infty \left|\vec{k}\right| \sin\left(\left|\vec{k}\right| z^0\right) \left(\int\limits_0^\pi e^{i\left|\vec{k}\right| R \cos\theta} \sin\theta \, d\theta\right) d\left|\vec{k}\right|.$$

To calculate the above integral over θ we make use of the variable change

$$\left|\vec{k}\right| R \cos\theta = \xi, \quad d\xi = -\left|\vec{k}\right| R \sin\theta \, d\theta,$$

and then

$$G(z) = \frac{\theta(z^0)}{(2\pi)^2} \int\limits_0^\infty |\vec{k}| \sin\left(|\vec{k}|z^0\right) \left(\int\limits_{-|\vec{k}|R}^{+|\vec{k}|R} \frac{e^{i\xi}}{|\vec{k}|R} d\xi \right) d|\vec{k}|$$

$$= \frac{2\,\theta(z^0)}{R\,(2\pi)^2} \int\limits_0^\infty \sin\left(|\vec{k}|z^0\right) \sin\left(|\vec{k}|R\right) d|\vec{k}|. \tag{11.10.195}$$

Next, using the well known trigonometric formula

$$\sin\alpha \sin\beta = \frac{1}{2}\left[\cos(\alpha - \beta) - \cos(\alpha + \beta)\right],$$

for $\alpha \equiv |\vec{k}|z^0$ and $\beta \equiv |\vec{k}|R$, the relation (11.10.195) can be written as

$$G(z) = \frac{\theta(z^0)}{R\,(2\pi)^2} \int\limits_0^\infty \left[\cos\left(|\vec{k}|z^0 - |\vec{k}|R\right) - \cos\left(|\vec{k}|z^0 + |\vec{k}|R\right) \right] d|\vec{k}|$$

$$= \frac{\theta(z^0)}{2R\,(2\pi)^2} \int\limits_{-\infty}^\infty \left[\cos\left(|\vec{k}|z^0 - |\vec{k}|R\right) - \cos\left(|\vec{k}|z^0 + |\vec{k}|R\right) \right] d|\vec{k}|,$$

$$\tag{11.10.196}$$

where, the last form of the above relation (the last equality in Eq. (11.10.196)) was written by virtue of the fact that the cosine function is an even function. Because the integration interval in Eq. (11.10.196) is symmetric and the sine function is odd, we can conveniently add it to construct necessary complex exponentials as follows:

$$G(z) = \frac{\theta(z^0)}{2R\,(2\pi)^2} \int\limits_{-\infty}^\infty \left[\cos\left(|\vec{k}|z^0 - |\vec{k}|R\right) + i\sin\left(|\vec{k}|z^0 - |\vec{k}|R\right) \right] d|\vec{k}|$$

$$- \frac{\theta(z^0)}{2R\,(2\pi)^2} \int\limits_{-\infty}^\infty \left[\cos\left(|\vec{k}|z^0 + |\vec{k}|R\right) + i\sin\left(|\vec{k}|z^0 + |\vec{k}|R\right) \right] d|\vec{k}|$$

$$= \frac{\theta(z^0)}{2R\,(2\pi)^2} \int\limits_{-\infty}^\infty \left[e^{i|\vec{k}|(z^0-R)} - e^{i|\vec{k}|(z^0+R)} \right] d|\vec{k}|$$

$$= \frac{2\pi\,\theta(z^0)}{2R\,(2\pi)^2} \left[\delta(z^0 - R) - \delta(z^0 + R) \right], \tag{11.10.197}$$

But, if $z^0 < 0$, then $\theta(z^0) = 0$ and $G(z)$ cancels. Thus, the Green's function in Eq. (11.10.197) can be non-zero only if $z^0 > 0$, in which case the second Dirac delta function in Eq. (11.10.197) also cancels, because $R > 0$, and thus $z^0 + R > 0$. Therefore we are left with

$$G(z) = \frac{\theta(z^0)}{4\pi R}\delta(z^0 - R),$$

or

$$G(z) = G_{\text{ret}}(x - x') = \frac{1}{4\pi R}\theta(x^0 - x'^0)\delta(x^0 - x'^0 - R). \qquad (11.10.198)$$

The *retarded Green's function* is also called *causal Green's function*, because the electromagnetic perturbation is produced at the moment t', which precedes in time the moment t of observation.

In a similar way, choosing the contour depicted in Fig. 11.79, one obtains the *advanced Green's function*:

$$G_{\text{adv}}(x - x') = \frac{1}{4\pi R}\theta\big[-(x^0 - x'^0)\big]\delta(x^0 - x'^0 + R)$$

$$= \frac{1}{4\pi R}\theta(x'^0 - x^0)\delta(x^0 - x'^0 + R). \qquad (11.10.199)$$

Using the relation

$$\delta(x^2 - a^2) = \frac{1}{2|a|}\big[\delta(x - a) + \delta(x + a)\big],$$

we can write

$$\delta\big[(x - x')^2\big] = \delta\big[(x^\mu - x'^\mu)(x_\mu - x'_\mu)\big] = \delta\big[(x^0 - x'^0)^2 - |\vec{r} - \vec{r}'|^2\big]$$

$$= \delta\big[(x^0 - x'^0 - R)(x^0 - x'^0 + R)\big]$$

$$= \delta\big[(x^0 - x'^0)^2 - R^2\big]$$

$$= \frac{1}{2R}\big[\delta(x^0 - x'^0 - R) + \delta(x^0 - x'^0 + R)\big].$$

$$(11.10.200)$$

Because the functions

$$\theta(x^0 - x'^0),$$

and

$$\theta\big[-(x^0 - x'^0)\big] = \theta(x'^{(0)} - x^{(0)})$$

select one or the other of the two terms in the right-hand side of relation (11.10.200), we have

$$\begin{cases} G_{\text{ret}}(x - x') = \dfrac{1}{2\pi}\theta(x^0 - x'^0)\delta\big[(x - x')^2\big], \\[3mm] G_{\text{adv}}(x - x') = \dfrac{1}{2\pi}\theta(x'^0 - x^0)\delta\big[(x - x')^2\big]. \end{cases} \qquad (11.10.201)$$

The unit step functions $\theta(x^0 - x'^0)$ and $\theta(x'^0 - x^0)$, which are not covariant, become covariant with respect to the proper orthochronous Lorentz transformations when taken in combination with the corresponding Dirac delta functions. Thus, Eq. (11.10.201) provides manifestly covariant expressions for the Green functions. The theta and delta functions in Eq. (11.10.201) show that the retarded and advanced Green functions are nonzero only inside the light hypercone ahead and behind the source, respectively.

The general solution of the non-homogeneous wave equation in vacuum,

$$\Box A^\mu = \mu_0 \, j^\mu,$$

satisfied by the four-potential $A^\mu(x)$ of the electromagnetic field can be written now with the help of the causal Green function as

$$A^\mu(x) = A_{\text{inc}}^\mu(x) + \mu_0 \int G_{\text{ret}}(x - x') \, j^\mu(x') \, d^4x', \qquad (11.10.202)$$

or by means of the advanced Green function as

$$A^\mu(x) = A_{\text{eme}}^\mu(x) + \mu_0 \int G_{\text{adv}}(x - x') \, j^\mu(x') \, d^4x', \qquad (11.10.203)$$

where A_{inc}^μ and A_{eme}^μ are two general solutions (corresponding to the two situations) of the homogeneous wave equation. In Eq. (11.10.202) appears the retarded Green's function. In this case, in the limit $x^0 \to -\infty$, the integral over the sources vanishes, assuming that the sources are localized in space and time (there are no sources at infinity). We see then that the four-potential A_{inc}^μ of the free field has the significance of *incident* or *incoming* four-potential, specified at $x^0 \to -\infty$. Similarly, in Eq. (11.10.203) with the advanced Green function, the solution A_{eme}^μ is the asymptotic *emerging* or *outgoing* four-potential, specified at $x^0 \to +\infty$. The radiation fields are defined as the difference between the emerging and incoming fields, that is

$$A_{\text{rad}}^\mu(x) = A_{\text{eme}}^\mu(x) - A_{\text{inc}}^\mu(x) = \mu_0 \int G(x - x') \, j^\mu(x') \, d^4x', \qquad (11.10.204)$$

where

$$G(x - x') = G_{\text{ret}}(x - x') - G_{\text{adv}}(x - x'). \qquad (11.10.205)$$

12

General Theory of Relativity

12.1 Problem No. 84

Find the geodesics lines corresponding to the metric

$$ds^2 = \frac{1}{t^2} \left(dt^2 - dx^2 \right). \tag{12.1.1}$$

Solution

Following the usual procedure, we apply the Euler-Lagrange equation for the dynamical variable x:

$$\frac{\partial f}{\partial x} - \frac{d}{ds} \left(\frac{\partial f}{\partial \dot{x}} \right) = 0, \tag{12.1.2}$$

where

$$f = \frac{1}{t^2} \left(\dot{t}^2 - \dot{x}^2 \right), \tag{12.1.3}$$

fulfilling the condition $f = 1$. Since f does not explicitly depend on x, the integration gives

$$\frac{\dot{x}}{t^2} = A = \text{const.} \tag{12.1.4}$$

Observing that

$$ds = \frac{1}{t} \left[1 - \left(\frac{dx}{dt} \right)^2 \right]^{1/2} dt, \tag{12.1.5}$$

we then have

$$\dot{x} = \frac{dx}{dt} \frac{dt}{ds} = \frac{dx}{dt} \frac{t}{\left[1 - \left(\frac{dx}{dt} \right)^2 \right]^{1/2}}.$$

Using Eq. (12.1.4) and separating variables, we can also write

$$\frac{dx}{dt} = \frac{At}{\sqrt{1 + A^2 t^2}},$$

and, by integrating the equation and a convenient arrangement of terms,

$$\frac{x^2}{a^2} - \frac{t^2}{a^2} = 1, \quad a \equiv \frac{1}{A}. \tag{12.1.6}$$

DOI: 10.1201/9781003402602-12

Giving values to the constant A, we obtain the geodesics in the shape of *equilateral hyperbolas, tangent to the light cone.*

12.2   Problem No. 85

At the point $\theta = \theta_0$, $\varphi = \varphi_0$ on the surface of the two-dimensional sphere $\left(ds^2 = d\theta^2 + \sin^2\theta\, d\varphi^2\right)$ we have $\vec{A} = \vec{u}_\theta$. Write the form of vector $\vec{A}$ as a result of a parallel transport along the circle $\theta = \theta_0$, as well as its magnitude after transport.

Solution

By definition, a tensor Q (of any order and/or variance) suffers a parallel transport if its absolute differential is zero

$$DQ = dQ + \delta Q = 0.$$

For an arbitrary contravariant four-vector A^μ, we have

$$DA^\mu = dA^\mu + \Gamma^\mu_{\nu\lambda}A^\nu\, dx^\lambda = 0,$$

or

$$u^\lambda\left(\frac{\partial A^\mu}{\partial x^\lambda} + \Gamma^\mu_{\nu\lambda}A^\nu\right) = 0, \quad u^\lambda = \frac{dx^\lambda}{ds}.$$

If a vector tangent to a curve is parallel transported, then the curve is called *autoparallel.* For $A^\mu = u^\mu$, it follows that

$$\frac{du^\mu}{ds} + \Gamma^\mu_{\nu\lambda}u^\nu u^\lambda = 0,$$

which are the differential equations of the geodesic lines. If the connection coefficients are the Christoffel symbols of the second kind, the manifold is Riemannian. Therefore, *the autoparallel curves of a Riemannian space are geodesics of that space.*

Let us now transpose this general formalism to our two-dimensional case. We then have

$$\nabla_\varphi A^i = A^i_{,\varphi} + \Gamma^i_{k\varphi}A^k = 0, \quad i,k = 1,2. \tag{12.2.7}$$

With $x^1 = \theta$ and $x^2 = \varphi$ in $ds^2 = d\theta^2 + \sin^2\theta\, d\varphi^2$ it follows that in our case f has the expression

$$f = \dot{\theta}^2 + \dot{\varphi}^2\sin^2\theta.$$

So, the Euler-Lagrange equations

$$\frac{\partial f}{\partial \theta} - \frac{d}{ds}\left(\frac{\partial f}{\partial \dot{\theta}}\right) = 0,$$

and

$$\frac{\partial f}{\partial \varphi} - \frac{d}{ds}\left(\frac{\partial f}{\partial \dot{\varphi}}\right) = 0,$$

give

$$\begin{cases} \Gamma_{22}^1 \equiv \Gamma_{\varphi\varphi}^\theta = -\sin\theta\cos\theta, \\ \Gamma_{12}^2 \equiv \Gamma_{\theta\varphi}^\varphi = \cot\theta, \end{cases}$$

and thus Eq. (12.2.7) yield

$$\begin{cases} 0 = A_{,\varphi}^\theta - \sin\theta\cos\theta\, A^\varphi, \\ 0 = A_{,\varphi}^\varphi + \cot\theta\, A^\theta. \end{cases} \tag{12.2.8}$$

If we take now the partial derivative of Eq. (12.2.8)$_1$ with respect to φ, and then use the Eq. (12.2.8)$_2$ we obtain

$$A_{,\varphi\varphi}^\theta = \sin\theta\cos\theta A_{,\varphi}^\varphi = -\cos^2\theta\, A^\theta, \tag{12.2.9}$$

i.e., a linear harmonic oscillator-type equation,

$$A_{,\varphi\varphi}^\theta + \cos^2\theta\, A^\theta = 0. \tag{12.2.10}$$

As is well known, the solution of this equation – viewed as a second-order ordinary differential equation for the unknown A^θ, the independent variable being φ – is

$$A^\theta = C_1\cos(\varphi\cos\theta) + C_2\sin(\varphi\cos\theta),$$

allowing to determine A^φ via Eq. (12.2.8)$_1$; it is given by

$$A^\varphi = \frac{1}{\sin\theta}\left[-C_1\sin(\varphi\cos\theta) + C_2\cos(\varphi\cos\theta)\right]. \tag{12.2.11}$$

To fix the (so far, arbitrary) constants of integration C_1 and C_2, we use the boundary conditions: at $\varphi = 0$, $\vec{A} = \vec{u}_\theta$, $A^\theta = 1$, $A^\varphi = 0$. Then $C_1 = 1$, $C_2 = 0$, and thus

$$\begin{cases} A^\theta = \cos(\varphi\cos\theta), \\ A^\varphi = -\dfrac{1}{\sin\theta}\sin(\varphi\cos\theta), \end{cases}$$

which leads to

$$\vec{A} = \cos(\varphi\cos\theta)\vec{u}_\theta - \frac{1}{\sin\theta}\sin(\varphi\cos\theta)\vec{u}_\varphi. \tag{12.2.12}$$

So, after the transport ($\varphi = 0 \to \varphi = 2\pi$), $\vec{A}$ becomes

$$\vec{A}_{\varphi=2\pi} = \cos(2\pi\cos\theta)\vec{u}_\theta - \frac{1}{\sin\theta}\sin(2\pi\cos\theta)\vec{u}_\varphi \neq \vec{u}_\theta, \tag{12.2.13}$$

but the magnitude of the vector remains the same. Indeed,

$$
\begin{aligned}
(A_\mu A^\mu)_{\varphi=2\pi} &= g_{\mu\nu} A^\mu A^\nu = g_{11} A^1 A^1 + g_{22} A^2 A^2 \\
&= g_{\theta\theta} A^\theta A^\theta + g_{\varphi\varphi} A^\varphi A^\varphi = \cos^2(2\pi\cos\theta) \\
&\quad + \sin^2\theta \left[\frac{1}{\sin^2\theta} \sin^2(2\pi\cos\theta) \right] \\
&= 1 = (A_\mu A^\mu)_{\varphi=0}.
\end{aligned}
\tag{12.2.14}
$$

12.3 Problem No. 86

Show that the covariant four-divergence of Einstein's tensor

$$
G_{\mu\nu} = R_{\mu\nu} - \frac{1}{2} R\, g_{\mu\nu}
$$

is zero.

Solution

Using the second Bianchi identity

$$
\nabla_\sigma R^\rho{}_{\mu\nu\lambda} + \nabla_\nu R^\rho{}_{\mu\lambda\sigma} + \nabla_\lambda R^\rho{}_{\mu\sigma\nu} = 0,
\tag{12.3.15}
$$

and recalling the Riemann tensor properties, we interchange the indices ν and λ in the first terms, then multiply by $g^{\mu\nu}$. The result is

$$
-g^{\mu\nu} R^\rho{}_{\mu\lambda\nu\,;\,\sigma} + g^{\mu\nu} R^\rho{}_{\mu\lambda\sigma\,;\,\nu} + g^{\mu\nu} R^\rho{}_{\mu\sigma\nu\,;\,\lambda} = 0.
$$

Multiply now by δ^λ_ρ and use the fact that the covariant derivative of the metric tensor is zero. We then have

$$
-(g^{\mu\nu} R_{\mu\nu})_{;\sigma} + (g^{\mu\nu} R_{\mu\sigma})_{;\nu} + (g^{\mu\nu} R^\lambda{}_{\mu\sigma\nu})_{;\lambda} = 0.
$$

The last term is the covariant four-divergence of the mixed Ricci tensor. Indeed,

$$
g^{\mu\nu} R^\lambda{}_{\mu\sigma\nu} = g^{\mu\nu} g^{\lambda\kappa} R_{\kappa\mu\sigma\nu} = g^{\mu\nu} g^{\lambda\kappa} R_{\mu\kappa\nu\sigma} = g^{\lambda\kappa} R^\nu{}_{\kappa\nu\sigma} = g^{\lambda\kappa} R_{\kappa\sigma} = R^\lambda{}_\sigma.
$$

Thus, we have obtained that

$$
-R_{;\sigma} + R^\nu{}_{\sigma\,;\,\nu} + R^\lambda{}_{\sigma\,;\,\lambda} = 0,
$$

or

$$
\nabla_\nu \left(R^\nu_\sigma - \frac{1}{2} R\, \delta^\nu_\sigma \right) = \nabla_\nu G^\nu_\sigma = 0,
\tag{12.3.16}
$$

and the problem is solved.

12.4   Problem No. 87

Determine the elementary space-like distance in a uniformly rotating coordinate system.

Solution

Let us start by elucidating the notions of space-like and time-like intervals in General Relativity.

Consider two time-like separated close events, that happen at the same point of space. Choosing x^1, x^2, x^3 as space coordinates and $x^0 = ct$ as time coordinate, it follows from the problem statement that $dx^1 = dx^2 = dx^3 = 0$, and thus the metric is

$$ds^2 = c^2 d\tau^2 = g_{\mu\nu} dx^\mu dx^\nu = g_{00}(dx^0)^2.$$

This means that the proper time separating the two events is

$$d\tau = \frac{1}{c}\sqrt{g_{00}}\, dx^0. \tag{12.4.17}$$

Recall that in special relativity the elementary space-like distance dl can be determined as the interval between two close events which take place at the same moment, by choosing $dx^0 = 0$. In general relativity this procedure cannot be used, because at different points the proper time τ is differently connected to x^0.

Suppose that a light signal emitted at point $B(x^i + dx^i)$, $i = 1, 2, 3$, is intercepted at the neighbouring point $A(x^i)$, and then transmitted back on the same path. Since the metric is isotropic, we have

$$g_{ik} dx^i dx^k + 2g_{i0} dx^i dx^0 + g_{00}(dx^0)^2 = 0.$$

The roots of this equation − viewed as an algebraic quadratic equation for the unknown quantity (dx^0) − are

$$(dx^0)_1 = -\frac{1}{g_{00}}\left[g_{i0} dx^i + \sqrt{(g_{i0}g_{k0} - g_{ik}g_{00})dx^i dx^k}\right],$$

$$(dx^0)_2 = -\frac{1}{g_{00}}\left[g_{i0} dx^i - \sqrt{(g_{i0}g_{k0} - g_{ik}g_{00})dx^i dx^k}\right],$$

and correspond to the propagation of the light signal in the two directions, between the points A and B of given coordinates. In Fig. 12.1 the solid lines are the world lines of the points A and B, while the dashed lines represent the world lines of the light signals. The "time" difference between the emission and reception of the light signal at the same point is

$$(dx^0)_2 - (dx^0)_1 = \frac{2}{g_{00}}\sqrt{(g_{i0}g_{k0} - g_{ik}g_{00})dx^i dx^k}.$$

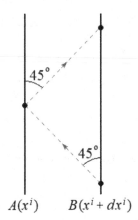

$$A(x^i) \qquad B(x^i + dx^i)$$

FIGURE 12.1
World lines of the points A and B (solid lines) and of the light signals (dashed lines).

The "true" elementary proper time is then

$$d\tau = \frac{1}{c}\sqrt{g_{00}}\left[(dx^0)_2 - (dx^0)_1\right] = \frac{2}{c\sqrt{g_{00}}}\sqrt{(g_{i0}g_{k0} - g_{ik}g_{00})\,dx^i\,dx^k},$$

while the elementary proper distance $dl = c\,d\tau/2$ is

$$(dl)^2 = \left(\frac{g_{i0}g_{k0}}{g_{00}} - g_{ik}\right)dx^i dx^k = \gamma_{ik}dx^i dx^k, \ i,k = 1,2,3, \qquad (12.4.18)$$

where

$$\gamma_{ik} = \frac{g_{i0}g_{k0}}{g_{00}} - g_{ik}, \ i,k = 1,2,3, \qquad (12.4.19)$$

is the *three-dimensional metric tensor*.

Now we can give the solution to the problem. Let us perform the transition from the inertial frame S' to another frame S, uniformly rotating around $Oz \equiv O'z' \equiv Oz'$-axis. The geometry of the problem suggest the use of cylindrical coordinates. Let r', φ', z', t and r, φ, z, t be the coordinates in the two frames, respectively. Then, in the inertial frame S' the metric is

$$ds^2 = c^2 dt^2 - dr'^2 - r'^2 d\varphi'^2 - dz'^2.$$

But $r' = r, z' = z, \varphi' = \varphi + \omega t$, so that

$$ds^2 = g_{\mu\nu}dx^\mu dx^\nu = (c^2 - \omega^2 t^2)dt^2 - 2\omega r^2\,d\varphi\,dt - dz^2 - r^2 d\varphi^2 - dr^2.$$

Thus,

$$\begin{cases} g_{00} = 1 - \dfrac{\omega^2 r^2}{c^2}, \\[2mm] g_{11} = g_{33} = -1, \\[2mm] g_{22} = -r^2, \\[2mm] g_{20} = -\dfrac{\omega r^2}{c}. \end{cases}$$

In the rotating frame S, the space-like distance element is worked out by using Eqs. (12.4.18) and (12.4.19). Choosing $x^0 = ct$, $x^1 = r$, $x^2 = \varphi$, $x^3 = z$, we obtain

$$\begin{cases} \gamma_{11} = g_{11} = -1, \\[2mm] \gamma_{33} = g_{33} = -1, \\[2mm] \gamma_{22} = \dfrac{(g_{20})^2}{g_{00}} - g_{22} = r^2 \left(1 - \dfrac{\omega^2 r^2}{c^2}\right)^{-1}. \end{cases}$$

The solution is then given by the expression

$$(dl)^2 = \gamma_{ik} dx^i dx^k = dr^2 + dz^2 + \frac{r^2 d\varphi^2}{1 - \dfrac{\omega^2 r^2}{c^2}}. \tag{12.4.20}$$

Observation: In the inertial (fixed) frame S', the ratio of a circle's circumference (with its centre on the axis of rotation) to its radius in the $z = \text{const.}$ plane is 2π, while in the non-inertial (rotating) frame S we have

$$L = \int_0^{2\pi} \frac{R d\varphi}{\sqrt{1 - \dfrac{\omega^2 R^2}{c^2}}} = \frac{2\pi R}{\sqrt{1 - \dfrac{\omega^2 R^2}{c^2}}}, \tag{87.5}$$

that is,

$$\frac{L}{R} > 2\pi. \tag{12.4.21}$$

12.5   Problem No. 88

Show that, irrespective of its mass, a body cannot orbit a Schwarzschild black hole at a distance smaller than $r_{\min}^{(\text{stable})} = 3r_S$, on a stable orbit, or $r_{\min}^{(\text{unstable})} = 3r_S/2$, on an unstable orbit.

Solution

The trajectory of a particle of mass m, moving in the spherically symmetric gravitational field described by the Schwarzschild metric can be found by means of the relativistic Hamilton-Jacobi equation

$$g^{\mu\nu}\frac{\partial S}{\partial x^\mu}\frac{\partial S}{\partial x^\nu} - m^2 c^2 = 0. \qquad (12.5.22)$$

The components of the metric tensor are straightforwardly inferred from the form of the Schwarzschild metric

$$ds^2 = g_{\mu\nu}dx^\mu dx^\nu = c^2\left(1 - \frac{r_S}{r}\right)dt^2$$

$$- \left(1 - \frac{r_S}{r}\right)^{-1} dr^2 - r^2 d\theta^2 - r^2 \sin^2\theta\, d\varphi^2, \qquad (12.5.23)$$

where $x^0 = ct$, $x^1 = r$, $x^2 = \theta$, $x^3 = \varphi$ and $r_S = 2GM/c^2$ is the Schwarzschild radius of the central body of mass M, which creates the gravitational field. As the field has central symmetry, the motion takes place in a plane which contains the source of the field (this is a general result, valid for any central field). We shall choose this plane as defined by the condition $\theta = \pi/2$. Then, the metric given by Eq. (12.5.23) becomes

$$ds^2 = c^2\left(1 - \frac{r_S}{r}\right)dt^2 - \left(1 - \frac{r_S}{r}\right)^{-1} dr^2 - r^2 d\varphi^2, \qquad (12.5.24)$$

from where

$$\begin{cases} g_{00} = 1 - \dfrac{r_S}{r}, \\[2mm] g_{11} \equiv g_{rr} = -\dfrac{1}{1 - \dfrac{r_S}{r}}, \\[2mm] g_{33} \equiv g_{\varphi\varphi} = -r^2. \end{cases} \qquad (12.5.25)$$

As $g_{\mu\lambda}g^{\nu\lambda} = \delta^\nu_\mu$, we have also

$$\begin{cases} g^{00} = \left(1 - \dfrac{r_S}{r}\right)^{-1}, \\[2mm] g^{11} \equiv g^{rr} = -\left(1 - \dfrac{r_S}{r}\right), \\[2mm] g^{33} \equiv g^{\varphi\varphi} = -\dfrac{1}{r^2}, \end{cases} \qquad (12.5.26)$$

such that the Hamilton-Jacobi equation (12.5.22) becomes

$$\left(1 - \frac{r_S}{r}\right)^{-1}\frac{1}{c^2}\left(\frac{\partial S}{\partial t}\right)^2 - \left(1 - \frac{r_S}{r}\right)\left(\frac{\partial S}{\partial r}\right)^2 - \frac{1}{r^2}\left(\frac{\partial S}{\partial \varphi}\right)^2 - m^2 c^2 = 0. \qquad (12.5.27)$$

Using the method of the separation of variables and taking into account that the system is conservative (the Hamiltonian does not depend explicitly on time), while the variable φ is cyclic, we shall seek the solution of Eq. (12.5.27) in the form

$$S \equiv S(t, r, \varphi) = -E_0 t + S_r(r) + L\varphi, \qquad (12.5.28)$$

where E_0 is the total (conserved) energy of the particle of mass m, while L is the angular momentum of the particle with respect to the centre of symmetry (and which is as well conserved in the case of a central force field). Introducing Eq. (12.5.28) into Eq. (12.5.27) we find the following expression for the radial part $S_r(r)$ of the action:

$$S_r(r) = \frac{1}{c} \int \frac{dr}{r} \sqrt{r^2 E_0^2 \left(1 - \frac{r_S}{r}\right)^{-2} - c^2 \left(m^2 c^2 r^2 + L^2\right) \left(1 - \frac{r_S}{r}\right)^{-1}}.$$

$$(12.5.29)$$

In Hamilton-Jacobi formalism, the dependence $r = r(t)$ is given by the equation

$$\frac{\partial S}{\partial E_0} = \text{const.}, \qquad (12.5.30)$$

while the trajectory of the particle is determined by the equation

$$\frac{\partial S}{\partial L} = \text{const.} \qquad (12.5.31)$$

Taking into account the relations (12.5.28) and (12.5.29), from Eq. (12.5.30) follows immediately that

$$ct = E_0 \int \left(1 - \frac{r_S}{r}\right)^{-1} \left[(rE_0)^2 - c^2 \left(1 - \frac{r_S}{r}\right) \left(m^2 c^2 r^2 + L^2\right)\right]^{-\frac{1}{2}} r\, dr.$$

$$(12.5.32)$$

With the notation

$$U_{\text{eff}}(r) \equiv \frac{c}{r} \sqrt{\left(1 - \frac{r_S}{r}\right) \left(m^2 c^2 r^2 + L^2\right)}$$

$$= mc^2 \sqrt{\left(1 - \frac{r_S}{r}\right) \left(1 + \frac{L^2}{m^2 c^2 r^2}\right)}, \qquad (12.5.33)$$

the dependence $r = r(t)$ – given in the integral form by the relation (12.5.32) – can be written in differential form as follows:

$$\frac{dr}{dt} = c \left(1 - \frac{r_S}{r}\right) \sqrt{1 - \left[\frac{U_{\text{eff}}(r)}{E_0}\right]^2}. \qquad (12.5.34)$$

This equation indicates that the function $U_{\text{eff}} = U_{\text{eff}}(r)$ plays the role of an *effective potential energy*, in the sense that, by analogy with the classical theory, the condition $E_0 \geq U_{\text{eff}}(r)$ establishes the intervals of values of the radial coordinate r within which the motion of the particle is allowed. In Fig. 12.2 we represented four curves of variation of the *reduced effective potential energy* U_{eff}/mc^2 as a function of the ratio r/r_S, corresponding to as many values of the angular momentum L.

The minima of the function $U_{\text{eff}}(r)$ correspond to the stable orbits of the particle, while the maxima correspond to unstable orbits. The values of the

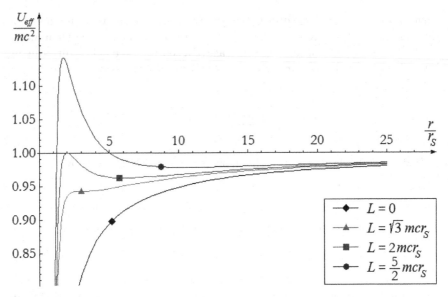

FIGURE 12.2
Reduced effective potential energy U_{eff}/mc^2 for various values of the angular momentum L of the particle.

radii of the circular orbits, as well as the corresponding values of the constant quantities E_0 and L are determined by the following equation system:

$$\begin{cases} U_{\text{eff}}(r) = E_0, \\ \dfrac{dU_{\text{eff}}}{dr} = 0, \end{cases} \tag{12.5.35}$$

which leads to the following relations that have to be satisfied simultaneously:

$$E_0 = cL\sqrt{\frac{2}{r}\frac{r - r_S}{r r_S}}, \tag{12.5.36}$$

and

$$\frac{r}{r_S} = \frac{L^2}{m^2 c^2 r_S^2}\left(1 \pm \sqrt{1 - \frac{3m^2 c^2 r_S^2}{L^2}}\right). \tag{12.5.37}$$

The "+" sign in relation (12.5.37) corresponds to the stable orbits $\left(\dfrac{d^2 U_{\text{eff}}}{dr^2} > 0\right)$, while the sign "−", corresponds to the unstable ones $\left(\dfrac{d^2 U_{\text{eff}}}{dr^2} < 0\right)$. The closest stable orbit from the centre of symmetry is char-

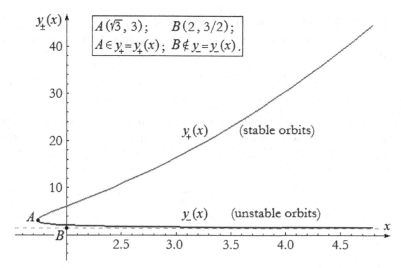

FIGURE 12.3

Dependence of $r/r_S \equiv y$ on $L/mcr_S \equiv x$. The upper branch gives the radius for stable, and the lower for unstable orbits. The point B does not belong to the curve $y_- = y_-(x)$, but to the horizontal asymptote.

acterized by the following parameters:

$$
\begin{cases}
r_{\min}^{\text{stable}} = 3r_S = \dfrac{6GM}{c^2}, \\[2mm]
L\big|_{r=r_{\min}^{\text{stable}}} = \sqrt{3}\, mcr_S = \dfrac{2\sqrt{3}\, GmM}{c}, \\[2mm]
E_0\big|_{r=r_{\min}^{\text{stable}}} = \dfrac{2}{3}\sqrt{2}\, mc^2,
\end{cases}
\tag{12.5.38}
$$

and it corresponds to the point A on Fig. 12.3. The smallest value for the radius of an unstable orbit is $r_{\min}^{\text{unstable}} = 3r_S/2$ and it is obtained in the limit $L \to \infty$, $E_0 \to \infty$ (point B corresponding to the horizontal asymptote figured with dotted line in Fig. 12.3). The upper branch in Fig. 12.3 gives the radius for stable, and the lower for unstable orbits. Note that point B does not belong to the curve $y_- = y_-(x)$, but to the horizontal asymptote.

With the notations

$$
\frac{L}{mcr_S} \equiv x, \qquad \frac{r}{r_S} \equiv y,
\tag{12.5.39}
$$

the relation (12.5.37) is written as

$$
y_\pm = x\left(x \pm \sqrt{x^2 - 3}\right).
\tag{12.5.40}
$$

The graphic representations of $y_\pm = y_\pm(x)$, given by the relations

$$
y_+ = x\left(x + \sqrt{x^2 - 3}\right) \quad \text{and} \quad y_- = x\left(x - \sqrt{x^2 - 3}\right)
\tag{12.5.41}
$$

are shown in Fig. 12.3, in which the branch $y_+ = y_+(x)$ corresponds to stable orbits, while the branch $y_- = y_-(x)$ corresponds to the unstable ones. Moreover,

$$\lim_{x\to\infty} y_-(x) = \lim_{x\to\infty} x\left(x - \sqrt{x^2 - 3}\right) = \frac{3}{2}, \qquad (12.5.42)$$

which justifies the value $r_{\min}^{\text{unstable}} = 3r_S/2$.

We note that the obvious difference compared to the classical case is that in a Newtonian gravitational field there are stable circular orbits at any distance from the centre of force, the radii of these orbits being given by the relation

$$r = \frac{1}{GM}\left(\frac{L}{m}\right)^2.$$

A

Appendix A: Elements of Tensor Calculus

A.1 n-dimensional Spaces

A n-dimensional space, S_n, is a set of elements, called *points*, which are in bi-univocal and bi-continuous correspondence with the values of n real variables $x^1, x^2, ..., x^n$. The variables x^1, x^2, ..., x^n are called *coordinates*. A system of given coordinates, x^1, x^2, ..., x^n, corresponds to a (figurative) point in S_n, and reciprocally. The number of coordinates defines the space *dimension* (n, in our case).

Let us have another coordinate system, x'^1, x'^2, ..., x'^n in S_n and consider the transformation

$$x^i = x^i\left(x'^k\right), \quad i, k = \overline{1, n}. \tag{A.1.1}$$

If the functional determinant (Jacobian) of transformation given by Eq. (A.1.1),

$$D = \frac{\left(x^1, x^2, ..., x^n\right)}{\left(x'^1, x'^2, ..., x'^n\right)} \tag{A.1.2}$$

is different from zero, the transformation is locally *reversible* or *bi-univocal* and we also have

$$x'^k = x'^k\left(x^i\right), \quad i, k = \overline{1, n}, \tag{A.1.3}$$

which means that x'^1, x'^2, ..., x'^n may constitute, in turn, a coordinate system in S_n. In other words, the choice of the coordinate set is not unique.

A transformation of the type given by Eq. (A.1.1) or Eq. (A.1.3) is called *coordinate transformation*. The two transformations are the reverse of each other. If the variables x^i are linear with respect to x'^k, the transformation in Eq. (A.1.1) is called *linear* or *affine*. In this case, the transformation in Eq. (A.1.3) is also linear.

If in the space S_n the law defining the distance between two points is given, the space is called *metric* or *metrized*, while the expression of the squared arc element $ds^2 \equiv (ds)^2$ is called *metric*. Otherwise, the space is called *amorphous*.

Let $f(x^1, x^2, ..., x^n)$ be a coordinate function. If transformation in Eq. (A.1.1) results in

$$f(x^1, x^2, ..., x^n) = f(x'^1, x'^2, ..., x'^n), \tag{A.1.4}$$

we say that $f(x^1, x^2, ..., x^n)$ is an *invariant*.

DOI: 10.1201/9781003402602-A

A.2   Contravariant and Covariant Vectors

From Eq. (A.1.3) we obtain by differentiation

$$dx'^k = \frac{\partial x'^k}{\partial x^i} dx^i = \bar{x}_i^k dx^i, \qquad (A.2.5)$$

where we used the notation

$$\bar{x}_k^i \equiv \frac{\partial x'^k}{\partial x^i}, \qquad (A.2.6)$$

and we also took into account the summation convention over repeated indices in the same product (Einstein's convention).

A system of quantities, A^i $(A^1, A^2, ..., A^n)$ which, as a result of transformation in Eq. (A.1.1) obey the rule given by Eq. (A.2.5) for the coordinate differentials, that is the law

$$A'^k = \bar{x}_i^k A^i, \quad i, k = \overline{1, n}, \qquad (A.2.7)$$

form a *first-order contravariant tensor* or a *n-dimensional contravariant vector*. The quantities A^i, $i = \overline{1, n}$, are the components of the contravariant vector in x^i variables, while A'^i are the components of the same vector in variables x'^i. Observing that

$$dx^i = \frac{\partial x^i}{\partial x'^k} dx'^k = \underline{x}_k^i dx'^k, \qquad (A.2.8)$$

where we entered the notation

$$\underline{x}_k^i \equiv \frac{\partial x^i}{\partial x'^k}, \qquad (A.2.9)$$

we will define the law of transformation of the components of a n-dimensional contravariant vector under the inverse transformation as

$$A^i = \underline{x}_k^i A'^k. \qquad (A.2.10)$$

The quantities $\bar{x}_i^k$ and $\underline{x}_i^k$ are also the elements of the direct and, respectively, reverse transformation matrix.

Let us now consider the function $f(x^i)$, $i = \overline{1, n}$, and perform the partial derivatives

$$\frac{\partial f}{\partial x'^i} = \frac{\partial f}{\partial x^k} \frac{\partial x^k}{\partial x'^i} = \underline{x}_i^k \frac{\partial f}{\partial x^k}, \quad i, k = \overline{1, n}. \qquad (A.2.11)$$

A system of n quantities B_i, $i = \overline{1, n}$, which at transformation in Eq. (A.1.1) transform as partial derivatives in Eq. (A.2.11) of a scalar function, that is according to the law

$$B_i' = \underline{x}_i^k B_k, \qquad (A.2.12)$$

defines a *first-order covariant tensor* or a *n-dimensional covariant vector*. The reverse transformation is

$$B_i = \overline{x}_k^i B_i'. \tag{A.2.13}$$

In the Euclidean space with orthogonal Cartesian coordinates (see **Appendix B**) we have

$$\overline{x}_i^k = \underline{x}_k^i = a_{ik},$$

therefore the notions of contravariant and covariant vector coincide.

Let us show that the summed product between a contravariant and a covariant vector is an invariant. We have

$$A'^k B_k' = \overline{x}_l^k \underline{x}_k^m A^l B_m = \delta_l^m A^l B_m = A^m B_m, \tag{A.2.14}$$

which proves the statement.

A.3 Second-Order Tensors

A system of n^2 quantities, T^{ik}, form a *second-order contravariant tensor* if when changing coordinates it transforms like the product $A^i B^k$, that is according to the law

$$T'^{ik} = \overline{x}_l^i \overline{x}_m^k T^{lm}; \quad T^{ik} = \underline{x}_l^i \underline{x}_m^k T'^{lm}. \tag{A.3.15}$$

A system of n^2 quantities, T_{ik}, form a *second-order covariant tensor* if when changing coordinates it transforms like the product $A_i B_k$, that is according to the law

$$T'_{ik} = \underline{x}_i^l \underline{x}_k^m T_{lm}; \quad T_{ik} = \overline{x}_i^l \overline{x}_k^m T'_{lm}. \tag{A.3.16}$$

A system of n^2 quantities, V_k^i which when changing coordinates it transforms like the product $A^i B_k$, that is according to the law

$$V'^i_{\cdot k} = \overline{x}_l^i \underline{x}_k^m V^l_{\cdot m}; \quad V^i_{\cdot k} = \underline{x}_l^i \overline{x}_k^m V'^l_{\cdot m}, \tag{A.3.17}$$

form a *second-order mixed tensor* (the point in front of the lower index k in $V'^i_{\cdot k}$ indicates that the first index of the tensor $V'^i_{\cdot k}$ is i, and k is the second index of the tensor; it should be noted that the order of the indices is very important when studying the symmetry property (the property of being symmetric or antisymmetric) of tensors. In addition, we point out that often this point is not shown, the order of the indices being sub-understood). Such a tensor is the Kronecker symbol δ_k^i. Indeed,

$$\delta'^i_k = \overline{x}_l^i \underline{x}_k^m \delta_m^l = \frac{\partial x'^i}{\partial x^l} \frac{\partial x^l}{\partial x'^k} = \frac{\partial x'^i}{\partial x'^k} = \delta_k^i. \tag{A.3.18}$$

In relation (A.3.17) the first index is i. This index can be lowered by means of the (covariant) metric tensor g_{ik} (see the next paragraph), on the left side of k, in the place indicated by the point below i. Also, the index k can be raised on the right side of i by means of a similar procedure (but using the contravariant fundamental tensor g^{ik}). We will not use the point in the text, while the first index will be specified only when necessary. If the tensor is symmetric, the index order does not matter. An example in this respect is the Kronecker symbol (see Eq. (A.3.18)).

Using the properties of the first- and second-order tensors, let us show that the product $T^{ik}A_iB_k$, $i = \overline{1,n}$, is invariant to a coordinate transformation. Indeed, we have

$$T'^{ik}A'_iB'_k = \overline{x}^i_l\overline{x}^k_m\underline{x}^p_i\underline{x}^q_kT^{lm}A_pB_q = \delta^p_l\delta^q_mT^{lm}A_pB_q = T^{pm}A_pB_m. \quad \text{(A.3.19)}$$

A.4   The Fundamental Metric Tensor

Consider an Euclidean space E_m and let $y_1, y_2, ..., y_m$ be the Cartesian coordinates of a point P in this space. (The *Euclidean* or *homaloidal space* is the space in which the Euclid's axioms are valid; such a space is without curvature, or *flat*). The square of the (infinitesimal) distance between the point P and a neighbouring point P', also called the *metric* of the variety E_m, is

$$(ds)^2 \equiv ds^2 = dy_jdy_j, \quad j = \overline{1,m}. \quad \text{(A.4.20)}$$

Consider in E_m a variety (sub-space) R_n, with $n < m$ and let $x^1, x^2, ..., x^n$ be the coordinates of a point in R_n. Obviously,

$$y_j = y_j(x^1, x^2, ..., x^n), \quad j = \overline{1,m}. \quad \text{(A.4.21)}$$

The metric of the variety R_n then is

$$ds^2 = \frac{\partial y_j}{\partial x^i}\frac{\partial y_j}{\partial k^k}dx^idx^k = g_{ik}dx^idx^k, \quad \text{(A.4.22)}$$

where we denoted

$$g_{ik}(x^1, x^2, ..., x^n) \equiv \frac{\partial y_j}{\partial x^i}\frac{\partial y_j}{\partial x^k}, \quad j = \overline{1,m}, \; i,k = \overline{1,n}. \quad \text{(A.4.23)}$$

The quadratic form given by Eq. (A.4.23) is positively defined, and according to Eq. (A.3.19) is invariant to a coordinate transformation. Since dx^i and dx^k are contravariant vectors, it follows that g_{ik} defined by Eq. (A.4.23) represent the components of a symmetric covariant tensor called *fundamental metric tensor* or, in short, *metric tensor*. Since (we will justify this statement at the end of the Sect. § **A.7**),

$$dx_i = g_{ik}dx^k, \quad \text{(A.4.24)}$$

we can still write

$$ds^2 = dx_i dx^i, \quad i = \overline{1, n}.$$

The relation (A.4.24) is valid for the components of any vector A^k, therefore

$$A_i = g_{ik} A^k, \quad i, k = \overline{1, n}. \tag{A.4.25}$$

Relations (A.4.25) can be considered as a system of n linear algebraic equations for unknowns $A^1, A^2, ..., A^n$. Solving this system according to Cramer's rule, we obtain

$$A^k = g^{kj} A_j, \quad j, k = \overline{1, n}, \tag{A.4.26}$$

where

$$g^{kj} = \frac{G_{kj}}{g} \tag{A.4.27}$$

are the components of the *contravariant metric tensor*. Here by G_{ik} the algebraic complement of the element g_{ik} in the determinant $g = \det(g_{ik})$ has been denoted.

Since $G_{jk} = G_{kj}$, we have $g^{kj} = g^{jk}$, and Eq. (A.4.25) yields

$$A_i = g_{ik} A^k = g_{ik} g^{kj} A_j, \tag{A.4.28}$$

and, consequently,

$$g_{ik} g^{jk} = g_i^j = \delta_i^j. \tag{A.4.29}$$

Using the fundamental metric tensors g_{ik} and g^{ik}, according to Eqs. (A.4.26) and (A.4.27), a contravariant vector can be obtained from a covariant vector, and reciprocally. In an Euclidean space $g_{ik} = \delta_{ik}$, so that

$$A_i = \delta_{ik} A^k = A^i,$$

which means that, in this case, the contravariant and covariant components are confused, *i.e.*, they are identical. That is why, for convenience, the lower indices are exclusively used (see Eq. (A.4.21)).

A.5   Tensors of Higher Order

The higher-order tensors (third-order, fourth-order, etc.) are analogously defined. For exemple, on a n-dimensional space, the system of n^{m+s} elements $T_{k_1 k_2 ... k_m}^{j_1 j_2 ... j_s}$ (from the set $\mathbb{R}$ of real numbers) which, when the coordinates change, transform according to the law

$$T_{k_1 k_2 ... k_m}^{\prime j_1 j_2 ... j_s} = \underset{k_1}{x}^{p_1} \underset{k_2}{x}^{p_2} ... \underset{k_m}{x}^{p_m} \overline{x}_{i_1}^{j_1} \overline{x}_{i_2}^{j_2} ... \overline{x}_{i_s}^{j_s} T_{p_1 p_2 ... p_m}^{i_1 i_2 ... i_s}, \tag{A.5.30}$$

where the first indices are the ones of contravariance, form a tensor of order (or rank) $m + s$, m times covariant and s times contravariant. In general, the number of *significant indices* (in the Sect. § A.8 we will present the classification of indices together with a rule − in the form of a *sui-generis* algorithm − of verifying the *a priori* validity of any formula in theoretical physics, which is the analogous rule of that used in phenomenological physics and that is called the *dimensional analysis*) gives the tensor order.

A.6 Operations with Tensors

A.6.1 Addition

The sum of two (or more) tensors can be defined only for tensors of the same rank and the same variance. For example, the sum of the tensors $U^{ij}{}_k$ and $V^{ij}{}_k$ is the tensor

$$T^{ij}{}_k = U^{ij}{}_k + V^{ij}{}_k, \tag{A.6.31}$$

where $T^{ij}{}_k$ is a tensor of the same type (variance) and order as the given tensors. The difference between two tensors is similarly defined because it always can be reduced to a sum.

A.6.2 Multiplication

Let $U^{i_1 i_2 \ldots i_p}_{j_1 j_2 \ldots j_q}$ and $V^{k_1 k_2 \ldots k_r}_{m_1 m_2 \ldots m_s}$ be two tensors of arbitrary types (variances) and orders. The product of the two tensors is defined as

$$T^{i_1 i_2 \ldots i_p k_1 k_2 \ldots k_r}_{j_1 j_2 \ldots j_q m_1 m_2 \ldots m_s} = U^{i_1 i_2 \ldots i_p}_{j_1 j_2 \ldots j_q} V^{k_1 k_2 \ldots k_r}_{m_1 m_2 \ldots m_s}, \tag{A.6.32}$$

being a tensor of order $p + q + r + s$, $p + r$ times contravariant, and $q + s$ times covariant.

A.6.3 Contraction

Consider a tensor (T) of order $p \geq 2$. Let us equal any two its indices, one lower and one upper, and perform summation over these indices (over their common value). The operation is called *contraction* (or *saturation*) with respect to the two indices, and the tensor obtained this way is called *once contracted tensor*. By a contraction, the tensor order decreases by two units. For example, by contraction of the indices i and k of the tensor T^{ij}_k, one obtains

$$T^{ij}_k \rightarrow T'^{ij}_i = \bar{x}^i_p \bar{x}^j_q \underline{x}^r_i T^{pq}_r = \delta^r_p \bar{x}^j_q T^{pq}_r = \bar{x}^j_q T^{pq}_p, \tag{A.6.33}$$

which is a contravariant vector. In general, if we equalize the first s indices of covariance to the first s indices of contravariance (meaning summation over the common value of any pair of two indices, one of covariance and the other of contravariance), we obtain the tensor $U^{i_{s+1} i_{s+2} \ldots i_p}_{j_{s+1} j_{s+2} \ldots j_q}$, which is $p - s$ times contravariant, and $q - s$ times covariant.

A.6.4   Raising and Lowering the Indices

Consider the tensor $T^i{}_{jkl}$, once contravariant and three times covariant. Multiplying this tensor by g_{is} and summing up over i, we have

$$g_{is}T^i{}_{jkl} = T_{sjkl}, \tag{A.6.34}$$

which is a covariant tensor, obtained by the convenient multiplication of $T^i{}_{jkl}$ with the fundamental metric tensor. We call this operation *lowering* of index i.

In a similar way, an index can be *raised*. In our case, for example, the raising of lower index j can be written as

$$g^{sj}T^i{}_{jkl} = T^{is}{}_{kl}. \tag{A.6.35}$$

In general, lowering/raising of an index of a tensor (T) is performed by a suitable multiplication with the metric tensor and summation over a pair of indices, one belonging to g_{ik} (or g^{ik}) and the other to the tensor (T). To raise/lower n indices, the operation must be repeated n times.

Observation: The raising/lowering operation does not modify the tensor order, but just its type (variance).

A.6.5   Symmetric and Antisymmetric Tensors

Consider the tensor $T^{i_1 i_2 ... i_p}_{j_1 j_2 ... j_q}$ with $p, q \geq 2$. If the tensor components do not change by an arbitrary permutation of a group of indices, either of covariance (and of covariance only) or of contravariance (and of contravariance only), we call the tensor *symmetric* in that group of indices (all indices being either of covariance or of contravariance, but not of different variances). If, for example, the above defined tensor (T) is symmetric in the first s covariant indices, $j_1, j_2, ..., j_s$, this property is denoted as $T^{i_1 i_2 ... i_p}_{(j_1 j_2 ... j_s) j_{s+1} ... j_q}$. The number of essentially distinct components of this tensor is given by $(C_n^s)\, n^{p+q-s}$, where (C_n^s) is the combination of n elements taken s at a time with repetition, where n is the space dimension, while s is the number of symmetry indices in the group mentioned above:

$$(C_n^s) = \frac{n(n+1)(n+2)\cdots(n+s-1)}{1\cdot 2\cdot 3\cdots s}. \tag{A.6.36}$$

For example, the second-order covariant tensor T_{ik} is symmetric if

$$T_{ik} = T_{ki}, \tag{A.6.37}$$

and, according to Eq. (A.6.36), it has $n(n+1)/2$ essentially distinct components (six in a space with three dimensions, ten in a space with four dimensions, etc.).

If the sign of the tensor (T) changes by permutation of any two indices from a group of indices, either of covariance or of contravariance (but never combined), the tensor is called *antisymmetric* with respect to that group of indices. Let (T) be antisymmetric in the first s covariant indices and denote by $[j_1 j_2 ... j_s]$ the group of these indices, that is $T^{i_1 i_2 ... i_p}_{[j_1 j_2 ... j_s] j_{s+1} ... j_q}$. We then have

$$T^{i_1 i_2 ... i_p}_{[j_1 j_2 ... j_s] j_{s+1} ... j_q} = (-1)^I T^{i_1 i_2 ... i_p}_{(j_1' j_2' ... j_s') j_{s+1} ... j_q}, \qquad (A.6.38)$$

where $(j_1' j_2' ... j_s')$ is an arbitrary permutation of the indices $(j_1 j_2 ... j_s)$, while I is the number of inversions introduced by the permutation with respect to the initial order. This category includes, for example, the Levi-Civita permutation symbol ε_{ijk}.

The number of essentially distinct components of an antisymmetric tensor (which satisfy the relation (A.6.38)) is given by $C_n^s \, n^{p+q-s}$, where C_n^s is the combination of n elements taken s at a time without repetition (normal/common combination),

$$C_n^s = \frac{n(n-1)(n-2)\cdots(n-s+1)}{1 \cdot 2 \cdot 3 \cdots s}. \qquad (A.6.39)$$

For example, the second-order covariant tensor A_{ik} is antisymmetric if

$$A_{ik} = -A_{ki}, \qquad (A.6.40)$$

and has $n(n-1)/2$ essentially distinct components (three in a three-dimensional space, six in a four-dimensional space, etc.).

The antisymmetry property $A_{ik} = -A_{ki}$, for $i = k$, leads to $A_{ii} = 0$ (no summation). In general, if the group of antisymmetry indices of a tensor contains two equal indices, the respective component of the tensor is zero.

More generally, a tensor $T^{i_1 i_2 ... i_p}_{j_1 j_2 ... j_q}$ is said to be *fully symmetric in s indices* – for instance in the group of first distinct indices of covariance $j_1, j_2, ..., j_s$ – if, for *any* permutation σ of these s indices,

$$T^{i_1 i_2 ... i_p}_{j_1' j_2' ... j_s' j_{s+1} ... j_q} \equiv T^{i_1 i_2 ... i_p}_{j_{\sigma(1)} j_{\sigma(2)} ... j_{\sigma(s)} j_{s+1} ... j_q} = T^{i_1 i_2 ... i_p}_{j_1 j_2 ... j_q},$$

and it is said to be *fully antisymmetric in s indices* – for instance in the group of first distinct indices of covariance $j_1, j_2, ..., j_s$ – if, for *any* permutation σ of these s indices,

$$T^{i_1 i_2 ... i_p}_{j_1' j_2' ... j_s' j_{s+1} ... j_q} \equiv T^{i_1 i_2 ... i_p}_{j_{\sigma(1)} j_{\sigma(2)} ... j_{\sigma(s)} j_{s+1} ... j_q} = s(\sigma) T^{i_1 i_2 ... i_p}_{j_1 j_2 ... j_q},$$

where $s(\sigma)$ is the signature of that permutation.

Here we have to take into account that any permutation σ of the first s natural numbers $\{1, ..., s\}$ can always be written as a succession of pairwise exchanges. The signature $s(\sigma)$ of a permutation is $+1$ if it can be factorized as an *even* number of pairwise exchanges, and -1 if it is an odd number – it turns out that this is a unique property, *i.e.*, the signature is not changed by factorizing a permutation in a different way.

A tensor which enjoys the property $T_i^k = 0$ for $i \neq k$ is called *diagonal tensor*. Such a tensor is, for example, the Kronecker symbol $\delta_i^k = g_i^k$. It is also called the *symmetric unit tensor of the second order* or the *second-order symmetric unit tensor*.

Observation: To determine the number of the essentially distinct components of a tensor, it is necessary to investigate whether there are no other relationships between its components, except those for symmetry or antisymmetry. For example, the symmetric tensor T_{ik} whose *trace* (or *spur*, i.e., the sum of the elements on the main diagonal) is zero, does not have 6 essentially distinct components, as we are tempted to say, but only 5 ($= 6-1$). In general, to find the number of essentially distinct components of a tensor, from the number of components given by the relationships $(C_n^s)\, n^{p+q-s}$ (for symmetry indices) and $C_n^s\, n^{p+q-s}$ (for antisymmetry indices) as many units must be subtracted as additional independent relations satisfy the components of that tensor.

A.6.6 Symmetrization and Alternation

The *symmetrization* of a tensor is one of the operations in tensor algebra that constructs a symmetric tensor (relative to a group of indices) from a given (random) tensor. Symmetrization always takes place over several upper or lower indices and never over combined indices. Thus, from the tensor $T_{j_1 j_2 \ldots j_q}^{i_1 i_2 \ldots i_p}$, a tensor which is symmetric in the first s indices of covariance can be formed by means of the operation

$$
\begin{aligned}
T_{(j_1 j_2 \ldots j_s) j_{s+1} \ldots j_q}^{i_1 i_2 \ldots i_p} &= \frac{1}{s!} \sum_{j_1', j_2', \ldots, j_s'} T_{j_1' j_2' \ldots j_s' j_{s+1} \ldots j_q}^{i_1 i_2 \ldots i_p} \\
&\equiv \frac{1}{s!} \sum_{\sigma \in S_s} T_{j_{\sigma(1)} j_{\sigma(2)} \ldots j_{\sigma(s)} j_{s+1} \ldots j_q}^{i_1 i_2 \ldots i_p},
\end{aligned}
\tag{A.6.41}
$$

where the sum is extended over all permutations of distinct indices j_1, j_2, ..., j_s. Here, S_s represents the set of the $s!$ permutations of distinct covariance indices j_1, j_2, ..., j_s, while σ designates the operation of permutation.

For example, the quantities

$$
T_{(ik)} = \left(T_{ik} + T_{ki} \right)/2
\tag{A.6.42}
$$

are the components of a second-order symmetric tensor.

The operation of transition from the random tensor $T_{j_1 j_2 \ldots j_q}^{i_1 i_2 \ldots i_p}$ to the antisymmetric (in the group of indices $j_1, j_2, \ldots, j_s$) tensor $T_{[j_1 j_2 \ldots j_s] j_{s+1} \ldots j_q}^{i_1 i_2 \ldots i_p}$ through

$$
\begin{aligned}
T_{[j_1 j_2 \ldots j_s] j_{s+1} \ldots j_q}^{i_1 i_2 \ldots i_p} &= \frac{1}{s!} \sum_{j_1', j_2', \ldots, j_s'} (-1)^I T_{j_1' j_2' \ldots j_s' j_{s+1} \ldots j_q}^{i_1 i_2 \ldots i_p} \\
&\equiv \frac{1}{s!} \sum_{\sigma \in S_s} s(\sigma) T_{j_{\sigma(1)} j_{\sigma(2)} \ldots j_{\sigma(s)} j_{s+1} \ldots j_q}^{i_1 i_2 \ldots i_p},
\end{aligned}
\tag{A.6.43}
$$

where the summation extends over all different permutations $(j'_1 j'_2 ... j'_s)$ of fixed and distinct indices $j_1, j_2, ..., j_s$, and I is the number of inversions from the permutation $(j'_1 j'_2 ... j'_s)$ with respect to the initial order, is called *alternation*. The tensor $T^{i_1 i_2 ... i_p}_{[j_1 j_2 ... j_q]}$ is called *alternated tensor* of the tensor $T^{i_1 i_2 ... i_p}_{j_1 j_2 ... j_q}$. If two of the distinct indices $j_1, j_2, ..., j_q$ are equal, the sums of Eq. (A.6.43) are zero. For example, the alternated tensor of T_{ik} is

$$T_{[ik]} = (T_{ik} - T_{ki})/2. \tag{A.6.44}$$

It can be checked that the two tensors defined above, namely $T^{i_1 i_2 ... i_p}_{(j_1 j_2 ... j_q)}$ and $T^{i_1 i_2 ... i_p}_{[j_1 j_2 ... j_q]}$ are, respectively, fully symmetric and fully antisymmetric — what is needed is the property that the signature of the composition of two permutations is the product of the signatures. These two tensors are called the *symmetric* and *antisymmetric parts* of $T^{i_1 i_2 ... i_p}_{j_1 j_2 ... j_q}$. In general $T^{i_1 i_2 ... i_p}_{j_1 j_2 ... j_q}$ is made of a symmetric part, an antisymmetric part, and a non-symmetric part.

Application. Any second-order tensor T_{ik} can be uniquely written as the sum between a symmetric tensor S_{ik} and an alternated one A_{ik}. Indeed,

$$T_{ik} = \frac{1}{2}(T_{ik} + T_{ki}) + \frac{1}{2}(T_{ik} - T_{ki}) = S_{ik} + A_{ik}. \tag{A.6.45}$$

In other words, any tensor of second order (and only of second order!) can be uniquely decomposed into a symmetric and an antisymmetric part,

$$T_{ik} = T_{(ik)} + T_{[ik]}.$$

Now let us look at the case of a third-order tensor, T_{ijk}. In this case there are six ($3! = 6$) permutations. The symmetric and antisymmetric parts of the tensor T_{ijk} are, respectively

$$T_{(ijk)} = \frac{1}{6}\left(T_{ijk} + T_{jki} + T_{kij} + T_{jik} + T_{kji} + T_{ikj}\right),$$

and

$$T_{[ijk]} = \frac{1}{6}\left(T_{ijk} + T_{jki} + T_{kij} - T_{jik} - T_{kji} - T_{ikj}\right).$$

Note that cyclic permutations of three indices

$$\{1, 2, 3\} \to \{2, 3, 1\} \text{ and } \{1, 2, 3\} \to \{3, 1, 2\}$$

can be obtained by two pairwise exchanges, hence have signature $+1$. We point out that in general,

$$T_{(ijk)} + T_{[ijk]} \neq T_{ijk},$$

which means that there is more to a rank-3 tensor — whether it is covariant or contravariant — than its symmetric part and its antisymmetric part. The same thing holds for all tensors of order greater than or equal to three.

A.7 Tensor Variance: An Intuitive Image

As we have seen, the *variance* of tensors can be of two types, namely, *covariance* and *contravariance*. Strictly speaking, one of the most direct and yet simple ways to define a covariant and/or contravariant first-order tensor is the one presented in Sect. § **A.2**; in the following, we wish to provide a more intuitive image about these concepts, and at the same time, to emphasize the necessity of introducing them.

For an easier presentation, we shall refer in the following to a three-dimensional space. Formally speaking, "between" the simplest type of coordinate system, namely the *Cartesian coordinate system* and the most general type of coordinate system, namely the *general curvilinear coordinate system*, there are two types of "intermediate" coordinate systems, namely:

1) *orthogonal curvilinear coordinate systems*: the coordinate axes are curvilinear, but at each point of space the vectors tangent to the axes form an orthogonal trihedron (see Fig. A.1.c);

2) *non-orthogonal rectilinear coordinate systems*: the coordinate axes are straight lines, but they form (between them) angles different from $\pi/2$ (see Fig. A.1.d).

The orthogonal curvilinear coordinate systems are discussed in **Appendix D**, where it is shown that the main effect of the curving of axes is the appearance of the Lamé coefficients. As we shall see, the non-orthogonality of the coordinate axes brings about the necessity of introducing the notion of variance of tensors.

To facilitate the graphical presentation, we shall mainly use a two-dimensional space (the generalization to three dimensions is trivial).

Let us consider in the Euclidean plane a vector $\vec{a}$, and express its components with respect to both an orthogonal and a non-orthogonal coordinate system (see Fig. A.2).

In the first system (Fig. A.2.a) we can write

$$\begin{cases} a_1 = \vec{a} \cdot \vec{u}_1, \\ a_2 = \vec{a} \cdot \vec{u}_2, \end{cases} \tag{A.7.46}$$

and if we denote

$$\begin{cases} \vec{a}_1 \equiv a_1 \vec{u}_1, \\ \vec{a}_2 \equiv a_2 \vec{u}_2, \end{cases}$$

then we have

$$\vec{a} = \vec{a}_1 + \vec{a}_2, \tag{A.7.47}$$

where $\vec{u}_1$ and $\vec{u}_2$ are the versors of the two axes, while a_1 and a_2 are the components of the vector $\vec{a}$ in this orthonormal basis (in other words, the orthogonal projections of the vector $\vec{a}$ on the coordinate axes).

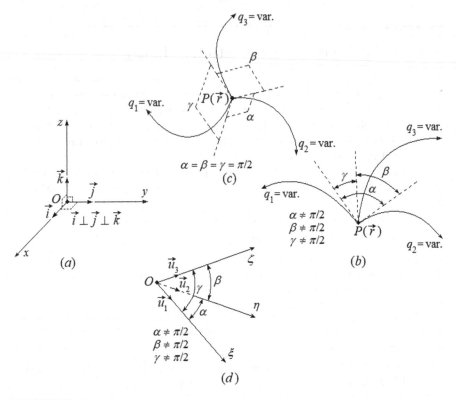

FIGURE A.1
Four types of coordinate system: (a) Cartesian (orthogonal rectilinear, (b) general curvilinear, (c) orthogonal curvilinear and (d) non-orthogonal rectilinear.

In the second system (Fig. A.2.b), we specify that there are not only one, but two possibilities of defining the components of the vector $\vec{a}$:

1) by orthogonal projection on the axes, leading to the components denoted by a_1 and a_2;

2) by drawing parallels to the axes through the "origin" and tip of the vector $\vec{a}$, leading to the components denoted by a'_1 and a'_2.

Thus, in the first case we can write

$$\begin{cases} a_1 = \vec{a} \cdot \vec{u}_1, \\ a_2 = \vec{a} \cdot \vec{u}_2, \end{cases} \tag{A.7.48}$$

but, with the notations

$$\begin{cases} \vec{a}_1 \equiv a_1 \vec{u}_1, \\ \vec{a}_2 \equiv a_2 \vec{u}_2, \end{cases}$$

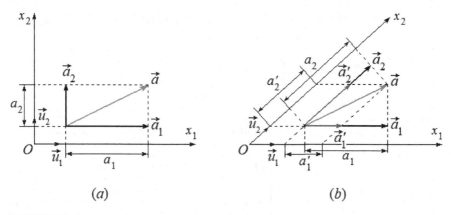

(a)   (b)

FIGURE A.2

Two types of rectilinear/straight coordinate axes: (a) orthogonal and (b) non-orthogonal.

a relation like that in Eq. (A.7.47) is not valid anymore, *i.e.*,

$$\vec{a} \neq \vec{a}_1 + \vec{a}_2. \tag{A.7.49}$$

In the second case, denoting

$$\begin{cases} \vec{a}\,'_1 \equiv a'_1 \vec{u}_1, \\ \vec{a}\,'_2 \equiv a'_2 \vec{u}_2, \end{cases} \tag{A.7.50}$$

we find that a relation like the one given by Eq. (A.7.47) remains valid, *i.e.*,

$$\vec{a} = \vec{a}\,'_1 + \vec{a}\,'_2, \tag{A.7.51}$$

but we cannot write anymore a relation of the type in Eq. (A.7.46), since in this case

$$\begin{cases} a'_1 \neq \vec{a} \cdot \vec{u}_1, \\ a'_2 \neq \vec{a} \cdot \vec{u}_2. \end{cases} \tag{A.7.52}$$

In other words, the components a'_1 and a'_2 are not the scalar products of the vector $\vec{a}$ with the versors of the coordinate axes.

We can make relations (A.7.49) and (A.7.52) valid at the same time, but for this we must first introduce a new basis in E_3, namely the dual basis of the existing one, *i.e.*, $\{\vec{u}_1, \vec{u}_2, \vec{u}_3\}$.

Let $\vec{u}$, $\vec{v}$ and $\vec{w}$ be three linearly independent vectors in E_3. By definition, they form a basis in E_3. Then, any vector $\vec{a}$ in E_3 can be written as

$$\vec{a} = \lambda \vec{u} + \mu \vec{v} + \nu \vec{w}, \tag{A.7.53}$$

where λ, μ and ν are three scalars, which are called the *components of the vector $\vec{a}$ in the basis* $\{\vec{u}, \vec{v}, \vec{w}\}$.

Let us now introduce three new vectors, denoted by $\vec{u}^*$, $\vec{v}^*$ and $\vec{w}^*$, simultaneously satisfying the following conditions:

$$\begin{cases} \vec{u}^* \cdot \vec{u} = 1, \\ \vec{u}^* \cdot \vec{v} = 0, \\ \vec{u}^* \cdot \vec{w} = 0, \end{cases} \quad \begin{cases} \vec{v}^* \cdot \vec{u} = 0, \\ \vec{v}^* \cdot \vec{v} = 1, \\ \vec{v}^* \cdot \vec{w} = 0, \end{cases} \quad \begin{cases} \vec{w}^* \cdot \vec{u} = 0, \\ \vec{w}^* \cdot \vec{u} = 0, \\ \vec{w}^* \cdot \vec{w} = 1. \end{cases} \qquad (A.7.54)$$

For instance, the vectors $\vec{u}^*$, $\vec{v}^*$ and $\vec{w}^*$ can be given by the relations

$$\begin{cases} \vec{u}^* = \dfrac{\vec{v} \times \vec{w}}{(\vec{u}, \vec{v}, \vec{w})}, \\[2mm] \vec{v}^* = \dfrac{\vec{w} \times \vec{u}}{(\vec{u}, \vec{v}, \vec{w})}, \\[2mm] \vec{w}^* = \dfrac{\vec{u} \times \vec{v}}{(\vec{u}, \vec{v}, \vec{w})}, \end{cases} \qquad (A.7.55)$$

where $(\vec{u}, \vec{v}, \vec{w}) = \vec{u} \cdot (\vec{v} \times \vec{w})$ is the scalar triple product of the three vectors.

From the definition, it follows that the vectors $\vec{u}^*$, $\vec{v}^*$ and $\vec{w}^*$ are also linearly independent, and thus they form a basis in E_3. Such vectors are called *dual* to the vectors of "original"/"direct" basis $\vec{u}$, $\vec{v}$ and $\vec{w}$, respectively, while the basis they form, *i.e.*, $\{\vec{u}^*, \vec{v}^*, \vec{w}^*\}$ is called *dual basis*. Moreover, one can easily check that the dual of a dual vector is the original vector, *i.e.*,

$$\begin{cases} \left(\vec{u}^*\right)^* = \vec{u}, \\ \left(\vec{v}^*\right)^* = \vec{v}, \\ \left(\vec{w}^*\right)^* = \vec{w}. \end{cases} \qquad (A.7.56)$$

Taking the scalar product of the vector given by Eq. (A.7.53) with the vectors $\vec{u}^*$, $\vec{v}^*$ and $\vec{w}^*$ of the dual basis and using Eq. (A.7.54), one can express the components of the vector $\vec{a}$ in the original basis as

$$\begin{cases} \lambda = \vec{a} \cdot \vec{u}^*, \\ \mu = \vec{a} \cdot \vec{v}^*, \\ \nu = \vec{a} \cdot \vec{w}^*. \end{cases} \qquad (A.7.57)$$

Going back to our problem, we notice that, using the vectors of the dual basis, we have

i) for the components a_1 and a_2 defined by Eq. (A.7.48), the relation (A.7.49) can also be valid, *i.e.*, we can write

$$\vec{a} = \vec{a}_1 + \vec{a}_2,$$

where this time

$$\begin{cases} \vec{a}_1 \equiv a_1 \vec{u}_1^*, \\ \vec{a}_2 \equiv a_2 \vec{u}_2^*, \end{cases} \qquad (A.7.58)$$

for, according to Eq. (A.7.57) we have

$$\begin{cases} a_1 = \vec{a} \cdot \left(\vec{u}_1^*\right)^* = \vec{a} \cdot \vec{u}_1, \\ a_2 = \vec{a} \cdot \left(\vec{u}_2^*\right)^* = \vec{a} \cdot \vec{u}_2, \end{cases}$$

in agreement with Eq. (A.7.48). This way, the relations (A.7.48) and (A.7.49) were agreed.

ii) for the components a_1' and a_2' defined by Eq. (A.7.50), we can write a relation like the one given by Eq. (A.7.52), by taking

$$\begin{cases} a_1' \equiv \vec{a} \cdot \vec{u}_1^*, \\ a_2' \equiv \vec{a} \cdot \vec{u}_2^*, \end{cases} \tag{A.7.59}$$

the relation (A.7.51) remaining valid since, according to Eq. (A.7.50) we have

$$\vec{a} = \vec{a}\,'_1 + \vec{a}\,'_2 = a_1'\vec{u}_1 + a_2'\vec{u}_2,$$

where, in agreement with Eq. (A.7.57) we have

$$\begin{cases} a_1' \equiv \vec{a} \cdot \vec{u}_1^*, \\ a_2' \equiv \vec{a} \cdot \vec{u}_2^*, \end{cases}$$

which are nothing but relationships (A.7.59).

In this way we made all the relations that previously were not valid simultaneously (in the sense that either one of them was valid and the other was not, or vice versa) to be put in agreement (*i.e.*, to be valid simultaneously).

The customary notations are different from the above, namely,

$$\begin{cases} a_1' \equiv a^1, \\ a_2' \equiv a^2, \end{cases} \qquad \begin{cases} \vec{u}_1^* \equiv \vec{u}^{\,1}, \\ \vec{u}_2^* \equiv \vec{u}^{\,2}. \end{cases} \tag{A.7.60}$$

In this way it is seen that in a non-orthogonal space, a vector $\vec{a}$ has not only one set, but two sets of components: one with *lower* indices,

$$a_1 = \vec{a} \cdot \vec{u}_1,$$
$$a_2 = \vec{a} \cdot \vec{u}_2, \tag{A.7.61}$$

and another with *upper* indices,

$$a^1 = \vec{a} \cdot \vec{u}^{\,1},$$
$$a^2 = \vec{a} \cdot \vec{u}^{\,2}. \tag{A.7.62}$$

The components with *lower* indices are called *covariant*, while the ones with *upper* indices *contravariant*. Besides, as can easily be seen from Fig. A.2.b, if the angle between the coordinate axes becomes a right angle ($\pi/2$

rads), *i.e.*, the coordinate system becomes orthogonal, then the two types of components coincide.

Returning to a three-dimensional non-orthogonal space, we find that in such a space we have not just one but two bases of interest: the "original" basis which is formed of covariant versors, $\{\vec{u}_1, \vec{u}_2, \vec{u}_3\}$, and its dual basis that is formed of contravariant versors, $\{\vec{u}^1, \vec{u}^2, \vec{u}^3\}$.

To conclude, in a non-orthogonal coordinate system, any vector has two sets of components: *covariant components*, by means of which one writes the vector in the dual, or contravariant basis, and *contravariant components*, used to write the vector in the "original"/"direct", or covariant basis.

For example, in the case of a three-dimensional non-orthogonal frame, the position/radius vector $\vec{r}$ of a point $P(\vec{r})$ can be written as follows:

1) in dual (contravariant) basis, using the covariant components x_1, x_2 and x_3, as

$$\vec{r} = x_1\vec{u}^1 + x_2\vec{u}^2 + x_3\vec{u}^3;$$

2) in the "original"/"direct" (covariant) basis, using the contravariant components x^1, x^2 and x^3, as

$$\vec{r} = x^1\vec{u}_1 + x^2\vec{u}_2 + x^3\vec{u}_3.$$

Clearly,

$$\begin{cases} d\vec{r} = dx_1\vec{u}^1 + dx_2\vec{u}^2 + dx_3\vec{u}^3, \\ d\vec{r} = dx^1\vec{u}_1 + dx^2\vec{u}_2 + dx^3\vec{u}_3, \end{cases}$$

therefore the metric of the space can be written in the following three ways:

$$ds^2 = \left|d\vec{r}\right|^2 = d\vec{r} \cdot d\vec{r}$$
$$= \begin{cases} \left(dx_1\vec{u}^1 + dx_2\vec{u}^2 + dx_3\vec{u}^3\right) \cdot \left(dx_1\vec{u}^1 + dx_2\vec{u}^2 + dx_3\vec{u}^3\right), \\ \left(dx^1\vec{u}_1 + dx^2\vec{u}_2 + dx^3\vec{u}_3\right) \cdot \left(dx^1\vec{u}_1 + dx^2\vec{u}_2 + dx^3\vec{u}_3\right), \\ \left(dx_1\vec{u}^1 + dx_2\vec{u}^2 + dx_3\vec{u}^3\right) \cdot \left(dx^1\vec{u}_1 + dx^2\vec{u}_2 + dx^3\vec{u}_3\right). \end{cases}$$

With the new notations, relations (A.7.54) can be written in a condensed manner as

$$\vec{u}^i \cdot \vec{u}_j = \delta^i_j, \tag{A.7.63}$$

hence

$$ds^2 = \left|d\vec{r}\right|^2 = d\vec{r} \cdot d\vec{r}$$
$$= \begin{cases} \left(dx_i\vec{u}^i\right) \cdot \left(dx_j\vec{u}^j\right) = \left(\vec{u}^i \cdot \vec{u}^j\right)dx_i dx_j, \\ \left(dx^i\vec{u}_i\right) \cdot \left(dx^j\vec{u}_j\right) = \left(\vec{u}_i \cdot \vec{u}_j\right)dx^i dx^j, \\ \left(dx_i\vec{u}^i\right) \cdot \left(dx^j\vec{u}_j\right) = \left(\vec{u}^i \cdot \vec{u}_j\right)dx_i dx^j. \end{cases} \tag{A.7.64}$$

If one denotes the scalar products of the contravariant and covariant versors by

$$\left(\vec{u}^i \cdot \vec{u}^j\right) \equiv g^{ij},$$
$$\left(\vec{u}_i \cdot \vec{u}_j\right) \equiv g_{ij}, \tag{A.7.65}$$

then the metric given by Eq. (A.7.64) can be written as

$$ds^2 = g^{ij} dx_i dx_j = g_{ij} dx^i dx^j = dx_i dx^i. \qquad \text{(A.7.66)}$$

Thus, we have arrived at the fundamental metric tensor, with contravariant components, g^{ij}, or covariant components, g_{ij}, and, at the same time we also have an intuitive picture of the "geometric meaning" of the components of the fundamental metric tensor.

Finally, from the last equality in Eq. (A.7.66) it follows that

$$dx_i = g_{ij} dx^j, \qquad \text{(A.7.67)}$$

which shows that the *lowering of indices* by means of the fundamental metric tensor appears naturally. Similarly, the equality

$$g^{ij} dx_i dx_j = dx_i dx^i$$

from Eq. (A.7.66) leads to

$$dx^i = g^{ij} dx_j, \qquad \text{(A.7.68)}$$

i.e., the *raising of indices* operation.

A.8   The Verification Method of *a priori* Validity of a Formula in Theoretical Physics

As is well known, the fastest way to check the correctness of a formula in *phenomenological physics* is *dimensional analysis*. Obviously, this method cannot fully assure us of the correctness of the formulae, but only of the correctness in terms of the units of measurement involved. Actually, it acts as a *sui-generis* filter in the sense that if a formula passes the dimensional test, then it is likely to be correct, but if that formula is not dimensionally correct, then it is definitely wrong. Dimensional analysis obviously has other applications, but this is not the place for such a discussion. The dimensional analysis used for this purpose is a rather minor application, but it is often used because it allows a quick check that a formula is written correctly.

In *theoretical physics* there is another "equivalent" test for a quick check of the correctness of a formula and it is called the *homogeneity analysis of indices*.

We draw the reader's attention to the fact that in reality it is not a question of a "replacement" of dimensional analysis with homogeneity analysis of indices, but of the fact that in phenomenological physics only dimensional analysis is done, while in theoretical physics both are done. It is just that in theoretical physics the homogeneity analysis of indices is done first. So, in this case the dimensional analysis specific to phenomenological physics must be complemented by the analysis of indices specifying the tensor quality (tensor order and tensor variance) of the quantities involved in the formula, the latter analysis taking precedence. Thus, in theoretical physics one first checks whether a formula is written correctly from the point of view of index homogeneity, and only afterwards does the dimensional analysis take place.

The peculiarity of formulas in theoretical physics is that they always contain quantities with indices, because in mathematical terms any physical quantity is designated by a tensor of a certain order and of a certain variance. Or this means indices! Just as in dimensional analysis we cannot add "kilograms to meters and get seconds or amperes", so in index analysis we can only add tensor quantities of the same order and the same variance. So here, in theoretical physics, the tensorial (and, more rarely, the spinorial) character of the quantities involved in the formulae takes precedence over the dimensions characterising those physical quantities.

Before indicating the concrete method of verifying the correctness of a formula in theoretical physics, it is absolutely necessary to make a small classification of the indices that can appear in a formula.

The indices of any physical tensor quantity can be of two kinds for each of the two classification criteria, namely variance and significance.

From the point of view of the *variance criterion*, the indices can be of two kinds, namely:

1) *covariance* indices and

2) *contravariance* indices.

From the point of view of the *significance criterion*, the indices are divided into:

1) *significant* indices (or *free* indices) and

2) *summation* indices (or *dummy* indices).

An index that is summed over is a summation index. It is also called a dummy index since any symbol can replace it without changing the meaning of the expression (provided that it does not collide with other index symbols in the same term).

An index that is not summed over is a free index and should appear only once per term. If such an index does appear, it usually also appears in every other term in an equation.

Most often in the formulas and calculations involved in theoretical physics (but not only, as this is equally true in mathematics, for example) the so-called *Einstein summation convention* is used. It is also known as the Einstein notation or Einstein summation notation and basically it is a notational convention that implies summation over a set of indexed terms in a formula, thus achieving brevity. As part of mathematics it is a notational subset of Ricci calculus; however, it is often used in physics applications that do not distinguish between tangent and cotangent spaces. It was introduced to physics by Albert Einstein in 1916. According to this convention, when an index variable appears twice in a single term and is not otherwise defined, it implies summation of that term over all the values of the summation index. Here are just two examples:

i) Einstein summation convention implies that instead of writing

$$\vec{a} \cdot \vec{b} = \sum_{i=1}^{3} a_i b_i = a_1 b_1 + a_2 b_2 + a_3 b_3,$$

we simply write

$$\vec{a} \cdot \vec{b} = a_i b_i.$$

Here we have only one summation (dummy) index, namely, the index i.

ii) Instead of writing

$$\sum_{\mu=0}^{3} \dot{x}^{\mu} \delta A_{\mu} = \sum_{\mu=0}^{3} \sum_{\nu=0}^{3} \sum_{\lambda=0}^{3} \Gamma_{\nu\lambda}^{\mu} A_{\mu} \dot{x}^{\nu} dx^{\lambda},$$

we will write

$$\dot{x}^{\mu} \delta A_{\mu} = \Gamma_{\nu\lambda}^{\mu} A_{\mu} \dot{x}^{\nu} dx^{\lambda}.$$

In this example, we have three summation (dummy) indices, namely, μ, ν and λ.

So, the Einstein summation convention involves omitting the writing of the symbol $\sum_{i=1}^{n}$ each time the summation from 1 to n is done over the index i, which necessarily *repeats itself*, *i.e.*, it appears twice: once as a covariance index and once as a contravariance index. Thus, when the variance of tensorial quantities is essential, the summation must be done diagonally. Of course, the value of n can vary from one case to another and in the same formula we can meet two or more sums that should be done up to different n.

The *homogeneity property of indices* "requires" that each term of a formula in theoretical physics has exactly the same free/significant indices (as number, quality/name, and variance), regardless of how many summation indices (which, as already mentioned, always occur in pairs of indices of different variances; in other words summation is always done "diagonally") any of the terms of that formula might contain.

To make it easier to understand, we give two examples below:

1) By definition, the Riemann-Christoffel curvature tensor in Einsteinian gravitational field theory is given by the formula

$$R_{\lambda\mu\nu}^{\rho} = \frac{\partial \Gamma_{\lambda\nu}^{\rho}}{\partial x^{\mu}} - \frac{\partial \Gamma_{\lambda\mu}^{\rho}}{\partial x^{\nu}} + \Gamma_{\lambda\nu}^{\sigma} \Gamma_{\sigma\mu}^{\rho} - \Gamma_{\lambda\mu}^{\sigma} \Gamma_{\sigma\nu}^{\rho}.$$

To check that this formula (and any other formula in theoretical physics) is written correctly in terms of index homogeneity, the following algorithm can be used:

i) first the simplest side of the formula is identified (most often the one containing the least number of terms, it is rarely necessary to analyse the term structure at this first stage);

$ii)$ in the simplest side of the formula we identify the simplest term (as a mode of writing, which means that we are interested in the number of tensor quantities that occur in that term, as well as the number of indices of these quantities);

$iii)$ in this simplest term of the simplest side of the analysed formula, the free/significant indices (as number, quality/name and variance) are identified;

$iv)$ the analysed formula is correctly written (i.e., the homogeneity property of indices is respected) if all terms of the formula, without exception, contain exactly the same free indices (as number, quality/name and variance).

In accordance with the algorithm presented, it is easily seen that the simplest side of the formula giving the Riemann tensor is the left side. Since this side has only one term, the second step of the algorithm is already completed. In agreement with the third step of the algorithm, in this term we identify free indices; there are four of them: ρ, λ, μ and ν − more precisely, all four indices of $R^{\rho}_{\lambda\mu\nu}$ are significant/free indices.

Thus, according to the fourth step of the algorithm, for the formula to be written correctly it is necessary that all its other terms contain four significant indices (neither more nor less), and that these are ρ, λ, μ and ρ (and no others) and in addition, all these four indices must have the same variance as in the simplest term of the formula, i.e., ρ must be a contravariance index, and the other three, i.e., λ, μ and ν must be covariance indices.

Let's see if this is so. The first term in the right side of the given formula has exactly the same significant indices, ρ is a contravariance index (as it should be), λ and ν are covariance indices (as they should be), and the index μ (which appears as a contravariance index of the position four-vector x^{μ}) is actually a covariance index, because the operator $\frac{\partial}{\partial x^{\mu}}$ is actually written in the form ∂_{μ} (here it must be taken into account that x^{μ} does not appear as a stand-alone tensor quantity, but as a quantity with respect to which it is derived; in other words, x^{μ} appears within an operator − moreover, it appears at the denominator − so that the index μ should not be seen as a contravariance index of the position four-vector, but as an index (here, of covariance) of the partial derivation operator ∂_{μ}, which it defines.

The second term in the right-hand side is very similar in structure to the first term in the same side, so this is also correct. The next two terms (and also the last two terms of the formula under consideration) are also very similar to each other, so we will only consider one of them. As can be seen, the term $\Gamma^{\sigma}_{\lambda\nu}\Gamma^{\rho}_{\sigma\mu}$ has five indices in total, including exactly the same four free indices as in the terms analysed above (namely ρ, λ, μ and ν − each with the required variance) but also the summation index σ. So this term is also written correctly, since the σ index is a dummy index. The analysis of this formula from the point of view of homogeneity of indices is therefore complete.

2) Finally, consider the following formula, also from Einsteinian gravitational field theory,

$$\frac{\partial T^{\nu}_{\mu}}{\partial x^{\nu}} = \frac{1}{2}\kappa \frac{\partial h_{\lambda\nu}}{\partial x^{\mu}} T^{\nu\lambda} - \frac{1}{2}\kappa \eta^{\nu\lambda} \frac{\partial h_{\lambda\nu}}{\partial x^{\sigma}} T^{\sigma}_{\mu},$$

and check whether or not it is written correctly in terms of index homogeneity.

Following the steps of the algorithm shown in the previous example, we identify the left side as the simplest. Since it has only one term, the second step of the algorithm is already completed. In this term we identify a single free index, namely the index μ, which is a covariance index.

So the formula is correctly written if each term of it has a single free index − in our case μ (and not another) − and, in addition, this is a covariance index in each term of the formula.

Indeed, the first term in the right side contains three indices, namely λ, ν and μ, but the indices λ and ν are dummy indices (they appear in pairs of indices of different variances: λ of $h_{\lambda\nu}$ is a covariance index, while the other λ in the pair, *i.e.*, λ of $T^{\nu\lambda}$ is a contravariance index; the same is true of the summation index ν), so this term too has only one significant index, it is − as it should be − μ itself (and not another), and in addition it is a covariance index, as is absolutely necessary for the formula to be correctly written. The second term in the right-hand side has four indices, namely ν, λ, σ and μ, but all are dummy indices, except μ, which is a covariance index. In conclusion, the formula is written correctly.

B

Appendix B: Tensors in 3D Euclidean Space

B.1 Cartesian Coordinates

A three-dimensional Euclidean space, E_3, is a system of points which are in a bi-univocal and bi-continuous correspondence with a system of three coordinates x_1, x_2 and x_3. If the three coordinates define an ordered system of three reciprocally orthogonal directions, they are called *Cartesian coordinates*. The three directions are also called *coordinate axes*. The Cartesian system of coordinates is defined by three *versors* (*unit vectors*) $\vec{u}_1$, $\vec{u}_2$ and $\vec{u}_3$, which satisfy the condition

$$\vec{u}_1 \cdot (\vec{u}_2 \times \vec{u}_3) \neq 0$$

and form an *orthonormal basis*. Let a_i, $i = 1, 2, 3$, be the components of the vector $\vec{a}$ in the orthonormal basis $\vec{u}_i$, $i = 1, 2, 3$. Note that algebraically a vector is actually a first-order tensor. Using the Einstein summation convention, we will write the analytical expression of the vector $\vec{a}$ in the form

$$\vec{a} = a_i \vec{u}_i. \tag{B.1.1}$$

A vector which defines the position of some point in E_3 is denoted by $\vec{r}(x_1, x_2, x_3)$. It is called *position vector* and its analytical expression is

$$\vec{r} = x_i \vec{u}_i. \tag{B.1.2}$$

Scalar product. Let $\vec{a}$ and $\vec{b}$ be any two vectors. The *scalar product* of the two vectors is defined by

$$\vec{a} \cdot \vec{b} = (a_i \vec{u}_i) \cdot (b_k \vec{u}_k) = a_i b_k (\vec{u}_i \cdot \vec{u}_k) = a_i b_i = ab \cos(\widehat{\vec{a}, \vec{b}}), \tag{B.1.3}$$

because

$$\vec{u}_i \cdot \vec{u}_k = \delta_{ik}. \tag{B.1.4}$$

The relation (B.1.4) can be regarded as a *sui-generis* "geometric" definition of the Kronecker symbol, which from a matrix point of view is nothing else than the unit matrix. Since here we have considered the three-dimensional Euclidean space E_3, we have

$$\delta_{ik} \equiv I_{3\times 3} = \begin{pmatrix} 1 & 0 & 0 \\ 0 & 1 & 0 \\ 0 & 0 & 1 \end{pmatrix}.$$

Strictly speaking, the algebraically correct name of the Kronecker symbol is the *second-order symmetric unit tensor* or[1] *second-rank symmetric unit tensor*. The symmetry property of δ_{ik} follows from its very geometric definition, since the scalar product of two vectors is symmetric:

$$\delta_{ik} = \vec{u}_i \cdot \vec{u}_k = \vec{u}_k \cdot \vec{u}_i = \delta_{ki}.$$

Vector product. By definition, the vector product of the vectors $\vec{a}$ and $\vec{b}$ is given by

$$\vec{a} \times \vec{b} = -\vec{b} \times \vec{a} = \varepsilon_{ijk} a_j b_k \vec{u}_i \equiv \begin{vmatrix} \vec{u}_1 & \vec{u}_2 & \vec{u}_3 \\ a_1 & a_2 & a_3 \\ b_1 & b_2 & b_3 \end{vmatrix}, \qquad \text{(B.1.5)}$$

the last "equality" involving formal writing in determinant form. The modulus of the vector $\vec{a} \times \vec{b}$ is

$$\left| \vec{a} \times \vec{b} \right| = |\vec{a}||\vec{b}| \sin(\widehat{\vec{a}, \vec{b}}) = ab \sin(\widehat{\vec{a}, \vec{b}}). \qquad \text{(B.1.6)}$$

We also have

$$\vec{u}_i \times \vec{u}_j = \varepsilon_{ijk} \vec{u}_k,$$
$$\vec{u}_s = \frac{1}{2} \varepsilon_{sij} \vec{u}_i \times \vec{u}_j, \qquad \text{(B.1.7)}$$

where

$$\varepsilon_{ijk} = \begin{cases} +1, & \text{cyclic permutations of } 1, 2, 3, \\ -1, & \text{non} - \text{cyclic permutations of } 1, 2, 3, \\ 0, & \text{at least two equal indices,} \end{cases} \qquad \text{(B.1.8)}$$

is the *Levi-Civita's permutation symbol* (see Sect. § **B.5**). It is easy to verify (this is only done by direct calculation) the property

$$\varepsilon_{ijk} \varepsilon_{lmn} = \begin{vmatrix} \delta_{il} & \delta_{im} & \delta_{in} \\ \delta_{jl} & \delta_{jm} & \delta_{jn} \\ \delta_{kl} & \delta_{km} & \delta_{kn} \end{vmatrix}. \qquad \text{(B.1.9)}$$

For $i = l$, we have

$$\varepsilon_{ijk} \varepsilon_{imn} = \delta_{jm} \delta_{kn} - \delta_{jn} \delta_{km}, \qquad \text{(B.1.10)}$$

---

[1]Although the notions of "order" and "rank" have different meanings, when it comes to tensors, the two terms are usually used with the same meaning.

and if $i = l$ and $j = m$ in Eq. (B.1.9), then it follows that

$$\varepsilon_{ijk}\varepsilon_{ijn} = 2\delta_{kn}. \tag{B.1.11}$$

Of course, if $i = l$, $j = m$ and $k = n$, then Eq. (B.1.9) gives

$$\varepsilon_{ijk}\varepsilon_{ijk} = 2\delta_{kk} = 6 = 3!. \tag{B.1.12}$$

Le us consider now three arbitrary vectors, $\vec{a}$, $\vec{b}$ and $\vec{c}$. Among other things, with these vectors can be defined the following two types of products:

1. Mixed product (or **scalar triple product**)

$$\vec{a} \cdot (\vec{b} \times \vec{c}) = \varepsilon_{ijk}a_i b_j c_k = \begin{vmatrix} a_1 & a_2 & a_3 \\ b_1 & b_2 & b_3 \\ c_1 & c_2 & c_3 \end{vmatrix} ; \tag{B.1.13}$$

in particular,

$$\vec{u}_i \cdot (\vec{u}_j \times \vec{u}_k) = \varepsilon_{ijk}, \tag{B.1.14}$$

relation that can be considered a *sui-generis* "geometric" definition of the *Levi-Civita symbol*.

2. Vector triple product

$$\vec{a} \times (\vec{b} \times \vec{c}) = (\vec{a} \cdot \vec{c})\vec{b} - (\vec{a} \cdot \vec{b})\vec{c}, \tag{B.1.15}$$

formula which is known as the *triple product expansion*, or *Lagrange's formula*, although the latter name is also used for several other formulae.

B.2   Orthogonal Coordinate Transformations

Transition from a three-orthogonal reference frame $S(Oxyz)$ to another one $S'(O'x'y'z')$ can be achieved in three fundamental ways:
  – translation of coordinate axes;
  – rotation of coordinate axes;
  – mirroring operation.

The first two types of transformations do not modify the *orientation* of the frame S' with respect to S (*proper transformations*), while the mirroring (e.g., $x' = -x$, $y' = y$, $z' = z$) changes the frames orientation (the right frame passes to a left one and vice-versa – *improper transformation*). Obviously, there are also possible combinations of the above operations, with the observation that two proper transformations yield a proper transformation (e.g., translation + rotation), while a proper transformation accompanied by an improper one (e.g., translation + mirroring) lead to an improper transformation.

Let S and S' have the same origin, and let $\vec{u}_i$ and $\vec{u}_i'$ be the two orthonormal bases. Expressing the position vector $\vec{r}$ in the two bases, we have

$$\vec{r} = x_i \vec{u}_i = x_i' \vec{u}_i', \qquad (B.2.16)$$

from which we deduce

$$\begin{cases} x_i' = a_{ik} x_k; & a_{ik} = \vec{u}_i' \cdot \vec{u}_k, \\ x_i = a_{ki} x_k'; & a_{ki} = \vec{u}_k' \cdot \vec{u}_i. \end{cases} \qquad (B.2.17)$$

We also have

$$\vec{u}_i' = a_{ik} \vec{u}_k; \quad \vec{u}_i = a_{ki} \vec{u}_k'. \qquad (B.2.18)$$

The coefficients a_{ik} form the *matrix of the orthogonal transformation* given by Eq. (B.2.17)$_1$, and a_{ki} — the *matrix of the inverse transformation* given by Eq. (B.2.17)$_2$.

The invariance of the square of the distance between two points

$$r^2 = x_i x_i = x_k' x_k',$$

allows us to deduce the *orthogonality condition*,

$$a_{ik} a_{im} = \delta_{km}, \quad i, k, m = 1, 2, 3. \qquad (B.2.19)$$

Aplying the rule of determinants multiplication, one finds

$$\det(a_{ik}) = \pm 1. \qquad (B.2.20)$$

On the other side,

$$\det(a_{ik}) = \vec{a}_1 \cdot (\vec{a}_2 \times \vec{a}_3) = \varepsilon_{ijk} \, a_{1i} a_{2j} a_{3k}, \qquad (B.2.21)$$

that is

$$\varepsilon_{ijk} \, a_{1i} a_{2j} a_{3k} = \pm 1, \qquad (B.2.22)$$

or

$$\varepsilon_{ijk} \, a_{li} a_{mj} a_{nk} = \pm \varepsilon_{lmn} = \varepsilon_{lmn} \det(a_{ik}). \qquad (B.2.23)$$

Here the plus sign corresponds to the case when the two frames have the same orientation, and the minus sign to the case when the frames have different orientations.

When changing the orthonormal basis (see Eqs. (B.2.17) and (B.2.18)), we have

$$\begin{aligned} \vec{p} + \vec{q} &= p_i \vec{u}_i + q_i \vec{u}_i = p_i a_{mi} \vec{u}_m' + q_i a_{mi} \vec{u}_m' \\ &= p_m \vec{u}_m' + q_m \vec{u}_m' = \vec{p}' + \vec{q}'; \\ \vec{p} \cdot \vec{q} &= p_i q_i = \left(a_{mi} p_m' \right) \left(a_{si} q_s' \right) = \delta_{ms} p_m' q_s' = p_s' q_s' = \vec{p}' \cdot \vec{q}'; \\ \vec{p} \times \vec{q} &= \varepsilon_{ijk} \, p_j q_k \vec{u}_i = \varepsilon_{ijk} \, a_{mj} a_{sk} a_{li} p_m' q_s' \vec{u}_l' \\ &= \left(\vec{p}' \times \vec{q}' \right) \det(a_{ik}); \\ \vec{p} \cdot \left(\vec{q} \times \vec{r} \right) &= \varepsilon_{ijk} \, p_i q_j r_k = \varepsilon_{ijk} \, a_{li} a_{mj} a_{nk} p_l' q_m' r_n' \\ &= \vec{p}' \cdot \left(\vec{q}' \times \vec{r}' \right) \det(a_{ik}). \end{aligned} \qquad (B.2.24)$$

Therefore, the operations of addition (subtraction) and scalar product of two vectors do not modify when changing the base, while the vector and scalar triple products change their sign or not, depending on whether or not the two bases have the same orientation.

We call *scalars of the first species* or *scalar invariants* (or, simply, *scalars*) those scalars whose sign — when passing from one base to another base — does not depend on the relative orientation of bases (e.g., temperature, mass, mechanical work, electric charge, etc.). The scalars whose sign depend on the relative orientation of bases are *scalars of the second species* or *pseudoscalars* (examples: the force moment with respect to an axis, the flux of a magnetic field etc.).

We call *polar vectors* (or, simply, *vectors*) those vectors which change their sign under an improper transformation (e.g., the velocity of a particle, the electric field intensity, the gradient of a scalar invariant, etc.). The vectors which do not change their sign under such a transformation are called *vectors of the second species* or *axial vectors* or, still, *pseudovectors* (e.g., the vector product of a two polar vectors, magnetic induction, the curl of a polar vector, angular velocity, etc.). For example, at the spatial inversion $x_i' = -x_i$, for a polar vector we have $A_i' = -A_i$, while for a pseudovector we have $B_i' = B_i$.

B.3   Elements of Vector Analysis

B.3.1   Scalar and Vector Fields

If to each point, $P \in D$ $(D \subset E_3)$ corresponds a value of a scalar $\varphi(P)$, we say that in D was defined a *scalar field*. If the position in D of point P is given by the position vector $\vec{r}$, the scalar field is usually written as $\varphi(\vec{r})$. In physics, in general we have $\varphi(\vec{r}(t), t)$.

If to any point, $P \in D$ $(D \subset E_3)$ one can associate a vector quantity, $\vec{A}(P)$, we will say that in D was defined a *vector field*, which is usually denoted by $\vec{A}(\vec{r})$. In general, we have $\vec{A}(\vec{r}(t), t)$.

A scalar field, φ (or vector field $\vec{A}$) is called *stationary* if φ (respectively, $\vec{A}$) do not explicitly depend on time, *i.e.*, $\varphi = \varphi(\vec{r}(t))$ (respectively $\vec{A} = \vec{A}(\vec{r}(t))$). Otherwise, the field is called *non-stationary*.

In Classical Electrodynamics the following **types of integrals** are frequently met:

a) *Curvilinear integral:* $\int_{P_1}^{P_2} \vec{a} \cdot d\vec{s}$, along a curve C, taken between the points P_1 and P_2 that belong to the curve C, where $\vec{a}$ is a vector with its application point on the curve, and $d\vec{s}$ is a vectorized element of the curve C. This integral is called *circulation* of the vector $\vec{a}$ along the curve C, from P_1 to P_2. If C is a closed curve, circulation is denoted as $\oint_{(C)} \vec{a} \cdot d\vec{s}$.

b) *Double integral:* $\int_{(S)} \vec{a} \cdot d\vec{S} = \int_{(S)} \vec{a} \cdot \vec{n}\, dS$, where $\vec{a}$ is a vector with its application point on the surface S, $d\vec{S}$ is a vectorized surface element, and $\vec{n}$ is the versor of the exterior normal to $d\vec{S}$. The integral is called *flux* of the vector $\vec{a}$ through the surface S. If S is a closed surface, the integral is denoted as $\oint_{(S)} \vec{a} \cdot d\vec{S}$.

c) *Triple integral:* $\int_{(D)} \vec{a}\, d\tau$, where the domain $D \subset E_3$ has the volume V, $d\tau$ is a volume element, and the origin of the vector $\vec{a}$ is in any point of D.

The above definitions remain obviously valid, with the necessary specifications, if $\vec{a}$ is a *vector field*.

B.3.2 First-Order Differential Vector "Operators"

Let's first explain why we put the word *operators* in the title of this sub-section in quotation marks. The reason for this is that, in reality, the *gradient, divergence* and *curl* − which will be briefly presented below − are not operators, but are the result of the action of the "nabla" operator (or Hamilton's operator, as it is also called) on various (scalar or vector) fields. Thus, the only operator used in this section is the "nabla" operator,

$$\nabla \equiv \vec{i}\frac{\partial}{\partial x} + \vec{j}\frac{\partial}{\partial y} + \vec{k}\frac{\partial}{\partial z} = \vec{u}_i\frac{\partial}{\partial x_i}, \tag{B.3.25}$$

and with its help we can define the three notions already mentioned, namely the gradient of a scalar (and only scalar) field, the divergence of a vector (and only vector) field and the curl of a vector (and only vector) field.

First of all it should be noted that "nabla" is not an ordinary operator, as it is both an operator and a vector, in other words, we could call it a "vector operator". For this reason, some authors even prefer to write nabla either in bold font (as vectors are usually written) or even with an arrow over it (as we have chosen to write vectors in this book), *i.e.*, $\vec{\nabla}$. However, we will write the nabla vector operator in a simple way, as in Eq. (B.3.25), *i.e.*, neither with the arrow above nor with bold font.

Being also a vector, we can take both the scalar and cross product of ∇ with other vectors, but being simultaneously *and* an operator, even the scalar product of two ordinary vectors is commutative, the situation is not the same with ∇, because, given the polar vector $\vec{a}$,

$$\vec{a} \cdot \nabla \neq \nabla \cdot \vec{a},$$

since the first quantity, *i.e.*, $\vec{a} \cdot \nabla$ is still an operator, but $\nabla \cdot \vec{a}$ is a scalar finite quantity which is quite the divergence of the vector (field) $\vec{a}$.

Since the vector product of two vectors is anticommutative, the more so

$$\vec{a} \times \nabla \neq \nabla \times \vec{a}.$$

Before moving on to briefly discuss gradient, divergence and curl, we point out that in this sub-section we will not be consistent in notation, in the sense that we will use both nabla operator notation and literal notation for gradient, divergence and curl at the same time, precisely to emphasize that both notations are equally good. Specifically, sometimes we write $\nabla \varphi$ and sometimes we write grad φ, for gradient, sometimes we write $\nabla \cdot \vec{A}$ and sometimes we write div$\vec{A}$, for divergence, and sometimes we write $\nabla \times \vec{A}$ and sometimes we write curl$\vec{A}$, for curl.

i) Gradient. Consider the scalar field $\varphi(\vec{r})$, where $\varphi(\vec{r})$ is a function of class C^1 in $D \subset E_3$. The vector field

$$\vec{A} = \operatorname{grad} \varphi = \frac{\partial \varphi}{\partial x_i} \vec{u}_i = \nabla \varphi, \quad i = 1, 2, 3, \tag{B.3.26}$$

where the Cartesian[2] coordinates were used, is called the *gradient* of the scalar field $\varphi(\vec{r})$. The vector field $\vec{A}$ defined by Eq. (B.3.26) is called *conservative*.

As the name suggests, the gradient of a scalar quantity is used in physics (and not only) whenever we need to express the fact that, that quantity varies "gradually" (e.g., the pressure in a fluid, the temperature of the water in a pool, the concentration of ions in a solution, etc.). This is exactly what the gradient does: it shows us in which direction the respective scalar quantity varies the fastest (indicated by the versor of the vector which is the gradient of that quantity) and how rapidly that variation is occurring (indicated by the modulus of the vector that is the gradient of that quantity).

Equipotential surfaces. Consider the fixed surface

$$\varphi(x, y, z) = K(= \text{const.}). \tag{B.3.27}$$

Observing that

$$d\varphi = \operatorname{grad} \varphi \cdot d\vec{r} \equiv \nabla \varphi \cdot d\vec{r} = 0, \tag{B.3.28}$$

we deduce that in each point of the surface $\varphi(x, y, z) = const.$, the vector $\nabla \varphi$ has the direction of the normal to the surface. Giving values to the constant K, we obtain a family of surfaces called *equipotential surfaces* or *level surfaces*. Equation (B.3.28) is called the *equation of equipotential surfaces* of scalar quantity $\varphi(\vec{r})$. For instance, if the Cartesian coordinates are used, then Eq. (B.3.28) writes explicitly as follows:

$$\frac{\partial \varphi}{\partial x_i} dx_i = \frac{\partial \varphi}{\partial x} dx + \frac{\partial \varphi}{\partial y} dy + \frac{\partial \varphi}{\partial z} dz = 0.$$

---

[2]Note that using different types of coordinates (for instance, spherical coordinates or cylindrical coordinates – to give just two examples) the gradient writes differently (has different forms).

Field lines. Consider the stationary vector field $\vec{A}(\vec{r})$ and a spatial curve C parametrically given by the equations $x_i = x_i(s)$, $i = 1, 2, 3$. If at any point of the curve C the field $\vec{A}(\vec{r})$ is tangent to the curve, the curve is called *line of the vector field* $\vec{A}(\vec{r})$. The differential equations of the field lines are explicitly obtained from the relation

$$\vec{A} \times d\vec{r} = 0, \qquad (B.3.29)$$

(where $d\vec{r}$ is a vectorized element of the field line), which is called the *equation of field lines* or *equation of force lines* of the vector field $\vec{A}(\vec{r})$. For instance, in Cartesian coordinates, Eq. (B.3.29) is written as follows:

$$\frac{dx}{A_x(x, y, z)} = \frac{dy}{A_y(x, y, z)} = \frac{dz}{A_z(x, y, z)},$$

while in spherical coordinates it looks like this

$$\frac{dr}{A_r(r, \theta, \varphi)} = \frac{r d\theta}{A_\theta(r, \theta, \varphi)} = \frac{r \sin \theta d\varphi}{A_\varphi(r, \theta, \varphi)}.$$

Directional derivative. Let us project the vector $\nabla \varphi$ on the direction defined by versor $\vec{u}$ of the line element $d\vec{r}$, whose modulus is $|d\vec{r}| \equiv ds$. Since $d\vec{r} = \vec{u} |d\vec{r}| = \vec{u} \, ds$, where ds is a line element on the direction defined by $\vec{u}$, we have

$$(\text{grad} \, \varphi) \cdot \vec{u} = \frac{d\varphi}{d\vec{r}} \cdot \frac{d\vec{r}}{ds} = \frac{d\varphi}{ds} = (\text{grad} \, \varphi)_u. \qquad (B.3.30)$$

The derivative $\frac{d\varphi}{ds}$ given by Eq. (B.3.30) is called *directional derivative* of the scalar field $\varphi(\vec{r})$ on the direction $\vec{u}$.

In particular, if $\vec{u}$ coincides with the versor $\vec{n}$ of the exterior normal to surface in Eq. (B.3.27), we get

$$(\text{grad} \, \varphi) \cdot \vec{n} = \frac{d\varphi}{dn} \geq 0, \qquad (B.3.31)$$

meaning that the gradient is oriented along the normal to the equipotential surfaces and its positive sense is the sense of increasing the field.

ii) Divergence. Taking the scalar product of vector operator $\nabla = \vec{u}_i \frac{\partial}{\partial x_i}$ with the vector field $\vec{A}(\vec{r})$, we obtain the *divergence* of $\vec{A}(\vec{r})$ in Cartesian[3] coordinates:

$$\text{div} \vec{A}(\vec{r}) \equiv \nabla \cdot \vec{A} = \frac{\partial A_i}{\partial x_i} \equiv \partial_i A_i \equiv A_{i,i}, \quad i = 1, 2, 3. \qquad (B.3.32)$$

In physics divergence is used whenever we want to express that a physical field has or does not have sources. If the divergence is positive, then the source is called a *spring*, and if the divergence is negative, then the source is called a

---

[3]The same observation applies as made in the footnote when discussing the gradient.

well. A field that has no sources (such as the magnetic field in the macrocosm − where the existence of the magnetic monopole has not been discovered) will by default have zero divergence,

$$\text{div}\vec{B} \equiv \nabla \cdot \vec{B} = 0. \tag{B.3.33}$$

In this case, the field lines close in on themselves and the field is called a *source-free* or *solenoidal* field. On the contrary, in the case of a vector field with sources, the field lines leave from positive or enter the negative sources. For example, in the case of the electrostatic field, we have

$$\nabla \cdot \vec{E} = \frac{1}{\epsilon_0}\rho,$$

where ρ is the spatial density of sources (the spatial density of electrical charge).

Divergence is also used to specify that a vector field is or isn't uniform, *i.e.*, if it has the same or, respectively, different values at different points in space or, mathematically speaking, its components aren't ($A_i = $ const., $i = 1, 2, 3$) or, respectively, are functions of coordinates ($A_i = A_i(x_i)$, $i = 1, 2, 3$). For uniform vector fields the divergence is zero, while non-uniform vector fields have a non-zero divergence.

iii) Curl. By curl of a vector field $\vec{A}(\vec{r})$ we mean the vector field $\vec{B}$ obtained from $\vec{A}$ as a result of the operation $\vec{B} = \nabla \times \vec{A}$. Analytically, in Cartesian[4] coordinates, the curl of $\vec{A}$ is expressed by

$$\text{curl}\,\vec{A} \equiv \nabla \times \vec{A} = \vec{u}_i\varepsilon_{ijk}\frac{\partial A_k}{\partial x_j} \equiv \vec{u}_i\varepsilon_{ijk}\partial_j A_k \equiv \vec{u}_i\varepsilon_{ijk}A_{k,j},$$

$$\left(\text{curl}\,\vec{A}\right)_s = \varepsilon_{sjk}\frac{\partial A_k}{\partial x_j} \equiv \varepsilon_{sjk}\partial_j A_k \equiv \varepsilon_{sjk}A_{k,j}. \tag{B.3.34}$$

In physics the curl is used to specify whether a vector field has or does not have vortices. For example, in fluid physics, whether or not the flow of a fluid is laminar is expressed mathematically by canceling or not canceling the curl of the vector field of velocities of that fluid. One even defines a vector, suggestively called the "vortex vector", by the relation

$$\Omega \equiv \frac{1}{2}\text{curl}\,\vec{v} = \frac{1}{2}\nabla \times \vec{v}.$$

Obviously, if $\Omega = 0$ means that the flow of respective fluid is laminar, and if $\Omega \neq 0$, then the fluid flow is turbulent.

A vector field $\vec{A}$ which satisfies the relation

$$\text{curl}\,\vec{A} \equiv \nabla \times \vec{A} = 0, \tag{B.3.35}$$

is called *irrotational* or *vortex free* or, still, *non-swirling*. Conversely, if $\nabla \times \vec{A} \neq 0$, then the field is called *eddy* or *turbulent* or *with vortices* or even *swirling*.

---

[4]The same observation applies as made in the footnote when discussing the gradient and divergence.

B.3.3 Second-Order Differential Operators

As can be seen, this time we have not put the word *operators* in quotation marks in the title of this sub-section (as in the previous sub-section), because second-order differential operators are genuine operators (in the full sense of the word).

Using the nabla vector operator, two new genuine operators which are second-order differential operators can be defined, namely, the *Laplacian* (or the *Laplace operator*) and the *d'Alembertian* (or the *d'Alembert operator*).

If we were to be very rigorous and go somewhat outside the narrow framework of the elementary mathematics we use in Classical Electrodynamics (which mathematics, however, turns out to be more than enough), in fact, both operators mentioned above are particular cases of the Laplace-Beltrami operator.

In differential geometry, the Laplace-Beltrami operator is a generalization of the Laplace operator to functions defined on submanifolds in Euclidean space and, even more generally, on Riemannian and pseudo-Riemannian manifolds. In Minkowski space the Laplace-Beltrami operator becomes the d'Alembert operator.

The Laplace operator is a second-order differential operator in the $n-$dimensional Euclidean space, defined as the divergence ($\nabla\cdot$) of the gradient ($\nabla\varphi$). It is usually denoted by the symbols $\nabla\cdot\nabla$, ∇^2 or Δ. In a Cartesian coordinate system, the Laplacian is given by the sum of second partial derivatives of the function with respect to each independent variable. In other coordinate systems, such as cylindrical and spherical coordinates, the Laplacian has different, more complicated (but very useful) forms.

Informally, the Laplacian $\Delta\varphi(P)$ of a scalar function φ at a point P measures by how much the average value of φ over small spheres or balls centered at P deviates from $\varphi(P)$.

Thus if φ is a twice-differentiable real-valued function, then the Laplacian of φ is the real-valued function defined by

$$\Delta\varphi = \nabla^2\varphi = \nabla \cdot \nabla\varphi, \tag{B.3.36}$$

or, in the literal notation,

$$\Delta\varphi = \nabla^2\varphi = \text{div}(\text{grad }\varphi),$$

where the latter notations (in both formulas above) derive from formally writing

$$\nabla = \left(\frac{\partial}{\partial x_1}, \frac{\partial}{\partial x_2}, \dots, \frac{\partial}{\partial x_n} \right).$$

Explicitly, the Laplacian of scalar function/field φ is thus the sum of all the unmixed second partial derivatives in the Cartesian coordinates x_i,

$$\Delta\varphi = \sum_{i=1}^{n} \frac{\partial^2\varphi}{\partial x_i^2} = \frac{\partial^2\varphi}{\partial x_i \partial x_i}. \tag{B.3.37}$$

As a second-order differential operator, the Laplace operator maps C^k functions to C^{k-2} functions for $k \geq 2$. It is a linear operator

$$\Delta : C^k(\mathbb{R}^n) \to C^{k-2}(\mathbb{R}^n),$$

or more generally, an operator

$$\Delta : C^k(\mathrm{M}) \to C^{k-2}(\mathrm{M}),$$

for any open set $\mathrm{M} \subseteq \mathbb{R}^n$.

By customizing Eq. (B.3.37) for $n = 3$, we get

$$\Delta\varphi = \sum_{i=1}^{3} \frac{\partial^2 \varphi}{\partial x_i^2} = \frac{\partial^2 \varphi}{\partial x_i \partial x_i} = \frac{\partial^2 \varphi}{\partial x^2} + \frac{\partial^2 \varphi}{\partial y^2} + \frac{\partial^2 \varphi}{\partial z^2}. \tag{B.3.38}$$

This explicit relation for the 3D-Laplacian in Cartesian coordinates can be obtained from the defining relation (B.3.36); indeed, taking into account that in Cartesian coordinates $\vec{u}_i = \mathrm{const.}, \forall i = \overline{1,3}$, we have

$$\Delta = \mathrm{div}(\mathrm{grad}) = \nabla \cdot \nabla = \left(\vec{u}_i \frac{\partial}{\partial x_i} \right) \cdot \left(\vec{u}_j \frac{\partial}{\partial x_j} \right) = \left(\vec{u}_i \cdot \vec{u}_j \right) \frac{\partial}{\partial x_i} \left(\frac{\partial}{\partial x_j} \right)$$

$$= \delta_{ij} \left(\frac{\partial^2}{\partial x_i \partial x_j} \right) = \frac{\partial^2 \varphi}{\partial x_i \partial x_i} = \frac{\partial^2 \varphi}{\partial x^2} + \frac{\partial^2 \varphi}{\partial y^2} + \frac{\partial^2 \varphi}{\partial z^2},$$

i.e., exactly the Eq. (B.3.38) which gives the Laplacian in Cartesian coordinates. For information only, we also give here the expression of the Laplacian in spherical coordinates; it is written as

$$\Delta\varphi = \frac{1}{r^2} \left\{ \frac{\partial}{\partial r} \left(r^2 \frac{\partial \varphi}{\partial r} \right) + \frac{1}{\sin\theta} \left[\frac{\partial}{\partial \theta} \left(\sin\theta \frac{\partial \varphi}{\partial \theta} \right) + \frac{1}{\sin\theta} \frac{\partial^2 \varphi}{\partial \varphi^2} \right] \right\}.$$

Regarding the other second-order differential operator which is the d'Alembert operator, as stated above, in fact it is nothing else than the Laplace-Beltrami operator written in Minkowski space. Indeed, anticipating a bit — the Euclidian-complex representation of Minkowski space will be presented in Sect. § C.1 — in this representation of Minkowski space, the fourth (temporal) coordinate is chosen as[5] $x_4 = ict$ (the other three coordinates obviously being $x_1 = x$, $x_2 = y$ and $x_3 = z$), and then we have

$$\Box \equiv \Delta_4 = \sum_{i=1}^{4} \frac{\partial^2}{\partial x_i \partial x_i} = \frac{\partial^2}{\partial x_1^2} + \frac{\partial^2}{\partial x_2^2} + \frac{\partial^2}{\partial x_3^2} + \frac{\partial^2}{\partial x_4^2}$$

$$= \frac{\partial^2}{\partial x^2} + \frac{\partial^2}{\partial y^2} + \frac{\partial^2}{\partial z^2} - \frac{1}{c^2} \frac{\partial^2}{\partial t^2}$$

$$= \Delta_3 - \frac{1}{c^2} \frac{\partial^2}{\partial t^2} \equiv \Delta - \frac{1}{c^2} \frac{\partial^2}{\partial t^2}.$$

---

[5] Here, in relation $x_4 = ict$, $i^2 = -1$, *i.e.*, it is the complex/imaginary i.

Note that in the hyperbolic representation of Minkowski space (which is equipped with the metric of signature -2), the four coordinates are chosen in the form $x^0 = ct$, $x^1 = x$, $x^2 = y$ and $x^3 = z$, and the d'Alembertian is written as follows:

$$\Box = \frac{1}{c^2}\frac{\partial^2}{\partial t^2} - \frac{\partial^2}{\partial x^2} - \frac{\partial^2}{\partial y^2} - \frac{\partial^2}{\partial z^2}. \tag{B.3.39}$$

Of course, in other coordinates – like spherical or cylindrical, for example – the d'Alembert operator has more complicated expressions.

We conclude this sub-section by pointing out that in Classical Electrodynamics equations containing a vector field under the d'Alembertian can be encountered. For example, in the Lorenz gauge, the second-order differential equation satisfied by the vector potential $\vec{A}(\vec{r}, t)$ of the magnetic field is written as

$$\Box \vec{A}(\vec{r}, t) = \mu_0 \vec{j}(\vec{r}, t), \tag{B.3.40}$$

where the d'Alembert operator is given by Eq. (B.3.39). Strictly speaking, the d'Alembert operator is defined only for a scalar field (being the divergence of the gradient, while the gradient is always applied only to a scalar field). However, the above vector relation is more than tolerated, and it should be understood in the sense that there we have in fact three scalar relations (obtained by projecting the above vector relation on the three axes of coordinates); for instance, in Cartesian coordinates relation (B.3.40) should be understood as follows:

$$\Box A_x(\vec{r}, t) = \mu_0 j_x(\vec{r}, t),$$
$$\Box A_y(\vec{r}, t) = \mu_0 j_y(\vec{r}, t),$$
$$\Box A_z(\vec{r}, t) = \mu_0 j_z(\vec{r}, t).$$

B.3.4 Fundamental theorems

Consider a domain $D \subset E_3$ bordered by the surface S and a vector field $\vec{A}$ of class C^1 in D and C^0 in $\overline{D}$ (the domain formed by the totality of the points of D, together with the points of the surface S). It is shown that

$$\oint_{(S)} \vec{A} \cdot \vec{n}\, dS = \int_D \operatorname{div}\vec{A}\, d\tau, \tag{B.3.41}$$

which is the mathematical expression of the *Green-Gauss-Ostrogradski theorem* (or simpler, *Green-Gauss theorem*, or, more often, *divergence theorem*). Here $\vec{n}$ is the versor of the exterior normal to S.

Let us contract the surface S so that the domain D becomes smaller and smaller. At the limit, we have

$$\operatorname{div}\vec{A} = \lim_{\Delta\tau\to 0}\frac{1}{\Delta\tau}\oint_S \vec{A}\cdot\vec{n}\, dS. \tag{B.3.42}$$

This relation can be considered as the definition of divergence of a vector field at a point. It is useful in that expresses the divergence *independently* of the coordinate system (it is an intrinsic definition relation). If $\operatorname{div}\vec{A} > 0$, we call that point a *positive source* (or *spring*), while if $\operatorname{div}\vec{A} < 0$, it is said that we have a *negative source* (or *well*).

Consider now a surface S which leans on the closed curve C and a vector field $\vec{A}$ of class C^1 on both the surface S and closed curve C. In this case the following relation is valid

$$\oint_{(C)} \vec{A} \cdot \vec{dr} = \int_{(S)} \left(\nabla \times \vec{A} \right) \cdot \vec{n} \, dS, \qquad (\text{B.3.43})$$

called the *Stokes-Ampère theorem*. Admitting that $\vec{n}$ is the versor of exterior normal to S, the direction of travel on C is given by the right hand screw rule (by convention, if a right hand screw is counter-clockwise screwed, it advances in the direction of $\vec{n}$). In particular, if $\vec{A} = \operatorname{grad}\varphi$, we have

$$\oint_{(C)} \vec{A} \cdot \vec{dr} = \oint_{(C)} d\varphi = 0, \qquad (\text{B.3.44})$$

because the integral of a total differential, over a closed curve, is always zero.

By shrinking the surface element until it becomes small enough so as $\nabla \times \vec{A}$ not to vary appreciably inside it, we have (at the limit):

$$\operatorname{curl}_n \vec{A} = \vec{n} \cdot \operatorname{curl}\vec{A} = \lim_{\Delta S \to 0} \frac{1}{\Delta S} \oint_{(C)} \vec{A} \cdot \vec{dr}, \qquad (\text{B.3.45})$$

which can serve as intrinsic definition relation of the curl at a point.

B.3.5 Some Consequences of the Green-Gauss-Ostrogradski and Stokes-Ampère Theorems

Consider $\vec{A} = \vec{e}\,\varphi$, where φ is a scalar field of class C^1, and $\vec{e}$ is a constant vector. Using Eq. (B.3.41) we then have

$$\oint_{(S)} \varphi \, d\vec{S} = \int_{(D)} \operatorname{grad}\varphi \, d\tau, \qquad (\text{B.3.46})$$

and, by means of Eq. (B.3.43),

$$\oint_{(C)} \varphi \, \vec{dl} = \int_{(S)} \left(\vec{n} \times \operatorname{grad}\varphi \right) dS. \qquad (\text{B.3.47})$$

If in Eq. (B.3.41) we choose $\vec{A} = \vec{e} \times \vec{B}$, where $\vec{e}$ is a constant vector, and

$\vec{B}$ a vector field of class C^1 in D, we obtain

$$\oint_{(S)} \vec{n} \times \vec{B} \, dS = \int_{(D)} \operatorname{curl} \vec{B} \, d\tau. \tag{B.3.48}$$

Let now the vector field $\vec{A}$ be of the form $\vec{A} = \psi \operatorname{grad} \varphi$, where $\psi \in C^1(D)$ and $\varphi \in C^2(D)$. From Eq. (B.3.41) we then have

$$\int_{(D)} \left(\operatorname{grad} \psi \cdot \operatorname{grad} \varphi + \psi \Delta \varphi \right) d\tau = \oint_{(S)} \psi \frac{\partial \varphi}{\partial n} dS, \tag{B.3.49}$$

where Δ is the *Laplace's operator* (the *Laplacian*). Switching between them the functions φ and ψ in Eq. (B.3.49), then subtracting the result from Eq. (B.3.49), we have

$$\int_{(D)} \left(\psi \Delta \varphi - \varphi \Delta \psi \right) d\tau = \oint_{(S)} \left(\psi \frac{\partial \varphi}{\partial n} - \varphi \frac{\partial \psi}{\partial n} \right) dS, \tag{B.3.50}$$

where, this time, both φ and ψ are twice continuously differentiable in D. The relations (B.3.49) and (B.3.50) are known as the *Green's first identity* and *Green's second identity*, respectively.

B.3.6 Useful Formulas

The scalar and vector fields on which are applied the "operators" grad, div and curl sometimes intervene as products of scalar and/or vector functions. Other times we meet repeated operations (like div $(\operatorname{grad} \varphi)$, grad $(\operatorname{curl} \vec{A})$, etc.).

Here are some formulas frequently used in Electrodynamics (and not only):

$$\operatorname{grad}(\varphi \, \psi) = \varphi \operatorname{grad} \psi + \psi \operatorname{grad} \varphi. \tag{B.3.51}$$

$$\operatorname{div}(\varphi \vec{A}) = \varphi \operatorname{div} \vec{A} + \vec{A} \, (\operatorname{grad} \varphi). \tag{B.3.52}$$

$$\operatorname{curl}(\varphi \vec{A}) = \varphi(\operatorname{curl} \vec{A}) + (\operatorname{grad} \varphi) \times \vec{A}. \tag{B.3.53}$$

$$\operatorname{div}(\vec{A} \times \vec{B}) = \vec{B} \cdot (\operatorname{curl} \vec{A}) - \vec{A} \cdot (\operatorname{curl} \vec{B}). \tag{B.3.54}$$

$$\operatorname{grad}(\vec{A} \cdot \vec{B}) = \vec{A} \times (\operatorname{curl} \vec{B}) + \vec{B} \times (\operatorname{curl} \vec{A})$$
$$+ (\vec{A} \cdot \nabla)\vec{B} + (\vec{B} \cdot \nabla)\vec{A}. \tag{B.3.55}$$

$$\operatorname{curl}(\vec{A} \times \vec{B}) = \vec{A} \operatorname{div} \vec{B} - \vec{B} \operatorname{div} \vec{A} + (\vec{B} \cdot \nabla)\vec{A} - (\vec{A} \cdot \nabla)\vec{B}. \tag{B.3.56}$$

$$\operatorname{div}(\operatorname{curl} \vec{A}) = 0. \tag{B.3.57}$$

$$\operatorname{curl}(\operatorname{grad} \varphi) = 0. \tag{B.3.58}$$

$$\text{div}(\text{grad}\,\varphi) = \Delta\varphi; \quad \Delta\cdot \equiv \frac{\partial^2\cdot}{\partial x_i \partial x_i} \equiv \partial_i\partial_i\cdot \equiv (\cdot)_{,ii}. \tag{B.3.59}$$

$$\text{curl}(\text{curl}\,\vec{A}) = \text{grad}(\text{div}\,\vec{A}) - \Delta\vec{A}. \tag{B.3.60}$$

Let $\vec{r}$ be the position vector of some point P, with respect to the origin of the Cartesian orthogonal reference frame $Oxyz$. We have

$$\begin{cases} \text{grad}\,r = \dfrac{\vec{r}}{r} = \vec{u}_r; \quad |\text{grad}\,r| = 1; \quad \text{div}\,\vec{r} = 3; \\[2mm] \text{curl}\,\vec{r} = 0; \quad \Delta\left(\dfrac{1}{r}\right)_{r\neq 0} = 0; \quad \nabla\left(\dfrac{1}{r}\right) = -\dfrac{\vec{r}}{r^3}. \end{cases} \tag{B.3.61}$$

If $\vec{A}$ is a constant vector, using Eqs. (B.3.54)–(B.3.56) and (B.3.61) we obtain

$$\text{div}\left(\vec{A}\times\vec{r}\right) = 0; \tag{B.3.62}$$

$$\text{grad}\left(\vec{A}\cdot\vec{r}\right) = \vec{A}; \tag{B.3.63}$$

$$\text{curl}\left(\vec{A}\times\vec{r}\right) = 2\vec{A}. \tag{B.3.64}$$

Given the fields $\varphi(r)$ and $\vec{A}(r)$, where $r = |\vec{r}|$, the validity of the following relations can be proved:

$$\begin{cases} \text{grad}\,\varphi(r) = \vec{u}_r\,\varphi', \\[1mm] \text{div}\,\vec{A}(r) = \vec{u}_r\cdot\vec{A}', \\[1mm] \text{curl}\,\vec{A}(r) = \vec{u}_r\times\vec{A}'; \end{cases} \tag{B.3.65}$$

where $\varphi' = \dfrac{d\varphi}{dr}$, $\vec{A}' = \dfrac{d\vec{A}}{dr}$, and $\vec{u}_r = \dfrac{\vec{r}}{r}$.

B.4   Second-Order Cartesian Tensors

An ordered system of three quantities, A_i, $i = 1, 2, 3$, that when changing the basis transforms according to the law (see Eq. (B.2.17)):

$$A'_i = a_{ik}A_k, \quad i, k = 1, 2, 3, \tag{B.4.66}$$

where the directional cosines a_{ik} satisfy the orthogonality condition given by Eq. (B.2.19), form an *first-order orthogonal affine tensor* (an *orthogonal affine vector*). The quantities A_i, $i = 1, 2, 3$, are called the *components* of the considered orthogonal affine vector.

A system of $3^2 = 9$ quantities T_{ik}, $i, k = 1, 2, 3$, that when changing the basis transforms as the product A_iB_k, that is according to the law

$$T'_{ik} = a_{ij}a_{km}T_{jm} \tag{B.4.67}$$

is called *second-order orthogonal affine tensor*.

The quantities T_{ik}, $i,k = 1,2,3$, are called the *components* of the considered second-order orthogonal affine tensor.
If in Eq. (B.4.67) we make $i = k$ and take into account the orthogonality condition given by Eq. (B.2.19), we have

$$T'_{ii} = a_{ij}a_{im}T_{jm} = \delta_{jm}T_{jm} = T_{mm} = T_{ii}. \tag{B.4.68}$$

The sum $T_{ii} = T_{11} + T_{22} + T_{33}$ is called the *trace* of the tensor T_{ik} and is written as $\mathrm{Tr}\,(T_{ik})$; sometimes the name *spur* (from German) is encountered – in this case it is written as $\mathrm{Sp}\,(T_{ik})$. According to Eq. (B.4.68), the trace of a second-order orthogonal affine tensor is invariant (remains unchanged) when changing the coordinates according to the law given by Eq. (B.2.17).
All the definitions and properties of the tensors met in **Appendix A** remain, obviously, valid. Consider, in this respect, the antisymmetric tensor A_{ik} and let us denote its three independent components by

$$A_{12} \equiv A_3, \; A_{23} \equiv A_1, \; A_{31} \equiv A_2,$$

or, in a condensed form

$$A_{ij} = \varepsilon_{ijk}A_k, \; A_i = \frac{1}{2}\varepsilon_{ijk}A_{jk}, \; i,j,k = 1,2,3. \tag{B.4.69}$$

The ordered system of quantities (numbers) A_i in Eq. (B.4.69) form a pseudovector, called *associated pseudovector* to the antisymmetric tensor A_{ik}. Such a situation is found, for example, in the case of the vector product of two vectors. In view of Eq. (B.3.57), we have

$$c_i = \left(\vec{a} \times \vec{b}\right)_i = \varepsilon_{ijk}a_jb_k = \frac{1}{2}\varepsilon_{ijk}c_{jk}, \tag{B.4.70}$$

where we denoted by c_{jk} the antisymmetric tensor

$$c_{jk} = a_jb_k - a_kb_j, \tag{B.4.71}$$

and took into account that the double sum over j and k of the product between the symmetric tensor $\frac{1}{2}(a_jb_k + a_kb_j)$ and the completely antisymmetric tensor ε_{ijk} is zero:

$$\varepsilon_{ijk}\frac{1}{2}(a_jb_k + a_kb_j) = 0. \tag{B.4.72}$$

Therefore, the "vector" associated with a second-order antisymmetric tensor is actually a pseudovector. The antisymmetric tensor A_{ik} and the pseudovector $A_i = \frac{1}{2}\varepsilon_{ijk}A_{jk}$ are called *dual* to each other, since the inverse relationship also occurs: given a vector of components A_k, $k = 1,2,3$, the second-order antisymmetric pseudotensor A_{ij}, $i,j = \overline{1,3}$, is associated with it through relation (B.4.69)$_1$.

B.5   Higher-Order Cartesian Tensors

A p-order tensor (T) in E_3 is a system of 3^p quantities called the *components* of the p-order tensor (T), which, under an orthogonal coordinate change, transforms according to the law

$$T'_{i_1 i_2 \ldots i_p} = a_{i_1 j_1} a_{i_2 j_2} \ldots a_{i_p j_p} T_{j_1 j_2 \ldots j_p},$$

$$i_1, i_2, \ldots, i_p; \ j_1, j_2, \ldots, j_p = 1, 2, 3. \qquad \text{(B.5.73)}$$

A p-order pseudotensor (T^*) in E_3 is a system of 3^p quantities called the *components* of the p-order pseudotensor (T^*), which, under an orthogonal coordinate change, transforms according to the law

$$T^{*'}_{i_1 i_2 \ldots i_p} = \big[\det(a_{ik}) \big] a_{i_1 j_1} a_{i_2 j_2} \ldots a_{i_p j_p} T^*_{j_1 j_2 \ldots j_p},$$

$$i_1, i_2, \ldots, i_p; \ j_1, j_2, \ldots, j_p = 1, 2, 3. \qquad \text{(B.5.74)}$$

Thus, under a proper orthogonal transformation $\big(\det(a_{ik}) = +1 \big)$, the pseudotensors transform like the tensors, while under an improper transformation $\big(\det(a_{ik}) = -1 \big)$, there is a sign difference. Comparing Eqs. (B.5.74) and (B.2.23), we deduce that the Levi-Civita permutation symbol defined by Eq. (B.1.8) is actually a pseudotensor. It is called *third-order completely anti-symmetric unit pseudotensor*. Under an orthogonal coordinate transformation, the Levi-Civita symbol transforms according to the law

$$\varepsilon'_{ijk} = \big[\det(a_{ik}) \big] a_{il} a_{jm} a_{kn} \, \varepsilon_{lmn}. \qquad \text{(B.5.75)}$$

Using Eq. (B.2.23), we have

$$\varepsilon'_{ijk} = \big[\det(a_{ik}) \big]^2 \varepsilon_{ijk} = \varepsilon_{ijk}, \qquad \text{(B.5.76)}$$

meaning that the components of the pseudotensor ε_{ijk} do not depend on the choice of the basis.

C

Appendix C: Tensors in Minkowski Space

C.1 Euclidean-Complex Representation

Let $x_1 = x$, $x_2 = y$, $x_3 = z$, $x_4 = ict$ be the coordinates of an event in Minkowski space. The metric then is

$$-ds^2 = dx^\mu dx_\mu = g_{\mu\nu}dx^\mu dx^\nu = dx_1^2 + dx_2^2 + dx_3^2 + dx_4^2$$
$$= dx^2 + dy^2 + dz^2 - c^2 dt^2, \qquad (C.1.1)$$

which means

$$g_{\mu\nu} = \delta_{\mu\nu}, \qquad (C.1.2)$$

corresponding to a pseudo-Euclidean space. Such a representation of Minkowski space is called *Euclidean-complex representation*.

The coordinates x_μ, $\mu = 1, 2, 3, 4$, of an event form a *four-vector* (or *quadrivector*), which is called the *position four-vector*. The *space components* x_1, x_2, x_3 are denoted by x_i, $i = 1, 2, 3$, and the *time component* by x_4. Under a change of coordinates the components of the position (or radius) four-vector transform according to

$$x'_\mu = a_{\mu\nu}x_\nu, \quad \mu, \nu = 1, 2, 3, 4. \qquad (C.1.3)$$

If x'_μ are the coordinates of the same event, but determined in the inertial frame S', moving along $Ox \equiv O'x'$-axis, with velocity V with respect to S, then $a_{\mu\nu}$ are elements of the *Lorentz transformation* matrix

$$A = \begin{pmatrix} \Gamma & 0 & 0 & i\dfrac{V}{c} \\ 0 & 1 & 0 & 0 \\ 0 & 0 & 1 & 0 \\ -i\dfrac{V}{c} & 0 & 0 & \Gamma \end{pmatrix}, \qquad (C.1.4)$$

while Eq. (C.1.3) represents, in condensed form, the Lorentz transformation

$$
\begin{cases}
t' = \Gamma\left(t - \dfrac{V}{c^2}x\right), \\[2mm]
x' = \Gamma\left(x - Vt\right), \\[2mm]
y' = y, \\[2mm]
z' = z.
\end{cases}
\tag{C.1.5}
$$

A system of four quantities A_μ, $\mu = 1, 2, 3, 4$, which transform like the coordinates, that is according to the law

$$
A'_\mu = a_{\mu\nu} A_\nu, \quad \mu, \nu = 1, 2, 3, 4,
\tag{C.1.6}
$$

form a *four-vector* (or a *quadrivector*). In Euclidean-complex representation the space components of a four-vector are real, and the time component is pure imaginary.

A *second-order four-tensor* transforms as the product $A_\mu B_\nu$, that is according to the law

$$
T'_{\mu\nu} = a_{\mu\lambda} a_{\nu\rho} T_{\lambda\rho}, \quad \lambda, \rho, \mu, \nu = 1, 2, 3, 4.
\tag{C.1.7}
$$

In the same way, one can define higher-order four-tensors (of order three, four, etc.).

In the Theory of Special Relativity, a special role is played by the *totally antisymmetric fourth-order unit four-pseudotensor* $\epsilon_{\mu\nu\lambda\rho}$. It is defined as being $+1$, -1 or 0, as indices are even, odd, or repeated-index permutation of $1, 2, 3, 4$, respectively. The quantities $\epsilon_{\mu\nu\lambda\rho}$ form a *four-pseudotensor* (sometimes called *four-axial tensor*) because they exhibit a four-tensor behaviour under translations, rotations, and special Lorentz transformations, but are not invariant under parity inversions.

It can be shown that

$$
\epsilon_{\mu\nu\lambda\rho}\epsilon_{\sigma\kappa\xi\zeta} =
\begin{vmatrix}
\delta_{\mu\sigma} & \delta_{\mu\kappa} & \delta_{\mu\xi} & \delta_{\mu\zeta} \\
\delta_{\nu\sigma} & \delta_{\nu\kappa} & \delta_{\nu\xi} & \delta_{\nu\zeta} \\
\delta_{\lambda\sigma} & \delta_{\lambda\kappa} & \delta_{\lambda\xi} & \delta_{\lambda\zeta} \\
\delta_{\rho\sigma} & \delta_{\rho\kappa} & \delta_{\rho\xi} & \delta_{\rho\zeta}
\end{vmatrix}.
\tag{C.1.8}
$$

It then results

$$
\begin{aligned}
\epsilon_{\mu\nu\lambda\rho}\epsilon_{\mu\nu\sigma\kappa} &= 2!\left(\delta_{\lambda\sigma}\delta_{\rho\kappa} - \delta_{\lambda\kappa}\delta_{\rho\sigma}\right), \\
\epsilon_{\mu\nu\lambda\rho}\epsilon_{\mu\nu\lambda\sigma} &= 3!\,\delta_{\rho\sigma} = 6\,\delta_{\rho\sigma}, \\
\epsilon_{\mu\nu\lambda\rho}\epsilon_{\mu\nu\lambda\rho} &= 4! = 24.
\end{aligned}
\tag{C.1.9}
$$

With the help of $\epsilon_{\mu\nu\lambda\rho}$ one can define the *dual* of an antisymmetric tensor. If $A_{\mu\nu}$ is a second-order antisymmetric four-tensor, then $A_{\mu\nu}$ and the second-order four-pseudotensor $\tilde{A}_{\mu\nu} = \frac{1}{2}\epsilon_{\mu\nu\lambda\rho}A_{\lambda\rho}$ are called *dual* to each other. Similarly, the third-order antisymmetric four-pseudotensor $A_{\mu\nu\lambda} = \epsilon_{\mu\nu\lambda\rho}A_\rho$ and the four-vector A_ρ are dual to each other.

There are four possible types of manifolds that can be embedded in the four-space, which means that there exist four types of integrals:

1) *Line integral*, when the integration is performed along a curve, with the arc element dx_μ as element of integration.

2) *Integral over a two-dimensional surface.*

In E_3, as surface element, one uses $d\tilde{S}_i$. This is the integration differential and it is the dual of the second-order antisymmetric tensor dS_{ik}:

$$d\tilde{S}_i = \frac{1}{2!}\epsilon_{ijk}\, dS_{jk}.$$

From the geometric point of view, $d\tilde{S}_i$ is a vector orthogonal to the considered surface element and having the same modulus as the elementary area of that surface element.

In the four-dimensional space,

$$d\tilde{S}_{\mu\nu} = \frac{1}{2}\epsilon_{\mu\nu\lambda\rho}dS_{\lambda\rho}, \quad \mu,\nu,\lambda,\rho = 1,2,3,4, \tag{C.1.10}$$

therefore the dual of the second-order four-tensor $dS_{\mu\nu}$ is a second-order four-pseudotensor and, geometrically, it represents a surface element equal to $dS_{\mu\nu}$ and orthogonal to it.

3) *Integral over a three-dimensional hypersurface.* In three dimensions, the volume element is constructed as the mixed product of three arc elements corresponding to three coordinate directions which intersect at a point. In four-dimensions, as hypersurface element one defines the three-order antisymmetric four-tensor $dS_{\mu\nu\lambda}$, together with its dual $d\tilde{S}_\mu$:

$$\begin{cases} d\tilde{S}_\mu = \dfrac{1}{3!}\epsilon_{\mu\nu\lambda\rho}\, dS_{\nu\lambda\rho}, \\[2mm] dS_{\mu\nu\lambda} = \epsilon_{\mu\nu\lambda\rho}\, d\tilde{S}_\rho. \end{cases} \tag{C.1.11}$$

Geometrically, the four-vector $d\tilde{S}_\mu$ is orthogonal to the hypersurface element, and has the modulus equal to the "area" of this element. In particular, $dS_0 = dx\,dy\,dz$ is the projection of the hypersurface element on the hyperplane $x_0 = $ const.

4) *Integral over the four-dimensional hypervolume.* The "volume" element in this case is

$$d\Omega = dx_1\, dx_2\, dx_3\, dx_4 = dS_\mu\, dx_\mu \text{ (no summation over } \mu), \tag{C.1.12}$$

where the hypersurface element dS_μ is orthogonal to the arc element dx_μ.

Using these notions, one can generalize the divergence theorem (or the Green-Gauss-Ostrogradski's theorem) and the Stokes-Ampère theorem in Minkowski space. In view of Eq. (C.1.12), we have

$$\oint A_\mu dS_\mu = \int \frac{\partial A_\mu}{\partial x_\mu}d\Omega, \tag{C.1.13}$$

which generalizes in four dimensions the divergence (Green-Gauss) theorem. Formally, the integral extended over a hypersurface can be transformed into an integral over the four-domain enclosed by the hypersurface by substituting

$$dS_\mu \to d\Omega \, \frac{\partial}{\partial x_\mu}, \quad \mu = 1, 2, 3, 4. \tag{C.1.14}$$

In a similar way, an integral over a two-dimensional surface, of element $dS_{\mu\nu} = dx_\mu \, dx_\nu$, can be transformed according to

$$\int \frac{\partial A_\mu}{\partial x_\nu} dS_{\mu\nu} = \int \frac{\partial A_\mu}{\partial x_\nu} \, dx_\mu \, dx_\nu = \oint A_\mu dx_\mu, \tag{C.1.15}$$

meaning that the circulation along a closed curve in four-dimensional space can be transformed into an integral over the two-dimensional surface bounded by the curve, by substituting

$$dx_\mu \to dS_{\mu\nu} \frac{\partial}{\partial x_\nu}. \tag{C.1.16}$$

After some calculation handling in Eq. (C.1.15), one obtains

$$\oint A_\mu dx_\mu = \frac{1}{2} \int \left(\frac{\partial A_\nu}{\partial x_\mu} - \frac{\partial A_\mu}{\partial x_\nu} \right) dS_{\mu\nu}. \tag{C.1.17}$$

One can also establish a formula connecting an integral over a two-dimensional surface, and the boundary three-dimensional surface. As an example, if $A_{\mu\nu}$ is a second-order antisymmetric four-tensor, we may write

$$\int \frac{\partial A_{\mu\nu}}{\partial x_\nu} dS_\mu = \frac{1}{2} \int \left(\frac{\partial A_{\mu\nu}}{\partial x_\nu} dS_\mu + \frac{\partial A_{\nu\mu}}{\partial x_\mu} dS_\nu \right)$$
$$= \frac{1}{2} \int \left(\frac{\partial A_{\mu\nu}}{\partial x_\nu} dS_\mu - \frac{\partial A_{\mu\nu}}{\partial x_\mu} dS_\nu \right).$$

If one denotes

$$d\tilde{S}_{\mu\nu} \to \frac{1}{2} \left(dS_\mu \frac{\partial}{\partial x_\nu} - dS_\nu \frac{\partial}{\partial x_\mu} \right), \tag{C.1.18}$$

it follows that

$$\int \frac{\partial A_{\mu\nu}}{\partial x_\nu} dS_\mu = \int A_{\mu\nu} d\tilde{S}_{\mu\nu}. \tag{C.1.19}$$

C.2   Hyperbolic Representation

Although the Euclidean-complex representation has the advantage of not having to distinguish between covariant and contravariant components of tensors, it is nowadays less and less used. Unlike the hyperbolic representation, in the Euclidean-complex representation Lorentz boosts get the geometric interpretation of rotations in Minkowski space, but at the cost of considering imaginary rotation angles. Therefore, the hyperbolic representation is preferred, because in this case the angles that allow writing Lorentz boosts as rotations in Minkowski space are real.

As we saw in the previous paragraph, in Euclidean-complex representation the coordinates of an event are chosen in the form $x_1 = x$, $x_2 = y$, $x_3 = z$, $x_4 = ict$, but the same event (like any other) can be also defined in Minkowski space by the choice $x^0 = ct$, $x^1 = x$, $x^2 = y$, $x^3 = z$, in which case the metric

$$ds^2 = dx_\mu \, dx^\mu = g_{\mu\nu} dx^\mu dx^\nu = c^2 dt^2 - dx^2 - dy^2 - dz^2 \qquad (C.2.20)$$

gives

$$\begin{cases} g_{00} = +1, \\ g_{11} = g_{22} = g_{33} = -1, \\ g_{\mu\nu} = 0, \ \text{if } \mu \neq \nu. \end{cases} \qquad (C.2.21)$$

A system of coordinates in a pseudo-Euclidean space (*e.g.*, Minkowski space-time) in which the line element has the form

$$ds^2 = \sum e_\mu^2 dx_\mu^2,$$

where $e_\mu = \pm 1$, is a *Galilean coordinate system*. So, above we have a Galilean coordinate system, while this representation of Minkowski space is called *hyperbolic*.

Relations (C.2.21) show that in such a representation one makes distinction between contravariance and covariance indices. For example, in case of a four-vector,

$$A_\mu = g_{\mu\nu} A^\nu \Rightarrow \begin{cases} A_0 = g_{00} A^0 = A^0, \\ A_i = g_{ii} A^i = -A^i, \quad \text{(no summation).} \end{cases} \qquad (C.2.22)$$

The square of a four-vector is

$$A^\mu A_\mu = A^0 A_0 + A^i A_i = A^0 A^0 - A^i A^i = \text{invariant.} \qquad (C.2.23)$$

In Minkowski space, the components of a contravariant four-vector transform according to Eq. (A.2.7), where the transformation matrix is

$$(\overline{x}_\mu^\nu) \equiv \Lambda^\nu{}_\mu = \begin{pmatrix} \Gamma & -\frac{V}{c}\Gamma & 0 & 0 \\ -\frac{V}{c}\Gamma & \Gamma & 0 & 0 \\ 0 & 0 & 1 & 0 \\ 0 & 0 & 0 & 1 \end{pmatrix}. \qquad (C.2.24)$$

If, for example, the relative motion of frames S and S' takes place along

$Ox \equiv O'x'$-axis, then the contravariant components of A^μ transform according to

$$\begin{cases} A'^0 = \Gamma\left(A^0 - \dfrac{V}{c}A^1\right), \\[2mm] A'^1 = \Gamma\left(-\dfrac{V}{c}A^0 + A^1\right), \\[2mm] A'^2 = A^2, \\[2mm] A'^3 = A^3, \end{cases} \tag{C.2.25}$$

while the covariant components obey the rule

$$\begin{cases} A'_0 = \Gamma\left(A_0 + \dfrac{V}{c}A_1\right), \\[2mm] A'_1 = \Gamma\left(\dfrac{V}{c}A_0 + A_1\right), \\[2mm] A'_2 = A_2, \\[2mm] A'_3 = A_3. \end{cases} \tag{C.2.26}$$

In view of Eqs. (C.2.21) and (A.4.27), we can write

$$g_{\mu\nu} = g^{\mu\nu}, \quad \mu, \nu = 0, 1, 2, 3. \tag{C.2.27}$$

In Galilean coordinates, both contravariant and covariant components of a four-vector are *real*.

Let us have a look over the relations written in previous paragraph in Euclidean-complex representation. At the beginning, we choose the contravariant Levi-Civita permutation symbol as

$$\epsilon^{0123} = +1. \tag{C.2.28}$$

It then follows that

$$\epsilon_{\mu\nu\lambda\rho} = g_{\mu\sigma}g_{\nu\kappa}g_{\lambda\xi}g_{\rho\zeta}\epsilon^{\sigma\kappa\xi\zeta} = -\epsilon^{\mu\nu\lambda\rho}, \tag{C.2.29}$$

because irrespective of the order of the four different indices, the product of the four metric tensors is -1. Then we have

$$\epsilon^{\mu\nu\lambda\rho}\epsilon_{\sigma\kappa\xi\zeta} = - \begin{vmatrix} \delta^\mu_\sigma & \delta^\mu_\kappa & \delta^\mu_\xi & \delta^\mu_\zeta \\ \delta^\nu_\sigma & \delta^\nu_\kappa & \delta^\nu_\xi & \delta^\nu_\zeta \\ \delta^\lambda_\sigma & \delta^\lambda_\kappa & \delta^\lambda_\xi & \delta^\lambda_\zeta \\ \delta^\rho_\sigma & \delta^\rho_\kappa & \delta^\rho_\xi & \delta^\rho_\zeta \end{vmatrix}. \tag{C.2.30}$$

Summation over two, three and four pairs of indices gives

$$\begin{cases} \epsilon^{\mu\nu\lambda\rho}\epsilon_{\sigma\kappa\lambda\rho} = -2!\,\delta^{\mu\nu}_{\sigma\kappa} = -2\left(\delta^\mu_\sigma\delta^\nu_\kappa - \delta^\mu_\kappa\delta^\nu_\sigma\right), \\[2mm] \epsilon^{\mu\nu\lambda\rho}\epsilon_{\sigma\nu\lambda\rho} = -3!\,\delta^\mu_\sigma = -6\,\delta^\mu_\sigma, \\[2mm] \epsilon^{\mu\nu\lambda\rho}\epsilon_{\mu\nu\lambda\rho} = -4! = -24. \end{cases} \tag{C.2.31}$$

If $A^{\mu\nu}$ is a second-order antisymmetric four-tensor, its dual is the second-order four-pseudotensor

$$\tilde{A}^{\mu\nu} = \frac{1}{2}\epsilon^{\mu\nu\lambda\rho}A_{\lambda\rho}.$$

The product

$$\tilde{A}^{\mu\nu}A_{\mu\nu}$$

is a pseudoscalar. In the same way, we observe that the antisymmetric third-order four-pseudotensor $\epsilon^{\mu\nu\lambda\rho}A_\rho$ and the four-vector A^μ are dual to each other.

The symbol of partial derivative $\frac{\partial}{\partial x^\mu}$ is a covariant four-vector, while the symbol $\frac{\partial}{\partial x_\mu}$ is a contravariant four-vector.

Consider now, as we already did in the Euclidean-complex representation of Minkowski space, the four possible types of integrals and the relations between them.

1) *Line integral*, with the arc element dx^μ.

2) *Integral over a two-dimensional surface*, with the surface element

$$d\tilde{S}^{\mu\nu} = \frac{1}{2}\epsilon^{\mu\nu\lambda\rho}dS_{\lambda\rho}, \tag{C.2.32}$$

which, geometrically, is an area element orthogonal (and quantitatively equal) to $dS_{\mu\nu}$.

3) *Integral over a three-dimensional hypersurface*, of surface element

$$\begin{cases} d\tilde{S}^\mu = -\frac{1}{3!}\epsilon^{\mu\nu\lambda\rho}dS_{\nu\lambda\rho}, \\ dS_{\mu\nu\lambda} = \epsilon_{\mu\nu\lambda\rho}dS^\rho, \end{cases} \tag{C.2.33}$$

such as

$$d\tilde{S}^0 = -dS_{123} = dS^{123}, \text{ etc.} \tag{C.2.34}$$

The four-vector $d\tilde{S}^\mu$ has its modulus equal to the area of the corresponding hypersurface element, being orthogonal to it.

4) *Integral over a four-dimensional domain*, the elementary hypervolume being

$$d\Omega = dx^0 dx^1 dx^2 dx^3 = dx^\mu dS_\mu \quad \text{(no summation)}, \tag{C.2.35}$$

where the line element dx^μ and the hypersurface element dS_μ are orthogonal.

The divergence (Green-Gauss) theorem in this representation is

$$\oint A^\mu dS_\mu = \int \frac{\partial A^\mu}{\partial x^\mu}d\Omega, \tag{C.2.36}$$

while the Stokes-Ampère theorem takes the form

$$\oint A_\mu dx^\mu = \frac{1}{2}\int\left(\frac{\partial A_\nu}{\partial x^\mu} - \frac{\partial A_\mu}{\partial x^\nu}\right)dS^{\mu\nu}. \tag{C.2.37}$$

Finally, the generalization of Eq. (C.1.19) in the hyperbolic representation of Minkowski space is

$$\int A^{\mu\nu}d\tilde{S}_{\mu\nu} = \frac{1}{2}\int\left(\frac{\partial A^{\mu\nu}}{\partial x^\nu}dS_\mu - \frac{\partial A^{\mu\nu}}{\partial x^\mu}dS_\nu\right) = \int\frac{\partial A^{\mu\nu}}{\partial x^\nu}dS_\mu. \tag{C.2.38}$$

C.3   Representation in Curvilinear Coordinates

One can represent the Minkowski space in the most general manner in a system of general curvilinear coordinates. These considerations are especially useful in the study of the gravitational field which, according to General Relativity, manifests itself in the curvature of space-time. However, locally, the space-time is flat (of Minkowski type) and the coordinate systems are in their turn locally defined. In general curvilinear coordinates, the metric tensor $g_{\mu\nu}$ depends on the coordinates.

Let us first express the law of transformation of the Levi-Civita symbol when one passes from the Galilean coordinates x^μ to an arbitrary set of general curvilinear coordinates,

$$x'^\mu = x'^\mu(x^\nu), \quad \mu, \nu = 0, 1, 2, 3.$$

According to the rule of transformation, we have

$$\epsilon'_{\mu\nu\lambda\rho} = \frac{\partial x^\sigma}{\partial x'^\mu} \frac{\partial x^\kappa}{\partial x'^\nu} \frac{\partial x^\xi}{\partial x'^\lambda} \frac{\partial x^\zeta}{\partial x'^\rho} \epsilon_{\sigma\kappa\xi\zeta}, \tag{C.3.39}$$

where $\epsilon_{\sigma\kappa\xi\zeta}$ is defined in the Galilean coordinates x^μ, and $\epsilon'_{\mu\nu\lambda\rho}$ in the curvilinear coordinates x'^μ.

If A_μ^ν, $\mu, \nu = 0, 1, 2, 3$, is an arbitrary second-order mixed four-tensor, it can be shown that

$$A_\mu^\sigma A_\nu^\kappa A_\lambda^\xi A_\rho^\upsilon \epsilon_{\sigma\kappa\xi\upsilon} = A \, \epsilon_{\mu\nu\lambda\rho}, \tag{C.3.40}$$

where $A = \det(A_\mu^\nu)$. Relation (C.3.40) is a generalization in four dimensions of Eq. (B.2.23). Then we may write

$$\epsilon'_{\mu\nu\lambda\rho} = \det\left(\frac{\partial x^\mu}{\partial x'^\nu}\right) \epsilon_{\mu\nu\lambda\rho} = \frac{1}{J} \epsilon_{\mu\nu\lambda\rho},$$

where

$$J = \frac{\partial(x'^0, x'^1, x'^2, x'^3)}{\partial(x^0, x^1, x^2, x^3)} \tag{C.3.41}$$

is the functional determinant of the transformation $x^\mu \to x'^\mu$.

Using the transformation rule, we have also

$$g'_{\mu\nu} = \frac{\partial x^\lambda}{\partial x'^\mu} \frac{\partial x^\rho}{\partial x'^\nu} \eta_{\lambda\rho},$$

where $\eta_{\lambda\rho} = \text{diag}\,(1, -1, -1, -1)$ is the *Minkowski metric tensor*. If we take the determinant of the above relation, we find

$$g = \frac{1}{J^2} \det(\eta_{\mu\nu}),$$

where $g = \det(g'_{\mu\nu})$. Since $\det(\eta_{\mu\nu}) = -1$, we have

$$J = \frac{1}{\sqrt{-g}}. \tag{C.3.42}$$

We then define the antisymmetric unit four-tensor in curvilinear coordinates by

$$\delta_{\mu\nu\lambda\rho} = \sqrt{-g}\, \epsilon_{\mu\nu\lambda\rho}. \tag{C.3.43}$$

The transformation rule of the contravariant components $\epsilon^{\mu\nu\lambda\rho}$ is found in a similar way:

$$\epsilon'^{\mu\nu\lambda\rho} = \frac{\partial x'^\mu}{\partial x^\sigma} \frac{\partial x'^\nu}{\partial x^\kappa} \frac{\partial x'^\lambda}{\partial x^\xi} \frac{\partial x'^\rho}{\partial x^\zeta} \epsilon^{\sigma\kappa\xi\zeta} = J\,\epsilon^{\mu\nu\lambda\rho},$$

that is

$$\delta^{\mu\nu\lambda\rho} = \frac{1}{\sqrt{-g}}\, \epsilon^{\mu\nu\lambda\rho}, \tag{C.3.44}$$

with $\epsilon'^{\mu\nu\lambda\rho} = \delta^{\mu\nu\lambda\rho}$. In view of Eqs. (C.3.43) and (C.3.44), relation (C.2.29) yields

$$\delta_{\mu\nu\lambda\rho} = g\,\delta^{\mu\nu\lambda\rho}. \tag{C.3.45}$$

If $g = -1$, we find the Galilean formula (C.2.29).

Let us now write the transformation rule of the four-dimensional elementary volume. In Galilean coordinates,

$$d\Omega = dx^0 dx^1 dx^2 dx^3$$

is an invariant. In the curvilinear coordinates, x'^μ, the element of four-volume is

$$d\Omega' = J d\Omega.$$

Since the four-volume must be an invariant, in the new coordinates x'^μ not $d\Omega'$, but $\sqrt{-g}\, d\Omega'$ has to be used as integration (hyper)volume element:

$$d\Omega \to \frac{1}{J}\, d\Omega' = \sqrt{-g}\, d\Omega'. \tag{C.3.46}$$

If, as a result of integration over Ω of the quantity $\sqrt{-g}\, \varphi$, with φ a scalar, one obtains an invariant, then $\sqrt{-g}\, \varphi$ is called a *scalar density*. In the same way are defined the notions of *vector density* $\sqrt{-g}\, A^\mu$ and *second-order tensor density* $\sqrt{-g}\, T^{\mu\nu}$, respectively.

The elementary three-dimensional surface is

$$\sqrt{-g}\, dS_\mu = -\frac{1}{3!}\epsilon_{\mu\nu\lambda\rho}\sqrt{-g}\, dS^{\nu\lambda\rho} = -\frac{1}{3!}\delta_{\mu\nu\lambda\rho}\, dS^{\nu\lambda\rho}. \tag{C.3.47}$$

Similarly, the two-dimensional surface element is

$$\sqrt{-g}\, d\tilde{S}_{\mu\nu} = \frac{1}{2!}\sqrt{-g}\epsilon_{\mu\nu\lambda\rho}\, dS^{\lambda\rho} = \frac{1}{2!}\delta_{\mu\nu\lambda\rho}\, dS^{\lambda\rho}. \tag{C.3.48}$$

C.4   Differential Operators in General Curvilinear Coordinates

In Special Relativity the fundamental equations of conservation involve the vector or tensor four-divergence "operators", written in terms of the usual derivatives. On curved space-times the usual partial derivatives with respect to coordinates have to be replaced by *covariant derivatives*. Here are the most important differential "operators", expressed in curvilinear coordinates.

As an auxiliary step, let us calculate the derivative

$$\frac{\partial}{\partial x^\nu}(\sqrt{-g}) = -\frac{1}{2\sqrt{-g}}\frac{\partial g}{\partial x^\nu} = \frac{\sqrt{-g}}{2g}\frac{\partial g}{\partial x^\nu}. \qquad (C.4.49)$$

We have

$$\frac{\partial g}{\partial x^\nu} = \frac{\partial}{\partial x^\nu}\begin{vmatrix} g_{00} & \cdots & g_{03} \\ \vdots & \ddots & \vdots \\ g_{30} & \cdots & g_{33} \end{vmatrix}$$

$$= \begin{vmatrix} \frac{\partial g_{00}}{\partial x^\nu} & \cdots & \frac{\partial g_{03}}{\partial x^\nu} \\ \vdots & \ddots & \vdots \\ g_{30} & \cdots & g_{33} \end{vmatrix} + \ldots + \begin{vmatrix} g_{00} & \cdots & g_{03} \\ \vdots & \ddots & \vdots \\ \frac{\partial g_{30}}{\partial x^\nu} & \cdots & \frac{\partial g_{33}}{\partial x^\nu} \end{vmatrix}$$

$$= \frac{\partial g_{0\sigma}}{\partial x^\nu}G^{0\sigma} + \ldots + \frac{\partial g_{3\sigma}}{\partial x^\nu}G^{3\sigma} = \frac{\partial g_{\rho\sigma}}{\partial x^\nu}G^{\rho\sigma}, \qquad (C.4.50)$$

where $G^{\rho\sigma}$ is the algebraic complement of the element $g_{\rho\sigma}$ in the determinant g. By means of Eq. (A.4.29), we also have

$$\frac{\partial}{\partial x^\nu}\left(\sqrt{-g}\right) = \frac{\sqrt{-g}}{2}\frac{G^{\rho\sigma}}{g}g_{\rho\sigma,\nu} = \frac{\sqrt{-g}}{2}g^{\rho\sigma}g_{\rho\sigma,\nu},$$

which gives

$$\frac{1}{2}g^{\rho\sigma}g_{\rho\sigma,\nu} = \frac{1}{\sqrt{-g}}\frac{\partial}{\partial x^\nu}\left(\sqrt{-g}\right). \qquad (C.4.51)$$

On the other hand, if in

$$\Gamma^\mu_{\nu\lambda} = \frac{1}{2}g^{\mu\sigma}\left(\frac{\partial g_{\sigma\lambda}}{\partial x^\nu} + \frac{\partial g_{\nu\sigma}}{\partial x^\lambda} - \frac{\partial g_{\lambda\nu}}{\partial x^\sigma}\right)$$

we set $\mu = \lambda$, one obtains

$$\Gamma^\lambda_{\nu\lambda} = \frac{1}{2}g^{\sigma\lambda}g_{\sigma\lambda,\nu} + \frac{1}{2}g^{\sigma\lambda}\left(\frac{\partial g_{\nu\sigma}}{\partial x^\lambda} - \frac{\partial g_{\lambda\nu}}{\partial x^\sigma}\right). \qquad (C.4.52)$$

Since the metric tensor is symmetric, and the expression in parentheses is antisymmetric in the same indices (σ and λ), the last term vanishes. Relations (C.4.51) and (C.4.52) then yield

$$\Gamma^{\lambda}_{\nu\lambda} = \frac{1}{\sqrt{-g}} \frac{\partial}{\partial x^{\nu}} \left(\sqrt{-g} \right). \tag{C.4.53}$$

This formula is of help in defining the covariant four-divergence, four-dimensional gradient, four-dimensional curl, and four-dimensional operators Laplacian and d'Alembertian in general curvilinear coordinates.

C.4.1 Four-Divergence

Let us consider the contravariant vector A^{ν}, and take its covariant derivative

$$\nabla_{\mu} A^{\nu} \equiv A^{\nu}_{;\mu} = A^{\nu}_{,\mu} + \Gamma^{\nu}_{\mu\lambda} A^{\lambda}. \tag{C.4.54}$$

Setting now $\nu = \mu$ and using Eq. (C.4.53), we find

$$A^{\mu}_{;\mu} = A^{\mu}_{,\mu} + \Gamma^{\mu}_{\mu\lambda} A^{\lambda} = A^{\mu}_{,\mu} + \frac{1}{\sqrt{-g}} \frac{\partial}{\partial x^{\lambda}} \left(\sqrt{-g} \right) A^{\lambda},$$

or, if one suitably changes the summation index in the last term,

$$A^{\mu}_{;\mu} = \frac{1}{\sqrt{-g}} \frac{\partial}{\partial x^{\mu}} \left(\sqrt{-g} A^{\mu} \right). \tag{C.4.55}$$

This is the *covariant four-divergence* of contravariant vector A^{μ}.

Let us now consider the contravariant second-order four-tensor $A^{\mu\lambda}$, and take its covariant derivative:

$$A^{\mu\lambda}_{;\nu} = A^{\mu\lambda}_{,\nu} + \Gamma^{\lambda}_{\sigma\nu} A^{\mu\sigma} + \Gamma^{\mu}_{\sigma\nu} A^{\sigma\lambda}. \tag{C.4.56}$$

Setting now $\lambda = \nu$ and using Eq. (C.4.53), we still have

$$\nabla_{\nu} A^{\mu\nu} \equiv A^{\mu\nu}_{;\nu} = A^{\mu\nu}_{,\nu} + \Gamma^{\nu}_{\sigma\nu} A^{\mu\sigma} + \Gamma^{\mu}_{\sigma\nu} A^{\sigma\nu}$$

$$= A^{\mu\nu}_{,\nu} + \frac{1}{\sqrt{-g}} \frac{\partial}{\partial x^{\nu}} \left(\sqrt{-g} \right) A^{\mu\nu} + \Gamma^{\mu}_{\sigma\nu} A^{\sigma\nu}.$$

In the second term on the r.h.s. we made a convenient change of summation indices. Therefore

$$A^{\mu\nu}_{;\nu} = \frac{1}{\sqrt{-g}} \frac{\partial}{\partial x^{\nu}} \left(\sqrt{-g} A^{\mu\nu} \right) + \Gamma^{\mu}_{\sigma\nu} A^{\sigma\nu}, \tag{C.4.57}$$

which is the *covariant four-divergence* of contravariant second-order four-tensor $A^{\mu\nu}$.

If the second-order four-tensor $A^{\mu\nu}$ is antisymmetric, and recalling that $\Gamma^{\mu}_{\sigma\nu}$ is symmetric in the lower indices, in this case we are left with

$$A^{\mu\nu}_{;\nu} = \frac{1}{\sqrt{-g}} \frac{\partial}{\partial x^{\nu}} \left(\sqrt{-g} A^{\mu\nu} \right). \tag{C.4.58}$$

C.4.2   Four-Dimensional Gradient

Consider the scalar function Φ. Its covariant derivative reduces, obviously, to the usual derivative, which is a covariant vector:

$$\Phi_{;\nu} = \Phi_{,\nu} \equiv A_\nu. \tag{C.4.59}$$

The contravariant components of A_ν are

$$A^\mu = g^{\mu\nu} A_\nu = g^{\mu\nu} \frac{\partial \Phi}{\partial x^\nu}. \tag{C.4.60}$$

C.4.3   Four-Dimensional d'Alembertian

Introducing Eq. (C.4.60) into Eq. (C.4.55), we obtain the *d'Alembertian* of Φ:

$$A^\mu_{;\mu} = \Box\,\Phi = \frac{1}{\sqrt{-g}} \frac{\partial}{\partial x^\mu} \left(\sqrt{-g} g^{\mu\nu} \frac{\partial \Phi}{\partial x^\nu} \right). \tag{C.4.61}$$

This is the most straightforward method to write the d'Alembertian (or the Laplacian) in any coordinate system.

The d'Alembertian is also known as the *Laplace-Beltrami operator*. Rigorously speaking, the Laplace-Beltrami operator given by Eq. (C.4.61) is the generalization of the Laplacian to an elliptic operator defined on a Riemannian manifold, the "usual" d'Alembertian

$$\Box = \frac{1}{v^2} \frac{\partial^2}{\partial t^2} - \Delta\,,$$

being the particular form of the Laplace-Beltrami operator in the Minkowski space.

C.4.4   Four-Dimensional Curl

Consider the covariant four-vector A_ν and form the covariant antisymmetric second-order four-tensor

$$F_{\mu\nu} = A_{\nu;\mu} - A_{\mu;\nu}.$$

But

$$A_{\nu;\mu} - A_{\mu;\nu} = \left(A_{\nu,\mu} - \Gamma^\lambda_{\mu\nu} A_\lambda \right) - \left(A_{\mu,\nu} - \Gamma^\lambda_{\nu\mu} A_\lambda \right) = A_{\nu,\mu} - A_{\mu,\nu}.$$

Consequently,

$$F_{\mu\nu} = A_{\nu;\mu} - A_{\mu;\nu} = A_{\nu,\mu} - A_{\mu,\nu}. \tag{C.4.62}$$

The quantities in Eq. (C.4.62) represent a four-dimensional curl, generalizing the notion of three-dimensional curl. One observes that the quantities in Eq. (C.4.62) do not depend on the metric.

D

Appendix D: Curvilinear Coordinates in the Physical Space

D.1 Elements of Arc Length, Area and Volume

Let $\vec{r}$ be the radius-vector of some point P and x^i, $i = 1, 2, 3$, its Cartesian coordinates. Suppose that there exist three real independent parameters x'^i, so that

$$x^i = x^i(x'^k), \quad i, k = 1, 2, 3. \tag{D.1.1}$$

To be (at least locally) reversible, *i.e.*, to have

$$x'^i = x'^i(x^k), \quad i, k = 1, 2, 3, \tag{D.1.2}$$

as well, it is necessary that the determinant of the Jacobian matrix be nonvanishing,

$$J = \det\left[\frac{\partial(x^1, x^2, x^3)}{\partial(x'^1, x'^2, x'^3)}\right] \neq 0. \tag{D.1.3}$$

If we fix the values of two parameters, say x'^2 and x'^3, we obtain the curve $x'^1 = variable$. In the same way one can obtain the curves $x'^2 = variable$ and $x'^3 = variable$. Thus, through each point of space pass three coordinate curves. The parameters x'^i are called *curvilinear coordinates* of the point P.

If at the point P (or any other point) the vectors

$$\vec{e}_i = \frac{\partial \vec{r}}{\partial x'^i}, \quad i = 1, 2, 3,$$

tangent to the three coordinate curves form a right orthogonal trihedron, then x'^1, x'^2, x'^3 form an *orthogonal coordinate system*.

D.1.1 Element of Arc Length

An elementary displacement of the point P is written as

$$d\vec{r} = \frac{\partial \vec{r}}{\partial x'^i}\, dx'^i = \vec{e}_i\, dx'^i. \tag{D.1.4}$$

DOI: 10.1201/9781003402602-D

Condition expressed by Eq. (D.1.3) shows that the three vectors $\vec{e}_1$, $\vec{e}_2$ and $\vec{e}_3$ are linearly independent, therefore they form a *basis*. Indeed,

$$(\vec{e}_1, \vec{e}_2, \vec{e}_3) = \left(\frac{\partial \vec{r}}{\partial x'^1}, \frac{\partial \vec{r}}{\partial x'^2}, \frac{\partial \vec{r}}{\partial x'^3} \right) = J,$$

and the determinant of this matrix is non-zero.

The squared arc element (the metric) is

$$ds^2 = d\vec{r} \cdot d\vec{r} = \left(\vec{e}_i dx'^i \right) \cdot \left(\vec{e}_k dx'^k \right) = g_{ik}\, dx'^i\, dx'^k, \qquad (D.1.5)$$

where g_{ik} is the covariant metric tensor.

If we fix x'^2 and x'^3, we obtain the elementary arc length on the coordinate curve $x'^1 = variable$:

$$(d_1 s)^2 = g_{11} \left(dx'^1 \right)^2,$$

that is

$$(d_1 s) = \sqrt{g_{11}} \left(dx'^1 \right).$$

In a similar way, we find two more relations. Therefore

$$\begin{cases} (d_1 s) = \sqrt{g_{11}} \left(dx'^1 \right), \\ (d_2 s) = \sqrt{g_{22}} \left(dx'^2 \right), \\ (d_3 s) = \sqrt{g_{33}} \left(dx'^3 \right). \end{cases} \qquad (D.1.6)$$

The elementary arc length is then

$$d\vec{s} = \left(\sqrt{g_{11}}\, dx'^1 \right) \vec{u}_1 + \left(\sqrt{g_{22}}\, dx'^2 \right) \vec{u}_2 + \left(\sqrt{g_{33}}\, dx'^3 \right) \vec{u}_3, \qquad (D.1.7)$$

where $\vec{u}_1$, $\vec{u}_2$ and $\vec{u}_3$ are the unit vectors (versors) of the basis vectors $\vec{e}_1$, $\vec{e}_2$ and $\vec{e}_3$, respectively.

D.1.2   Area Element

The area element constructed, for example, on the length elements $d_1\vec{s}$ and $d_2\vec{s}$, is

$$\begin{cases} d\vec{S}_3 = d_1\vec{s} \times d_2\vec{s} = \dfrac{\partial \vec{r}}{\partial x'^1} \times \dfrac{\partial \vec{r}}{\partial x'^2} dx'^1\, dx'^2, \\ \left| d\vec{S}_3 \right| = \left| \vec{e}_1 \times \vec{e}_2 \right| dx'^1\, dx'^2. \end{cases} \qquad (D.1.8)$$

Thus, $d\vec{S}_3$ is orthogonal to the plane determined by $\vec{e}_1$ and $\vec{e}_2$ (but not necessarily pointing in the $\vec{e}_3$ direction).

D.1.3   Volume Element

The volume element is found by taking the scalar triple product of the three line elements $d_1\vec{s}$, $d_2\vec{s}$ and $d_3\vec{s}$, i.e.,

$$d\tau = \left(d_1\vec{s}, d_2\vec{s}, d_3\vec{s} \right) = \left(\vec{e}_1, \vec{e}_2, \vec{e}_3 \right) dx'^1\, dx'^2\, dx'^3 = \sqrt{g}\, dx'^1\, dx'^2\, dx'^3. \qquad (D.1.9)$$

D.2 First-Order Differential Operators in Curvilinear Coordinates

All formulas obtained in **Appendix C** for the "operators" divergence, gradient and curl are, obviously, valid also in three dimensions. Omitting the "prime" superscript for coordinates, we re-write the specified formulas:

$$\text{div } \vec{A} = \frac{1}{\sqrt{g}} \frac{\partial}{\partial x^i} \left(\sqrt{g} A^i \right), \quad i = 1, 2, 3, \tag{D.2.10}$$

$$(\text{grad } \Phi)_i = \frac{\partial \Phi}{\partial x^i}, \qquad (\text{grad } \Phi)^i = g^{ik} \frac{\partial \Phi}{\partial x^k}, \tag{D.2.11}$$

$$\Delta \Phi = \frac{1}{\sqrt{g}} \frac{\partial}{\partial x^i} \left(\sqrt{g} g^{ik} \frac{\partial \Phi}{\partial x^k} \right). \tag{D.2.12}$$

To express the curl, we consider the antisymmetric tensor (see Eq. (C.4.62)):

$$F_{ik} = A_{k,i} - A_{i,k}, \tag{D.2.13}$$

and let B_i be its associated dual (see Eq. (C.3.43)):

$$B_i = \frac{1}{2} \sqrt{g} \, \epsilon_{ijk} F^{jk}. \tag{D.2.14}$$

Therefore, the covariant and contravariant components of curl $\vec{A}$ are

$$\left(\text{curl } \vec{A} \right)_i = \frac{1}{2} \sqrt{g} \, \epsilon_{ijk} F^{jk},$$

$$\left(\text{curl } \vec{A} \right)^i = \frac{1}{2} \frac{1}{\sqrt{g}} \epsilon^{ijk} F_{jk}. \tag{D.2.15}$$

D.2.1 Differential Operators in Terms of Orthogonal Components

If the basis vectors $\vec{e}_1$, $\vec{e}_2$ and $\vec{e}_3$ form an orthogonal trihedron, the curvilinear coordinates are called *orthogonal*. In such a coordinate system the metric tensor g_{ik} is *diagonal*:

$$g_{ik} = \begin{cases} g_{ii}, & i = k \text{ (no summation)}, \\ 0, & i \neq k. \end{cases} \tag{D.2.16}$$

In curvilinear orthogonal coordinates one usually utilizes the *orthogonal* (or *physical*) components, instead of contravariant and covariant components of vectors and tensors. To find the transformation relations between the contravariant (covariant) components of a vector $\vec{A}$ and its orthogonal components, we represent the vector in the basis $\{\vec{e}_k\}_{k=1,2,3}$, then take the dot product by the unit vector of the coordinate curve on which the projection takes place. Denoting by $A_{(i)}$ the orthogonal components, we then have

$$A_{(i)} = \vec{A} \cdot \vec{u}_i = A^k \vec{e}_k \cdot \vec{u}_i = A^k \sqrt{g_{kk}} \, \vec{u}_k \cdot \vec{u}_i = A^k \sqrt{g_{kk}} \, \delta_{ik}$$

$$= \sqrt{g_{ii}} \, A^i = \sqrt{g_{ii}} \, g^{ii} A_i = \frac{1}{g_{ii}} \sqrt{g_{ii}} \, A_i = \frac{1}{\sqrt{g_{ii}}} A_i \text{ (no summation)},$$

or

$$\begin{cases} A^i = \dfrac{1}{\sqrt{g_{ii}}} A_{(i)}, \\ A_i = \sqrt{g_{ii}} \, A_{(i)} \,. \end{cases} \tag{D.2.17}$$

Usind Eq. (D.2.17), we are now able to write the differential "operators" in Eqs. (D.2.10)−(D.2.15) in curvilinear orthogonal coordinates:

$$\text{div } \vec{A} = \frac{1}{\sqrt{g}} \left[\frac{\partial}{\partial x^1} \left(\sqrt{\frac{g}{g_{11}}} A_{(1)} \right) + \frac{\partial}{\partial x^2} \left(\sqrt{\frac{g}{g_{22}}} A_{(2)} \right) + \frac{\partial}{\partial x^3} \left(\sqrt{\frac{g}{g_{33}}} A_{(3)} \right) \right],$$

or, since $g = g_{11} \, g_{22} \, g_{33}$,

$$\text{div } \vec{A} = \frac{1}{\sqrt{g}} \left[\frac{\partial}{\partial x^1} \left(\sqrt{g_{22}g_{33}} \, A_{(1)} \right) + \frac{\partial}{\partial x^2} \left(\sqrt{g_{33}g_{11}} \, A_{(2)} \right) \right.$$
$$\left. + \frac{\partial}{\partial x^3} \left(\sqrt{g_{11}g_{22}} \, A_{(3)} \right) \right]. \tag{D.2.18}$$

Also,

$$(\text{grad } \Phi)_i = \frac{\partial \Phi}{\partial x^i} = \sqrt{g_{ii}} \, (\text{grad } \Phi)_{(i)} \,,$$

$$(\text{grad } \Phi)^i = g^{ii} \frac{\partial \Phi}{\partial x^i} = \frac{1}{\sqrt{g_{ii}}} (\text{grad } \Phi)_{(i)} \,,$$

which are, in fact, one and the same relation,

$$(\text{grad } \Phi)_{(i)} = \frac{1}{\sqrt{g_{ii}}} \frac{\partial \Phi}{\partial x^i} \,. \tag{D.2.19}$$

Similarly,

$$\Delta \Phi = \frac{1}{\sqrt{g}} \left[\frac{\partial}{\partial x^1} \left(\sqrt{\frac{g_{22}g_{33}}{g_{11}}} \frac{\partial \Phi}{\partial x^1} \right) + \frac{\partial}{\partial x^2} \left(\sqrt{\frac{g_{33}g_{11}}{g_{22}}} \frac{\partial \Phi}{\partial x^2} \right) \right.$$
$$\left. + \frac{\partial}{\partial x^3} \left(\sqrt{\frac{g_{11}g_{22}}{g_{33}}} \frac{\partial \Phi}{\partial x^3} \right) \right]. \tag{D.2.20}$$

Finally,

$$(\text{curl } \vec{A})_{(i)} = \frac{1}{\sqrt{g_{ii}}} (\text{curl } \vec{A})_i = \frac{1}{2} \frac{1}{\sqrt{g_{ii}}} \epsilon_{ijk} F^{jk} \text{ (no summation over } i). \tag{D.2.21}$$

We shall present the detailed calculation of one component, the other two being obtained by cyclic permutations. For example,

$$(\text{curl } \vec{A})_{(1)} = \sqrt{g_{22}g_{33}}\, F^{23} = \sqrt{g_{22}g_{33}}\, g^{22}g^{33} F_{23}$$

$$= \frac{1}{\sqrt{g_{22}g_{33}}} \left(\frac{\partial A_3}{\partial x^2} - \frac{\partial A_2}{\partial x^3} \right)$$

$$= \frac{1}{\sqrt{g_{22}g_{33}}} \left[\frac{\partial}{\partial x^2} \left(\sqrt{g_{33}}\, A_{(3)} \right) - \frac{\partial}{\partial x^3} \left(\sqrt{g_{22}}\, A_{(2)} \right) \right].$$

The orthogonal components of curl $\vec{A}$ therefore are

$$\begin{cases} (\text{curl } \vec{A})_{(1)} = \dfrac{1}{\sqrt{g_{22}g_{33}}} \left[\dfrac{\partial}{\partial x^2} \left(\sqrt{g_{33}}\, A_{(3)} \right) - \dfrac{\partial}{\partial x^3} \left(\sqrt{g_{22}}\, A_{(2)} \right) \right], \\[2mm] (\text{curl } \vec{A})_{(2)} = \dfrac{1}{\sqrt{g_{33}g_{11}}} \left[\dfrac{\partial}{\partial x^3} \left(\sqrt{g_{11}}\, A_{(1)} \right) - \dfrac{\partial}{\partial x^1} \left(\sqrt{g_{33}}\, A_{(3)} \right) \right], \\[2mm] (\text{curl } \vec{A})_{(3)} = \dfrac{1}{\sqrt{g_{11}g_{22}}} \left[\dfrac{\partial}{\partial x^1} \left(\sqrt{g_{22}}\, A_{(2)} \right) - \dfrac{\partial}{\partial x^2} \left(\sqrt{g_{11}}\, A_{(1)} \right) \right]. \end{cases} \qquad \text{(D.2.22)}$$

Observation: Sometimes one uses the notation

$$\begin{cases} \sqrt{g_{11}} = h_1, \\ \sqrt{g_{22}} = h_2, \\ \sqrt{g_{33}} = h_3. \end{cases}$$

The quantities h_i, $i = 1, 2, 3$, are called *Lamé's coefficients*.

D.3 Differential Operators in Spherical and Cylindrical Coordinates

D.3.1 Spherical Coordinates

In spherical coordinates, the squared arc element (the metric) of the Euclidean three-dimensional space is given by

$$ds^2 = dr^2 + r^2\, d\theta^2 + r^2 \sin^2\theta\, d\varphi^2. \qquad \text{(D.3.23)}$$

We choose $x^1 = r$, $x^2 = \theta$ and $x^3 = \varphi$, and obtain the components of the metric tensor:

$$\begin{cases} g_{11} = \dfrac{1}{g^{11}} = 1, \\[2mm] g_{22} = \dfrac{1}{g^{22}} = r^2, \\[2mm] g_{33} = \dfrac{1}{g^{33}} = r^2 \sin^2\theta, \end{cases} \qquad \text{(D.3.24)}$$

which allow us to write the differential operators in spherical coordinates, as follows

$$\operatorname{div}\vec{A} = \frac{1}{r^2 \sin\theta}\left[\frac{\partial}{\partial r}\left(r^2 \sin\theta A_r\right) + \frac{\partial}{\partial\theta}\left(r \sin\theta A_\theta\right) + \frac{\partial}{\partial\varphi}\left(r A_\varphi\right)\right], \quad \text{(D.3.25)}$$

$$\operatorname{grad}\Phi = \frac{\partial\Phi}{\partial r}\vec{u}_r + \frac{1}{r}\frac{\partial\Phi}{\partial\theta}\vec{u}_\theta + \frac{1}{r\sin\theta}\frac{\partial\Phi}{\partial\varphi}\vec{u}_\varphi, \quad \text{(D.3.26)}$$

$$\Delta\Phi = \frac{1}{r^2}\left\{\frac{\partial}{\partial r}\left(r^2\frac{\partial\Phi}{\partial r}\right) + \frac{1}{\sin\theta}\left[\frac{\partial}{\partial\theta}\left(\sin\theta\frac{\partial\Phi}{\partial\theta}\right) + \frac{1}{\sin\theta}\frac{\partial^2\Phi}{\partial\varphi^2}\right]\right\}, \quad \text{(D.3.27)}$$

$$\operatorname{curl}\vec{A} = \frac{1}{r\sin\theta}\left[\frac{\partial}{\partial\theta}\left(A_\varphi \sin\theta\right) - \frac{\partial A_\theta}{\partial\varphi}\right]\vec{u}_r + \frac{1}{r\sin\theta}\left[\frac{\partial A_r}{\partial\varphi} - \frac{\partial}{\partial r}\left(r A_\varphi \sin\theta\right)\right]\vec{u}_\theta$$
$$+ \frac{1}{r}\left[\frac{\partial}{\partial r}\left(r A_\theta\right) - \frac{\partial A_r}{\partial\theta}\right]\vec{u}_\varphi, \quad \text{(D.3.28)}$$

where $\vec{u}_r$, $\vec{u}_\theta$ and $\vec{u}_\varphi$ are the unit vectors of the three reciprocally orthogonal directions r, θ and φ, respectively.

D.3.2 Cylindrical Coordinates

In cylindrical coordinates, the metric (the squared arc element) of the Euclidean three-dimensional space is given by

$$ds^2 = d\rho^2 + \rho^2 d\varphi^2 + dz^2. \quad \text{(D.3.29)}$$

Choosing $x^1 = \rho$, $x^2 = \varphi$ and $x^3 = z$, we have

$$\begin{cases} g_{11} = \dfrac{1}{g^{11}} = 1, \\[2mm] g_{22} = \dfrac{1}{g^{22}} = \rho^2, \\[2mm] g_{33} = \dfrac{1}{g^{33}} = 1. \end{cases} \quad \text{(D.3.30)}$$

In view of Eq. (D.3.30), the differential operators in cylindrical coordinates are

$$\operatorname{div}\vec{A} = \frac{1}{\rho}\left[\frac{\partial}{\partial\rho}\left(\rho A_\rho\right) + \frac{\partial A_\varphi}{\partial\varphi} + \frac{\partial}{\partial z}\left(\rho A_z\right)\right], \quad \text{(D.3.31)}$$

$$\operatorname{grad}\Phi = \frac{\partial\Phi}{\partial\rho}\vec{u}_\rho + \frac{1}{\rho}\frac{\partial\Phi}{\partial\varphi}\vec{u}_\varphi + \frac{\partial\Phi}{\partial z}\vec{k}, \quad \text{(D.3.32)}$$

$$\Delta\Phi = \frac{1}{\rho}\left[\frac{\partial}{\partial\rho}\left(\rho\frac{\partial\Phi}{\partial\rho}\right) + \frac{1}{\rho}\frac{\partial^2\Phi}{\partial\varphi^2} + \rho\frac{\partial^2\Phi}{\partial z^2}\right], \quad \text{(D.3.33)}$$

$$\operatorname{curl} \vec{A} = \left(\frac{1}{\rho} \frac{\partial A_z}{\partial \varphi} - \frac{\partial A_\varphi}{\partial z} \right) \vec{u}_\rho + \left(\frac{\partial A_\rho}{\partial z} - \frac{\partial A_z}{\partial \rho} \right) \vec{u}_\varphi$$

$$+ \left(\frac{\partial A_\varphi}{\partial \rho} + \frac{1}{\rho} A_\varphi - \frac{1}{\rho} \frac{\partial A_\rho}{\partial \varphi} \right) \vec{k}, \text{(D.3.34)}$$

where $\vec{u}_\rho$, $\vec{u}_\varphi$ and $\vec{k}$ are the unit vectors of the three reciprocally orthogonal directions ρ, φ and z, respectively.

E

Appendix E: Dirac's Delta Distribution

E.1 Definition and Generalities

As it is known, the spatial density of an electric charge distribution formed by a single point electric charge is zero everywhere, except for the point where the charge is located, where it is infinite. This is an example of *distribution*. While a "regular" *function*, $y(x)$, is defined as a correspondence between a set B of values of y (usually real or complex numbers) corresponding to some values of x, that belong to another set A of real or complex numbers (more exactly, a function from a set A to a set B is an assignment of an element of B to each element of A; the set A is called the domain of the function and the set B is called the codomain of the function) a *distribution* is a *functional* characterized by the fact that to each function makes it correspond a number (to each function $y = y(x)$ is assigned a number). Analogically speaking, a functional is still "sort of" a function, but this time the set A is a set of functions (in the general case it is a linear/vector space), and the set B is a set of scalars (from the field of scalars over which that vector space is defined). So simplistically speaking, a functional is a correspondence between a vector space and a set of numbers (scalars). For example, being given the function $f(x)$, a distribution can be defined by attaching to $f(x)$ the value

$$\int_{-\infty}^{+\infty} f(x)y(x)dx,$$

which is a number (the result of a definite integral is always a number).

Dirac's δ distribution can be defined by means of some functions which, at the limit, tend towards *delta*. Consider, for example, a function $\delta(x,\alpha)$ depending on the variable x and a parameter $\alpha > 0$, which satisfies the conditions

$$\begin{cases} \lim_{\alpha \to 0} \delta(x,\alpha) = \begin{cases} 0, & \text{for} \quad x \neq 0, \\ \infty, & \text{for} \quad x = 0, \end{cases} \\ \lim_{\alpha \to 0} \int_{-\infty}^{+\infty} \delta(x,\alpha)dx = 1. \end{cases} \qquad \text{(E.1.1)}$$

DOI: 10.1201/9781003402602-E

Let us denote

$$
\begin{cases}
\delta(x) \equiv \lim_{\alpha \to 0} \delta(x, \alpha), \\
\int_{-\infty}^{+\infty} \delta(x)dx \equiv \lim_{\alpha \to 0} \int_{-\infty}^{+\infty} \delta(x, \alpha)dx = 1.
\end{cases}
\tag{E.1.2}
$$

We draw attention to the fact that, in general, the uniform convergence of $\delta(x, \alpha)$ for $x = 0$ and $\alpha \to 0$ is not achieved, so that the operations of integration and limit cannot be reversed; relationship $(E.1.2)_2$ is, therefore, a simple notation. In the same sense must be understood the integral

$$
\int_{-\infty}^{+\infty} f(x)\delta(x)dx = \lim_{\alpha \to 0} \int_{-\infty}^{+\infty} f(x)\delta(x, \alpha)dx,
\tag{E.1.3}
$$

where $f(x)$ is a continuous function throughout its domain of definition.

Let $[a, b]$ be a closed interval on x-axis and x_0 some point on this axis. It then follows that

$$
\begin{cases}
\delta(x - x_0) = \begin{cases} 0, & x \neq x_0, \\ \infty, & x = x_0, \end{cases} \\
\int_a^b \delta(x - x_0)dx = \begin{cases} 1, & x_0 \in [a, b], \\ 0, & x_0 \notin [a, b]. \end{cases}
\end{cases}
\tag{E.1.4}
$$

If we extend the limits of integration between $-\infty$ and $+\infty$, then we have

$$
\int_{-\infty}^{+\infty} \delta(x - x_0)dx = 1.
\tag{E.1.5}
$$

Besides,

$$
\int_{-\infty}^{+\infty} f(x)\delta(x - x_0)dx = f(x_0),
\tag{E.1.6}
$$

which is sometimes referred to as the *sifting property* or the *sampling property* of Dirac's delta function. The delta function is said to "sift out" the value at $x = x_0$.

If instead of x we consider the time t, it follows that the effect of convolving a function $f(t)$ with the time-delayed Dirac's delta $\delta_{t_0}(t) = \delta(t - t_0)$ is to time-delay $f(t)$ by the same amount:

$$
\begin{aligned}
(f * \delta_{t_0})(t) &= \int_{-\infty}^{+\infty} f(\tau)\delta(t - t_0 - \tau)d\tau \\
&= \int_{-\infty}^{+\infty} f(\tau)\delta(\tau - (t - t_0))d\tau = f(t - t_0),
\end{aligned}
$$

where we used (in advance − see Eq. (E.1.16)) the property of Dirac's delta distribution of being even.

Returning to the x variable, for $x_0 = 0$, we therefore have

$$\int_{-\infty}^{+\infty} f(x)\delta(x)dx = f(0), \tag{E.1.7}$$

which is a formula with which we can specify the value of any function f at the origin.

E.1.1   Examples

a) Let $\delta(x, \alpha)$ be of the form

$$\delta(x, \alpha) = \frac{1}{\pi} \frac{\alpha}{\alpha^2 + x^2}, \quad \text{with } \alpha > 0. \tag{E.1.8}$$

We have

$$\lim_{\alpha \to 0} \delta(x, \alpha) = \lim_{\alpha \to 0} \frac{1}{\pi} \frac{\alpha}{\alpha + \frac{x^2}{\alpha}} = \begin{cases} 0, & x \neq 0, \\ \infty, & x = 0. \end{cases}$$

$$\int_{-\infty}^{+\infty} \delta(x, \alpha)dx = \frac{1}{\pi} \int_{-\infty}^{+\infty} \frac{\alpha}{\alpha^2 + x^2}dx = \frac{1}{\pi} \left(\arctan \frac{x}{\alpha} \right) \bigg|_{-\infty}^{+\infty} = 1.$$

b) Consider

$$\delta(x, \alpha) = \begin{cases} \dfrac{1}{\alpha}, & x \in \left[-\dfrac{\alpha}{2}, +\dfrac{\alpha}{2} \right], \\ 0, & x \notin \left[-\dfrac{\alpha}{2}, +\dfrac{\alpha}{2} \right]. \end{cases} \tag{E.1.9}$$

We have

$$\int_{-\infty}^{+\infty} \delta(x, \alpha)dx = \frac{1}{\alpha} \int_{-\frac{\alpha}{2}}^{+\frac{\alpha}{2}} f(x)dx.$$

By applying the average theorem, we can write

$$\int_{-\frac{\alpha}{2}}^{+\frac{\alpha}{2}} f(x)dx = f(\xi) \int_{-\frac{\alpha}{2}}^{+\frac{\alpha}{2}} dx = \alpha f(\xi),$$

where $\xi \in \left[-\frac{\alpha}{2}, +\frac{\alpha}{2} \right]$. Therefore,

$$\lim_{\alpha \to 0} \int_{-\infty}^{+\infty} f(x)\delta(x, \alpha)dx = \int_{-\infty}^{+\infty} f(x)\delta(x)dx = f(0),$$

since at the limit $\alpha \to 0$ the interval $\left[-\frac{\alpha}{2}, +\frac{\alpha}{2} \right]$ reduces to a point, namely the origin. This way, we have shown that the function given by Eq. (E.1.9) verifies the relation (E.1.7).

c) Let us choose

$$\delta(x, \alpha) = \frac{1}{\alpha\sqrt{\pi}} e^{-x^2/\alpha^2}. \tag{E.1.10}$$

We then have

$$\frac{1}{\alpha\sqrt{\pi}} \int_{-\infty}^{+\infty} e^{-x^2/\alpha^2}\,dx = \frac{1}{\sqrt{\pi}} \int_{-\infty}^{+\infty} e^{-t^2}\,dt = 1,$$

therefore

$$\lim_{\alpha\to 0} \frac{1}{\alpha\sqrt{\pi}} e^{-x^2/\alpha^2} = \delta(x).$$

E.1.2   Properties of Dirac's Delta Distribution

1) Let $y_1(x)$ and $y_2(x)$ be two known distributions and a and b two constants. By definition, the distribution $y = ay_1 + by_2$ will be expressed as

$$\int_{-\infty}^{+\infty} \big[ay_1 + by_2\big]\,dx = a \int_{-\infty}^{+\infty} f(x)y_1(x)\,dx + b \int_{-\infty}^{+\infty} f(x)y_2(x)\,dx. \quad \text{(E.1.11)}$$

2) Let $y(x)$ be a function with the property $y(\pm\infty) = 0$. We have

$$\int_{-\infty}^{+\infty} f(x)\frac{dy}{dx}\,dx = \big[fy\big]_{-\infty}^{+\infty} - \int_{-\infty}^{+\infty} \frac{df}{dx} y\,dx = - \int_{-\infty}^{+\infty} \frac{df}{dx} y\,dx. \quad \text{(E.1.12)}$$

If $y(x)$ is a known distribution, then relation (E.1.12) is considered as definition of the derivative of this distribution, dy/dx, with respect to its variable, x. For example, the derivative $\frac{d}{dx}\big[\delta(x)\big]$ is expressed by

$$\int_{-\infty}^{+\infty} f(x)\frac{d}{dx}\big[\delta(x)\big]\,dx = - \int_{-\infty}^{+\infty} \frac{df}{dx}\delta(x)\,dx = - \left[\frac{df(x)}{dx}\right]_{x=0}. \quad \text{(E.1.13)}$$

In general,

$$\int_{-\infty}^{+\infty} f(x)\delta^{(n)}(x)\,dx = (-1)^n \left[f^{(n)}(x)\right]_{x=0}. \quad \text{(E.1.14)}$$

3) If y is a function of x, and x a function of t, that is we have a dependence of the form $y\big(x(t)\big)$, then

$$\int f(t)y\big(x(t)\big)\,dt = \int f\big(t(x)\big)y(x)\frac{dt}{dx}\,dx. \quad \text{(E.1.15)}$$

This relation is considered as definition of the distribution $y\big(x(t)\big)$, where the distribution $y(x)$ is known, and $x(t)$ is some function.

4) Let us calculate

$$\int_{-\infty}^{+\infty} f(x)\delta(-x)\,dx = - \int_{-\infty}^{+\infty} f(-y)\delta(y)\,dy$$

$$= f(0) = + \int_{-\infty}^{+\infty} f(x)\delta(x)\,dx,$$

where we have used Eq. (E.1.7). Since this relation is true for any $f(x)$, it follows that

$$\delta(-x) = \delta(x), \qquad (\text{E.1.16})$$

meaning that Dirac's delta distribution is *even*.

5) Another example is offered by the change of variable $y = ax$, where the constant a can be either positive or negative. If $a > 0$, we have

$$\int_{-\infty}^{+\infty} f(x)\delta(ax)dx = \int_{-\infty}^{+\infty} f\left(\frac{y}{a}\right)\delta(y)\frac{dy}{a}$$

$$= \frac{1}{a}f(0) = \int_{-\infty}^{+\infty} f(x)\frac{\delta(x)}{a}dx.$$

If $a < 0$, by means of Eq. (E.1.16) we find $\delta(ax) = \delta(-ax)$, therefore in this case $-a$ is positive. The last relation then gives

$$\delta(ax) = \frac{1}{|a|}\delta(x). \qquad (\text{E.1.17})$$

6) In three dimensions, corresponding to variables x, y and z, the Dirac's delta distribution is defined as

$$\delta(\vec{r} - \vec{r}_0) = \delta(x - x_0)\delta(y - y_0)\delta(z - z_0) = \begin{cases} 0, & \vec{r} \neq \vec{r}_0, \\ \infty, & \vec{r} = \vec{r}_0, \end{cases} \qquad (\text{E.1.18})$$

$$\int_D \delta(\vec{r} - \vec{r}_0)d\vec{r} = \begin{cases} 1, & P_0(x_0, y_0, z_0) \in D, \\ 0, & P_0(x_0, y_0, z_0) \notin D, \end{cases} \qquad (\text{E.1.19})$$

$$\int_D f(\vec{r})\delta(\vec{r} - \vec{r}_0)d\vec{r} = \begin{cases} f(\vec{r}_0), & \vec{r}_0 \in D, \\ 0, & \vec{r}_0 \notin D, \end{cases} \qquad (\text{E.1.20})$$

or, if we extend the integration domain to the entire three-dimensional space,

$$\int_{-\infty}^{+\infty} f(\vec{r})\delta(\vec{r} - \vec{r}_0)d\vec{r} = f(\vec{r}_0), \qquad (\text{E.1.21})$$

where $d\vec{r} = dx\,dy\,dz$.

7) Let $f(\vec{r}) \equiv f(x, y, z)$ be a function of coordinates. Its Fourier transform is

$$F(\vec{k}) = \frac{1}{(2\pi)^{3/2}} \int f(\vec{r})e^{-i\vec{k}\cdot\vec{r}}d\vec{r}, \qquad (\text{E.1.22})$$

together with the inverse Fourier transform

$$f(\vec{r}) = \frac{1}{(2\pi)^{3/2}} \int F(\vec{k})e^{i\vec{k}\cdot\vec{r}}d\vec{k}, \qquad (\text{E.1.23})$$

where $d\vec{k} = dk_x dk_y dk_z$. Replacing Eq. (E.1.22) in Eq. (E.1.23), we have

$$f(\vec{r}) = \int f(\vec{r}') \left[\frac{1}{(2\pi)^3} \int e^{i\vec{k}\cdot(\vec{r}-\vec{r}')} \right] d\vec{r}'.$$

Using Eq. (E.1.21), it follows that

$$\delta(\vec{r} - \vec{r}') = \frac{1}{(2\pi)^3} \int e^{i\vec{k}\cdot(\vec{r}-\vec{r}')} d\vec{k}, \qquad \text{(E.1.24)}$$

expressing the *Fourier transform of the three-dimensional Dirac's delta distribution*.

8) Let us suppose that the function $\varphi(x)$ is such that the equation $\varphi(x) = 0$ has n (only) simple roots: $x_1, x_2, ..., x_n$, that is

$$\varphi(x_i) = 0, \ \varphi'(x_i) \neq 0, \ i = 1, 2, ..., n.$$

Then we can write

$$\int_{-\infty}^{+\infty} f(x)\delta[\varphi(x)] \, dx = \sum_{i=1}^{n} \int_{a_{i-1}}^{a_i} f(x)\delta[\varphi(x)] \, dx,$$

where

$$a_0 < x_1 < a_1 < x_2 < a_2 < \cdots < a_{n-1} < x_n < a_n.$$

Making use of the fact that $\varphi(x)$ is differentiable at x_i, and ε is very small (in the limit it tends to zero), and using Eq. (E.1.17), we have

$$\int_{a_{i-1}}^{a_i} f(x)\delta[\varphi(x)] \, dx = \int_{x_i-\varepsilon}^{x_i+\varepsilon} f(x)\delta[\varphi'(x_i)(x - x_i)] \, dx = \frac{f(x_i)}{|\varphi'(x_i)|},$$

therefore

$$\int_{-\infty}^{+\infty} f(x)\delta[\varphi(x)] \, dx = \sum_{i=1}^{n} \frac{f(x_i)}{|\varphi'(x_i)|} = \int_{-\infty}^{+\infty} f(x) \sum_{i=1}^{n} \frac{\delta(x - x_i)}{|\varphi'(x_i)|} dx$$

resulting in

$$\delta[\varphi(x)] = \sum_{i=1}^{n} \frac{\delta(x - x_i)}{|\varphi'(x_i)|}. \qquad \text{(E.1.25)}$$

In particular, if

$$\varphi(x) = x^2 - a^2,$$

we have

$$\delta(x^2 - a^2) = \frac{\delta(x + a) + \delta(x - a)}{2|a|}. \qquad \text{(E.1.26)}$$

E.2   The Origin of Dirac's Delta Distribution and Its Connection to Heaviside's Function

Using the *delta* distribution, we will write the spatial density of an electric charge distribution consisting of a single point electric charge q, located at the point $P_0(\vec{r}_0)$, as being

$$\rho(\vec{r}) = q\,\delta(\vec{r} - \vec{r}_0). \tag{E.2.27}$$

Therefore, we can formulate the following interpretations:

a) $\delta(x)$ is the linear density of an electric charge distribution, which consists of a single point charge of value $+1$ (measured in Coulombs), located at the origin.

b) $\delta(x - x_0)$ is the linear density of an electric charge distribution consisting of a single point charge of value $+1$, located at the point of coordinate x_0.

c) $\delta(\vec{r} - \vec{r}_0)$ is the spatial density of an electric charge distribution which consists of a single point charge of value $+1$, located at the point $P_0(\vec{r}_0)$.

Contrary to the opinion of some who believe and claim that the Dirac delta distribution has its "origin" in electrostatics, being introduced in order to express mathematically the density of discrete electric charge distributions, in reality the "birth" of Dirac δ distribution is related to Quantum Mechanics.

Since the volume of a geometric point is zero, it appears more than obvious that the spatial density of an electric charge distribution composed by a single electric point charge, located at the point $P_0(\vec{r}_0)$ is given by

$$\rho(\vec{r}) = \begin{cases} 0, & \vec{r} \neq \vec{r}_0, \\ \infty, & \vec{r} = \vec{r}_0, \end{cases} \tag{E.2.28}$$

which is indeed very similar to relation (E.1.18) that defines the δ distribution. We specify that actually the Dirac δ distribution is defined by two relations, namely (E.1.18) together with (E.1.19) (for the three-dimensional case).

Since in our case the total electric charge contained in the considered domain D, is the electric point charge q itself, that is

$$\int_{(D)} \rho(\vec{r})\,d\vec{r} = q, \tag{E.2.29}$$

then both relations (E.2.28) and (E.2.29) are automatically satisfied if the spatial density of this electric charge distribution is defined by Eq. (E.2.27).

This way, the two relations which define the Dirac's delta distribution, namely Eqs. (E.1.18) and (E.1.19), necessarily result. Indeed, since q is finite, according to Eqs. (E.2.27) and (E.2.28) we have

$$\delta(\vec{r} - \vec{r}_0) = \begin{cases} 0, & \vec{r} \neq \vec{r}_0, \\ \infty, & \vec{r} = \vec{r}_0, \end{cases}$$

which is the first of the two defining relations of δ, while according to Eqs. (E.2.29) and (E.2.27), it necessarily follows that

$$\int_{(D)} \delta(\vec{r} - \vec{r}_0)\, d\vec{r} = 1, \quad \text{if } P_0(\vec{r}_0) \in D,$$

and

$$\int_{(D)} \delta(\vec{r} - \vec{r}_0)\, d\vec{r} = 0, \quad \text{if } P_0(\vec{r}_0) \notin D,$$

that is exactly the second definition relation of Dirac's δ distribution.

Even if these things appear as very obvious and even necessary (making some believe that this is the "origin" of δ), in fact the Dirac's delta distribution was first introduced by the French physicist Paul Adrien Maurice Dirac (that's why this distribution is named after Dirac) as a substitute of the Kronecker symbol for the continuous spectra. Therefore, the Dirac's δ distribution is nothing else but the "equivalent" of δ_{ij} (always used when we consider discrete quantities; in fact, the indices i and j always take only integer values) for the case when the physical quantities have a continuous variation.

As well known, the fact that the basis vectors of a vector space form an orthonormal set (*i.e.*, their "length" equals 1 and they are reciprocally orthogonal) is mathematically expressed by a relation which necessarily involves the use of the Kronecker symbol:

$$(v_i, v_k) = \delta_{ik}, \tag{E.2.30}$$

where (v_i, v_k) designates the scalar product of the vectors v_i and v_k, each of them of "length" 1. Obviously, the indices i and k take values from 1 to n, where n is the vector space dimension. One of the simplest examples concerns the three-dimensional Euclidean space E_3, where three-orthogonal reference frames are considered, such as the three-orthogonal Cartesian frame x, y and z. In this case, the basis vectors are the three versors of coordinate axes $\vec{i}$, $\vec{j}$ and $\vec{k}$. In order to use the Einstein's summation convention, we will introduce the notations

$$x \to x_1, \quad \vec{i} \to \vec{u}_1,$$
$$y \to x_2, \quad \vec{j} \to \vec{u}_2,$$
$$z \to x_3, \quad \vec{k} \to \vec{u}_3.$$

Obviously, $|\vec{u}_i| = 1, \forall i = 1, 2, 3$, and

$$\vec{u}_i \cdot \vec{u}_k = \delta_{ik}, \quad i, k = 1, 2, 3. \tag{E.2.31}$$

A somewhat similar situation appears in the non-relativistic quantum mechanics (as an example) where the quantum state of a physical system is described by the so-called "wave function" of the system, which is nothing else but the solution of Schrödinger's equation.

Within the operatorial formalism of the Quantum Mechanics, the Schrödinger's equation is an eigenvalue equation for the Hamilton operator,

$$\hat{H}\psi_k = E_k\psi_k, \quad \text{(no summation)},$$

where $\{\psi_k\}_{k\in\mathbb{N}}$ represents the set of eigenvectors of $\hat{H}$, and $\{E_k\}_{k\in\mathbb{N}}$ represents the set of eigenvalues of $\hat{H}$, also called *Hamilton's operator spectrum* or *energy spectrum*.

These eigenvectors are nothing else than the wave functions of the studied physical system, but in the framework of the mathematical formalism specific to operatorial quantum mechanics, they are elements (vectors) of a Hilbert space, more exactly, the Hilbert space formed by the state vectors of the physical quantum system under study.

From the point of view of mathematical analysis, these vectors of the Hilbert space of state vectors, $\{\psi_k\}_{k\in\mathbb{N}}$, are nothing but complex functions of real variables satisfying certain boundary, continuity and differentiability conditions, specific to each particular problem.

Usually, in Quantum Mechanics, the scalar product operation with these vectors is defined as

$$(\psi_j, \psi_k) \equiv \int \psi_j^* \psi_k \, dq, \tag{E.2.32}$$

where q designates the set of all coordinates which serve to completely describe the quantum state of the system, and $*$ means the complex conjugation.

When the spectrum of certain operator is *discrete*, one says that the states corresponding to the studied physical system are *quantified*; there is *no quantification* when the spectrum is *continuous*, and the quantities characterizing the state of the system take continuous values in a certain interval/domain.

In the case of quantification, the *orthogonality* condition of the state vectors of the system writes as follows:

$$(\psi_j, \psi_k) = c\,\delta_{jk}, \tag{E.2.33}$$

where δ_{jk} is the Kronecker's symbol, and c a complex constant. In general, the state vectors are normalized to unity, condition that in quantum mechanics writes in the form

$$(\psi_k, \psi_k) = 1, \quad \text{(no summation)}, \tag{E.2.34}$$

which actually means that

$$\int |\psi_k|^2 dq = 1, \quad \forall\, k = \overline{1,n}, \tag{E.2.35}$$

where the integral extends over the whole space. In this situation (when the state vectors are orthogonal to each other and normalized to unity), the condition expressed by Eq. (E.2.33) (with $c = 1$) represents the *orthonormality* condition of the state vectors, namely

$$(\psi_j, \psi_k) = \delta_{jk}, \quad \forall\, j, k = \overline{1,n}. \tag{E.2.36}$$

The problem Dirac was asking himself was how to write the condition expressed by Eq. (E.2.36) when the states of the physical system are no longer quantized, but the spectrum is continuous. This means that the state functions can no longer be indexed by means of the natural numbers j or k, but they will have to be indexed by means of an index that takes values continuously.

If, for example, f is a physical quantity which takes values continuously (it has a continuous spectrum), then the corresponding "eigenfunction" can be written as ψ_f where, this time, the index f varies continuously. For the case of a continuous spectrum, the relation corresponding to the orthonormality relation (E.2.36) was written by Dirac as

$$\left(\psi_f,\ \psi_{f'}\right) = \delta(f' - f), \tag{E.2.37}$$

which actually means

$$\int \psi_f^* \psi_{f'}\, dq = \delta(f' - f). \tag{E.2.38}$$

We note that instead of indices j and k taking discrete values, indices f and f' take continuous values, and instead of δ_{jk}, Dirac wrote $\delta(f' - f)$, keeping the same notation (the Greek letter δ) but "adapting" it to the case of the continuous spectrum.

The condition expressing the *principle of superposition of quantum states*, which is a first rank principle in Quantum Mechanics ("the chief positive principle of Quantum Mechanics"), is written for the discrete spectrum as

$$\psi = \sum_k c_k \psi_k, \tag{E.2.39}$$

where ψ is the wave function for an arbitrary quantum state of the system, the sum expands over all possible values of the discrete index k, and c_k are complex constants. These constants have a very important physical significance: $|c_k|^2$ determines the probability of the corresponding value F_k of the physical quantity F when the physical system is in the quantum state described by the wave function ψ.

The analogue of the relation (E.2.39) in the continuous spectrum case is written as follows:

$$\psi(q) = \int c_f \psi_f(q) df, \tag{E.2.40}$$

where the integral extends over the whole range of values taken by the continuous variable f.

As in the case of the discrete spectrum, also in the case of the continuous spectrum it can be easily shown that the coefficients c_f in relation (E.2.40) are given by the expression

$$c_f = \left(\psi_f(q),\ \psi(q)\right) = \int \psi_f^*(q)\psi(q) dq. \tag{E.2.41}$$

Replacing Eq. (E.2.40) into Eq. (E.2.41), we obtain

$$c_f = \int c_{f'} \left(\int \psi_f^* \psi_{f'} dq \right) df'. \tag{E.2.42}$$

This relation has to be identically satisfied (*i.e.*, must be true for any coefficient c_f).

Considering Dirac's notation introduced by Eq. (E.2.38), according to Eq. (E.2.42) we must have

$$c_f = \int c_{f'} \delta(f' - f) df'. \tag{E.2.43}$$

Obviously, for the relation (E.2.43) to take place it is necessary that

$$\int \delta(f' - f) df' = 1. \tag{E.2.44}$$

But for the relation (E.2.42) to be true it is necessary first of all that the coefficient of $c_{f'}$ from the integrand, that is the integral $\int \psi_f^* \psi_{f'} dq$ "noted" by Dirac with $\delta(f' - f)$, to be zero for all $f' \neq f$ and, obviously, in order that the integral over f' not to be canceled, it is necessary that for $f' \neq f$ this coefficient (that is $\int \psi_f^* \psi_{f'} dq = \delta(f' - f)$) be infinite. Indeed, the only quantity which multiplied by zero can give a non-zero result is obviously infinity, because in mathematical analysis infinity multiplied by zero is a quantity called *indeterminate form*, and this indeterminate form is most often used in physics where it is necessary to obtain a non-zero quantity by multiplication from zero. Indeed, being an indeterminate form, we can give (assign) it, in a proper way, any value we want (actually any value we need). In our case, the manner in which the function $\delta(f' - f)$ becomes infinite for $f' - f = 0$ is determined by the fact that we must have

$$\int c_{f'} \delta(f' - f) df' = c_f,$$

which in the limit $f' \to f$ leads to Eq. (E.2.44).

This way, we are led to the following two properties of this symbol denoted by Dirac as $\delta(f' - f)$:

$$1) \ \delta(f' - f) = \begin{cases} 0, & f' \neq f, \\ \infty, & f' = f; \end{cases}$$

$$2) \ \int \delta(f' - f) \, df' = 1.$$

These are the very *relations of definition* of this "quantity" which at present bears the name of the one who introduced it, namely *Dirac's δ distribution*.

Therefore, using notations specific to mathematical analysis, the Dirac's delta distribution has the following properties

$$\delta(x) = \begin{cases} 0, & x \neq 0, \\ \infty, & x = 0; \end{cases} \tag{E.2.45}$$

$$\int_{-\infty}^{+\infty} \delta(x)\, dx = 1. \tag{E.2.46}$$

We can take as integration limits any values (diferent from $-\infty$ and $+\infty$) so that the point $x = 0$ lies between them. In addition, if $f(x)$ is a function continuous at the point $x = 0$, then according to Eqs. (E.2.45) and (E.2.46) we must have

$$\int_{-\infty}^{+\infty} f(x)\delta(x)\, dx = f(0). \tag{E.2.47}$$

Indeed, when x runs the entire range from $-\infty$ to $+\infty$, $\delta(x)$ is 0 for all non-zero values of x and so, in this case the integral in Eq. (E.2.47) will be zero; the only possibility for the integral in Eq. (E.2.47) to be non-zero is for x to be zero, in which case we are left with $f(0)$, because f(0) being constant goes out under the integral and the remaining integral equals 1 according to Eq. (E.2.46).

The relation (E.2.47) can be written in a more general form, namely

$$\int f(x)\delta(x - a)\, dx = f(a), \tag{E.2.48}$$

where the integration interval must include the point $x = a$, and the function $f(x)$ must be continuous at that point. The above discussion obviously leads to

$$\delta(-x) = \delta(x), \tag{E.2.49}$$

that is the Dirac's delta distribution is *even*.

In addition to the above discussed properties of the Dirac's δ distribution, in the following we will present another method of introducing it, which is very intuitive and easy to understand. As we will see, this new procedure puts into evidence the natural connection of the Dirac's δ with another special function frequently used in physics (and not only), namely the Heaviside's function.

The *Heaviside step function*, or the *unit step function*, usually denoted by θ, is a step function, named after Oliver Heaviside (1850–1925), the value of which is 0 for negative arguments and 1 for positive ones.

In mathematical analysis a function $f : \mathbb{R} \to \mathbb{R}$ is called a *step function* (or *staircase function*) if it can be written as

$$f(x) = \sum_{i=0}^{n} \rho_i \chi_{I_i}(x), \quad \forall x \in \mathbb{R}, \tag{E.2.50}$$

where $n \geq 0$, ρ_i are real numbers, I_i are intervals, and $\chi_I(x)$ is the *indicator function* of I, i.e.,

$$\chi_I(x) = \begin{cases} 0, & x \notin I, \\ 1, & x \in I. \end{cases} \tag{E.2.51}$$

Besides, the intervals $\{I_i\}_{i=\overline{0,n}}$ have the following two properties:

1) $\{I_i\}_{i=\overline{0,n}}$ are pairwise disjoint: $I_i \cap I_j = \emptyset$, for $i \neq j$,

2) the union of all intervals is the entire real line: $\bigcup\limits_{i=0}^{n} I_i = \mathbb{R}$.

Therefore, this new way of defining the Dirac's delta distribution makes use of the Heaviside function, which, in its turn, plays a very important role in many applications, not only in physics, but also in functional analysis, game theory, electronics, biochemistry, neuroscience, etc.

The Heaviside function transposes/models mathematically, for example, the action of a switch in an electric circuit. Indeed, omitting the transient regime, before closing the switch the current intensity in the circuit is zero, and after closing the switch the current intensity gets a certain value, which is determined by the circuit elements. This situation can be mathematically described as follows:

$$i(t) = \begin{cases} 0, & t \leq 0, \\ I, & t > 0, \end{cases} \tag{E.2.52}$$

supposing that we turned on the switch at the "initial" moment $t_0 = 0$. Relation (E.2.52) can also be written as

$$i(t) = I\,\theta(t), \tag{E.2.53}$$

where

$$\theta(t) = \begin{cases} 0, & t \leq 0, \\ 1, & t > 0, \end{cases} \tag{E.2.54}$$

is just the *Heaviside step function*. This function can also be defined in two more ways, namely:

1) as an *indicator function*,

$$\theta(x) = 1_{x>0}; \tag{E.2.55}$$

2) as the derivative of the *ramp function*:

$$\theta(x) = \frac{d}{dx}\left(\max\{x,0\}\right), \quad \text{for } x \neq 0. \tag{E.2.56}$$

The Heaviside step function was originally developed in operational calculus for the solution of differential equations, where it represents a signal that switches on at a specified time and stays switched on indefinitely. Oliver Heaviside, who developed the operational calculus as a tool in the analysis of telegraphic communications, represented the function as **1**.

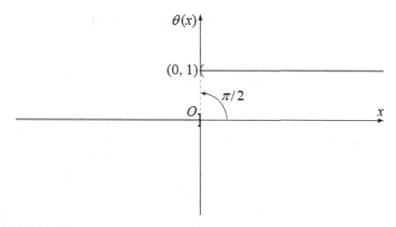

FIGURE E.1
Graphical representation of the unit step function (Heaviside function).

In the above example, if the switch is not acted at the moment $t_0 = 0$, but an arbitrary moment of time $t_0 \neq 0$, then we can write

$$i(t) = \begin{cases} 0, & t \leq t_0, \\ I, & t > t_0, \end{cases} \tag{E.2.57}$$

which can be written by means of the Heaviside step function as

$$i(t) = I\,\theta(t - t_0), \tag{E.2.58}$$

where

$$\theta(t - t_0) = \begin{cases} 0, & t - t_0 \leq 0, \\ 1, & t - t_0 > 0, \end{cases} \tag{E.2.59}$$

which is equivalent to

$$\theta(t - t_0) = \begin{cases} 0, & t \leq t_0, \\ 1, & t > t_0. \end{cases} \tag{E.2.60}$$

In the following presentation we will put into evidence the connection between the Heaviside's function (which, as we have seen, can be considered as a particular step function, namely the unit step function, being also denoted by $H(x)$, or $u(x)$, or $\mathbf{1}(x)$, or even $1(x)$) and the Dirac's δ distribution. To this end, we will appeal to the graphic representation of $\theta(x)$ as shown in Fig. E.1.

As can be observed in Fig. E.1, the function $\theta(x)$ is not continuous at the point $x = 0$, which is a discontinuity of the first kind (or a jump discontinuity), since at this point $\theta(t)$ has a finite jump (a single limit does not exist because the one-sided limits exist and are finite, but are not equal):

$$\begin{cases} \lim\limits_{\substack{x \to 0 \\ x < 0}} \theta(x) \equiv \theta(x - 0) = 0, \\ \lim\limits_{\substack{x \to 0 \\ x > 0}} \theta(x) \equiv \theta(x + 0) = 1. \end{cases} \tag{E.2.61}$$

Since the Heaviside step function is not continuous at the point $x = 0$, it is not differentiable either in the strict sense of "orthodox" mathematical analysis. For any other value of the argument x, except for the value $x = 0$, the function $\theta(x)$ is continuous, therefore differentiable, and we can write

$$\frac{d}{dx}\,\theta(x) = \begin{cases} 0, & x < 0, \\ 0, & x > 0. \end{cases} \tag{E.2.62}$$

As one observes in Fig. E.1, in reality, at the point $x = 0$ the function $\theta(x)$ makes a sudden jump from 0 to 1, the function "graph" (drawn in the figure with a dotted line) being a "vertical straight line segment", meaning a straight line segment which makes with the x-axis an angle of 90°.

As well-known, the geometric significance of the derivative of a function at one point is the tangent of the angle between the x-axis and the line tangent to the graph of the function at that point (in other words, the slope of the tangent line is equal to the derivative of the function at the marked point). In our case, at the point $x = 0$ this angle is $\pi/2$, and $\tan\frac{\pi}{2} = \infty$. Thus, the value of the derivative of $\theta(x)$ at the point $x = 0$ is actually infinite, and that is way it's said that at the point $x = 0$ the function $\theta(x)$ is not differentiable. But extending the meaning of the notion of differentiability, we can write

$$\frac{d}{dx}\,\theta(x) = \begin{cases} 0, & x \neq 0, \\ \infty, & x = 0, \end{cases} \tag{E.2.63}$$

because, for $x \neq 0$ the function $\theta(x)$ being constant (0 for $x < 0$, and 1 for $x > 0$) has zero derivative, while at the point $x = 0$, in agreement with previous specifications, its derivative is ∞.

In view of Eqs. (E.2.45) and (E.2.63), we can write

$$\frac{d}{dx}\theta(x) \equiv \frac{d\theta}{dx} = \delta(x), \tag{E.2.64}$$

meaning that the derivative of the Heaviside step function (defined on the whole real axis) is just the Dirac's δ distribution. But this statement can also be made and vice-versa, this being another definition of the Dirac's δ distribution, namely that the *Dirac's δ distribution equals the derivative of the Heaviside step function*:

$$\delta(x) = \frac{d}{dx}\theta(x).$$

Integrating this relation from $-\infty$ to $+\infty$, we obtain

$$\int\limits_{-\infty}^{+\infty} \delta(x)dx = \int\limits_{-\infty}^{+\infty} \frac{d}{dx}\theta(x)\,dx = \int\limits_{-\infty}^{+\infty} d\theta(x)$$

$$= \theta(x)\Big|_{-\infty}^{+\infty} = \theta(+\infty) - \theta(-\infty) = 1 - 0 = 1,$$

therefore

$$\int\limits_{-\infty}^{+\infty} \delta(x)dx = 1. \tag{E.2.65}$$

Taking into account Eq. (E.2.64), relations (E.2.63) and (E.2.65) represent the relations of definition of the Dirac's delta distribution, known from the beginning of this appendix, but now with a higher level of understanding.

The small generalization provided by Eq. (E.2.60) can also be extended to Dirac's δ distribution, in the sense that we can write

$$\delta(x - x_0) = \frac{d}{dx}\theta(x - x_0), \tag{E.2.66}$$

and, obviously,

$$\int\limits_{-\infty}^{+\infty} \delta(x - x_0)\,dx = 1. \tag{E.2.67}$$

Returning to the integral of $\delta(x)$, performed close to formula (E.2.65) in text, let us observe that the result is the same for any integration interval that contains the point $x = 0$, because $\theta(a) = 0, \forall a \in (-\infty, 0)$ and $\theta(b) = 1, \forall b \in (0, +\infty)$, so that

$$\int_a^b \delta(x)dx = \int_a^b d\theta(x) = \theta(x)\Big|_a^b = \theta(b) - \theta(a) = 1 - 0 = 1. \tag{E.2.68}$$

Now it is easier to understand why the integral of $\delta(x)$, extended over any interval containing $x = 0$, equals 1. More than that, in view of the generalization given by Eqs. (E.2.66) and (E.2.67), we can state that the integral of $\delta(x - x_0)$ extended over any interval containing the point x_0, is also equal to 1.

Therefore, the Heaviside step function can be considered as the integral of Dirac's delta distribution. It is even written sometimes under the form

$$\theta(x) = \int\limits_{-\infty}^{x} \delta(y)dy,$$

even if this kind of writing is not entirely correct from the point of view of ordinary mathematical analysis (being also senseless, from the same point of view) for the value $x = 0$ of the argument of the Heaviside step function.

However, depending on the formalism chosen to make sense of the integrals containing the Dirac's delta distribution, this operation can be made to make sense and even be legitimate. A concrete example of this is given by considering the Heaviside step function as a cumulative distribution function of a random variable which is *almost surely zero*. In probability theory and statistics, the cumulative distribution function of a real-valued random variable (or just distribution function of such variable), evaluated at x, is the probability that the random variable will take a value less than or equal to x. Besides, the meaning of "almost surely" is related to probability 1. Indeed, in probability theory, an event is said to happen *almost surely* if it happens with probability 1. In other words, the set of possible exceptions may be non-empty, but it has probability 0. The concept is analogous to the concept of "almost everywhere" in measure theory, for instance.

We conclude our remarks on Dirac's δ distribution by returning to one of its most important and widely used properties, namely the *sifting property* (or *sampling property*). As we have already seen (see relation (E.1.6)), in the one-dimensional case it is written as follows:

$$\int_{-\infty}^{+\infty} f(x)\delta(x - x_0)dx = f(x_0),$$

the most important particular case being

$$\int_{-\infty}^{+\infty} f(x)\delta(x)dx = f(0).$$

As the name implies, this property is used to "sift out" or to "filter" (or even to "extract", in some sense) from the set of all the values of the function $f(x)$ only one, namely, the value of f at point x_0, *i.e.*, $f(x_0)$. This property can be understood very easily and even in two ways. A first way to understand the sifting property given by Eq. (E.2.48) or, equivalently, Eq. (E.1.6), appeals to the very definition of Dirac's δ distribution. If we refer to the form in Eq. (E.1.6) of this property, note that $\delta(x-x_0)$ falling under the integral is zero for all but one value of x (the integrating variable), *i.e.*, for $x = x_0$, in which case $\delta(x - x_0)$ takes the value ∞, but as we restrict the integration interval, which was originally $x \in (-\infty, +\infty)$, to a very small interval around x_0, which in the limit tends to zero, it is obvious that $x \to x_0$, and the relation (E.1.6) retains its validity, for, as we have seen, it does not matter how large the integration interval is, as long as it contains the point $x = x_0$.

But at the limit $x \to x_0$, $f(x) \to f(x_0)$ and since $f(x_0)$ is a constant, it falls outside the integral, and according to the second relation in Dirac's δ definition, namely Eq. (E.1.4)$_2$, the remaining integral has the value 1. So, looking at the sifting property expressed by Eq. (E.1.6) from the point of view of a "passing to the limit process", when the integration interval $[a, b]$ which contains $x = x_0$, tends to zero, it appears more than obvious.

The second way of understanding the sifting property of the Dirac's δ distribution appeals to the real meaning of this distribution, as nothing more than the "analogue" in the continuous case of Kronecker symbol δ_{jk} in the discrete case. Obviously, as is well known, the sums in the discrete case pass into corresponding integrals in the continuous case.

One of the most used properties of δ_{jk} is in fact also a sifting property (or, in other words, a "filtering" property), and it is often used to determine the values of some constant coefficients appearing with summation indices in a given (single or multiple) sum. To understand exactly what this is all about, below we'll give a concrete example.

Often, the general solution of a problem of mathematical physics equations (*e.g.*, the solution of the Poisson equation or the Laplace equation) can be determined by the Fourier method (the method of separation of variables), in which case the solution is expressed as an infinite (simple or double) sum of the form (for instance):

$$V(x, y, z) = \sum_{j=1}^{\infty} \sum_{k=1}^{\infty} A_{jk} f(\lambda_j x) g(\mu_k y) h(\nu_{jk} z), \qquad (E.2.69)$$

where, depending on the symmetry of the problem the functions f, g and h (considered here as functions of Cartesian variables x, y and respectively z, but they may have other coordinates as arguments, as required by the symmetry of the problem) or only part of them (*e.g.*, only f and g or only f) form an orthogonal set of functions. An example is the Laplace equation in Cartesian coordinates x, y, z, in which case the functions f and g can be a sine, and the function h can be a hyperbolic sine (see **Problem No. 7** in the first chapter of the book):

$$V(x, y, z) = \sum_{j=1}^{\infty} \sum_{k=1}^{\infty} A_{jk} \sin(\lambda_j x) \sin(\mu_k y) \sinh(\nu_{jk} z). \qquad (E.2.70)$$

In this situation, the set of constant quantities A_{jk} can be determined from the boundary condition $V(x, y, c) = V(x, y)$, where c is a constant and the function $V(x, y)$ is taken to be given/known. Making use of the orthogonality relations

$$\begin{cases} \int\limits_0^a \sin(\lambda_j x) \sin(\lambda_k x) dx = \dfrac{a}{2} \delta_{jk}, \\[4mm] \int\limits_0^b \sin(\mu_j y) \sin(\mu_k y) dy = \dfrac{b}{2} \delta_{jk}, \end{cases} \qquad (E.2.71)$$

the set of constants A_{jk} can be determined as follows. Note that the constant quantities A_{jk} cannot be directly removed from below the double sum because these quantities are indexed by the summation indices themselves (not a single constant, but an infinite set of constants). Basically multiply the solution given by Eq. (E.2.70) where the boundary condition $V(x, y, c) = V(x, y)$ is imposed, by $\sin(\lambda_i x) \sin(\mu_l y)$ and the obtained result is integrated over x and y, from 0 to a and from 0 to b, respectively, in order to use the orthogonality conditions expressed by Eq. (E.2.71). The following (in this case, double) sum will result in the right-hand member:

$$\frac{ab}{4} \sum_{j=1}^{\infty} \sum_{k=1}^{\infty} A_{jk} \delta_{ji} \delta_{kl} \sinh(\nu_{jk} c). \tag{E.2.72}$$

The summation over the indices j and k in the relation (E.2.72) is nothing but a *sui-generis* "filtering" (or sifting) operation, for out of the whole infinite double set of constant quantities A_{jk} only one remains, and the sum "disappears". Obviously, since of all the values of δ_{ji}, $j, i = 1, 2, 3, ...$, only one is non-zero (that for which $j = i$) and the same for δ_{kl} (only the term satisfying the condition $k = l$ is non-zero) it follows that after the double summation over j and k only one term remains, namely that for which $j = i$ and $k = l$, i.e.,

$$\frac{ab}{4} \sum_{j=1}^{\infty} \sum_{k=1}^{\infty} A_{jk} \delta_{ji} \delta_{kl} \sinh(\nu_{jk} c) = \frac{ab}{4} A_{il} \sinh(\nu_{il} c). \tag{E.2.73}$$

In other words, summing with the Kronecker symbol "filters" (in other words, "extracts") from the infinite discrete set of quantities $\{A_{jk}\}_{j,k=1,2,3,...}$, only one quantity, namely A_{il}. The sifting ("filtering") property of Dirac δ distribution is nothing more than the "translation" to the continuous case of this property (also being a "filtering" property) of the Kronecker symbol from the discrete case. Obviously, going from discrete to continuous, sums are replaced by integrals, and discrete quantities (such as A_{jk}) are replaced by continuously varying quantities (such as the function $f(x)$). The analogy is therefore quite obvious.

F

Appendix F: Green's Function

F.1 Solving Differential Equations by the Green's Function Method

Many physical phenomena are described by equations of the type

$$Lf(\vec{r}) = u(\vec{r}), \qquad (F.1.1)$$

where L is a linear differential operator, $f(\vec{r})$ – an unknown function, and $u(\vec{r})$ – a given/known function. If the function $G(\vec{r}, \vec{r}')$ is the solution of the equation

$$LG(\vec{r}, \vec{r}') = \delta(\vec{r} - \vec{r}'), \qquad (F.1.2)$$

then the solution of the equation (F.1.1) is given by

$$f(\vec{r}) = \int G(\vec{r}, \vec{r}')u(\vec{r}')d\vec{r}', \qquad (F.1.3)$$

and $G(\vec{r}, \vec{r}')$ is called the *Green's function* attached to the equation (F.1.1).

The electrodynamic phenomena which we refer to in the applications are described by non-homogeneous, second-order partial differential equations of the form

$$Lf(x) = s(x), \qquad (F.1.4)$$

where the linear operator L is a differential operator, mostly of second order; for instance, it can be: the Laplacian Δ, the d'Alembertian $\Box$, etc., x is a multidimensional variable $x = (x_1, x_2, ..., x_n)$, where most often $n = 4$, and the function $s(x)$ is called the *source density*. If $f(x)$ is a vector function, then $s(x)$ is also a vector function. Such equations are, for example, the equations satisfied by the electrodynamic potentials $V(\vec{r}, t)$ and $\vec{A}(\vec{r}, t)$, when the role of the source density, $s(\vec{r}, t)$, is played by the electric charge spatial density $\rho(\vec{r}, t)$ and conduction current density $\vec{j}(\vec{r}, t)$, respectively.

The solution of Eq. (F.1.4) can formally be found by applying the inverse operator L^{-1} to the left of the equation, in which case the searched solution is written symbolically as

$$f(x) = L^{-1}s(x). \qquad (F.1.5)$$

Using the sifting property of Dirac's δ distribution, we can write

$$s(x) = \int\limits_{(D')} s(x')\delta(x - x')dx', \qquad (F.1.6)$$

and replacing this result in Eq. (F.1.5) and taking into account that L acts only on the variable x, we have

$$f(x) = \int\limits_{(D')} s(x')\big[L^{-1}\delta(x - x')\big]dx'. \qquad (F.1.7)$$

Let us note the quantity in the square bracket in Eq. (F.1.7) by $G(x, x')$,

$$L^{-1}\delta(x - x') \equiv G(x, x'), \qquad (F.1.8)$$

and call it the *Green's function* of our problem (the problem defined by Eq. (F.1.4)). Then, according to Eq. (F.1.7), the solution of the problem takes the following form:

$$f(x) = \int\limits_{(D')} G(x, x')s(x')dx', \qquad (F.1.9)$$

i.e., it has exactly the same form as the solution given by Eq. (F.1.3), thereby justifying the form of this solution.

Thus, since the function $s(x)$ is known/given, the solution of Eq. (F.1.4) is determined by the Green's function of the problem. In other words, once the corresponding/appropriate Green's function $G(x, x')$ is found, the solution of the problem is "automatically" known, the only remaining operation to perform being the calculation of the integral in Eq. (F.1.9).

Let's now find out the equation that the Green's function must satisfy; to this end the operator L must be applied to the left of the Eq. (F.1.8), thus obtaining the following equation:

$$LG(x, x') = \delta(x - x'). \qquad (F.1.10)$$

Equation (F.1.10) is very similar to the original equation (F.1.4), the only notable difference being that while the original equation has the $s(x)$ function as its source term, the equation satisfied by the Green's function has the Dirac's delta distribution as its source. Apparently, one might say that we got "from smoke into somother", but this is not at all the case, because while the initial source $s(x)$ may have a very complicated form, the Green's function source is one of the simplest and in addition has a number of very important and useful properties that make it very advantageous compared to the source $s(x)$. Moreover, since the Dirac's delta distribution has a very simple Fourier integral representation, Eq. (F.1.10) satisfied by the Green's function can often be solved using the Fourier transform method, which is an elegant and quick method.

We mention that the "definition relation" (F.1.8) can be generalized as

$$G(x, x') = L^{-1}\delta(x - x') + G_0(x, x'), \tag{F.1.11}$$

where the function $G_0(x, x')$ verifies the condition/equation

$$LG_0(x, x') = 0, \tag{F.1.12}$$

which is nothing more than the homogeneous equation attached to the non-homogeneous equation (F.1.10).

Since the Dirac's delta distribution $\delta(x - x')$ is even, according to Eq. (F.1.10) the function $G(x, x')$ is *symmetric*, *i.e.*,

$$G(x, x') = G(x', x).$$

The symmetry property of the Green's function expresses mathematically a very important phenomenological situation, namely: a source located at the point P' produces at the point P the same effect as that produced at P' if the source would be placed at P.

If we use as independent variables x, y, z and t, then Eq. (F.1.10) becomes

$$LG(\vec{r}, t; \vec{r}', t') = \delta(\vec{r} - \vec{r}')\delta(t - t') = \delta(x - x')\delta(y - y')\delta(z - z')\delta(t - t'), \tag{F.1.13}$$

consequently, the solution of the equation

$$Lf(\vec{r}, t) = s(\vec{r}, t) \tag{F.1.14}$$

is

$$f(\vec{r}, t) = \int G(\vec{r}, t; \vec{r}', t')s(\vec{r}', t')d\vec{r}' dt', \tag{F.1.15}$$

where $d\vec{r}' = dx' dy' dz'$. The problem of finding the solution of Eq. (F.1.14) reduces to determination of the suitable Green's function for the studied application.

Bibliography

[1] Abraham, M., Becker, R., *The Classical Theory of Electricity and Magnetism*, Blackie and Son Ltd., London (1960).

[2] Alexeev, A., *Recueil de Problèmes d'Électrodynamique*, Mir, Moscou (1980).

[3] Arfken, G., *Mathematical Methods for Physicists*, 3rd edn., Orlando, Academic Press (1985).

[4] Arfken, G., and Weber, H.J., *Mathematical Methods for Physicists*, 4th edn., Academic Press, New York (1995).

[5] Baldin, A.M., Gol'danskii, V.I., and Rozenthal, I.L., *Kinematics of Nuclear Reactions*, Pergamon Press, New York (1961).

[6] Barut, A.O., *Electrodynamics and Classical Theory of Fields and Particles*, Macmillan, New York (1964); Dover reprint (1980).

[7] Batygin, V.V., and Toptygin, I.N., *Problems in Electrodynamics*, 2nd edn., London, Academic Press (1978).

[8] Bergmann, P.G., *Introduction to the Theory of Relativity*, Prentice-Hall, Englewood Cliffs, NJ (1942); Dover reprint (1976).

[9] Binns, K. J., Lawrenson, P.J., and Trowbridge, C.W., *The Analytic and Numerical Solution of Electric and Magnetic Fields*, Wiley, New York (1992).

[10] Bleaney, B.I., Bleaney, B., *Electricity and Magnetism*, 3rd edn., Oxford University Press, Oxford (1989).

[11] Bohm, D., *The Special Theory of Relativity*, Benjamin, New York (1965); Addison-Wesley, Reading, MA (1989); Routledge, Abingdon, reprint edn. (1996).

[12] Bondi, H., *Relativity and Common Sense*, 2nd edn., Dover, New York (1980).

[13] Brau, C.A., *Modern Problems in Classical Electrodynamics*, Oxford University Press, Oxford (2003).

[14] Chaichian, M., Merches, I., Tureanu, A., *Mechanics: An Intensive Course*, Springer, Berlin, Heidelberg (2012).

[15] Chaichian, M., Merches, I., Radu, D., Tureanu, A., *Electrodynamics: An Intensive Course*, Springer-Verlag, Berlin, Heidelberg, ISBN: 978-3-642-17380-6, ISBN (eBook): 978-3-642-17381-3 (2016).

[16] Chakravarty, A.S., *Introduction to the Magnetic Properties of Solids*, Wiley, New York (1980).

[17] Chen, M., University of California, Berkeley, *Physics Problems with Solutions*, Englewood Cliffs, NJ, Prentice-Hall, Inc. (1974).

[18] Churchill, R.V., and Brown, J.W., *Fourier Series and Boundary Value Problems*, 5th edn., McGraw-Hill, New York (1993).

[19] Clemmow, P.C., *The Plane Wave Spectrum Representation of Electromagnetic Fields*, Pergamon Press, Oxford (1966); reissued in IEEE/OUP series (1996).

[20] Clemmow, P.C., and Dougherty, J.P., *Electrodynamics of Particles and Plasmas*, Addison- Wesley, Reading, MA (1969); reprinted (1990).

[21] Collin, R.E., *Field Theory of Guided Waves*, 2nd edn., IEEE Press, Piscataway, NJ (1991).

[22] Condon, E.U., and Odishaw, H., eds., *Handbook of Physics*, 2nd edn., McGraw-Hill, New York (1967).

[23] Cowling, T.G., *Magnetohydrodynamics*, 2nd edn., Hilger, London (1976).

[24] Craik, D.J., *Magnetism: Principles and Applications*, Wiley, New York (1995).

[25] Cronin, J., Greenberg, D., and Telegdi, V., *University of Chicago Graduate Problems in Physics*, Chicago, University of Chicago Press (1979).

[26] Einstein, A., *Relativity: The Special and the General Theory*, Broadway Books (1995).

[27] Fitzpatrick, R., *Maxwell's Equations and the Principles of Electromagnetism*, Jones and Bartlett Publishers, Massachusetts (2008).

[28] Fleich, D., *A Students Guide to Maxwell's Equations*, Cambridge University Press, Cambridge (2008).

[29] Friedman, B., *Principles and Techniques of Applied Mathematics*, Wiley, New York (1956); Dover reprint (1990).

[30] Fujimoto, M., *Physics of Classical Electromagnetism*, Springer, Berlin, Heidelberg (2007).

[31] Good Jr., R.H., and Nelson, T.J., *Classical Theory of Electric and Magnetic Fields*, Academic Press, New York (1971).

[32] Goldstein, H., *Classical Mechanics*, 2nd edn., Addison-Wesley, Reading, MA (1980).

[33] Gradshteyn, I.S., and Ryzhik, I.M., *Tables of Integrals, Series, and Products*, 4th edn., prepared by Yu. V. Geronimus and M. Yu. Tseytlin, corr. enlarged edn. by A. Jeffrey, Academic Press, New York (1980).

[34] Greiner, W., *Classical Electrodynamics*, Springer, Berlin, Heidelberg (1998).

[35] Haus, H.A., and Melcher, J.R., *Electromagnetic Fields and Energy*, Prentice-Hall, Englewood Cliffs, NJ (1989).

[36] Heald, M.A., and Marion, J.B., *Classical Electromagnetic Radiation*, 3rd edn., Saunders, Philadelphia (1995).

[37] Hehl, F.W., and Obukhov, Yu.N., *Foundations of Classical Electrodynamics: Charge, Flux, and Metric*, Birkhäuser, Boston, MA (2003).

[38] Ida, N., and Bastos, J.P.A., *Electromagnetics and Calculation of Fields*, Springer-Verlag, New York (1992).

[39] Jackson, J.D., *Classical Electrodynamics*, New York, John Wiley & Sons, Inc. (1975); 3rd edn., Wiley, New York (1998).

[40] Jauch, J.M., and Rohrlich, F., *The Theory of Photons and Electrons*, Springer, Berlin (1976).

[41] Jefimenko, O.D., *Electricity and Magnetism*, Appleton-Century Crofts, New York (1966); 2nd edn., Electret Scientific, Star City, WV (1989).

[42] Johnson, C.C., *Field and Wave Electrodynamics*, McGraw-Hill, New York (1965).

[43] Jones, D.S., *The Theory of Electromagnetism*, Pergamon Press, Oxford (1964).

[44] Jordan, E.C., and Balmain, K.G., *Electromagnetic Waves and Radiating Systems*, 2nd edn., Prentice-Hall, Englewood Cliffs, NJ (1968).

[45] Kilmister, C.W., *Special Theory of Relativity*, Pergamon Press, Oxford (1970).

[46] Kraus, J.D., *Antennas*, 2nd edn., McGraw-Hill, New York (1988).

[47] Kraus, J.D., and Fleisch, D.A., *Electromagnetics with Applications*, 5th edn., McGraw Hill, New York (1998).

[48] Landau, L.D., and Lifshitz, E.M., *Classical Theory of Fields*, Vol. 2 of Course of Theoretical Physics, 4th edn., Elmsford, New York, Pergamon Press (1975).

[49] Landau, L.D., Lifshitz, E.M., and Pitaevskii, L.P., *Electrodynamics of Continuous Media*, Vol. 8 of Course of Theoretical Physics, 2nd edn., Elmsford, New York, Pergamon Press (1984).

[50] Lavrentiev, M., and Chabat, B., *Méthodes de la théorie des functions d'une variable complexe*, Moscou, Mir (1972).

[51] Lightman, A.P., Press, W.H., Price, R.H., and Teukolski, S.A., *Problem Book in Relativity and Gravitation*, Princeton University Press, Princeton (1975).

[52] Lorentz, H.A., Einstein, A., Minkowski, H., and Weyl, H., *The Principle of Relativity: A Collection of Original Memoirs on the Special and General Theory of Relativity*, Dover, New York (1968).

[53] Lorraine, P., and Corson, D.R., *Electromagnetic Fields and Waves*, Freeman, New York (1970).

[54] Low, F.E., *Classical Field Theory: Electromagnetism and Gravitation*, Wiley, New York (1997).

[55] Marion, J.B., and Heald, M.A., *Classical Electromagnetic Radiation*, 2nd ed., New York, Academic Press (1980).

[56] Maxwell, J.C., *Elementary Treatise on Electricity*, Clarendon Press, Oxford (1888).

[57] Maxwell, J.C., *Treatise on Electricity and Magnetism*, 3rd edn. (1891), 2 vols., reprint, Dover, New York (1954).

[58] Melia, F., *Electrodynamics* (Chicago Lectures in Physics), University of Chicago Press, Chicago (2001).

[59] Merches, I., and Radu, D., *Analytical Mechanics − Solutions to Problems in Classical Physics*, CISP − CRC Press, Taylor & Francis Group, Boca Raton, London, New-York, ISBN: 978-1-4822-3939-3 (2015).

[60] Nasar, S.A., *2000 Solved Problems in Electromagnetics*, McGraw-Hill, New York (1992).

[61] Newbury, N., Newman, M., Ruhl, J., Staggs, S., and Thorsett, S., *Princeton Problems in Physics with Solutions*, Princeton, Princeton University Press (1991).

[62] Oppenheimer, J.R., *Lectures on Electrodynamics*, Gordon and Breach Science Publishers, Philadelphia (1970).

[63] Panofsky, W.K.H., and Phillips, M., *Classical Electricity and Magnetism*, 2nd edn., Dover Publications, New York (2005).

[64] Pauli, W., *Theory of Relativity*, Pergamon, London (1958).

[65] Pollack, G., and Stump, D., *Electromagnetism*, Addison Wesley, Boston (2001).

[66] Purcell, E.M., *Electricity and Magnetism*, Vol. 2 of Berkeley Physics Course, New York, McGraw-Hill (1965); 2nd edn., McGraw-Hill, New York (1985).

[67] Press, W.H., Flannery, B.P., Teukolsky, S.A., and Vetterling, W.T., *Numerical Recipes, The Art of Scientific Computing*, Cambridge University Press, Cambridge (1986).

[68] Ramo, S., Whinnery, J.R., and Van Duzer, T., *Fields and Waves in Communication Electronics*, 3rd edn., Wiley, New York (1994).

[69] Reitz, J.R., Milford, F.J., and Christy, R.W., *Foundations of Electromagnetic Theory*, 4th edn., Addison Wesley, Boston (2008).

[70] Resnick, R., *Introduction to Special Relativity*, Wiley, New York (1968).

[71] Rindler, W., *Special Relativity*, 2nd edn., Oliver and Boyd, Edinburgh (1966).

[72] Robinson, F.N.H., *Macroscopic Electromagnetism*, Pergamon Press, Oxford (1973).

[73] Rohrlich, F., *Classical Charged Particles*, Addison-Wesley, Reading, MA (1965, 1990).

[74] Sadiku, M.N.O., *Numerical Techniques in Electromagnetics*, CRC Press, Boca Raton, FL (1992).

[75] Sard, R.D., *Relativistic Mechanics*, Benjamin, New York (1970).

[76] Schwartz, M., *Principles of Electrodynamics*, Dover, New York (1987).

[77] Sidney, B.C., and Boris, E.N., *A Guide to Physics Problems, part 1, Mechanics, Relativity, and Electrodynamics*, Kluwer Academic Publishers, New York, Boston, Dordrecht, London, Moscow, eBook ISBN: 0-306-48400-5, Print ISBN: 0-306-44679-0 (1994).

[78] Smythe, W.R., *Static and Dynamic Electricity*, 3rd edn., New York, Hemisphere Publishing Corp. (1989).

[79] Sommerfeld, A., *Electrodynamics*, Academic Press, New York (1964).

[80] Strang, G., *Introduction to Applied Mathematics*, Wellesley-Cambridge Press, Wellesley, MA (1986).

[81] Synge, J.L., *Relativity, The Special Theory*, 2nd edn., Elsevier Science Ltd., Amsterdam (1980).

[82] Taylor, E.F., and Wheeler, J.A., *Spacetime Physics*, 2nd edn., Freeman, San-Francisco (1992).

[83] Titchmarsh, E.C., *Introduction to the Theory of Fourier Integrals*, 3rd edn., Oxford University Press, Oxford (1986).

[84] Ugarov, V.A., *Special Theory of Relativity*, Mir Publishers, Moscow (1979).

[85] Vanderlinde, J., *Classical Electromagnetic Theory*, Wiley, New York (1993).

[86] Watson, G.N., *Theory of Bessel Functions*, 2nd edn., Cambridge University Press, Cambridge (1952); reprinted (1995).

[87] Zhou, P.-B., *Numerical Analysis of Electromagnetic Fields*, Springer-Verlag, New York (1993).

Printed in the United States
by Baker & Taylor Publisher Services